Lecture Notes in Mathematics

A collection of informal reports and seminars
Edited by A. Dold, Heidelberg and B. Eckmann, Zürich

142

Klaus W. Roggenkamp

McGill University, Montreal

Lattices over Orders II

Springer-Verlag
Berlin · Heidelberg · New York 1970

PREFACE

This volume is a continuation of Volume I and reference is made to the statements in Volume I simply by number without quoting special theorems.

I would like to express my gratitude to Verena Huber-Dyson, who has read these notes carefully and who has made valuable improvements.

At this point I have to mention my wife Christa, who has patiently endured all my moods during the preparation of these notes, with the equanimity that only a wife has, and who has typed most of Vol. I and all of Vol. II for me.

There are more distinguished people who should have written these notes; however,

"nullus est liber tam malus, ut non aliqua parte prosit".

(Plinius sen.)

CONTENT

CHAPTER VI

MODULES OVER ORDERS, ONE-SIDED IDEALS OVER MAXIMAL ORDERS *)

§1 Local equivalence

We show that the question of reducibility and decomposition of lattices over semi-perfect orders can already be decided modulo the reduction by a sufficiently high power of the underlying prime ideal. Moreover, lattices are locally isomorphic if and only if they are isomorphic over the completions.

We shall use the following notation:

R with prime element π and quotient field K is the localization of a Dedekind domain at some maximal ideal;

A is a finite dimensional separable K-algebra;

Λ is an R-order in A;

$\underline{\underline{H}}(\Lambda)$ is the Higman ideal of Λ in R (cf. V, (3.1));

$\underline{\underline{H}}(\Lambda) = \pi^{s_1}R$ for some $s_1 \in \underline{\underline{N}}$;

$\underline{\underline{P}}_R(M)$ is the set of projective endomorphism in $\mathrm{End}_\Lambda(M) \cap R$ (cf. V, 2.1), $M \in {}_\Lambda\underline{\underline{M}}^o$;

$\underline{\underline{P}}_R(M) = \pi^{s(M)}R$ for some $s(M) \in \underline{\underline{N}}$.

If $X \in {}_\Lambda\underline{\underline{M}}^f$, then we denote by $\hat{X}$ the π-adic completion of X.

1.1 Theorem: Let $M, N \in {}_\Lambda\underline{\underline{M}}^o$. Then $M \cong_\Lambda N$ if and only if $M/\pi^s M \cong N/\pi^s N$ as $\Lambda/\pi^s\Lambda$-modules for some $s > s(M)$.

Proof: Obviously $M \cong N$ implies $M/\pi^s M \cong N/\pi^s N$ for every $s \in \underline{\underline{N}}$. Let us therefore assume $\varphi : \bar{M} \xrightarrow{\sim} \bar{N}$ is a $\bar{\Lambda}$-isomorphism, where "-" denotes reduction modulo π^s with $s > s(M)$. We put

$$\psi = \pi^{s-1}\chi\varphi : M \longrightarrow \bar{N},$$

where $\chi : M \longrightarrow \bar{M}$ is the canonical epimorphism. Then

$$\mathrm{Im}\,\psi = \pi^{s-1}N/\pi^s N \neq 0$$

*) In this chapter, prime ideals are always assumed to be different from zero.

by Nakayama's lemma. Moreover, since $s - 1 \geq s(M)$, $\psi \in \mathrm{Hom}_\Lambda(M,\bar{N})$ is a projective homomorphism (cf. V, 2.1, 2.2). (Observe $\mathrm{ext}^1_\Lambda(\psi,1_X) = \mathrm{ext}^1_\Lambda(\chi\psi,1_X)\mathrm{ext}^1_\Lambda(\pi^{s-1}1_M,1_X) = 0$ (cf. II, 5.5, 5.9).)

If $\rho : N \longrightarrow \bar{N}$ is the canonical epimorphism, then we can complete the following diagram (cf. V, 2.3)

$$\begin{array}{ccccccccc} & & & & & & M & & \\ & & & & & \sigma\swarrow & \downarrow\psi & & \\ 0 & \longrightarrow & \pi^s N & \xrightarrow{\iota} & N & \xrightarrow{\rho} & \bar{N} & \longrightarrow & 0, \end{array}$$

and $\mathrm{Im}\,\sigma \subset \pi^{s-1}N$. However, $\rho|_{\pi^{s-1}N} : \pi^{s-1}N \longrightarrow \pi^{s-1}N/\pi^s N$ is an essential epimorphism (cf. III, 7.2; IV, 2.6), and the relation $\sigma\rho|_{\pi^{s-1}N} = \psi$ implies $\mathrm{Im}\,\sigma = \pi^{s-1}N$. Since $\mathrm{rk}(M) = \mathrm{rk}(N)$ (rk = rank), we conclude that $\sigma : M \longrightarrow \pi^{s-1}N$ is an isomorphism. But $\pi^{s-1}N \cong N$, and the theorem is established. #

1.2 <u>Corollary</u>: Let $M,N \in {}_\Lambda\underline{\underline{M}}^o$. Then $M \cong_\Lambda N$ if and only if $\hat{M} \cong_{\hat{\Lambda}} \hat{N}$.

<u>Proof</u>: Since $\underline{\underline{P}}_{\hat{R}}(\hat{M}) = \hat{R} \otimes_R \underline{\underline{P}}_R(M)$ (cf. V, 4.4), $\underline{\underline{P}}_{\hat{R}}(\hat{M}) = \hat{\pi}^{s(M)}\hat{R}$. Obviously $M \cong N$ implies $\hat{M} \cong \hat{N}$. We assume $\hat{M} \cong \hat{N}$ and put $s = s(M) + 1$. Then $\hat{M}/\hat{\pi}^s\hat{M} \cong \hat{N}/\hat{\pi}^s\hat{N}$; but $\hat{M}/\hat{\pi}^s\hat{M} \cong M/\pi^s M$ (cf. I, 9.18) and we obtain the isomorphism $M/\pi^s M \cong N/\pi^s N$. Since $s > s(M)$, we apply (1.1) to conclude $M \cong N$. #

1.3 <u>Theorem</u> (Maranda [1], Higman [5]): Let $\hat{M} \in {}_{\hat{\Lambda}}\underline{\underline{M}}^o$. Then M is reducible if and only if $\hat{M} =_{\hat{R}} \hat{M}_1 \oplus \hat{M}_2$ (as $\hat{R}$-modules), $0 \neq \hat{M}_i$, $i = 1,2$, and $\hat{\Lambda}\hat{M}_1 \subset \hat{M}_1 + \hat{\pi}^s\hat{M}_2$, for some $s > 2s_1$. (We recall that $\underline{\underline{H}}(\hat{\Lambda}) = \hat{\pi}^{s_1}\hat{R}$.)

<u>Proof</u>: If $\hat{M}$ is reducible the statement is trivial, since then $\hat{M}$ contains an $\hat{R}$-pure submodule $\hat{N}$ of smaller rank (cf. proof of VI, 1.13), and thus $\hat{M} =_{\hat{R}} \hat{N} \oplus \hat{M}/\hat{N}$, $\hat{M}/\hat{N}$ being $\hat{R}$-projective.

<u>Conversely</u>, assume that $\hat{M} =_{\hat{R}} \hat{M}_1 \oplus \hat{M}_2$ is a non-trivial decomposition of $\hat{R}$-modules with $\hat{\Lambda}\hat{M}_1 \subset \hat{M}_1 + \hat{\pi}^s\hat{M}_2$, $s > 2s_1$. For $m \in \hat{M}$, $\lambda \in \hat{\Lambda}$ we denote by $[\lambda m]_i$ the part of λm which lies in $\hat{M}_i$, $i = 1,2$. The hypotheses

then imply that for $m_1 \in \hat{M}_1$, $[\lambda m_1]_2 \in \hat{\pi}^s\hat{M}_2$ and we shall write

$$[\lambda m_1]_2 = \hat{\pi}^s(m_1^{\psi_\lambda}),$$

where

$$\psi_\lambda : \hat{M}_1 \longrightarrow \hat{M}_2$$

is an $\hat{R}$-homomorphism satisfying

(i) $m_1^{\psi_{\lambda_1\lambda_2}} = [\lambda_1(m_1^{\psi_{\lambda_2}})]_2 + ([\lambda_2 m_1]_1)^{\psi_{\lambda_1}}$ for $m_1 \in \hat{M}_1$ and $\lambda_1, \lambda_2 \in \hat{\Lambda}$.

We put

$$\overline{M}_1 = (\hat{M}_1 + \hat{\pi}^s\hat{M}_2)/\hat{\pi}^s\hat{M} \text{ and } \overline{M}_2 = \overline{M}/\overline{M}_1,$$

where $\overline{M} = \hat{M}/\hat{\pi}^s\hat{M}$. Then both $\overline{M}_1$ and $\overline{M}_2$ are $\hat{\Lambda}$-modules. The map

$$\overline{\psi} : \hat{\Lambda} \longrightarrow \mathrm{Hom}_{\hat{R}}(\overline{M}_1, \overline{M}_2)$$
$$\lambda \longmapsto \overline{\psi}_\lambda,$$

where $\overline{\psi}_\lambda : m_1 + \hat{\pi}^s M_1 \longmapsto m_1^{\psi_\lambda} + \hat{\pi}^s\hat{M}_2$ is a derivation (cf. III, 4.3). (It should be observed that $\overline{M}_1 \cong \hat{M}_1/\hat{\pi}^s\hat{M}_1$ and $\overline{M}_2 \cong \hat{M}_2/\hat{\pi}^s\hat{M}_2$. We identify both structures and consider $\hat{M}_1/\hat{\pi}^s\hat{M}_1 = \overline{M}_1$ and $\hat{M}_2/\hat{\pi}^s\hat{M}_2 = \overline{M}_2$ as $\hat{\Lambda}$-modules.) That $\overline{\psi}$ is a derivation follows immediately from (i). From the properties of the Higman ideal (cf. V, 3.3) we conclude that

$$\hat{\pi}^{s_1}\overline{\psi} : \hat{\Lambda} \longrightarrow \mathrm{Hom}_{\hat{R}}(\overline{M}_1, \overline{M}_2)$$

is an inner derivation; i.e., there exists $\overline{\varphi} \in \mathrm{Hom}_{\hat{R}}(\overline{M}_1, \overline{M}_2)$ such that

$$\hat{\pi}^{s_1}(\overline{m}_1^{\overline{\psi}_\lambda}) = (\lambda\overline{m}_1)^{\overline{\varphi}} - \lambda(\overline{m}_1^{\overline{\varphi}})$$

for every $\lambda \in \hat{\Lambda}$, $\overline{m}_1 \in \overline{M}_1$. However, $\hat{M}_1$ is a projective $\hat{R}$-module, and hence we have an epimorphism

$$\mathrm{Hom}_{\hat{R}}(\hat{M}_1, \hat{M}_2) \longrightarrow \mathrm{Hom}_{\hat{R}}(\overline{M}_1, \overline{M}_2) \text{ (cf. IV, 3.7),}$$

and there exists $\varphi \in \mathrm{Hom}_{\hat{R}}(\hat{M}_1, \hat{M}_2)$ such that

(ii) $\hat{\pi}^{s_1}(m_1^{\psi_\lambda}) \equiv ([\lambda m_1]_1^{\varphi}) - [\lambda(m_1^{\varphi})]_2 \bmod (\hat{\pi}^s\hat{M}_2)$. We now consider the following $\hat{R}$-submodule $\hat{M}_1^{(1)}$ of $\hat{M}$:

$$\hat{M}_1^{(1)} = \{(m_1, \hat{\pi}^{s-s_1}(m_1^{\varphi})) : m_1 \in \hat{M}_1\}.$$

Then $\hat{M}_1^{(1)} \cong_{\hat{R}} \hat{M}_1$ under the map $m_1 \longmapsto (m_1, \hat{\pi}^{s-s_1}(m_1^{\varphi}))$. Obviously $\hat{M}_2$ is still an $\hat{R}$-complement of $\hat{M}_1^{(1)}$ in $\hat{M}$. We claim that

$$\hat{\Lambda}\hat{M}_1^{(1)} \subset \hat{M}_1^{(1)} + \hat{\pi}^{s+1}\hat{M}_2.$$

For $(m_1, \hat{\pi}^{s-s_1}m_1^{\varphi}) \in \hat{M}_1^{(1)}$ and for $\lambda \in \hat{\Lambda}$ we have

$$\lambda(m_1, \hat{\pi}^{s-s_1}m_1^{\varphi}) = ([\lambda m_1]_1 + \hat{\pi}^{s-s_1}[\lambda(m_1^{\varphi})]_1,$$
$$\hat{\pi}^{s}m_1^{\psi\lambda} + \hat{\pi}^{s-s_1}[\lambda(m_1^{\varphi})]_2);$$

for the terms in the second position we get:

$$\hat{\pi}^{s-s_1}(\hat{\pi}^{s_1}m_1^{\psi\lambda} + [\lambda(m_1^{\varphi})]_2) \equiv \hat{\pi}^{s-s_1}\{([\lambda(m_1^{\varphi})]_2 +$$
$$+ ([\lambda m_1]_1^{\varphi} - [\lambda(m_1^{\varphi})]_2)\} \bmod \hat{\pi}^{2s-s_1}\hat{M}$$

using (ii). Since $2s - s_1 \geq s + 1$, it suffices to show that

$$\hat{\pi}^{s-s_1}([\lambda m_1]_1)^{\varphi} - \hat{\pi}^{s-s_1}\langle [\lambda m_1]_1^{\varphi} + \hat{\pi}^{s-s_1}([\lambda(m_1^{\varphi})]_1)^{\varphi}\rangle \in \hat{\pi}^{s+1}\hat{M}_2.$$

But

$$\hat{\pi}^{2(s-s_1)}([\lambda(m_1^{\varphi})]_1)^{\varphi} \in \hat{\pi}^{s+1}\hat{M}_2, \text{ since } 2(s-s_1) \geq s+1 \ (s > 2s_1).$$

This proves the claim. Now, we start all over again with the pair $\hat{M}_1^{(1)}, \hat{M}_2$ etc. This way we construct a family of $\hat{R}$-modules $\hat{M}_1^{(i)}$ such that $\hat{M} =_R \hat{M}_1^{(i)} \oplus \hat{M}_2$ and such that $\overline{M}_1^{(i)} = \hat{M}_1^{(i)}/\hat{\pi}^{s+i}\hat{M}_1^{(i)}$ is a $\hat{\Lambda}$-module; moreover, we have a $\hat{\Lambda}$-homomorphism

$$\sigma_{01} : \overline{M}_1 \longleftarrow \overline{M}_1^{(1)} = (\hat{M}_1^{(1)} + \hat{\pi}^{s+1}\hat{M})/\hat{M}\hat{\pi}^{s+1},$$
$$\overline{m}_1 \longleftarrow (\overline{m}_1, \hat{\pi}^{s-s_1}\overline{m}_1^{\overline{\varphi}});$$

similarly for all i. Moreover, $\hat{\Lambda} = \varprojlim \hat{\Lambda}/\hat{\pi}^{i}\hat{\Lambda}$ and the family $\{\overline{M}_1^{(i)}, \sigma_{ij}\}$ satisfies the hypotheses of the projective limit (cf. I, § 9) and thus, $\varprojlim \overline{M}_1^{(i)} = \hat{M}_0$ is a $\hat{\Lambda}$-submodule of $\hat{M}$ of smaller $\hat{R}$-rank. This means that $\hat{M}$ is reducible. #

1.4 <u>Theorem</u> (Maranda [1] , Higman [5], Heller [1]: Let $\hat{M} \in {}_{\hat{\Lambda}}\underline{\underline{M}}^{o}$ and let $\underline{\underline{P}}_{\hat{R}}(\hat{M}) = \hat{\pi}^{s_0}\hat{R}$. Then $\hat{M}$ decomposes if and only if $\hat{M}/\hat{\pi}^{s}\hat{M}$

decomposes for some $s > s_0$.

Proof: Since $\bar{M} = \hat{M}/\hat{\pi}^s\hat{M} \cong \hat{R}/\hat{\pi}^s\hat{R} \otimes_{\hat{R}} \hat{M}$, and since $\hat{R}/\hat{\pi}^s\hat{R} \otimes_{\hat{R}} -$ is an additive functor, we need only to prove one direction. Let us assume that $\bar{M}$ decomposes. We consider the exact sequence

$$E : 0 \longrightarrow \hat{M} \xrightarrow{\ \varphi\ } \hat{M} \longrightarrow \bar{M} \longrightarrow 0,$$

where φ is multiplication by $\hat{\pi}^s$. Applying $\mathrm{Hom}_{\hat{\Lambda}}(\hat{M},-)$ to E, we obtain the exact sequence where φ_* is multiplication by $\hat{\pi}^s$,

$$0 \longrightarrow \mathrm{End}_{\hat{\Lambda}}(\hat{M}) \xrightarrow{\ \varphi_*\ } \mathrm{End}_{\hat{\Lambda}}(\hat{M}) \xrightarrow{\ \sigma\ } \mathrm{Hom}_{\hat{\Lambda}}(\hat{M},\bar{M}) \longrightarrow \mathrm{Ext}^1_{\hat{\Lambda}}(\hat{M},\hat{M}) \longrightarrow 0,$$

since $\mathrm{Im}\,\varphi_{**} = 0$, where $\varphi_{**} : \mathrm{Ext}^1_{\hat{\Lambda}}(\hat{M},\hat{M}) \longrightarrow \mathrm{Ext}^1_{\hat{\Lambda}}(\hat{M},\hat{M})$ is multiplication by $\hat{\pi}^s \in \underline{\underline{P}}_{\hat{R}}(\hat{M})$. Thus

$$\mathrm{Im}\,\sigma \cong \mathrm{End}_{\hat{\Lambda}}(\hat{M})/\hat{\pi}^s\mathrm{End}_{\hat{\Lambda}}(\hat{M}) \overset{\mathrm{def}}{=} \overline{\mathrm{End}_{\hat{\Lambda}}(\hat{M})}.$$

Since

$$\mathrm{Ext}^1_{\hat{\Lambda}}(\hat{M},\hat{M}) \cong \mathrm{Hom}_{\hat{\Lambda}}(\hat{M},\bar{M})/\mathrm{Im}\,\sigma,$$

and since $\hat{\pi}^{s_0}\mathrm{Ext}^1_{\hat{\Lambda}}(\hat{M},\hat{M}) = 0$, we conclude

$$\hat{\pi}^{s_0}\mathrm{Hom}_{\hat{\Lambda}}(\hat{M},\bar{M}) \subset \overline{\mathrm{End}_{\hat{\Lambda}}(\hat{M})}.$$

Moreover,

$$\mathrm{Hom}_{\hat{\Lambda}}(\hat{M},\bar{M}) \overset{\mathrm{nat}}{\cong} \mathrm{End}_{\hat{\Lambda}}(\bar{M}) \text{ (cf. IV, 3.7),}$$

and we find that

$$\hat{\pi}^{s_0}\mathrm{End}_{\hat{\Lambda}}(\bar{M}) \subset \overline{\mathrm{End}_{\hat{\Lambda}}(\hat{M})}.$$

Let $\bar{\varepsilon}$ be a non-trivial idempotent in $\mathrm{End}_{\hat{\Lambda}}(\bar{M})$; this exists since $\bar{M}$ decomposes. Thus we can find a $\hat{\Lambda}$-module $X \neq 0$ such that

$$\bar{M} = \bar{M}\bar{\varepsilon} \oplus X.$$

If $\hat{\pi}^{s_0}X = 0$, then $X \subset \hat{\pi}\bar{M}$ since $s > s_0$, and $\bar{M} = \bar{M}\bar{\varepsilon} + \hat{\pi}\bar{M}$ and Nakayama's lemma implies $\bar{M}\bar{\varepsilon} = \bar{M}$, a contradiction. Similarly one shows that $\hat{\pi}^{s_0}\bar{M}\bar{\varepsilon} \neq 0$. Consequently, $0 \neq \hat{\pi}^{s_0}\bar{\varepsilon} \in \hat{\pi}^{s_0}\mathrm{End}_{\hat{\Lambda}}(\bar{M}) \subset \overline{\mathrm{End}_{\hat{\Lambda}}(\hat{M})}$, and there exists $\alpha \in \mathrm{End}_{\hat{\Lambda}}(\hat{M})$ such that the following diagram is commutative

$$\begin{array}{ccc} \hat{M} & \xrightarrow{\ \alpha\ } & \hat{M} \\ \mathrm{can}\downarrow & & \downarrow\mathrm{can} \\ \bar{M} & \xrightarrow{\ \hat{\pi}^{s_0}\bar{\varepsilon}\ } & \hat{\pi}^{s_0}\bar{M}\bar{\varepsilon} \subset \bar{M}\ . \end{array}$$

In particular, Im $\alpha \subset \hat{\pi}^{s_0}\hat{M}$. This shows, that we can write $\alpha = \hat{\pi}^{s_0}\beta$ for some $\beta \in \mathrm{End}_{\hat{\Lambda}}(\hat{M})$. Then the diagram

$$\begin{array}{ccc} \hat{M} & \xrightarrow{\hat{\pi}^{s_0}\beta} & \hat{M} \\ \rho\downarrow & & \downarrow\rho \\ \overline{M} & \xrightarrow{\hat{\pi}^{s_0}\overline{\varepsilon}} & \overline{M} \end{array}$$

is commutative, where ρ is the canonical epimorphism; i.e., $\hat{\pi}^{s_0}(\beta\rho - \rho\overline{\varepsilon}) = 0$ and we conclude that

$$(\beta\rho - \rho\overline{\varepsilon})\Big|_{\hat{\pi}^{s_0}\hat{M}} = 0.$$

Thus we obtain the commutative diagram

$$\begin{array}{ccc} \hat{\pi}^{s_0}\hat{M} & \xrightarrow{\beta} & \hat{\pi}^{s_0}\hat{M} \\ \rho\downarrow & & \downarrow\rho \\ \hat{\pi}^{s_0}\overline{M} & \xrightarrow{\overline{\varepsilon}} & \hat{\pi}^{s_0}\overline{M}, \end{array}$$

and one finds readily, that $\overline{\varepsilon}\Big|_{\hat{\pi}^{s_0}\hat{M}} \in \mathrm{End}_{\hat{\Lambda}}(\hat{\pi}^{s_0}\overline{M})$ is a non trivial idempotent, and the above commutative diagram shows that

$$\overline{\varepsilon} \in \overline{\mathrm{End}_{\hat{\Lambda}}(\hat{\pi}^{s_0}\hat{M})}.$$

Since $\mathrm{End}_{\hat{\Lambda}}(\hat{\pi}^{s_0}\hat{M})$ is semi-perfect (cf. IV, 2.1), we can lift $\overline{\varepsilon}$ to a non-trivial idempotent ε of $\mathrm{End}_{\hat{\Lambda}}(\hat{\pi}^{s_0}\hat{M})$; i.e., $\hat{\pi}^{s_0}\hat{M}$ decomposes; but $\hat{\pi}^{s_0}\hat{M} \cong \hat{M}$ and so $\hat{M}$ decomposes. #

<u>Exercises §1</u>:

We keep the notation of §1.

1.) Let $\underline{\underline{H}}(\hat{\Lambda}) = \hat{\pi}^{s_1}\hat{R}$. Show that $\hat{M} \in {}_{\hat{\Lambda}}\underline{\underline{M}}^{o}$ decomposes if and only if $\hat{M} =_{\hat{R}} \hat{M}_1 \oplus \hat{M}_2$, where $\hat{M}_i$, $i=1,2$, are non-zero $\hat{R}$-modules such that $\hat{\Lambda}\hat{M}_1 \subset \hat{M}_1 + \hat{\pi}^{s}\hat{M}_2$, $\hat{\Lambda}\hat{M}_2 \subset \hat{M}_2 + \hat{\pi}^{s}\hat{M}_1$ for some $s > 2s_1$. (<u>Hint</u>: Use 1.3.)

2.) Assume that $\underline{\underline{H}}(\hat{\Lambda}) = \hat{R}$, and let $\hat{M}_1, \hat{M}_2 \in {}_{\hat{\Lambda}}\underline{\underline{M}}^{o}$ be irreducible. If $\varphi: \hat{M}_1 \longrightarrow \hat{M}_2$ is a non-zero map, show that $\varphi = \hat{\pi}^{s}\varphi_0$, where

$\varphi_0 : \hat{M}_1 \longrightarrow \hat{M}_2$ is an isomorphism and s is some non-negative integer. (__Hint__: Let s be the largest integer such that $\mathrm{Im}\,\varphi \subset \hat{\pi}^s \hat{M}_2$; then $\hat{\pi}^{-s}\varphi : \hat{M}_1 \longrightarrow \hat{M}_2$ is a non-zero map, which induces a non-zero map $\hat{M}_1/\hat{\pi}\hat{M}_1 \longrightarrow \hat{M}_2/\hat{\pi}\hat{M}_2$. Now use (1.1) and (1.3).)

4. Let $\underline{\underline{H}}(\Lambda) = R$ and show that for $M, N \in {}_{\Lambda}\underline{\underline{M}}^{o}$,

$$M \cong N \Longleftrightarrow KM \cong KN \Longleftrightarrow \overline{M} \cong \overline{N},$$

where "-" denotes reduction modulo π.

§2 Separable orders

Separable orders are shown to be maximal orders in algebras, that are unramified at all maximal ideals of R.

We keep the notation of §1.

2.1 **Lemma:** An $\hat{R}$-order $\hat{\Lambda}$ in $\hat{A}$ is separable if and only if $\bar{\Lambda} = \hat{\Lambda}/\hat{\pi}\hat{\Lambda}$ is a separable $\bar{R} = \hat{R}/\hat{\pi}$ $\hat{R}$-algebra.

Proof: Let us first assume that $\hat{\Lambda}$ is separable; then we have the splitting $\hat{\Lambda}^e$-sequence

$$0 \longrightarrow \operatorname{Ker}\hat{\varepsilon} \longrightarrow \hat{\Lambda}^e \xrightarrow{\hat{\varepsilon}} \hat{\Lambda} \longrightarrow 0 \quad (\text{cf. III, } 4.7),$$

which induces the split exact sequence

$$E_1 : 0 \longrightarrow \bar{R} \otimes_{\hat{R}} \operatorname{Ker}\hat{\varepsilon} \longrightarrow \bar{R} \otimes_{\hat{R}} \hat{\Lambda}^e \longrightarrow \bar{R} \otimes_{\hat{R}} \hat{\Lambda} \longrightarrow 0.$$

But $\bar{R} \otimes_{\hat{R}} \hat{\Lambda}^e \overset{\text{nat}}{\cong} \bar{\Lambda} \otimes_{\hat{R}} \hat{\Lambda}^{op}$, and the sequence

$$0 \longrightarrow \hat{\pi}\hat{\Lambda}^{op} \longrightarrow \hat{\Lambda}^{op} \longrightarrow \bar{\Lambda}^{op} \longrightarrow 0$$

gives rise to the exact sequence

$$\bar{\Lambda} \otimes_{\hat{R}} \hat{\pi}\hat{\Lambda}^{op} \xrightarrow{\sigma} \bar{\Lambda} \otimes_{\hat{R}} \hat{\Lambda}^{op} \longrightarrow \bar{\Lambda} \otimes_{\hat{R}} \bar{\Lambda}^{op} \longrightarrow 0.$$

Since $\operatorname{Im}\sigma = 0$, we get

$$\bar{R} \otimes_{\hat{R}} \hat{\Lambda}^e \cong \bar{\Lambda} \otimes_{\hat{R}} \bar{\Lambda}^{op} \cong \bar{\Lambda}^e.$$

Hence the splitting of the sequence E_1 shows that $\bar{\Lambda}$ is a projective $\bar{\Lambda}^e$-module, and thus $\bar{\Lambda}$ is a separable $\bar{R}$-algebra (cf. III, 4.7). As a consequence of this argument we see, that we have an exact sequence

$$E_2 : 0 \longrightarrow \hat{\Lambda}^e \xrightarrow{\varphi} \hat{\Lambda}^e \longrightarrow \bar{\Lambda}^e \longrightarrow 0,$$

where φ is multiplication by $\hat{\pi}$ - this does not need the fact that $\hat{\Lambda}$ is separable. Now we shall assume that $\bar{\Lambda}$ is a separable $\bar{R}$-algebra; we show that $\hat{\Lambda}$ is separable using the arguments employed already in the proof of (IV, 3.4). Since $\bar{\Lambda}$ is separable, we have $\operatorname{hd}_{\bar{\Lambda}^e}(\bar{\Lambda}) = 0$ ($\operatorname{hd}_{-}(-)$ is the homological dimension, cf. II, § 4). We have the unitary ring homomorphism $\hat{\Lambda}^e \longrightarrow \bar{\Lambda}^e$, $x \otimes y^{op} \longmapsto \bar{x} \otimes \bar{y}^{op}$, and from the change of rings theorem (II, 4.6) we get

$$\mathrm{hd}_{\hat{\Lambda}^e}(\bar{\Lambda}) \leq \mathrm{hd}_{\bar{\Lambda}^e}(\bar{\Lambda}) + \mathrm{hd}_{\hat{\Lambda}^e}(\bar{\Lambda}^e),$$

and the exact sequence E_2 implies (cf. II, 4.5)

$$\mathrm{hd}_{\hat{\Lambda}^e}(\bar{\Lambda}^e) = 1 + \mathrm{hd}_{\hat{\Lambda}^e}(\hat{\Lambda}^e).$$

Altogether this yields $\mathrm{hd}_{\hat{\Lambda}^e}(\bar{\Lambda}) \leq 1$. Now we continue as in the proof of (IV, 3.4), to conclude that $\hat{\Lambda}$ is a separable $\hat{R}$-order in $\hat{A}$. #

2.2 <u>Corollary</u>: Let $\hat{\Lambda}$ be a separable $\hat{R}$-order in $\hat{A}$. Then $\bar{\Lambda}$ is semi-simple and rad $\hat{\Lambda} = \hat{\pi}\hat{\Lambda}$.

<u>Proof</u>: This is an immediate consequence of (III, 5.9; IV, 2.6 and 2.1). #

2.3 <u>Corollary</u>: Let $\hat{\Lambda} = \hat{\Lambda}_1 \oplus \hat{\Lambda}_2$, where $\hat{\Lambda}_i$ is an $\hat{R}$-order in $\hat{A}_i$, i=1,2. Then $\hat{\Lambda}$ is separable if and only if $\hat{\Lambda}_1$ and $\hat{\Lambda}_2$ are separable.

<u>Proof</u>: $\hat{\Lambda}$ is separable if and only if $\bar{\Lambda}$ is separable (cf. 2.1) if and only if $\bar{\Lambda}_1$ and $\bar{\Lambda}_2$ are separable (cf. III, Ex. 6,7) if and only if $\hat{\Lambda}_1$ and $\hat{\Lambda}_2$ are separable. #

2.4 <u>Lemma</u>: Let $\hat{\Lambda}$ be a separable $\hat{R}$-order in $\hat{A}$. Then $\hat{M} \in {}_{\hat{\Lambda}}\underline{\underline{M}}^o$ is irreducible if and only if $\bar{M} \in {}_{\bar{\Lambda}}\underline{\underline{M}}^o$ is simple.

<u>Proof</u>: If $\hat{M}$ is reducible, $\bar{M}$ can not be simple. Conversely, assume that $\bar{M}$ is not simple. Since $\bar{M}$ is projective, it decomposes non-trivially. But then $\hat{M}$ decomposes, since $\hat{\Lambda}$ is separable; i.e., $\underline{\underline{H}}(\hat{\Lambda}) = \hat{R}$ (cf. 1.4). #

2.5 <u>Theorem</u>: If $\hat{\Lambda}$ is separable, then $\hat{\Lambda}$ is maximal. And if $\hat{L}$ is a finite separable extension field of $\hat{K}$ with maximal $\hat{R}$-order $\hat{\underline{\Omega}}$, then $\hat{\underline{\Omega}} \otimes_{\hat{R}} \hat{\Lambda}$ is a separable $\hat{\underline{\Omega}}$-order.

<u>Proof</u>: From (V, 3.4, 4.2) it follows that $\hat{\Lambda}$ is hereditary. Thus, to show that $\hat{\Lambda}$ is maximal, it suffices to assume that $\hat{A}$ is simple (cf. 2.3, IV, 4.3, 4.5). In view of (2.1 and IV, 3.7) separable $\hat{R}$-orders are

invariant under Morita equivalences. Since $\hat{\Lambda}$ is semi-perfect (IV, 2.1), we may assume that in the decomposition

$$\hat{\Lambda} = \oplus_{i=1}^{n} \hat{\Lambda} e_i$$

into indecomposable lattices, $\hat{\Lambda} e_i \not\cong \hat{\Lambda} e_j$ for $i \neq j$. If now $\hat{M}$ is an irreducible $\hat{\Lambda}$-lattice, then $\hat{M} \cong \hat{\Lambda} e_i$ for some i, since $\hat{\Lambda}$ is hereditary and because of the Krull-Schmidt theorem for projective $\hat{\Lambda}$-lattices (cf. III, 7.7). Because of the method of lifting idempotents, $\bar{\Lambda} = \hat{\Lambda}/\hat{\pi}\hat{\Lambda}$ is a direct sum of n skewfields (cf. 2.4, IV, 3.5), say $\bar{\Lambda} = \oplus_{i=1}^{n} \underline{\underline{k}}_i$. Comparing the dimensions, one finds $n = 1$. In fact, $\underline{\underline{k}}_i \cong \hat{\Lambda} e_i/\hat{\pi}\hat{\Lambda} e_i$, and if $\hat{A} = (\hat{D})_n$, $\hat{D}$ a skewfield over $\hat{K}$ of dimension m over $\hat{K}$, then $\dim_{\hat{R}}(\mathrm{End}_{\hat{\Lambda}}(\hat{M})) = m$, $\hat{R}$ being a principal ideal ring. However, $\hat{M}$ is projective and thus $\mathrm{End}_{\hat{\Lambda}}(\hat{M}/\hat{\pi}\hat{M}) \cong \mathrm{End}_{\hat{\Lambda}}(\hat{M})/\hat{\pi}\,\mathrm{End}_{\hat{\Lambda}}(\hat{M})$ (cf. proof of IV, 3.7); i.e., $\dim_{\bar{R}} \mathrm{End}_{\bar{\Lambda}}(\hat{M}/\hat{\pi}\hat{M}) = m$. On the other hand $\hat{M}/\hat{\pi}\hat{M} \cong \underline{\underline{k}}_1$ and $\mathrm{End}_{\hat{\Lambda}/\hat{\pi}\hat{\Lambda}}(\hat{M}/\hat{\pi}\hat{M}) \cong \underline{\underline{k}}_1$. But $\dim_{\bar{R}}(\hat{M}/\hat{\pi}\hat{M}) = m \cdot n$, $\hat{M}$ being irreducible. Thus $m = nm$ and $n = 1$; i.e., $\hat{\Lambda}$ is a hereditary $\hat{R}$-order in a skewfield $\hat{D}$. However, by (V, 4.10 and IV, 5.2) there are no non-maximal hereditary $\hat{R}$-orders in $\hat{D}$; i.e., $\hat{\Lambda}$ is maximal.

If now $\hat{L}$ is a finite separable extension of $\hat{K}$, then the maximal $\hat{R}$-order $\hat{\Omega}$ in $\hat{L}$ exists (cf. IV, 4.6) and the exact sequence

$$0 \longrightarrow \hat{\Omega} \otimes_{\hat{R}} \mathrm{Ker}\,\hat{\varepsilon} \longrightarrow \hat{\Omega} \otimes_{\hat{R}} \hat{\Lambda}^e \longrightarrow \hat{\Omega} \otimes_{\hat{R}} \hat{\Lambda} \longrightarrow 0$$

is split exact, $\hat{\Lambda}$ being separable. But $\hat{\Omega} \otimes_{\hat{R}} \hat{\Lambda}^e \cong (\hat{\Omega} \otimes_{\hat{R}} \hat{\Lambda})^e$ and thus $\hat{\Omega} \otimes_{\hat{R}} \hat{\Lambda}$ is separable. #

2.6 __Remark__: If $\hat{\Lambda}$ is a maximal $\hat{R}$-order in $\hat{A}$, then $\hat{\Omega} \otimes_{\hat{R}} \hat{\Lambda}$ need not be maximal, where $\hat{\Omega}$ is the maximal $\hat{R}$-order in some separable extension of $\hat{K}$ (cf. Ex. 2,2). In fact, if $\hat{\Lambda}$ stays maximal under all "extensions of the ground ring", then $\hat{\Lambda}$ is separable (cf. 2.8).

2.7 __Lemma__: Let $\hat{\Lambda}$ be a separable $\hat{R}$-order in $\hat{A}$. Then every maximal $\hat{R}$-order in $\hat{A}$ is separable.

__Proof__: We may assume $\hat{A}$ to be simple (cf. 2.3). Let $\hat{\Gamma}$ be a maximal

$\hat{R}$-order in $\hat{A}$. Since $\hat{\Lambda}$ is separable, it is maximal (cf. 2.4) and thus, $\hat{\Lambda}$ and $\hat{\Gamma}$ are conjugate (cf. IV, 5.8); i.e., $\hat{\Lambda} = a\hat{\Gamma}a^{-1}$ for some regular element $a \in \hat{A}$. Hence we obtain an $\bar{R}$-algebra isomorphism $\bar{\Lambda} \longrightarrow \bar{\Gamma}$, "-" denoting reduction modulo rad $\hat{R}$. Since $\bar{\Lambda}$ is separable, so is $\bar{\Gamma}$ (cf. Ex. 2,1). But then $\hat{\Gamma}$ is separable (cf. 2.1). #

2.8 __Theorem:__ Let $\hat{\Lambda}$ be an $\hat{R}$-order in the separable finite dimensional $\hat{K}$-algebra $\hat{A}$. Let $\hat{L}$ be a finite dimensional separable splitting field for $\hat{A}$ and let $\hat{\Omega}$ be the maximal $\hat{R}$-order in $\hat{L}$. Then $\hat{\Lambda}$ is separable if and only if $\hat{\Omega} \otimes_{\hat{R}} \hat{\Lambda}$ is maximal.

__Proof:__ By (III, 6.13) there exist finite separable splitting fields for $\hat{A}$; let $\hat{L}$ with maximal $\hat{R}$-order $\hat{\Omega}$ be one of them. Then $\hat{\Omega}$ is a free $\hat{R}$-module with a finite basis. If $\hat{\Lambda}$ is separable, $\hat{\Omega} \otimes_{\hat{R}} \hat{\Lambda}$ is maximal by (2.5). __Conversely__, assume that $\hat{\Omega} \otimes_{\hat{R}} \hat{\Lambda}$ is maximal. We __claim__ that $\hat{\Omega} \otimes_{\hat{R}} \hat{\Lambda} = \hat{\Lambda}'$ is separable. In view of (2.7) it suffices to show that some maximal $\hat{\Omega}$-order in $\hat{A}' = \hat{L} \otimes_{\hat{K}} \hat{A}$ is separable, and we may assume that $\hat{A}'$ is simple (cf. 2.3), say $\hat{A}' = (\hat{L})_n$ - $\hat{L}$ is a splitting field for $\hat{A}$. Then $\hat{\Gamma}' = (\hat{\Omega})_n$ is a maximal $\hat{R}$-order in $\hat{A}'$ and $\hat{\Gamma}'/\text{rad}\ \hat{\Omega}\cdot\hat{\Gamma}' =$ $= (\hat{\Omega}/\text{rad}\ \hat{\Omega})_n$ is a separable $\hat{\Omega}/\text{rad}\ \hat{\Omega}$-algebra and $\hat{\Gamma}'$ is a separable $\hat{\Omega}$-order (cf. 2.1). Now we turn to the original situation where $\hat{A}$ is a separable finite dimensional $\hat{K}$-algebra, and $\hat{\Omega} \otimes_{\hat{R}} \hat{\Lambda}$ is a separable $\hat{\Omega}$-order. Assume that for some $\hat{X} \in {}_{\hat{\Lambda}^e}\underline{\underline{M}}^f$, $\text{Ext}^1_{\hat{\Lambda}^e}(\hat{\Lambda},\hat{X}) \neq 0$. Then $\hat{\Omega} \otimes_{\hat{R}} \text{Ext}^1_{\hat{\Lambda}^e}(\hat{\Lambda},\hat{X}) \neq 0$, since $\hat{\Omega}$ is a free $\hat{R}$-module with a finite number of generators. In particular, $\hat{\Omega}$ is $\hat{R}$-flat and we apply (III, 1.2) to conclude

$$\hat{\Omega} \otimes_{\hat{R}} \text{Ext}^1_{\hat{\Lambda}^e}(\hat{\Lambda},\hat{X}) \cong \text{Ext}^1_{\hat{\Omega} \otimes_{\hat{R}} \hat{\Lambda}^e}(\hat{\Omega} \otimes_{\hat{R}} \hat{\Lambda}, \hat{\Omega} \otimes_{\hat{R}} \hat{X}) \neq 0.$$

But $\hat{\Omega} \otimes_{\hat{R}} \hat{\Lambda}^e \cong (\hat{\Omega} \otimes_{\hat{R}} \hat{\Lambda})^e$. Since $\hat{\Omega} \otimes_{\hat{R}} \hat{\Lambda}$ is separable, we have obtained a contradiction; i.e., $\hat{\Lambda}$ is separable. #

2.9 __Remark:__ The above proof shows that maximal orders in split algebras are separable, and in the next theorem we shall demonstrate that algebras, in which there exist separable orders, can not be "too far off" from being split.

2.10 <u>Theorem</u>: Assume that $\hat{R}/\mathrm{rad}\,\hat{R}$ is a finite field. $\hat{\Lambda}$ is a separable order in the simple separable $\hat{K}$-algebra $\hat{A}$ if and only if $\hat{\Lambda}$ is maximal and $\hat{A} = (\hat{C})_n$, where $\hat{C}$ is an unramified field extension of $\hat{K}$.

<u>Proof</u>: Let $\hat{A} = (\hat{D})_n$, where $\hat{D}$ is a separable skewfield over $\hat{K}$. If $\hat{\Lambda}$ is a separable $\hat{R}$-order in $\hat{A}$, then we may assume $\hat{\Lambda} = (\hat{\Omega})_n$, where $\hat{\Omega}$ is the unique maximal $\hat{R}$-order in $\hat{D}$ (cf. IV, 5.2; VI, 2.7). The condition $\mathrm{rad}\,\hat{\Lambda} = \hat{\pi}\hat{\Lambda}$, where $\hat{\pi}\hat{R} = \mathrm{rad}\,\hat{R}$ implies $\mathrm{rad}\,\hat{\Omega} = \hat{\pi}\hat{\Omega}$. In particular, we conclude that the center $\hat{\Sigma}$ of $\hat{\Omega}$ is unramified over $\hat{R}$ (i.e., $\mathrm{rad}\,\hat{\Sigma} = \hat{\pi}\hat{\Sigma}$). Since $\hat{R}$ has a finite residue class field, the same is true for $\hat{\Sigma}$, and we apply (IV, 6.7) to conclude that $\hat{\Omega} = \hat{\Sigma}$; i.e., $\hat{A} = (\hat{C})_n$, where the center $\hat{C}$ of $\hat{A}$ is unramified over $\hat{K}$.
<u>Conversely</u>, if $\hat{A} = (\hat{C})_n$, where $\hat{C}$ is an unramified field extension of $\hat{K}$, then one shows as in the proof of (2.8) that every maximal $\hat{R}$-order in $\hat{A}$ is separable. #

2.11 <u>Remark</u>: Using (2.10) one can characterize separable R-orders globally*) if we assume that R has finite residue class fields modulo the maximal ideals: An R-order Λ in the finite dimensional separable K-algebra A is separable if and only if

(i) Λ is maximal,

(ii) the centers of the simple components of A are unramified at every prime ideal $\underline{\underline{p}}$ of R,

(iii) A is unramified at every maximal ideal $\underline{\underline{p}}$ of R.

(For the definitions we refer to IV, 6.5.) We point out, that Λ can be separable in $A = (D)_n$, though D is not commutative (cf. 6.8).

---

*) i.e., R is any Dedekind domain.

Exercises §2:

We keep the notation of §2.

1.) Let $\hat{\Lambda}_1$ and $\hat{\Lambda}_2$ be maximal $\hat{R}$-orders in $\hat{A}$. Show that $\bar{\Lambda}_1$ is separable if and only if $\bar{\Lambda}_2$ is separable, where "-" denotes reduction modulo rad $\hat{R}$.

2.) Construct an example of a maximal $\hat{R}$-order $\hat{\Lambda}$ in $\hat{A}$ and a finite separable extension field $\hat{L}$ of $\hat{K}$ with maximal $\hat{R}$-order $\hat{\Omega}$ such that $\hat{\Omega} \otimes_{\hat{R}} \hat{\Lambda}$ is not maximal. (Hint: Use the results of IV, §6, in particular 6.16.)

3.) Prove the statements of (2.11).

4.) If $\hat{\Gamma}$ is a maximal $\hat{R}$-order in the simple separable $\hat{K}$-algebra $\hat{A}$, show that - up to isomorphism - there exists only one simple $\hat{\Gamma}$-module.

§3 The Krull-Schmidt theorem

It is shown that the Krull-Schmidt theorem is valid for lattices over orders which are semi-perfect. The Krull-Schmidt theorem is locally valid for projective lattices over commutative orders. Cancellation is allowed locally, and we prove a local analogue to the theorem of Noether-Deuring.

We keep the notation of § 1; in particular R is the localization of a Dedekind domain at a maximal ideal and by "$\hat{\ }$" we denote the corresponding completion.

3.1 Theorem (Borevich-Faddeev [1]; Reiner [6]): If Λ is an R-order in the separable finite dimensional K-algebra A, then the Krull-Schmidt theorem is valid for Λ-lattices if $\mathrm{End}_\Lambda(M)$ is semi-perfect for every indecomposable $M \in {}_\Lambda\underline{\underline{M}}^o$.

Proof: Because of (I, 4.10) it is enough to show that $\mathrm{End}_\Lambda(M)$ is completely primary if M is an indecomposable Λ-lattice. This means that we have to show $\mathrm{End}_\Lambda(M)/\mathrm{rad}\ \mathrm{End}_\Lambda(M)$ is a skewfield. However $\pi\,\mathrm{End}_\Lambda(M) \subset \mathrm{rad}\ \mathrm{End}_\Lambda(M)$ (cf. IV, 2.6) and $\mathrm{End}_\Lambda(M)/\mathrm{rad}\ \mathrm{End}_\Lambda(M)$ is a semi-simple $R/\pi R$-algebra. Thus it suffices to show that $\mathrm{End}_\Lambda(M)/\mathrm{rad}\ \mathrm{End}_\Lambda(M)$ does not have any non-trivial idempotents. However, this follows from the method of lifting idempotents, since $\mathrm{End}_\Lambda(M)$ is semi-perfect, and since $\mathrm{End}_\Lambda(M)$ does not have non-trivial idempotents, M being indecomposable. #

3.2 Corollary: The Krull-Schmidt theorem is valid for $\hat{\Lambda}$-lattices.

Proof: $\mathrm{End}_{\hat{\Lambda}}(\hat{M})$ is semi-perfect (cf. IV, 2.1) and we can apply (3.1). #

3.3 Corollary (Heller [1]): Assume that $A = \oplus_{i=1}^{n} (D_i)_{n_i}$ is the decomposition of A into simple K-algebras. If $\hat{D}_i$ is a skewfield $1 \le i \le n$, then the Krull-Schmidt theorem is valid for Λ-lattices.

Proof: In view of (3.2) and (1.2) it suffices to show that for $M \in {}_\Lambda\underline{\underline{M}}^o$, M is indecomposable if and only if $\hat{M} \in {}_{\hat{\Lambda}}\underline{\underline{M}}^o$ is indecomposable if all D_i are skewfields;

i.e., $\mathrm{End}_\Lambda(M)$ has no non-trivial idempotents if and only if $\mathrm{End}_{\hat\Lambda}(\hat M)$ has no non-trivial idempotents. Since $K \otimes_R \mathrm{End}_\Lambda(M)$ satisfies the same hypotheses as does A (cf. III, 5.5), it suffices to show that an R-order Λ in A is indecomposable as module if and only if $\hat\Lambda$ is indecomposable as module. Assume that $\hat\Lambda$ decomposes, say $\hat\Lambda = \hat\Lambda\hat e_1 \oplus \hat\Lambda\hat e_2$, where $\hat e_1$ and $\hat e_2$ are orthogonal idempotents. Because of the hypotheses on A, all idempotents in $\hat A$ come from idempotents in A; i.e., there exist idempotents e_1 and e_2 in A such that

$$\hat K \otimes_K Ae_i \cong \hat A\hat e_i, \ i=1,2.$$

According to (IV, 1.9) there exist Λ-lattices M_i such that $\hat M_i \cong \hat\Lambda\hat e_i$, $i=1,2$, and thus $\hat\Lambda \cong \hat M_1 \oplus \hat M_2$; now we use (1.2) to conclude $M_1 \oplus M_2 \cong \Lambda$; i.e., Λ decomposes. The other direction is obvious. #

3.4 <u>Remark</u>: The hypotheses of (3.3) are satisfied if $A = \oplus_{i=1}^{n} (K_i)_{n_i}$, where K_i are extension fields of K, which are unramified at π (i.e., if R_i is the maximal R-order in K_i, then $\mathrm{rad}\, R_i = \pi R_i$). Thus, in particular the Krull-Schmidt theorem is valid if K splits A. We remark also that under the assumptions of (3.3), Λ is semi-perfect. This follows readily from the proof.

3.5 <u>Corollary</u> (Reiner [6]): For an R-order Λ in A, cancellation is allowed in ${}_{\Lambda}\underline{\underline{M}}^{o}$; i.e., for -lattices M,N,X we have

$$M \oplus X \cong N \oplus X \text{ if and only if } M \cong N.$$

<u>Proof</u>: We have

$$M \oplus X \cong N \oplus X \Longleftrightarrow \hat M \oplus \hat X \cong \hat N \oplus \hat X \Longleftrightarrow \hat M \cong \hat N \Longleftrightarrow M \cong N$$

(cf. 1.2, 3.2). #

3.6 <u>Corollary</u> (Reiner [6]): For $M,N \in {}_{\Lambda}\underline{\underline{M}}^{o}$, we have

$$M^{(n)} \cong N^{(n)} \text{ if and only if } M \cong N,\ n \in \underline{\underline{N}}.$$

<u>Proof</u>: By (1.2), $M^{(n)} \cong N^{(n)}$ if and only if $\hat M^{(n)} \cong \hat N^{(n)}$. However,

the Krull-Schmidt theorem is valid for ${}_{\hat{\Lambda}}\underline{\underline{M}}^o$, and thus $\hat{M}^{(n)} \cong \hat{N}^{(n)}$ if and only if $\hat{M} \cong \hat{N}$ and another application of (1.2) shows $M \cong N$. #

3.7 Theorem (Roggenkamp [4]): Let A be a commutative separable algebra. For an R-order Λ, the Krull-Schmidt theorem is valid for the projective Λ-lattices.

Proof: Since A is commutative, every idempotent of Λ is central, and we may assume that Λ is indecomposable as Λ-lattice. In fact, if $\Lambda = \oplus_{i=1}^{n} \Lambda_i$ is the decomposition of Λ into indecomposable orders, then each Λ_i is indecomposable as module, A being commutative. If we have two decompositions of $M \in {}_{\Lambda}\underline{\underline{M}}^o$ into indecomposable lattices,

$$M = \oplus_{i=1}^{s} M_i = \oplus_{j=1}^{t} N_j,$$

then for each $1 \leq k \leq n$, we have

$$\Lambda_k M = \oplus_{i=1}^{s} \Lambda_k M_i = \oplus_{i=1}^{t} \Lambda_k N_j,$$

and since M_i and N_j are indecomposable, there exists for each i exactly one Λ_k such that $\Lambda_k M_i = M_i$ and $\Lambda_k M_{i'} = 0$ for $i \neq i'$. Thus, if the Krull-Schmidt theorem is valid for the projective Λ_i-lattices, $1 \leq i \leq n$, then it is valid for the projective Λ-lattices.

From now on Λ is an indecomposable commutative order, and it suffices to show that every indecomposable projective Λ-lattice is isomorphic to Λ. Let $\{e_i\}_{1 \leq i \leq n}$ be a complete set of non-equivalent primitive idempotents of A; we write

$$\hat{A}e_i = \oplus_{j=1}^{s_i} \hat{A}\hat{e}_{ij}, \quad 1 \leq i \leq n,$$

where $\{\hat{e}_{ij}\}_{\substack{1 \leq i \leq n \\ 1 \leq j \leq s_i}}$ are the non-equivalent primitive idempotents of $\hat{A}$. (Observe that A is commutative.) Let

$$\hat{\Lambda} = \oplus_{j=1}^{n} \hat{\Lambda}\,\hat{\varepsilon}_j,$$

where $\{\hat{\varepsilon}_j\}_{1 \leq j \leq n}$ are the non-equivalent primitive idempotents of $\hat{\Lambda}$. Now, for a projective indecomposable Λ-lattice P, we have

$$KP \cong \oplus_{i=1}^{n} Ae_i^{(\alpha_i)},$$

$$\hat{P} \cong \oplus_{j=1}^{n} \hat{\Lambda}\hat{\varepsilon}_j^{(\beta_j)}.$$

If $\beta_j > 0$ for all j, then Λ is a direct summand of P; in fact, $\beta_j > 0$ for all j, implies $\hat{P} \cong \hat{\Lambda} \oplus \hat{X}$, where $\hat{X}$ is a projective $\hat{\Lambda}$-lattice. However, $(\hat{K} \otimes_K KP)/\hat{\Lambda} \cong \hat{K} \otimes_{\hat{R}} \hat{X}$, and by (IV, 1.9) there exists a Λ-lattice M with $\hat{M} \cong \hat{X}$; i.e., $\hat{R} \otimes_R (\Lambda \oplus M) \cong \hat{P}$. Applying (1.2), we conclude $P \cong \Lambda \oplus M$ and Λ is a direct summand of P. Since P was assumed to be indecomposable, $P \cong \Lambda$. If $\beta_1, \ldots, \beta_m = 0$ for some $1 \leq m < n$ and $\beta_j > 0$ for $j > m$, we put $\hat{M} = \oplus_{j=1}^{m} \hat{\Lambda}\hat{\varepsilon}_j$ and $L = \oplus_{i'} Ae_{i'}$, where the sum is taken over all those i' for which $\alpha_{i'} = 0$. Then $\hat{K} \otimes_K L = \hat{K} \otimes_{\hat{R}} \hat{M}$, and there exists a Λ-lattice M_1, such that $\hat{M}_1 \cong \hat{M}$. However, $\hat{M}$ is a proper direct summand of $\hat{\Lambda}$ and thus M_1 is a proper direct summand of Λ, a contradiction. Thus $P \cong \Lambda$ and the Krull-Schmidt theorem is valid for the projective Λ-lattices. #

3.8 <u>Theorem</u> (Reiner-Zassenhaus [1]): Let K' be a finite dimensional separable extension field of K and let R' be the integral closure of R in K'. If $\underline{\underline{p}}'$ is a maximal ideal of R' containing $\pi R'$, $\pi R = \text{rad } R$, we let $R'_{\underline{\underline{p}}'}$ be the localization of R' at $\underline{\underline{p}}'$. For an R-order Λ in A, $R' \otimes_R \Lambda = \Lambda'$ is an R-order in $A' = K' \otimes_K A$ and if $M, N \in {}_{\Lambda}\underline{\underline{M}}^o$, then

$$M \cong_{\Lambda} N \text{ if and only if } R'_{\underline{\underline{p}}'} \otimes_R M \cong_{\Lambda'_{\underline{\underline{p}}'}} R'_{\underline{\underline{p}}'} \otimes_R N.$$

<u>Proof</u>: Since A is separable, A' is a separable K'-algebra and Λ' is an R'-order in A'. Obviously, $M \cong N$ implies $R'_{\underline{\underline{p}}'} \otimes_R M \cong R'_{\underline{\underline{p}}'} \otimes_R N$.

<u>Conversely</u>, R' is an R-order in K' and thus $\hat{R} \otimes_R R'$ is an $\hat{R}$-order in $\hat{K} \otimes_K K'$. However, if $\pi R' = \prod_{i=1}^{n} \underline{\underline{p}}_i^{\alpha_i}$ is the decomposition of $\pi R'$ into maximal ideals in R', then

$$\hat{R} \otimes_R R' \cong \oplus_{i=1}^{n} R'_i$$

(cf. IV, 5.9), and each R_1' has a unique maximal ideal $\hat{R} \otimes_R \underline{\underline{p}}_1 = \hat{\underline{\underline{p}}}_1$ and it follows from (IV, 2.2) that R_1' is complete with respect to the $\hat{\underline{\underline{p}}}_1$-adic topology; i.e.,

$$R_1' = \hat{R}'_{\underline{\underline{p}}_1} .$$

Since each $\hat{R}'_{\underline{\underline{p}}_1}$ is an $\hat{R}$-order, it is finitely generated as $\hat{R}$-module (cf. VI, 1.1). In particular, $\hat{R}'_{\underline{\underline{p}}'}$ is $\hat{R}$-free, say $\hat{R}'_{\underline{\underline{p}}'} \cong_{\hat{R}} \hat{R}^{(m)}$ for some m. Now $R'_{\underline{\underline{p}}'} \otimes_R M = R'_{\underline{\underline{p}}'} \otimes_R N$ implies $\hat{R}'_{\underline{\underline{p}}'} \otimes_R M \cong \hat{R}'_{\underline{\underline{p}}'} \otimes_R N$ and hence $\hat{M}^{(m)} \cong \hat{N}^{(m)}$ as $\hat{\Lambda}$-modules, since $\hat{R}'_{\underline{\underline{p}}'} \otimes_{\hat{R}} \hat{\Lambda} =_{\hat{\Lambda}} \hat{\Lambda}^{(m)}$. Now we apply (1.2) and (3.6) to conclude $M \cong N$. #

We remark that (3.8) is an integral version of the Noether-Deuring theorem, which states that for A-modules L_1, L_2 we have $L_1 \cong_A L_2$ if and only if $K' \otimes_K L_1 \cong K' \otimes_K L_2$ (cf. Ex. 3,2).

3.9 Remarks on the Krull-Schmidt theorem: We now assume that R is any Dedekind domain with quotient field K and Λ is an R-order in the separable finite dimensional K-algebra A. If R is not a principal ideal domain, then surely the Krull-Schmidt theorem cannot hold for Λ-lattices. For, if $\underline{\underline{a}}$ and $\underline{\underline{b}}$ are coprime (i.e., relatively prime) ideals in R, then

$$R \oplus \underline{\underline{a}}\,\underline{\underline{b}} \cong \underline{\underline{a}} \oplus \underline{\underline{b}} \quad \text{(cf. Ex. 3,1).}$$

But even if R is a principal ideal domain, then the Krull-Schmidt theorem does not hold for Λ-lattices in general (cf. Reiner [8], Ex. 3,3; 3,5). If $R^{\#}$ is the localization of R at some maximal ideal of R, then we have seen in (IV, 5.7) that the Krull-Schmidt theorem holds for $\Gamma^{\#}$-lattices in case $\Gamma^{\#}$ is a maximal order. It also holds for $\Lambda^{\#}$-lattices in case the hypotheses of (3.3) are satisfied and it also holds for projective $\Lambda^{\#}$-lattices in case A is commutative. However, these seem to be the only general cases, where the Krull-Schmidt theorem holds for ${}_{\Lambda^{\#}}\underline{\underline{M}}^{o}$ or for ${}_{\Lambda^{\#}}\underline{\underline{P}}^{f}$ (cf. Berman-Gudivok [1], Roggenkamp [4]). In the latter paper it is shown, that not even the non-equivalent primitive idempotents of $\Lambda^{\#}$ need to be unique; and even if they are unique, one still can construct examples, where there are projective indecomposable $\Lambda^{\#}$-lattices, the rank of which is strictly larger than the rank of $\Lambda^{\#}$. We shall see in (IX, 2.29) that the Krull-Schmidt theorem need not even hold for $\Lambda^{\#}$-lattices in case $\Lambda^{\#}$ is hereditary.

Exercises §3:

1.) Let R be a Dedekind domain and $\underline{\underline{a}}$ and $\underline{\underline{b}}$ coprime ideals; show that $\underline{\underline{a}} \oplus \underline{\underline{b}} \cong R \oplus \underline{\underline{a}}\,\underline{\underline{b}}$.

2.) Noether-Deuring theorem: Let A be a separable K-algebra and let K' be a separable extension field of K. If $A' = K' \otimes_K A$ and if $M, N \in {}_A\underline{\underline{M}}^f$, then

$$M \cong_A N \text{ if and only if } K' \otimes_K M \cong_{A'} K' \otimes_K N.$$

(Hint: Prove the theorem first when K' is a finite extension of K. For the general case, let K" be an extension of K' which contains a splitting field S for A with $[S : K] < \infty$.)

3.) Let R be the localization of a Dedekind domain at some maximal ideal. Construct an example of an R-order Λ in some separable K-algebra for which the Krull-Schmidt theorem is not valid for projective Λ-lattices. (Hint: Let $p \in \underline{\underline{Z}}$ be a rational prime number, and let K with ring of integers R be an algebraic number field such that $pR = \underline{\underline{p}}_1\underline{\underline{p}}_2$, where $\underline{\underline{p}}_1$ and $\underline{\underline{p}}_2$ are distinct prime ideals of R. If "$\hat{-}$" denotes the $p\underline{\underline{Z}}$-adic completion, then $\hat{R} = \hat{R}_1 \oplus \hat{R}_2$, where $\hat{R}_1$ and $\hat{R}_2$ are complete Dedekind domains. Let $A = (K)_2$; then $\hat{A} = (\hat{K}_1)_2 \oplus (\hat{K}_2)_2$, where $\hat{K}_i$ is the quotient field of $\hat{R}_i$, i=1,2. By $\hat{\pi}_i$ we denote the prime element of $\hat{R}_i$; i.e., $\hat{\pi}_i\hat{R}_i = \operatorname{rad} \hat{R}_i$, i=1,2. In $\hat{A}$ we consider the following $\hat{\underline{\underline{Z}}}_p$-order

$$\hat{\Lambda} = \hat{\Lambda}_1 \oplus \hat{\Lambda}_2 \quad \text{with}$$

$$\hat{\Lambda}_i = \left\{ \begin{pmatrix} \alpha & \beta \\ \hat{\pi}_i\gamma & \delta \end{pmatrix}, \alpha, \beta, \gamma, \delta \in \hat{R}_i \right\}, \quad i=1,2.$$

Then $\hat{\Lambda}$ has four irreducible non-isomorphic projective lattices. And if we put $\Lambda = \hat{\Lambda} \cap (K)_2$, then one can construct irreducible projective Λ-lattices M_1, M_2, N_1, N_2 such that $M_1 \oplus M_2 \cong N_1 \oplus N_2$ but $M_1 \not\cong N_1$ and $M_1 \not\cong N_2$.)

4.) Again R is the localization of a Dedekind domain with quotient field K, and K' a finite separable extension field. Let S be the integral closure of R in K'. If $\{\pi_i S\}_{1 \le i \le t}$ are the maximal ideals in S, then

$$\hat{K} \otimes_K K' = \oplus_{i=1}^t \hat{K}_i, \text{ where } \{\hat{K}_i\}_{1 \le i \le t}$$

are complete fields with rings of integers $\{\hat{R}_i\}$. Let $A = (K')_n$ and Λ an R-order A. Then $\hat{A} = \oplus_{i=1}^t (\hat{K}_i)_n$ and we put $\hat{A}_i = (\hat{K}_i)_n$.

Prove the following statements:

(i) For $P \in {}_{\Lambda}\underline{P}^f$, we have $\hat{K} \otimes_{\hat{R}} \hat{P} = \oplus_{i=1}^t \hat{L}_i$, $\hat{L}_i \in {}_{\hat{A}_i}\underline{M}^f$, and $(\hat{L}_i : \hat{K}_i) = d$ is independent of i.

(ii) If we have for some $\hat{P} \in {}_{\hat{\Lambda}}\underline{P}^f$

$$\hat{K} \otimes_{\hat{R}} \hat{P} = \oplus_{i=1}^t \hat{L}_i, \quad \hat{L}_i \in {}_{\hat{A}_i}\underline{M}^f,$$

and if $(\hat{L}_i : \hat{K}_i) = d$ is independent of i, then there exists $P_1 \in {}_{\Lambda}\underline{P}^f$ with $\hat{P}_1 \cong \hat{P}$.

(iii) Let for $P \in {}_{\Lambda}\underline{P}^f$,

$$\hat{P} = \oplus_{i=1}^s \hat{M}_i$$

be the decomposition of $\hat{P}$ into indecomposable submodules. Then P is decomposable if and only if there exists a proper subset $\{\hat{N}_j\}_{1 \le j \le s'}$

of $\{M_i\}_{1 \le i \le s}$ such that

$$\hat{K} \otimes_K (\oplus_{j=1}^{s'} \hat{N}_j) = \oplus_{i=1}^{t} \hat{L}_i, \ \hat{L}_i \in {}_{\hat{\Lambda}_i}\underline{\underline{M}}^f$$

and $(\hat{L}_i : \hat{K}_i) = d$ is independent of i.

5.) Use Ex. 4 to construct an example of an R-order Λ in a full matrix ring, where $1 \in \Lambda$ has two different decompositions into non-equivalent primitive idempotents in Λ. Also construct an example of an R-order Λ and an indecomposable projective lattice P with rank(P) > rank(Λ).

§4 The Jordan-Zassenhaus theorem

It is proved that an order in a semi-simple algebra A over an algebraic number field K has only finitely many non-isomorphic lattices that span a fixed A-module. The same statement is proved if K is a finite extension field of the field of the rational functions over a finite field.

In this section we assume first that K is an algebraic number field with ring of integers R and Λ is an R-order in the semi-simple K-algebra A. We remark that A is automatically separable since K has characteristic zero (cf. III, Ex. 6,4).

4.1 __Theorem:__ Let Γ be a maximal R-order in A and let $L \in {}_A\underline{\underline{M}}^f$ be given. Then the set $\underline{\underline{M}}_\Gamma(L) = \{M \in {}_\Gamma\underline{\underline{M}}^o : KM \cong L\}$ contains only finitely many, say $m_\Gamma(L)$, non-isomorphic Γ-lattices.

__Proof:__ Since Γ is hereditary, it suffices to prove the statement for a simple A-module L. If $A = \oplus_{i=1}^n A_i$ is the decomposition of A into simple K-algebras, then Γ decomposes accordingly, and there exists exactly one $1 \leq i \leq n$ such that $A_i L \neq 0$. Thus, we may assume that A is simple. If $M \in \underline{\underline{M}}_\Gamma(L)$, then M is irreducible, and we have a Morita equivalence between ${}_\Gamma\underline{\underline{M}}^o$ and ${}_\Omega\underline{\underline{M}}^o$, where $\Omega = \mathrm{End}_\Gamma(M)$ is a maximal order in a skewfield D over K (cf. IV, 5.4; III, 2.1). Moreover, the set $\underline{\underline{M}}_\Gamma(L)$ corresponds to the set of left Ω-ideals in D. Since R is the ring of algebraic integers in K, Ω can be viewed as a $\underline{\underline{Z}}$-order in the finite dimensional skewfield D over $\underline{\underline{Q}}$ (cf. IV, Ex. 4,3).

4.2 __Definition:__ Let I be a left Ω-ideal in Ω.*) __The norm of I__ is defined as

$$N(I) = \text{number of elements in } \Omega/I.$$

Then N(I) is finite, since Ω/I is a finitely generated $\underline{\underline{Z}}$-torsion module, and since $\underline{\underline{Z}}$ has finite residue rings modulo its non-zero ideals.

4.3 __Lemma:__ Let $\underline{\varphi}$ be a $\underline{\underline{Z}}$-linear transformation, mapping a fixed $\underline{\underline{Z}}$-basis

*) This means $KI = \Omega$ and $I \subseteq \Omega$.

of Ω onto a fixed $\underline{\underline{Z}}$-basis of the left Ω-ideal I in Ω. Then

$$N(I) = |\det(\underline{\underline{\varphi}})|,$$

where "$|\_|$" denotes the absolute value and $\underline{\underline{\varphi}}$ is considered as matrix.

<u>Proof</u>: This follows immediately from the invariant factor theorem for lattices over principal ideal domains and is left as an exercise (cf. Ex. 4,1). #

4.4 <u>Proposition</u>: There exists a real positive constant c depending only on Ω and D such that in any left Ω-ideal $I \subset \Omega$, there exists $0 \neq x \in I$ such that

$$N(\Omega x) \leq c \cdot N(I).$$

<u>Proof</u>: Let $\Omega = \oplus_{i=1}^{n} \underline{\underline{Z}}\,\omega_i$; we write $\underline{\underline{\varphi}}_{\omega_i}$ for the matrix induced by right multiplication with ω_i, relative to the basis $\{\omega_j\}_{1\leq j\leq n}$. If $\{X_i\}_{1\leq i\leq n}$ are real variables, then

$$\det(\sum_{i=1}^{n} X_i \underline{\underline{\varphi}}_{\omega_i}) = F(X_1,\dots,X_n)$$

is a homogeneous polynomial in the variables $X_1,\dots,X_n$. Hence there exists a real positive constant c that depends only on the matrices $\{\underline{\underline{\varphi}}_{\omega_i}\}_{1\leq i\leq n}$ such that

$$|F(X_1,\dots,X_n)| \leq c \cdot a^n, \text{ if } |X_i| \leq a, 1\leq i\leq n.$$

This c is the desired constant. In fact, the set

$$\{\sum_{i=1}^{n} \alpha_i \omega_i : \alpha_i \in \underline{\underline{Z}},\ 0 \leq \alpha_i \leq N(I)^{1/n}\}$$

contains more than N(I) different elements, and hence the difference of two of them must lie in I; i.e., I contains an element

$$x = \sum_{i=1}^{n} \beta_i \omega_i,\ \beta_i \in \underline{\underline{Z}},\ |\beta_i| \leq N(I)^{1/n},$$

and

$$N(\Omega x) \leq c \cdot N(I), \text{ by } (4.3).$$

Now we <u>return to the proof of (4.1)</u>: Let I be a left Ω-ideal. Then I is isomorphic to an integral ideal, and we may thus assume $I \subset \Omega$. According to (4.4) we may pick $0 \neq x \in I$ such that

$$N(\Omega x) \leq cN(I).$$

Then x is invertible in D, and if we put $I' = Ix^{-1}$, then $I' \cong_\Omega I$ and $\Omega x \subset I$ implies $\Omega \subset I'$. But then

$$|I' : \Omega| = |I : \Omega x| = |\Omega : \Omega x|/|\Omega : I| = N(\Omega x)/N(I) \leq c,$$

where $|X : Y|$ denotes the number of elements in X/Y. Thus in every isomorphism class of left Ω-ideals we have found a left Ω-ideal $I \supset \Omega$ with $|I : \Omega| \leq c$. Since c is a finite number, it suffices to show that for a fixed integer $0 < m \leq c$ there exist only finitely many Ω-ideals $I \supset \Omega$ with $|I : \Omega| = m$. But $|I : \Omega| = m$ implies

$$m\Omega \subset mI \subset \Omega.$$

Since $\Omega/m\Omega$ has only finitely many elements, there are only finitely many possibilities for mI. Hence there are only finitely many non-isomorphic left Ω-ideals. This completes the proof of (4.1). #

4.5 <u>Theorem</u> (Jordan-Zassenhaus; Zassenhaus [1]): Let R be the ring of algebraic integers in an algebraic number field K and Λ an R-order in the semi-simple K-algebra A. If $L \in {}_A\underline{\underline{M}}^f$, then there are only finitely many non-isomorphic Λ-lattices that span L; i.e., $m'_\Lambda(L) < \infty$.

<u>Proof</u>: Let Γ be a maximal R-order in A containing Λ, then $m_\Gamma(L) < \infty$ by (4.1), and if $M \in \underline{\underline{M}}_\Lambda(L)$, then $\Gamma M \in \underline{\underline{M}}_\Gamma(L)$, and it suffices to show that there are only finitely many non-isomorphic Λ-lattices M with $\Gamma M \cong M_o$, where M_o is a fixed Γ-lattice. Replacing M by an isomorphic module, we may assume $\Gamma M = M_o$. If we choose $0 \neq r \in R$ such that $r\Gamma \subset \Lambda$, then

$$rM_o \subset M \subset M_o,$$

and since M_o/rM_o is a finite abelian group, R having finite residue class rings, there are only finitely many possibilities for M. #

4.6 <u>Remark</u>: In the proof of (4.1) and (4.5) essentially two properties of R were used:

(i) R has finite class number,

(ii) R has finite residue class rings.

To derive (4.5) from (4.1), only (ii) is needed; and once (i) and (ii) are satisfied by a Dedekind domain R, one can derive (4.5) as soon as one can establish an analogue to (4.4). We shall see in (5.5), that if (i) is not satisfied, it can happen that there exists a Λ-lattice M_o and infinitely many non-isomorphic Λ-lattices M with $M_{\underline{\underline{p}}} \cong M_{o_{\underline{\underline{p}}}}$ for every maximal ideal $\underline{\underline{p}}$ of R. On the other hand, (ii) guarantees that for every $L \in {}_A\underline{\underline{M}}^f$, the number $m_\Lambda(L)$ is locally finite; i.e., $m_{\Lambda_{\underline{\underline{p}}}}(L) < \infty$ for every maximal ideal $\underline{\underline{p}}$ of R (cf. 4.9).

4.7 Theorem (Higman-MacLaughlin [1]): Let $\underline{\underline{k}}$ be a finite field and X an indeterminate over $\underline{\underline{k}}$. We put $R = \underline{\underline{k}}[X]$ and $K = \underline{\underline{k}}(X)$; then R is the polynomial ring over $\underline{\underline{k}}$ and K, the field of rational functions over $\underline{\underline{k}}$, is the quotient field of R. If Λ is an R-order in the separable finite dimensional K-algebra A, and if $L \in {}_A\underline{\underline{M}}^f$, then $m_\Lambda(L) < \infty$; i.e., there are only finitely many non-isomorphic Λ-lattices spanning L.

Proof: As follows from the remark and from the proof of (4.1), it suffices to prove the analogue of (4.4).

3.8 Lemma: Let K and R be as in (4.7), and let D be a finite dimensional separable skewfield over K and Ω a maximal R-order in D. For a left Ω-ideal $I \subset \Omega$, N(I) is defined as the number of elements in Ω/I. Then there exists a real positive constant c which depends only on Ω and D and an element $0 \neq \alpha \in I$ such that

$$N(\Omega\alpha) \leq cN(I).$$

Proof: R is a principal ideal ring, and it satisfies (i) and (ii) of (4.6). If $\underline{\underline{k}}$ has q elements, then $N(I) = q^t$, where t is a positive integer depending on I. We fix an R-basis for Ω, $\{\omega_i\}_{1\leq i\leq n}$, and write

$t = s \cdot n + u$ where $0 \leq u < n$. Then the set

$$\{\sum_{i=1}^{n} f_i \omega_i : f_i \in R, \deg(f_i) \leq s\} \quad (\deg f(X) = \text{degree } f(X))$$

contains $q^{(s+1)n} > q^t$ different elements, and hence two of them are congruent modulo I; i.e., there exist $\alpha \in I$,

$$\alpha = \sum_{i=1}^{n} f_i \omega_i, \ \deg(f_i) \leq s.$$

On the other hand, it is easily checked that

$$N(\Omega\beta) = q^{\deg(\det \underline{\varphi}_\beta)} \quad \text{for } 0 \neq \beta \in \Omega,$$

where $\underline{\varphi}_\beta$ is the matrix of right multiplication by β with respect to the basis $\{\omega_i\}_{1\leq i\leq n}$. Since the determinant of $\underline{\varphi}_\beta$ is a homogeneous function of degree n in f_i, where $\beta = \sum_{i=1}^{n} f_i \omega_i$, we conclude

$$\deg(\det \underline{\varphi}_\alpha) \leq n \cdot s + c \leq t + c,$$

where c is a constant depending only on Ω and D. Thus

$$N(\Omega\alpha) = q^{\deg(\det \underline{\varphi}_\alpha)} \leq q^{t+c} = q^c \cdot N(I). \quad \#$$

This also proves (4.7). #

4.9 <u>Lemma</u>: Let R be the localization of a Dedekind domain at some maximal ideal; put rad $R = \pi R$, and assume that $R/\pi R$ is a finite field. If Λ is an R-order in the separable K-algebra A, then for every $L \in {}_A\underline{\underline{M}}^f$, there are only finitely many non-isomorphic Λ-lattices which span L.

<u>Proof</u>: Let $\pi^{s_1}R = \underline{\underline{H}}(\Lambda)$, where $\underline{\underline{H}}(\Lambda)$ is the Higman ideal of Λ. Then for every $0 \neq r \in \underline{\underline{H}}(\Lambda)$, $r \cdot 1_M$ is a projective endomorphism of M (cf. V, 2, 4.2), and if we choose $s > s_1$, then $R/\pi^s R$ is a finite ring, and $\Lambda/\pi^s\Lambda$ is also a finite ring. If $M \in \underline{\underline{M}}_\Lambda(L)$; i.e., $KM \cong L$, then $M/\pi^s M$ is $R/\pi^s R$-free on $\dim_K(L)$ elements. However, there are only finitely many $\Lambda/\pi^s\Lambda$-modules with this property. Since $s > s_1$, $M \cong N$ if and only if $M/\pi^s M \cong N/\pi^s N$ (cf. 1.2), and $\underline{\underline{m}}_\Lambda(L) < \infty$. #

4.10 <u>Definition</u>: An $\underline{\underline{A}}$-field is a <u>global field</u>; i.e., it is either an algebraic number field or a finite algebraic extension of the field of rational functions over a finite field. By an $\underline{\underline{A}}$-field K with Dedekind domain R we understand a Dedekind domain R with quotient field K. In (4.5) and (4.7) we have shown that the Jordan-Zassenhaus theorem holds for R-orders in separable K-algebras, where K is an $\underline{\underline{A}}$-field with Dedekind domain R (cf. Ex. 4,3).

Exercises §4:

1.) Let Ω be a $\underline{\underline{Z}}$-order in a skewfield D of finite dimension over $\underline{\underline{Q}}$. Prove (4.3) and show $N(\Omega\alpha) = |\det(\underline{\underline{\varphi}}_\alpha)|$ for $\alpha \in \Omega$. Under the hypotheses of (4.8) show that

$$N(\Omega\alpha) = q^{\deg(\det \underline{\underline{\varphi}}_\alpha)} \quad \text{where } 0 \neq \alpha \in \Omega.$$

2.) (Roggenkamp [1]). Let K be a field and A a finite dimensional separable K-algebra. Let R = K[X] and K' = K(X), where X is an indeterminate over K. Show that every maximal R-order Γ in $A' = K' \otimes_K A$ is a principal ideal domain, in particular (4.1) is valid for Γ. (Hint: Show first that $\Gamma_o = R \otimes_K A$ is a maximal R-order in A', which is a principal ideal ring. To see this, choose a Galois extension K_1 of K which splits A (cf. III, 6.13; IV, 6.10). Then K_1A and K_1A' are split and have the same idempotents. Then $Gal(K_1:K) = Gal(K_1(X) : K(X))$. Use this to show that A and A' have the same idempotents, and that central (primitive) idempotents are preserved. Use this to show

$$A = \oplus_{i=1}^{n} (D_i)_{n_i} \Longrightarrow A' = \oplus_{i=1}^{n} (K(X) \otimes_K D_i)_{n_i},$$

where D_i are separable skewfields over K. Thus

$$R \otimes_K A \cong \oplus_{i=1}^{n} (R \otimes_K D_i)_{n_i}.$$

Now show that $R \otimes_K D_i$ is a principal ideal ring, and that $R \otimes_K A$ is maximal and a principal ideal ring. - If now Γ is any maximal R-order in A, then Γ and Γ_o are Morita equivalent.)

3.) Let K be an $\underline{\underline{A}}$-field with Dedekind domain R and Λ an R-order in the separable K-algebra A. Show that the Jordan-Zassenhaus theorem is valid for ${}_{\Lambda}\underline{\underline{M}}^{o}$. (*Hint*: If K is an algebraic number field with ring of integers R_o, then there exists a multiplicative system S in R_o such that $R = R_{o_S}$. If K is a finite extension of $\underline{\underline{k}}(X)$, $\underline{\underline{k}}$ a finite field, then the integral closure R_o of $\underline{\underline{k}}[X]$ in K is a Dedekind domain with quotient field K, and then there exists a multiplicative system S such that $R = R_{o_S}$. Observe that K is a separable extension of $\underline{\underline{k}}(X)$ and so R_o is a $\underline{\underline{k}}[X]$-order.)

§5 Irreducible lattices

It is shown that the absolutely irreducible Λ-lattices are in a one-to-one correspondence with lattices over a certain class of maximal orders. We prove that R-orders in a split algebra have finitely many non-isomorphic irreducible lattices whenever R is a local Dedekind domain.

Let R be a Dedekind domain with quotient field K and Λ an R-order in the separable finite dimensional K-algebra A.

5.1 Definition: By $\underline{\underline{Ir}}(\Lambda)$ we denote the set of isomorphism classes of irreducible Λ-lattices. For $M \in {}_{\Lambda}\underline{\underline{M}}^{o}$ we shall write $M \in\in \underline{\underline{Ir}}(\Lambda)$ to indicate that M is irreducible.

5.2 Lemma: Let $\{e_i\}_{1 \le i \le s}$ be the set of central primitive orthogonal idempotents of A. Then

$$\underline{\underline{Ir}}(\Lambda) = \dot{\bigcup_{i=1}^{s}} \underline{\underline{Ir}}(\Lambda e_i)$$

is the disjoint union of a finite number of sets.

Proof: $\Lambda_1 = \oplus_{i=1}^{s} \Lambda e_i$ is an R-order in A containing Λ (cf. IV, 4.4). We define a map

$$\Phi : \underline{\underline{Ir}}(\Lambda_1) \longrightarrow \underline{\underline{Ir}}(\Lambda),$$

$$({}_{\Lambda_1}M) \longmapsto ({}_{\Lambda}M),$$

where (X) denotes the isomorphism class of X and the subscript indicates whether M should be considered as Λ_1-lattice or as Λ-lattice. Φ is well-defined, since every Λ_1-isomorphism is a Λ-isomorphism and since a lattice M is irreducible if and only if KM is simple (cf. IV, 1.13). Moreover, Φ is injective since $Hom_{\Lambda_1}(M_1,M_2) = Hom_{\Lambda}(M_1,M_2)$ (cf. IV, 1.14). To show that Φ is surjective, let $M \in\in \underline{\underline{Ir}}(\Lambda)$. Then KM is simple, and there exists exactly one central idempotent, say e_1, such that $e_1M = M$ and $e_iM = 0, 2 \le i \le s$. Thus,

$\Lambda_1 M = \Lambda_1 e_1 M = \Lambda M = M$ and $M \in\in \underline{\underline{Ir}}(\Lambda_1)$. Thus, Φ is a bijection and

$$\underline{\underline{Ir}}(\Lambda) = \underline{\underline{Ir}}(\Lambda_1) = \dot{\bigcup_{i=1}^{s}} \underline{\underline{Ir}}(\Lambda e_i);$$

obviously this union is disjoint. #

5.3 __Remark:__ In view of (5.2) it suffices, for the computation of $\underline{\underline{Ir}}(\Lambda)$, to assume that A is a simple separable K-algebra.

5.4 __Lemma:__ Let Γ be a maximal R-order in the simple separable K-algebra A. For a fixed $M_o \in\in \underline{\underline{Ir}}(\Gamma)$, we put $\Omega = \mathrm{End}_\Gamma(M_o)$. If $\{I_\alpha\}_{\alpha \in S}$ are representatives of the different classes of left Ω-ideals in $K\Omega$, then

$$\underline{\underline{Ir}}(\Gamma) = \{(M_o \otimes_\Omega I_\alpha) : \alpha \in S\},$$

and $M_o \otimes_\Omega I_\alpha \cong M_o \otimes_\Omega I_\beta$ if and only if $\alpha = \beta$.

__Proof:__ By (IV, 5.5) M_o is a progenerator for ${}_\Gamma\underline{\underline{M}}^o$ and we have a Morita-equivalence between ${}_\Gamma\underline{\underline{M}}^o$ and ${}_\Omega\underline{\underline{M}}^o$, which preserves irreducible lattices (cf. IV, 3.7). Since

$$\underline{\underline{Ir}}(\Omega) = \{(I_\alpha) : \alpha \in S\}, \text{ we conclude}$$

$$\underline{\underline{Ir}}(\Gamma) = \{(M_o \otimes_\Omega I_\alpha) : \alpha \in S\},$$

and no two of these modules are isomorphic. (It should be observed that $M_o \otimes_\Omega I_\alpha = \mathrm{Hom}_\Omega(\mathrm{Hom}_\Omega(M_o,\Omega),I_\alpha)$ (cf. III, §2, 2.3.) #

5.5 __Remark:__ Assume that R has infinite class number, and let Γ be a maximal R-order in $A = (K)_n$. If $\{I_j\}_{j=1,2,\dots}$ are representatives of the different ideal classes in K, then $\underline{\underline{Ir}}(\Gamma)$ contains infinitely many elements; on the other hand, for every maximal ideal $\underline{\underline{p}}$ of R, $\underline{\underline{Ir}}(\Gamma_{\underline{\underline{p}}})$ contains exactly one element (cf. IV, 5.4).

5.6 __Notation:__

Λ = R-order in the simple separable K-algebra A,

$A = (D)_n$, D a finite dimensional separable skewfield over K,

$\{\Gamma_\beta\}_{\beta \in T}$ = distinct maximal R-orders in A containing Λ,

M_β = fixed irreducible Γ_β-lattice, $\beta \in T$,

$\Omega_\beta = \mathrm{End}_{\Gamma_\beta}(M_\beta)$, $\beta \in T$,

$\{I_\kappa^\beta\}_{\kappa \in S_\beta}$ = representatives of the distinct classes of left Ω_β-ideals in $K\Omega_\beta$.

$\underline{\underline{Ir}}(\Gamma_\beta) = \{(M_\beta \otimes_{\Omega_\beta} I_\kappa^\beta) : \kappa \in S_\beta\}$ (cf. 5.5).

5.7 __Theorem__ (Roggenkamp [6]): In the notation of (5.6) we have

(i) $|\underline{\underline{Ir}}(\Lambda)| \geq \sum_{\beta \in T} s_\beta$, where $|\underline{\underline{Ir}}(\Lambda)|$ denotes the number of elements in $\underline{\underline{Ir}}(\Lambda)$.

(ii) $\underline{\underline{Ir}}(\Lambda) = \dot{\bigcup}_{\beta \in T} \underline{\underline{Ir}}(\Gamma_\beta)$ if and only if $\mathrm{End}_\Lambda(M)$ is a maximal order for every $M \in\in \underline{\underline{Ir}}(\Lambda)$.

(iii) If (ii) holds, then

$$\underline{\underline{Ir}}(\Lambda) = \{(M_\beta \otimes_{\Omega_\beta} I_\kappa^\beta) : \beta \in T, \kappa \in S_\beta\};$$

moreover, $M_\beta \otimes_{\Omega_\beta} I_\kappa^\beta \cong M_{\beta'} \otimes_{\Omega_{\beta'}} I_{\kappa'}^{\beta'}$ if and only if $\beta = \beta'$ and $\kappa = \kappa'$.

For the __proof__ of (4.7) we shall establish the following statement.

5.8 __Proposition__: Let $M \in\in \underline{\underline{Ir}}(\Gamma_\alpha)$, $N \in\in \underline{\underline{Ir}}(\Gamma_\beta)$, $\alpha, \beta \in T$, $\alpha \neq \beta$; then ${}_\Lambda M \not\cong {}_\Lambda N$.

__Proof__: Assume that $\varphi : {}_\Lambda M \xrightarrow{\sim} {}_\Lambda N$ is a Λ-isomorphism. Then we make M into a Γ_β-lattice denoted by M_β:

$$\gamma_\beta \circ m = (\gamma_\beta(m\varphi))\varphi^{-1}, \ \gamma_\beta \in \Gamma_\beta, \ m \in M.$$

Since φ is Λ-linear, Λ acts as M and M_β in the same way. Since both $M \in {}_{\Gamma_\alpha}\underline{\underline{M}}^o$ and $M_\beta \in {}_{\Gamma_\beta}\underline{\underline{M}}^o$ are progenerators (cf. IV, 5.5), we have (cf. III, § 2)

$$\Omega_1 = \mathrm{End}_{\Gamma_\alpha}(M) \text{ and } \Gamma_\alpha = \mathrm{End}_{\Omega_1}(M);$$

$$\Omega_2 = \mathrm{End}_{\Gamma_\beta}(M_\beta) \text{ and } \Gamma_\beta = \mathrm{End}_{\Omega_2}(M_\beta).$$

However by (IV, 1.14),

$$\Omega_1 = \mathrm{End}_{\Gamma_\alpha}(M) = \mathrm{End}_\Lambda(M) = \mathrm{End}_\Lambda(M_\beta) = \mathrm{End}_{\Gamma_\beta}(M_\beta) = \Omega_2,$$

and M and M_β are the same $\Omega_1 = \Omega_2$-module. Thus

$$\Gamma_\alpha = \mathrm{End}_{\Omega_1}(M) = \mathrm{End}_{\Omega_2}(M_\beta) = \Gamma_\beta \text{ and } \alpha = \beta,$$

a contradiction. #

Now we turn to the <u>proof of 5.7</u>: Because of (5.4) and (5.8), the lattices

$$\{M_\beta \otimes_{\Omega_\beta} I_\kappa^\beta : \kappa \in S_\beta, \beta \in T\}$$

are non-isomorphic irreducible Λ-lattices; but this is the statement of (i).

(ii) If $\underline{\underline{\mathrm{Ir}}}(\Lambda) = \dot\bigcup_{\beta \in T} \underline{\underline{\mathrm{Ir}}}(\Gamma_\beta)$, then every irreducible Λ-lattice M is of the form $M_\beta \otimes_{\Omega_\beta} I_\kappa^\beta$ and thus $\mathrm{End}_\Lambda(M) \overset{\text{ring}}{\cong} \mathrm{End}_{\Gamma_\beta}(M_\beta \otimes_{\Omega_\beta} I_\kappa^\beta)$ which is a maximal order in $K\Omega_\beta$ (cf. IV, 5.5).

<u>Conversely</u>, assume that $\Omega_M = \mathrm{End}_\Lambda(M)$ is a maximal R-order for every $M \in\in \underline{\underline{\mathrm{Ir}}}(\Lambda)$. Then M is a progenerator for $\underline{\underline{M}}^o_{\Omega_M}$ (cf. IV, 5.5), and $\Gamma_M = \mathrm{End}_{\Omega_M}(M)$ is a maximal R-order in A, Ω_M being maximal. However, the elements of Λ act as Ω_M-endomorphisms on M and so $\Gamma_M \supset \Lambda$ and M is a left Γ_M-lattice. Thus $\Gamma_M = \Gamma_\beta$ for some $\beta \in T$ and $M \cong M_\beta \otimes_{\Omega_\beta} I_\kappa^\beta$ for some $\beta \in S_\beta$. With (i) we conclude $\underline{\underline{\mathrm{Ir}}}(\Lambda) = \dot\bigcup_{\beta \in T} \underline{\underline{\mathrm{Ir}}}(\Gamma_\beta)$.

(iii) is an immediate consequence of (5.8) and (ii). #

5.9 <u>Corollary</u>: If for every $M \in\in \underline{\underline{\mathrm{Ir}}}(\Lambda)$, $\mathrm{End}_\Lambda(M) = \Omega$ is independent of M, then Ω is maximal. If $\{I_\kappa\}_{\kappa \in S}$ are representatives of the different classes of left Ω-ideals in $K\Omega$, then

$$\underline{\underline{\mathrm{Ir}}}(\Lambda) = \{(M_\beta \otimes_\Omega I_\kappa) : \beta \in T, \kappa \in S\},$$

and there are $|T|\cdot|S|$ non-isomorphic irreducible Λ-lattices.

<u>Proof</u>: Since $M_\beta \in\in \underline{\underline{\mathrm{Ir}}}(\Lambda)$ for every $\beta \in T$, $\mathrm{End}_\Lambda(M_\beta) = \Omega_\beta = \Omega$ is maximal. The remainder of the statements follows from (5.8). #

5.10 <u>Lemma</u> (Maranda [2]): If Λ is an R-order in $A = (K')_n$ where K' is a finite separable extension of K and if $\Lambda \cap K' = R'$ is the maximal R-order in K', then (5.9) is applicable.

<u>Proof</u>: If $M \in\in \underline{\underline{\mathrm{Ir}}}(\Lambda)$, then $\mathrm{End}_A(KM) = K'$ and $R' \subset \mathrm{End}_\Lambda(M)$. However,

R' is maximal, and thus $\mathrm{End}_\Lambda(M) = R'$. #

5.11 <u>Lemma</u> (Roggenkamp [10]): Let R be a local Dedekind domain (i.e., R is the localization of some Dedekind domain at a maximal ideal $\underline{\underline{p}}$), and let Λ be an R-order in the separable finite dimensional K-algebra A. Then there are only finitely many distinct maximal R-orders in A containing Λ.

<u>Proof</u>: Let us denote by "^" the completion of R. Since the R-orders in A and the $\hat{R}$-orders in A are in a one-to-one correspondence, which preserves inclusions (cf. IV, 1.9), it suffices to show that there are only finitely many maximal $\hat{R}$-orders in $\hat{A}$ containing $\hat{\Lambda}$. Assume that $\{\hat{\Gamma}_i\}_{i=1,2,\ldots}$ is an infinite set of distinct maximal $\hat{R}$-orders containing $\hat{\Lambda}$. Then we have a descending chain of $\hat{\Lambda}$-lattices

$$\hat{\Gamma}_1 \supset \hat{\Gamma}_1 \cap \hat{\Gamma}_2 \supset \ldots \supset \bigcap_{i=1}^{n} \hat{\Gamma}_i \supset \ldots \supset \bigcap_{i=1,2,\ldots} \hat{\Gamma}_i \supset \hat{\Lambda}.$$

Since $\hat{\Gamma}_1$ is a noetherian $\hat{R}$-module, the above chain has to terminate; i.e., there exists $n_o \in \underline{\underline{N}}$ such that

$$\bigcap_{i=1}^{n_o} \hat{\Gamma}_i = \bigcap_{i=1}^{n_o+s} \hat{\Gamma}_i \text{ for } s = 1,2,\ldots .$$

We denote this order by $\hat{\Gamma}_o = \bigcap_{i=1}^{n_o} \hat{\Gamma}_i$, and conclude that $\hat{\Gamma}_o$ is contained in infinitely many maximal orders. If $\hat{A} = \oplus_{i=1}^{t} \hat{A}_i$ is the decomposition of $\hat{A}$ into simple $\hat{K}$-algebras, then $\hat{\Gamma}_o$, as the intersection of maximal orders, decomposes accordingly, say $\hat{\Gamma}_o = \oplus_{i=1}^{t} \hat{\Gamma}_{o_i}$. Since at least one $\hat{\Gamma}_{o_i}$ is contained in infinitely many maximal orders, we may assume that $\hat{A}$ is simple; even central simple, since $\hat{\Gamma}_{o_1}$ contains the maximal order in the center of $\hat{A}_1$. Thus we have the situation:

$\hat{A} = (\hat{D})_n$, where $\hat{D}$ is a central skewfield over $\hat{K}$, and we view $\hat{D}$ embedded into $(\hat{D})_n$ diagonally; this embedding is fixed in the sequel.

$\hat{\Gamma}_o = \bigcap_{i=1}^{n} \hat{\Sigma}_i$, where $\hat{\Sigma}_i$ are maximal $\hat{R}$-orders in $\hat{A}$.

$\hat{\Gamma}_o \subset \hat{\Gamma}_i$, $i=1,2\ldots$ where $\hat{\Gamma}_i$ are distinct maximal $\hat{R}$-orders.

$\hat{\Omega}$ is the unique maximal $\hat{R}$-order in $\hat{D}$ (cf. IV, 5.2). Let $\hat{\Gamma}$ be a maximal $\hat{R}$-order in $\hat{A}$. We put $\hat{\Omega}_{ij}(\hat{\Gamma}) = \{\omega \in \hat{D} : \omega$ occurs at the (i,j)-position of some $\gamma \in \hat{\Gamma}\}$.

5.12 <u>Claim</u>: $\hat{\Gamma}$ is uniquely determined by $\{\hat{\Omega}_{ij}(\hat{\Gamma})\}$. (We <u>remark</u> that this is true for any order which contains $\hat{\Omega}$ in the diagonal and the elements $\underline{\underline{E}}_{ii}$.)

<u>Proof</u>: Let $\underline{\underline{E}}_{ij}$ be the matrix in $\hat{A}$ with 1 at the (i,j)-position and zeros elsewhere. Then $\hat{\Gamma}\underline{\underline{E}}_{ii}$ is an irreducible $\hat{\Gamma}$-lattice, and $\text{End}_{\hat{\Gamma}}(\hat{\Gamma}\underline{\underline{E}}_{ii}) = \underline{\underline{E}}_{ii}\hat{\Gamma}\underline{\underline{E}}_{ii} = \hat{\Omega}$. Consequently, $\hat{\Gamma}\underline{\underline{E}}_{ii} \subset \text{End}_{\underline{\underline{E}}_{ii}\hat{\Gamma}\underline{\underline{E}}_{ii}}(\hat{\Gamma}\underline{\underline{E}}_{ii}) = \hat{\Gamma}$; i.e., $\underline{\underline{E}}_{ii} \in \hat{\Gamma}$, $1 \le i \le n$. Now, let $\omega \in \hat{\Omega}_{ij}(\hat{\Gamma})$; i.e., there exists $\gamma \in \hat{\Gamma}$ such that ω is at the (i,j)-position of γ. But then also $\underline{\underline{E}}_{ii}\gamma\underline{\underline{E}}_{jj} = \omega\underline{\underline{E}}_{ij} \subset \hat{\Gamma}$. Thus $\hat{\Omega}_{ij}(\hat{\Gamma})\underline{\underline{E}}_{ij} \subset \hat{\Gamma}$, and $\hat{\Gamma}$ is uniquely determined by $\{\hat{\Omega}_{ij}(\hat{\Gamma})\}_{1 \le i,j \le n}$. This shows also that $\hat{\Omega}_{ij}(\hat{\Gamma})$ is a two-sided $\hat{\Omega}$-ideal, and that $\hat{\Gamma}$ contains $\hat{\Omega}$ in the diagonal embedding. This proves the claim. #

Now, to continue with the lemma, we may assume that $\hat{\Sigma}_1 = (\hat{\Omega})_n$, since all maximal $\hat{R}$-orders in $\hat{A}$ are conjugate (cf. IV, 5.8). There exists a positive integer t such that $\hat{\Gamma}_o \supset \omega_o^t(\hat{\Omega})_n$, where $\omega_o\hat{\Omega} = \text{rad}\,\hat{\Omega}$. Consequently, $\hat{\Gamma}_i \supset \omega_o^t(\hat{\Omega})_n$, $i=1,2\ldots$. We shall show that this cannot happen for infinitely many $\hat{\Gamma}_i$. If it did, according to the claim, there would exist an index (k,l) and an infinite subset of maximal orders $\{\hat{\Gamma}'_j\}_{j=1,2\ldots} \subset \{\hat{\Gamma}_i\}_{i=1,2\ldots}$ such that

$$\hat{\Omega}_{kl}(\hat{\Gamma}'_j) = \omega_o^{-t_j}\hat{\Omega},$$

where $\{t_j\}$ is an infinite increasing chain of positive integers. We now choose j such that $t_j > 2t$. Because of the claim we have

$$\omega_o^{-t_j} \underline{\underline{E}}_{kl} \in \Gamma_j' \supset \omega_o^{t}(\hat{\Omega})_n.$$

Hence

$$\omega_o^{t} \underline{\underline{E}}_{1k} \omega_o^{-t_j} \underline{\underline{E}}_{kl} \omega_o^{t} \underline{\underline{E}}_{l1} \in \Gamma_j', \ 1 \le l \le n.$$

Thus $\omega_o^{-1}\underline{\underline{E}} \in \hat{\Gamma}_j'$; but this can not happen for a maximal order $\hat{\Gamma}$, since ω_o^{-1} is not integral in $\hat{D}$ (cf. Ex. 5,3). Thus we have obtained a contradiction. #

5.13 __Corollary__: Let R be a local Dedekind domain and A a finite dimensional K-algebra, which is split by K. If Λ is an R-order in A, then Λ has only finitely many non-isomorphic irreducible lattices.

__Proof__: According to (5.11) there are only finitely maximal R-orders in A containing Λ. Now the statement follows from (5.10) and (5.7). #

5.14 __Lemma__: Let R be a Dedekind domain and Λ an R-order in the separable finite dimensional K-algebra A. Assume that the Jordan-Zassenhaus theorem is valid for Λ-lattices. Then there are only finitely many maximal R-orders in A that contain Λ.

__Proof__: This is an immediate consequence of (5.7,1), since the Jordan-Zassenhaus theorem (4.5) asserts that $\mathrm{card}(\underline{\underline{Ir}}(\Lambda)) < \infty$. #

__Exercises §5__:

1.) Let $A = (\underline{\underline{Q}})_2 \oplus (\underline{\underline{Q}})_2$ and let

$$\Lambda = \left\{ \begin{pmatrix} a_1 & a_2 \\ pa_3 & a_4 \end{pmatrix}, \begin{pmatrix} a_1+pb_1 & b_2 \\ p^2b_3 & b_4 \end{pmatrix} \quad a_i, b_i \in \underline{\underline{Z}} \right\}, \ p \text{ a rational prime number.}$$

Compute the five irreducible representations of Λ explicitely.

2.) Let Λ be an R-order in $A = \oplus_{i=1}^{n} Ke_i$, where e_i are orthogonal idempotents. Compute $\underline{\underline{Ir}}(\Lambda)$.

3.) Let $\hat{R}$ be a complete Dedekind domain, and let $\hat{D}$ be a central skew-field over $\hat{K}$. If $\hat{\Omega}$ is the maximal $\hat{R}$-order in $\hat{D}$ and if $\underline{\underline{E}}$ is the n-dimensional identity matrix, show that $d\underline{\underline{E}}$ is integral over $\hat{R}$ if and only if $d \in \hat{\Omega}$. Thus no $\hat{R}$-order in $\hat{A}$ can contain elements of the form $d\underline{\underline{E}}$ with $d \notin \hat{\Omega}$. (Hint: Use IV, 6.2, 6.3 and IV, 1.4.)

4.) Let R be a local Dedekind domain and let Λ be an R-order in $A=(K)_n$. Then there are only finitely many non-isomorphic irreducible Λ-lattices. However in general, the Jordan-Zassenhaus theorem need not hold for irreducible lattices. Give an example! (Hint: Assume that R has an infinite residue class field; i.e., $R/\pi R$ is infinite, and let T be a separable finite dimensional extension field of K and let S be maximal R-order in T. Assume furthermore that $\pi S = \underline{\underline{P}}_1\underline{\underline{P}}_2$, where $\underline{\underline{P}}_1$ and $\underline{\underline{P}}_2$ are different prime ideals of T. Denote by "$\hat{\ }$" the π-adic completion. Then $\hat{T} = \hat{T}_1 \oplus \hat{T}_2$ and $\hat{S} = \hat{S}_1 \oplus \hat{S}_2$. Write

$$\hat{S}_1 = \oplus_{i=1}^{n} \hat{R}\omega_i^{(1)},\ \omega_1^{(1)} = 1,$$

$$\hat{S}_2 = \oplus_{i=1}^{n} \hat{R}\omega_i^{(2)},\ \omega_1^{(2)} = 1.$$

Assume $n \geq 3$. Then we consider the following $\hat{R}$-order in $\hat{T} = \hat{T}_1 \oplus \hat{T}_2$:

$$\hat{\Lambda} = \{(\textstyle\sum_{i=1}^{n} r_i\omega_i^{(1)}, \sum_{i=3}^{n} r_i'\omega_i^{(2)} + (r_1+\hat{\pi}r_1')\omega_1^{(2)} + (r_2+\hat{\pi}r_2')\omega_2^{(2)}) : r_i, r_i' \in \hat{R}\}.$$

Let $\hat{e}_1, \hat{e}_2$ be the central idempotents in $\hat{T}$ and put $\hat{M}_i = \hat{\Lambda}\hat{e}_i$. Then $\mathrm{Ext}^1_{\hat{\Lambda}}(\hat{M}_1, \hat{M}_2) \cong \hat{R}/\hat{\pi}\hat{R} \oplus \hat{R}/\hat{\pi}\hat{R}$. Show that among the exact sequences

$$0 \longrightarrow \hat{M}_2 \longrightarrow \hat{X} \longrightarrow \hat{M}_1 \longrightarrow 0$$

there are infinitely many with non-isomorphic middle term

(cf. Reiner [9], Roggenkamp [10]).
Then $\hat{\Lambda}$ satisfies (5.13).
Consider $\Lambda = T \cap \hat{\Lambda}$ and show that Λ has infinitely many non-isomorphic irreducible lattices; thus (5.13) becomes false if one drops the hypothesis that K splits A.)

§6 Infinite primes and algebras over $\underline{\underline{A}}$-fields

Infinite primes and totally definite quaternion algebras are defined; we quote without proof, that an algebra over an $\underline{\underline{A}}$-field is split if and only if it is split at all finite and infinite primes and the norm theorem.

We recall that an $\underline{\underline{A}}$-field K is either an algebraic number field, and then we say that K is an $\underline{\underline{A}}$-field of characteristic zero, or K is a finite algebraic extension of $\underline{\underline{k}}(X)$, where $\underline{\underline{k}}$ is a finite field and X is an ideterminate over $\underline{\underline{k}}$; then we say that K is an $\underline{\underline{A}}$-field of characteristic $p > 0$, where p is the characteristic of $\underline{\underline{k}}$. We take R to be a Dedekind domain with quotient field K.

6.1 Definition of infinite primes in algebraic number fields: Let K be an $\underline{\underline{A}}$-field of characteristic zero; then K is a separable extension of $\underline{\underline{Q}}$ and there exists a primitive element $\alpha \in K$ such that $K = \underline{\underline{Q}}(\alpha)$. Let $f(X) = \min_{K/\underline{\underline{Q}}}(\alpha, X)$. Since $\underline{\underline{C}}$ - the field of complex numbers - is algebraically closed,

$$f(X) = \prod_{i=1}^{n}(X - \alpha_i),\ \alpha_i \in \underline{\underline{C}},\ \alpha_1 = \alpha,\ \alpha_i \neq \alpha_j.$$

With every α_i, $1 \leq i \leq n$, we may associate an embedding

$$\varphi_i : K \longrightarrow \underline{\underline{C}},$$

$$\varphi_i : \underline{\underline{Q}}(\alpha) \longrightarrow \underline{\underline{Q}}(\alpha_i),$$

$$\sum_{j=1}^{n} a_j \alpha^j \longmapsto \sum_{j=1}^{n} a_j \alpha_i^j,\ a_j \in \underline{\underline{Q}}.$$

We now define archimedian valuations on K:

$$\nu_\infty^i : K \longrightarrow \underline{\underline{R}},$$

$$k \longrightarrow |k^{\varphi_i}|,$$

where $\underline{\underline{R}}$ is the field of real numbers and "$|\ -\ |$" is the ordinary absolute value. Then $\nu_\infty^i = \nu_\infty^j$ if and only if α_i and α_j are complex conjugate elements. Thus, if f(X) has n_1 real roots and $2n_2$ complex roots, then there are exactly $n_1 + n_2$ different archimedian valuations,

all extending the absolute value on $\underline{\underline{Q}}$. The valuation ν_∞^i is called <u>real</u> if α_i is real; otherwise ν_∞^i is called <u>imaginary</u>.
Since the prime ideals are in one-to-one correspondence with the non-archimedian valuations, one calls the archimedian valuations <u>infinite primes</u> and denotes them by $\underline{\underline{p}}_\infty$. $\underline{\underline{p}}_\infty$ is called a <u>real infinite prime</u> if the corresponding valuation is real and an <u>imaginary infinite prime</u> otherwise. The prime ideals are then called <u>finite primes</u>. The finite and infinite primes form the set of all <u>primes of K</u>. The <u>completion of K at the infinite prime</u> $\underline{\underline{p}}_\infty$ is the completion $\hat{K}_{\underline{\underline{p}}_\infty}$ of K with respect to the valuation ν_∞. Thus

$$\hat{K}_{\underline{\underline{p}}_\infty} = \underline{\underline{C}} \text{ if } \underline{\underline{p}}_\infty \text{ is imaginary}$$

and

$$\hat{K}_{\underline{\underline{p}}_\infty} = \underline{\underline{R}} \text{ if } \underline{\underline{p}}_\infty \text{ is real.}$$

6.2 <u>Lemma</u> (Frobenius): If D is a non-commutative finite dimensional skewfield over $\underline{\underline{R}}$, then D is the skewfield of real quaternions $\underline{\underline{H}}(\underline{\underline{R}})$.

<u>Proof</u>: There is only one finite dimensional extension field of $\underline{\underline{R}}$, namely $\underline{\underline{C}} = \underline{\underline{R}}(i), i^2 = -1$. If the center of D is different from $\underline{\underline{R}}$, then the center of D is $\underline{\underline{C}}$; but $\underline{\underline{C}}$ is algebraically closed and so there are no finite dimensional skewfields over $\underline{\underline{C}}$ (cf. III, 6.2), and hence D must be a central skewfield over $\underline{\underline{R}}$. Since any maximal subfield of D must be isomorphic to $\underline{\underline{C}}$, $(D : \underline{\underline{R}}) = 4$ (cf. III, 6.5), and we may assume that D contains $\underline{\underline{C}} = \underline{\underline{R}}(i)$ as maximal subfield. We have an automorphism

$$\begin{aligned} \sigma : \underline{\underline{C}} &\longrightarrow \underline{\underline{C}}, \\ \underline{\underline{R}}(i) &\longrightarrow \underline{\underline{R}}(-i), \\ i &\longmapsto -i, \end{aligned}$$

and according to (III, 6.6) this automorphism which is $\underline{\underline{R}}$-linear is given by conjugation with an element $j_o \in D$; i.e., $j_o i j_o^{-1} = -i$. Then $j_o \notin \underline{\underline{R}}(i)$ and so $D = \underline{\underline{R}}(i, j_o)$, since $(D : \underline{\underline{R}}) = 4$. However, $j_o^2 i j_o^{-2} = i$

shows that $j_0^2 \in \text{center}(D) = \underline{\underline{R}}$. But $j_0 \notin \underline{\underline{R}}$ and so $\underline{\underline{R}}(j_0) \cong \underline{\underline{C}}$. This implies that j_0^2 has to be a negative element in $\underline{\underline{R}}$, say $j_0^2 = -r^2$, $0 \neq r \in \underline{\underline{R}}$. We put $j = j_0 r^{-1}$ and $ij = k$. Then it is easily checked that

$$i^2 = j^2 = k^2 = -1; \; ij = -ji = k; \; jk = -kj = i; \; ki = -ik = j.$$

Moreover, the elements $1,i,j,k$ form an $\underline{\underline{R}}$-basis for D. Thus $D = \underline{\underline{H}}(\underline{\underline{R}})$. #

6.3 **Remark:** $\underline{\underline{C}}$ is a splitting field for $\underline{\underline{H}}(\underline{\underline{R}})$ and we have a realization of $\underline{\underline{H}}(\underline{\underline{R}})$ as (2x2)-matrices over $\underline{\underline{C}}$:

$$\underline{\underline{I}} = \begin{pmatrix} 1 & 0 \\ 0 & -1 \end{pmatrix}, \quad \underline{\underline{J}} = \begin{pmatrix} 0 & 1 \\ -1 & 0 \end{pmatrix}, \quad \underline{\underline{K}} = \begin{pmatrix} 0 & 1 \\ 1 & 0 \end{pmatrix},$$

and $\underline{\underline{H}}(\underline{\underline{R}}) = \underline{\underline{R}}(\underline{\underline{I}},\underline{\underline{J}},\underline{\underline{K}})$ (cf. Ex. 6,1).

6.4 **Remark:** If K is an algebraic number field, then the finite and infinite primes are in one-to-one correspondence with the equivalence classes of non-archimedian and archimedian valuations, and the non-archimedian valuations are in one-to-one correspondence with the prime ideals in R. With each valuation ν of K we may associate a unique topology $\underline{\underline{T}}_\nu$ and we may form the completion $\hat{K}_\nu$ of K with respect to this topology. Then we have an embedding

$$\varphi_\nu : K \longrightarrow \hat{K}_\nu ,$$

and the valuation ν is archimedian if and only if $\hat{K}_\nu$ is either $\underline{\underline{C}}$ or $\underline{\underline{R}}$; i.e., if and only if ν corresponds to an infinite prime. If now K is an $\underline{\underline{A}}$-field of characteristic $p > 0$, then similar considerations as above show that every valuation of K is non-archimedian, since we can not embed K into a field of characteristic zero. Consequently all valuations are induced from the $\underline{\underline{p}}$-adic topologies, where $\underline{\underline{p}}$ is a prime ideal of R. Thus in this case there do not exist infinite primes. (For the details of these considerations we refer to Bourbaki [2 , Ch. 6].) If K is an $\underline{\underline{A}}$-field, then we talk about the set of all finite and infinite primes, and it should be noted that in case char $K > 0$ the set of infinite primes is empty.

6.5 <u>Definition</u>: Let K be an $\underline{\underline{A}}$-field and A a central simple K-algebra. By $\underline{\underline{S}}_1$ we denote the set of all finite and infinite primes of K. For $\underline{\underline{p}} \in \underline{\underline{S}}_1$ we define the <u>ramification index of A at</u> $\underline{\underline{p}}$ as $m_{\underline{\underline{p}}}$, where

$$\hat{K}_{\underline{\underline{p}}} \otimes_K A = (\hat{D}_{\underline{\underline{p}}})_{n_{\underline{\underline{p}}}},$$

and $(\hat{D}_{\underline{\underline{p}}} : \hat{K}_{\underline{\underline{p}}}) = m^2_{\underline{\underline{p}}}$, $\hat{K}_{\underline{\underline{p}}}$ being the completion of K at $\underline{\underline{p}}$ (cf. III, 6.5, 6.10). We say that A is <u>unramified</u> at $\underline{\underline{p}} \in \underline{\underline{S}}_1$ if $m_{\underline{\underline{p}}} = 1$.

6.6 <u>Theorem</u>: Let A be a central simple algebra over an $\underline{\underline{A}}$-field K. Then A is unramified at almost all finite and infinite primes of K.

<u>Proof</u>: There are only finitely many infinite primes (cf. 6.1). Let Γ be a maximal R-order in A. Then $\hat{\Gamma}_{\underline{\underline{p}}}$, where $\underline{\underline{p}}$ is a finite prime of K, is separable if and only if $\underline{\underline{p}}$ does not divide the Higman ideal $\underline{\underline{H}}(\Gamma)$ (cf. V, 3.4). However, there are only finitely many prime divisors of $\underline{\underline{H}}(\Gamma)$. Thus $\hat{\Gamma}_{\underline{\underline{p}}}$ is separable for all but a finite number of prime ideals, and it follows from (2.10) that $\hat{A}_{\underline{\underline{p}}}$ has to be unramified if $\hat{\Gamma}_{\underline{\underline{p}}}$ is separable. (This follows also from V, Ex. 3,3 and from the results in IV, § 6.) Hence A is unramified at all but a finite number of primes.#

6.7 <u>Lemma</u>: Let K be an $\underline{\underline{A}}$-field of characteristic zero and let A be a central simple K-algebra. If $\underline{\underline{p}}_\infty$ is an infinite imaginary prime of K, then A is unramified at $\underline{\underline{p}}_\infty$. If $\underline{\underline{p}}_\infty$ is real, then $m_{\underline{\underline{p}}_\infty} = 1$ or $m_{\underline{\underline{p}}_\infty} = 2$.

<u>Proof</u>: This is an immediate consequence of (III, 6.2; VI, 6.1, 6.2). #

We quote without proof the following theorem from class-field theory (cf. Weil [1, Ch. XI, § 2, Theorem 2]).

6.8 <u>Theorem</u> (Hasse [2]): Let K be an $\underline{\underline{A}}$-field and A a central simple K-algebra. Then K is a splitting field for A if and only if A is unramified at every finite and infinite prime of K. Moreover, if A is ramified at some prime, then it is ramified at at least two primes. As a consequence of (6.8) one can obtain the following theorem of Eichler (cf. Weil [1, Ch. XI, Proposition 3]).

6.9 Theorem (Eichler [1,2]): Let A be a central simple algebra over an $\underline{\underline{A}}$-field K. Then an element $0 \neq k \in K$ is the reduced norm of a regular element $a \in A$; i.e., $\mathrm{Nrd}_{A/K}(a) = k$, if and only if k is positive at all infinite primes of K at which A is ramified (i.e., $\nu_{\underline{\underline{p}}_\infty}(k) > 0$ for every $\underline{\underline{p}}_\infty$ at which A is ramified.)

We remark that $St_K(A) = \{\alpha \in K, 0 \neq \alpha$ is positive at every ramified infinite real prime of $K\}$ is called the ray modulo the ramified infinite real primes.

We quote without proof one more theorem of Hasse (cf. Albert [1, Ch. IV, §11, Theorem 27]).

6.10 Theorem (Hasse [2]): Let A be a central simple algebra over an $\underline{\underline{A}}$-field with $(A : K) = n^2$. If E is a splitting field for A with $(E : K) = n$, then E is isomorphic to a maximal subfield of A.

6.11 Lemma: Let A be a central simple algebra over an $\underline{\underline{A}}$-field K. If f(X) is an irreducible polynomial over $\hat{K}_{\underline{\underline{p}}}$, the degree of which is a multiple of the ramification index $m_{\underline{\underline{p}}}$ of A, then for every root ω of f(X), $\hat{K}_{\underline{\underline{p}}}(\omega)$ is a splitting field for $\hat{A}_{\underline{\underline{p}}}$.

Proof: If $\underline{\underline{p}}$ is infinite the statement is clear since $K_{\underline{\underline{p}}}(\omega) = \underline{\underline{C}}$ if the degree of f(X) is larger than 1. If $\underline{\underline{p}}$ is finite, then $E = \hat{K}_{\underline{\underline{p}}}(\omega)$ is an extension field of $\hat{K}_{\underline{\underline{p}}}$, the degree of which over $K_{\underline{\underline{p}}}$ is a multiple of $m_{\underline{\underline{p}}}$, and the statement follows from (IV, 6.17). #

6.12 Definition: Let K be an $\underline{\underline{A}}$-field. A central simple K-algebra A is called a <u>totally definite quaternion algebra</u>, if $(A : K) = 4$ and A is ramified at every infinite prime of K.

6.13 <u>Remark</u>: If K is an $\underline{\underline{A}}$-field of characteristic $p > 0$, then there do not exist totally definite quaternion algebras, since K has no infinite primes (cf. 6.4). If A is an algebraic number field, and A is a totally definite quaternion algebra then every infinite prime $\underline{\underline{p}}_\infty$ of K has to be real and $A_{\underline{\underline{p}}_\infty} = \underline{\underline{H}}(\underline{\underline{R}})$ (cf. 6.2).

In the theory of lattices over orders, the totally definite quaternion algebras play an exceptionally bad role. In fact for an order Λ in a totally definite quaternion algebra there exist Λ-lattices M,N,X, where X is a direct summand of $M^{(n)}$ for some $n \in \underline{\underline{N}}$, such that $M \oplus X \cong N \oplus X$, but $M \not\cong N$. However, Jacobinski has shown that such a "misbehavior" can not happen if A "does not involve" totally definite quaternion algebras (cf. VIII, §5).

<u>Exercises §6</u>:

1.) Show that the $\underline{\underline{R}}$-algebra A, spanned by the matrices $1, \underline{\underline{I}}, \underline{\underline{J}}, \underline{\underline{K}}$ of (6.3) is isomorphic to $\underline{\underline{H}}(\underline{\underline{R}})$, and that $\underline{\underline{H}}(\underline{\underline{R}})$ is a skewfield.

2.) Show that the rational quaternion algebra $\underline{\underline{H}}(\underline{\underline{Q}})$ (cf. III, Ex. 5.9) is totally definite quaternion algebra. Which elements of $\underline{\underline{Q}}$ can occur as reduced norms?

§7 A theorem of Eichler on algebras that are not totally definite quaternion algebras

This section is devoted to the proof of Theorem (7.2) below; this theorem plays a key role in Jacobinski's theory of genera and in the proof of his cancellation theorem (VII, VIII).

We keep the notation of §6; K is always an $\underline{\underline{A}}$-field and R is a Dedekind domain with quotient field K.

7.1 Definition: Let Λ an R-order in the separable finite dimensional K-algebra A. We say that $M \in {}_{\Lambda}\underline{\underline{M}}^{o}$ satisfies Eichler's condition if none of the simple components of $\mathrm{End}_A(KM)$ is a totally definite quaternion algebra.

We point out here, that every Λ-lattice M satisfies Eichler's condition if K is an $\underline{\underline{A}}$-field of characteristic $p > 0$.

The following theorem - Swan's formulation of a theorem of Eichler - is most important in the proof of Jacobinski's cancellation theorem (VIII, 5.1). We present here a proof due to R.G. Swan (unpublished).

7.2 Theorem (Eichler [1],[2], Swan): Let Λ be an R-order in the separable finite dimensional K-algebra A. Let $M \in {}_{\Lambda}\underline{\underline{M}}^{o}$ satisfy Eichler's condition. Then there exists a finite set of prime ideals $\underline{\underline{S}}(M)$ such that, given

(i) a simple left Λ-module U with $\mathrm{ann}_R(U) = \underline{\underline{p}}_o \notin \underline{\underline{S}}(M)$,

(ii) an ideal $\underline{\underline{a}}$ of R with $(\underline{\underline{a}}, \underline{\underline{p}}_o) = 1$,

(iii) two epimorphisms

$$\psi, \varphi : M \longrightarrow U,$$

then there exists a Λ-automorphism τ of M such that

$$\tau|_{\mathrm{Ker}\,\varphi} : \mathrm{Ker}\,\varphi \xrightarrow{\sim} \mathrm{Ker}\,\psi \text{ is an isomorphism,}$$

moreover for every $m \in M$,

$$m\tau - m \in \underline{\underline{a}}M.$$

The proof of this theorem will occupy the remainder of this section, and during the course of the proof we shall need some formal lemmata which we shall establish first.

7.3 Approximation theorem (Eichler [1]): Let K be an algebraic number field with ring of integers R. Let $k \in K$ and $0 < c \in \mathbb{Q}$ be given, and assume that $k^{(1)},\dots,k^{(s)}$ are conjugates of k such that the corresponding conjugate fields $K^{(i)}$ are real (i.e., they are subfields of $\mathbb{R}$). If $s < (K : \mathbb{Q})$, then for every set $\{\varepsilon_i\}_{1\le i\le s}$ of arbitrary preassigned positive numbers there exists $r \in R$ such that

$$|k^{(i)} - cr^{(i)}| < \varepsilon_i, 1\le i\le s,$$

where " $|-|$ " denotes the absolute value.

Proof: For $0 < x \in \mathbb{R}$ we denote by $[x]$ the largest non-negative integer $\le x$, and for $x < 0$ we put $[x] = -[-x]$; and $[0] = 0$. (Observe that this definition is different from the usual one!)

Claim 1: There exists $w \in R$ such that

$$|w^{(i)}| < \varepsilon_i/c.$$

Proof: We observe that (7.3) is only interesting if $s \geq 1$, and we shall assume that. Let $\{\alpha_i\}_{1\le i\le n}$ be an integral basis of K over $\mathbb{Q}$; i.e., $\alpha_i \in R$. Let $\{\alpha_i^{(j)}\}_{1\le j\le n}$ be the conjugates of α_i, $1\le i\le n$, numbered in such a way that for $1\le j\le s$, $\alpha_i^{(j)} \in K^{(j)}$, where $K^{(j)}$, $1\le j\le s$, are the real fields of the hypothesis. Since K is a separable extension of $\mathbb{Q}$ and since $\det(\alpha_i^{(j)})$ is the discriminant of the basis $\{\alpha_i\}_{1\le i\le n}$ (cf. Ex. 7,1), we have $\det(\alpha_i^{(j)}) \neq 0$ (cf. III, 6.18). By Minkovski's lemma for complex numbers (cf. Ex. 7,3) there exist $z_1,\dots,z_n \in \mathbb{Z}$, not all zero such that

$$0 < |\sum_{i=1}^{n} \alpha_i^{(1)} z_i| < \varepsilon_1/c.$$

It should be noted that we have used heavily the fact $s < n$. We put $w = \sum_{i=1}^{n} \alpha_i z_i \in R$. #

Claim 2: Let $w \in R$ be as in Claim 1. Then $r = [k^{(1)}/c\, w^{(1)}]\, w \in R$ and we have

$$| k^{(1)} - cr^{(1)} | < \varepsilon_1 .$$

The proof consists in a straight forward computation.

Now we shall prove (7.3) by induction on s. Assume that we have found $x \in R$ such that

$$| k^{(i)} - cx^{(i)} | < \varepsilon_i/2, \quad 1 \leq i \leq s-1.$$

If $| k^{(s)} - cx^{(s)} | < \varepsilon_s$, we are done. Thus, we can assume $|(k^{(s)} - cx^{(s)})/\varepsilon_s| \geq 1$ and hence $[(k^{(s)} - cx^{(s)})/\varepsilon_s] \neq 0$.

Denoting by D the discriminant of the integral basis $\{\alpha_i\}_{1 \leq i \leq n}$, we apply once more Minkowski's lemma (Ex. 7.3): There exists $0 \neq x_o \in R$ such that

$$|x_o^{(i)}| < \varepsilon_i/2c \left|\left[\sqrt{|D|}(k^{(s)} - cx^{(s)})/\varepsilon_s\right]\right|^{-1} = \eta_i, \quad 1 \leq i \leq s-1,$$

$$|x_o^{(s)}| < \varepsilon_s/c = \eta_s,$$

$$|x_o^{(j)}| \leq \eta_j, \quad s+1 \leq j \leq n,$$

where $\{\eta_j\}_{s+1 \leq j \leq n}$ are real positive numbers such that $\eta_j = \eta_{j'}$, if $K^{(j)}$ and $K^{(j')}$ are complex conjugate fields, and such that $\prod_{i=1}^{n} \eta_i = \sqrt{|D|}$. Since $0 \neq x_o$ is integral, we have $1 \leq \prod_{i=1}^{n} | x_o^{(i)} |$; thus

$$| x_o^{(s)} | > \varepsilon_s/c\sqrt{|D|} .$$

We put

$$r = x + [(k^{(s)} - cx^{(s)})/cx_o^{(s)}]x_o.$$

Then r is integral, and we have for $1 \leq i \leq s-1$,

$$|k^{(i)} - cr^{(i)}| \leq |k^{(i)} - cx^{(i)}| + c|[(k^{(s)} - cx^{(s)})/cx_o^{(s)}]x_o^{(i)}|$$

$$< \varepsilon_1/2 + c|[\sqrt{|D|}\,(k^{(s)} - cx^{(s)})/\varepsilon_s]\,x_o^{(i)}| < \varepsilon_1,$$

and

$$|k^{(s)} - cr^{(s)}| = |k^{(s)} - cx^{(s)} - c\,[(k^{(s)} - cx^{(s)})/cx_o^{(s)}]x_o^{(s)}| < \varepsilon_s.$$

Thus r has the desired properties. #

7.4 Lemma: Let $\underline{k}$ be a finite field and let $2 \leq m \in \underline{Z}$ be given. Then there exists a polynomial

$$f(X) = X^m + c_{m-1}X^{m-1} + \ldots + c_1X + (-1)^m \in \underline{k}[X],$$

which has no root in $\underline{k}$.

Proof: If $(\underline{k} : 1) = q$, then there are q^{m-1} polynomials of the above form f(X). If f(X) has a zero in $\underline{k}$, then we can write

$$f(X) = (X - c)(X^{m-1} + c'_{m-2}X^{m-2} + \ldots + c^{-1}(-1)^{m-1}),$$

where $0 \neq c \in \underline{k}$. However, there are $q^{m-2}(q - 1)$ polynomials of this form. Hence the statement follows since $q^{m-2}(q - 1) < q^{m-1}$. #

7.5 Lemma: Let $\hat{K}$ and $\hat{R}$ be the completion of an $\underline{A}$-field K with Dedekind domain R at some prime ideal $\underline{p}$ of R. Then there exist irreducible polynomials of the form

$$X^m + \alpha_{m-1}X^{m-1} + \ldots + \alpha_1X + (-1)^m \in K[X],$$

which are arbitrarily close to $(X - 1)^m$ in the $\underline{p}$-adic topology.

For the proof we refer to Weil [1, Ch. XI, §3, Lemma 2]. #

7.6 Lemma: Let $\hat{K}$ and $\hat{R}$ be the completion of an $\underline{A}$-field K with R, $\hat{\pi}\hat{R} = \mathrm{rad}\,\hat{R}$. If f(X) is a monic irreducible polynomial in $\hat{R}[X]$, then there exists an exponent s such that every polynomial $g(X) \in \hat{K}[X]$ with

$$g(X) \equiv f(X) \bmod (\hat{\pi}^s \hat{R}[X])$$

is also irreducible.

<u>Proof</u>: Since $f(X)$ is a monic irreducible polynomial in $\hat{R}[X]$, $\hat{\Lambda} = \hat{R}[X]/(f(X))$ is an $\hat{R}$-order in the field $\hat{A} = \hat{K}[X]/(f(X))$. To prove (7.3) it suffices to show that there exists $s \in \underline{\underline{N}}$ such that $f(X)$ is irreducible modulo $\hat{\pi}^s$. Assume that there exists a decomposition

$$f(X) \equiv h_1(X)h_2(X) \bmod (\hat{\pi}^s \hat{R}[X]);$$

then we may choose $h_1(X), h_2(X) \in \hat{R}[X]$ to be monic (cf. Gauss' lemma I, Ex. 7.6). We put $\hat{M}_1 = \hat{R}[X]/(h_1(X))$ and $\hat{M}_2 = \hat{R}[X]/(h_2(X))$. Then $\hat{\Lambda} =_{\hat{R}} \hat{M}_1 \oplus \hat{M}_2$ as $\hat{R}$-lattices and

$$\hat{\Lambda} \hat{M}_1 \subset \hat{M}_1 + \hat{\pi}^s \hat{M}_2,$$

since $f(X)$ is reducible modulo $\hat{\pi}^s$. If s is large enough, we can apply (1.3) to conclude that $\hat{\Lambda}$ is reducible; i.e., $f(X)$ is reducible. Thus $f(X)$ must be irreducible modulo $\hat{\pi}^s$ for sufficiently large s. #

Now we <u>turn to the proof of (7.2)</u>:

We <u>recall the notation</u>: K is an $\underline{\underline{A}}$-field with Dedekind domain R, A is a finite dimensional separable K-algebra and Λ an R-order in A. $M \in {}_{\Lambda}\underline{\underline{M}}^o$ satisfies Eichler's condition; i.e., none of the simple components in $\mathrm{End}_A(KM)$ is a totally definite quaternion algebra.

7.7 <u>Reduction of the proof of (7.2) to the case where Λ is a maximal R-order</u>. Let us assume that Eichler's theorem is true for maximal R-orders in A, and let Γ be a maximal R-order in A containing Λ. Then $\Gamma M \in {}_{\Gamma}\underline{\underline{M}}^o$ also satisfies Eichler's condition, since $\mathrm{End}_A(K\Gamma M) = \mathrm{End}_A(KM)$. According to (7.2) there exists a finite set of prime ideals $\underline{\underline{S}}(\Gamma M)$ for which the statement is true for ΓM. Set $\underline{\underline{S}}(\Gamma/\Lambda) = \{\underline{\underline{p}} : \underline{\underline{p}}$ a prime ideal in R such that $\Gamma_{\underline{\underline{p}}} \neq \Lambda_{\underline{\underline{p}}}\}$. Then $\underline{\underline{S}}(\Gamma/\Lambda)$ is finite by (IV, 1.8), and we put

$$\mathfrak{S}(M) = \mathfrak{S}(\Gamma M) \cup \mathfrak{S}(\Gamma/\Lambda).$$

Assume now that the following data are given:

(i) U, a simple left Λ-module with $\operatorname{ann}_R(U) = \mathfrak{p}_0 \notin \mathfrak{S}(M)$,

(ii) $\mathfrak{a}$, an ideal of R with $(\mathfrak{a}, \mathfrak{p}_0) = 1$,

(iii) $\varphi, \psi : M \longrightarrow U$ are two epimorphisms.

Then U is also a simple Γ-module. In fact, U is a $\Lambda/\mathfrak{p}_0\Lambda$ -module: but

$$\Lambda/\mathfrak{p}_0\Lambda \cong \Lambda_{\mathfrak{p}_0}/\mathfrak{p}_0\Lambda_{\mathfrak{p}_0} = \Gamma_{\mathfrak{p}_0}/\mathfrak{p}_0\Gamma_{\mathfrak{p}_0} \cong \Gamma/\mathfrak{p}_0\Gamma,$$

since $\mathfrak{p}_0 \notin \mathfrak{S}(\Gamma/\Lambda)$. Thus we may extend φ and ψ uniquely to epimorphisms

$$\varphi', \psi' : \Gamma M \longrightarrow U.$$

We put $\mathfrak{b} = \{r \in R : r\Gamma \subset \Lambda\}$ and $\mathfrak{c} = \mathfrak{a}\mathfrak{b}$. Then $(\mathfrak{c}, \mathfrak{p}_0) = 1$, since $(\mathfrak{a}, \mathfrak{p}_0) = 1$ and $(\mathfrak{b}, \mathfrak{p}_0) = 1$. We have assumed (7.2) to be true for ΓM, and so there exists an automorphism τ' of ΓM such that

$$\tau'|_{\operatorname{Ker}\varphi'} : \operatorname{Ker}\varphi' \xrightarrow{\sim} \operatorname{Ker}\psi' \text{ with}$$

$$m' - m'\tau' \in \mathfrak{c}\Gamma M \text{ for every } m' \in \Gamma M.$$

We put $\tau = \tau'|_M$. Then for every $m \in M$, $m\tau - m \in \mathfrak{a}\mathfrak{b}\Gamma M \subset \mathfrak{a}M$, and since τ' is an automorphism of ΓM,

$$(m\tau - m)\tau'^{-1} = m - m\tau'^{-1} \in \mathfrak{a}\mathfrak{b}\Gamma M\tau'^{-1} \subset \mathfrak{a}M;$$

thus $\tau'^{-1}|_M$ is the inverse to τ and τ is an automorphism of M with $m\tau - m \in \mathfrak{a}M$ for every $m \in M$. Moreover, $\operatorname{Ker}\psi' \cap M = \operatorname{Ker}\psi$, and thus

$$\tau|_{\operatorname{Ker}\varphi} : \operatorname{Ker}\varphi \xrightarrow{\sim} \operatorname{Ker}\psi$$

is an isomorphism. #

7.8 <u>Reduction of the proof of (7.2) to the case where Γ is a maximal R-order in a central simple algebra.</u> Let Γ be a maximal R-order in the separable K-algebra A and let $M \in {}_{\Gamma}\mathfrak{M}^{o}$ satisfy Eichler's condition. We decompose $A = \oplus_{i=1}^{n} A_i$ into simple K-algebras. Γ and M decompose accordingly: $\Gamma = \oplus_{i=1}^{n} \Gamma_i$, $M = \oplus_{i=1}^{n} M_i$ (where certain M_i may be zero).

If (7.2) is true for the Γ_i-lattice M_i, which obviously satisfies Eichler's condition, then we take $\underline{\underline{S}}(M) = \bigcup_{i=1}^{n} \underline{\underline{S}}(M_i)$, and the epimorphisms φ and ψ decompose into

$$\varphi_i, \psi_i : M_i \longrightarrow U.$$

Since U is a simple Γ_i-module for exactly one i, φ_i, ψ_i are different from zero for exactly one i, say i = 1. Then we can find an automorphism τ_1 of M_1, which has the desired properties for M_1; but then $\tau_1 \oplus (\oplus_{i=2}^{n} 1_{M_i}) = \tau$ has the desired properties for M. Thus, we may assume that Γ is a maximal R-order in a simple separable K-algebra A. Let K' be the center of A. Then K' is again an $\underline{\underline{A}}$-field, since it is a finite extension of K, and R' is the integral closure of R in K'. Since for every prime ideal $\underline{\underline{p}}'$ of R', $R \cap \underline{\underline{p}}'$ is a prime ideal of R it suffices to prove (7.2) in case A is a central simple K-algebra.

7.9 Proof of (7.2) for a maximal R-order Γ in a central simple K-algebra A. We have the situation:

A = central simple K-algebra,

Γ = maximal R-order in A,

$M \in {}_{\Gamma}\underline{\underline{M}}^{o}$ satisfies Eichler's condition,

$B = \mathrm{End}_A(KM)$ is a central simple K-algebra, since A contains only one class of simple modules,

$\Omega = \mathrm{End}_{\Gamma}(M)$ is a maximal R-order in B (cf. IV, 5.5),

$m^2 = (B : K)$.

7.10 Lemma: If m = 1, then (7.2) holds for every finite set of prime ideals of R.

Proof: If m = 1, then Ω = R and Γ is separable. M is an irreducible Γ-lattice. If

$$\varphi : M \longrightarrow U, \; \mathrm{ann}_R U = \underline{\underline{p}}_o,$$

is a non-zero epimorphism of M onto a simple Γ-module, then we also have an epimorphism

$$\overline{\varphi}: M/\underline{\underline{p}}_o M \longrightarrow U.$$

But $M/\underline{\underline{p}}_o M \cong \hat{M}_{\underline{\underline{p}}_o}/\underline{\underline{p}}_o \hat{M}_{\underline{\underline{p}}_o}$, and according to (1.3), $M/\underline{\underline{p}}_o M$ is simple, since $\hat{M}_{\underline{\underline{p}}_o}$ is irreducible. Thus $\overline{\varphi}$ is an isomorphism. Now, given the data of (7.2), we conclude that $\overline{\varphi}$ and $\overline{\psi}$ are isomorphisms and Ker φ = Ker ψ = $= \underline{\underline{p}}_o M$. #

7.11 <u>We may thus assume $m \geq 2$</u>.

7.12 <u>Lemma</u>: There are only finitely many non-conjugate maximal R-orders in B, say $\Omega_1, \ldots, \Omega_s$.

<u>Proof</u>: If Ω_1 is a maximal R-order B, then $\Omega\Omega_1$ is a left Ω-lattice. However, the Jordan-Zassenhaus theorem is valid for Ω-lattices (cf. 4.5, 4.7, 4.10). However $\Omega\Omega_1 \cong_\Omega \Omega\Omega_2$ if and only if there exists a regular element $a \in B$ such that $\Omega\Omega_1 a = \Omega\Omega_2$. Comparing the right orders we find $\Omega_2 = a^{-1}\Omega_1 a$, Ω_1 and Ω_2 being maximal; i.e., Ω_1 and Ω_2 are conjugate. Consequently, there are only finitely many non-conjugate maximal R-orders in B, say $\Omega_1, \ldots, \Omega_s$. #

7.13 <u>Notation</u>:

(i) Let $0 \neq r_1 \in R$ be such that $r_1\Omega_i \subset \Omega$, $1 \leq i \leq s$.

(ii) $\underline{\underline{S}}(M) = \{\underline{\underline{p}}, \underline{\underline{p}}$ a prime ideal of R such that $\underline{\underline{p}} | r_1 R$ or B is ramified at $\underline{\underline{p}}$ (cf. 6.5)$\}$. Since we have a Morita equivalence between A and B we observe that A and B are simultaneously ramified or unramified at a prime $\underline{\underline{p}}$ of K.

Assume now that the data of (7.2) are given with respect to $\underline{\underline{S}}(M)$; i.e., a simple left Λ-module U with $\underline{\underline{p}}_o = \mathrm{ann}_R(U) \notin \underline{\underline{S}}(M)$, an ideal $\underline{\underline{a}}$ with $(\underline{\underline{a}}, \underline{\underline{p}}_o) = 1$, and two epimorphisms $\varphi, \psi: M \longrightarrow U$.

(iii) Let $0 \neq r_2 \in R$ be such that $r_2 \in \underline{\underline{a}}$ and $(r_2, \underline{\underline{p}}_o) = 1$.

(iv) Since $\overline{\Omega} = \mathrm{End}_{\Gamma/\underline{\underline{p}}_o\Gamma}(M/\underline{\underline{p}}_o M)$ is a finite ring, it contains only finitely many units. Let $u_1, \ldots, u_t$ be preimages of these units in Ω.

Then $\{u_i\}_{1\le i\le t}$ are units in $\Omega_{\underline{\underline{p}}_o}$, as follows from an application of Nakayama's lemma.

Let $0 \neq r_3 \in R$ be such that $r_3 u_i^{-1} \in \Omega, 1\le i\le s$ and $(r_3, \underline{\underline{p}}_o) = 1$. The latter condition can be satisfied since $u_i^{-1} \in \Omega_{\underline{\underline{p}}_o}$.

(v) Set $r = r_1 r_2 r_3$.

(vi) $\underline{\underline{S}}_o = \{\underline{\underline{p}} : \underline{\underline{p}}$ a prime ideal in R, $\underline{\underline{p}} \in \underline{\underline{S}}(M)$ or $\underline{\underline{p}}$ divides $r\}$.

7.14 Theorem (Eichler [2]): There exists a polynomial

$$f(X) = X^m + a_{m-1}X^{m-1} + \dots + a_1 X + (-1)^m \in R[X] \;^{*)}$$

satisfying the following conditions:

(i) $f(X)$ modulo $\underline{\underline{p}}_o$ has no root in $\bar{R} = R/\underline{\underline{p}}_o$.

(ii) $f(X) \equiv (X-1)^m \bmod (r^m R)$.

(iii) $f(X)$ is irreducible over $\hat{K}_{\underline{\underline{p}}}$ for any prime ideal $\underline{\underline{p}}$ of R at which B is ramified.

(iv) $f(X)$ has no root in $\hat{K}_{\underline{\underline{p}}\,\infty}$ for any infinite prime of K at which B is ramified.

Proof:

We first prove (7.14) locally and then we globalize it.

Let us pick monic polynomials $f_{\underline{\underline{p}}}(X) \in \hat{R}_{\underline{\underline{p}}}[X]$ of degree m with constant term $(-1)^m$ such that

(i) $f_{\underline{\underline{p}}_o}(X)$ has no root in $R/\underline{\underline{p}}_o$ (cf. 6.4),

(ii) for all $\underline{\underline{p}}$ in $\underline{\underline{S}}_o$, $f_{\underline{\underline{p}}}(X)$ is irreducible in $\hat{K}_{\underline{\underline{p}}}$ and

$$f_{\underline{\underline{p}}}(X) \equiv (X-1)^m \bmod ((\underline{\underline{p}}R_{\underline{\underline{p}}})^{r(\underline{\underline{p}})+1}),$$

where $r(\underline{\underline{p}})$ is the exact power of $\underline{\underline{p}}$ dividing $r^m R$.

*) We recall that $m^2 = (B : K)$.

This second condition can be satisfied, since we can find $f_{\underline{\underline{p}}}(X) \in \hat{K}_{\underline{\underline{p}}}[X]$, irreducible, monic and with constant term $(-1)^m$ satisfying the second congruence (cf. 7.5). But then it follows from the proof of (IV, 6.2) that $f_{\underline{\underline{p}}}(X) \in \hat{R}_{\underline{\underline{p}}}[X]$.

(i) and (ii) do not conflict, since $\underline{\underline{p}}_o$ does not divide r (7.13).

By the Chinese remainder theorem (I, 7.7), we can find a monic polynomial $f(X) \in R[X]$ with constant term $(-1)^m$ such that

$$f(X) \equiv f_{\underline{\underline{p}}_o}(X) \bmod \underline{\underline{p}}_o \hat{R}_{\underline{\underline{p}}_o}[X],$$

$$f(X) \equiv f_{\underline{\underline{p}}}(X) \bmod (\underline{\underline{p}}\hat{R}_{\underline{\underline{p}}}[X])^{s_1(\underline{\underline{p}})},$$

where $s_1(\underline{\underline{p}}) = r(\underline{\underline{p}}) + 1$ is such that (7.6) applies, for all $\underline{\underline{p}} \in \underline{\underline{S}}_o$ (cf. 7.13).

We have thus found a polynomial $f(X) \in R[X]$ such that

7.15 (i) $f(X)$ modulo $\underline{\underline{p}}_o$ has no root in $R/\underline{\underline{p}}_o$,

(ii) $f(X) \equiv (X-1)^m \bmod (r^m R)$,

(iii) $f(X)$ is irreducible over $\hat{K}_p$ for every prime ideal of R at which B is ramified, and

$f(X)$ is monic and has constant term $(-1)^m$.

To prove the rest of the theorem; i.e., (iv), we establish:

7.16 <u>Lemma</u>: If K is an algebraic number field we can determine $f(X) \in R[X]$ such that it satisfies (7.15) and such that $f(X)$ has no root is $\hat{K}_{\underline{\underline{p}}\infty}$ for every infinite prime $\underline{\underline{p}}_\infty$ at which B is ramified.

We <u>remark</u>, that this is the crucial point, where Eichler's condition is needed.

<u>Proof</u>: <u>Case 1: m > 2</u>. If m is odd, then $(B : K) = m^2$ is odd, and B can not be ramified at any infinite prime. Hence we may assume <u>m</u> to be <u>even</u>. Then $f(X)$ has even degree. If $K^{(1)}, \dots, K^{(s)}$ are the real conjugate fields of K, then $f^{(i)}(X)$ - the i-th conjugate of $f(X)$ - is positive for large real x, $1 \le i \le s$; and for x = 0, we have $f(0)^{(i)} > 0$. Now we choose $0 < z \in \underline{\underline{Z}}$ such that

(i) $\underline{\underline{p}}^{r(\underline{\underline{p}})}$ divides $z \cdot R$ for every $\underline{\underline{p}} \in \underline{\underline{S}}_o$ and $\underline{\underline{p}}_o$ divides $z \cdot R$.

(ii) $(f(x) + zx^2)^{(i)} > 0$ for every $x \in \underline{\underline{R}}$. Then $f(X) + zX^2$ can not have a root in any $\hat{K}_{\underline{\underline{p}}_\infty}$ for real $\underline{\underline{p}}_\infty$, since this would mean, that $(f(X) + zX^2)^{(i)}$ has a real root. In particular, for every root ω of $f(X) + zX^2$ and for every prime $\underline{\underline{p}}_\infty$ at which B is ramified, $\hat{K}_{\underline{\underline{p}}_\infty}(\omega) = \underline{\underline{C}}$, and thus ω generates a splitting field for $\hat{B}_{\underline{\underline{p}}_\infty}$. In addition, since $m > 2$, and by the choice z, $f(X) + zX^2$ satisfies the same congruences as $f(X)$.

Case 2: m = 2. Since M satisfies Eichler's condition, B is not ramified at all infinite primes of K; say B is ramified at $\underline{\underline{p}}_{\infty 1},\dots,\underline{\underline{p}}_{\infty s}$, and let $K^{(1)},\dots,K^{(s)}$ be the real conjugate fields corresponding to these infinite primes. Then $(K : \underline{\underline{Q}}) > s$ and we may apply the approximation theorem (7.3) as follows: Let

$$f(X) = X^2 + \alpha X + 1$$

and choose $0 < c \in \underline{\underline{Z}}$ such that $\underline{\underline{p}}^{r(\underline{\underline{p}})}$ divides $c \cdot R$ for every prime $\underline{\underline{p}} \in \underline{\underline{S}}_o$ and $\underline{\underline{p}}_o$ divides cR. According to (7.3) we can find $y \in R$ (here it should be observed that R is a localization of the algebraic integers in K, cf. Ex. 4,3) such that

$$|\alpha^{(i)} - cy^{(i)}| < 2,\ 1 \leq i \leq s.$$

Now the polynomial

$$X^2 + (\alpha - cy)X + 1$$

also satisfies (7.15) and its roots are

$$x_{1,2} = -(\alpha - cy)/2 \pm \sqrt{((\alpha - cy)^2 - 4)/4}.$$

But because of the choice of y, this can never lie in any of the real fields $K^{(1)},\dots,K^{(s)}$.

This proves (7.15). #

7.17 Lemma: If f(X) satisfies (7.15), then every root of f(X) generates a splitting field for B.

Proof: At the finite primes $\underline{\underline{p}}$ at which B is ramified, f(X) is irreducible in $\hat{K}_{\underline{\underline{p}}}$, since all these primes are contained in $\underline{\underline{S}}_o$. Thus $\hat{K}_{\underline{\underline{p}}}(\omega)$ is a splitting field for $\hat{B}_{\underline{\underline{p}}}$, if ω is a root of f(X) (cf. 6.11). But f(X) is also irreducible at the infinite primes at which B is ramified; i.e., $\hat{K}_{\underline{\underline{p}}_\infty}(\omega)$ is a splitting field for $\hat{B}_{\underline{\underline{p}}_\infty}$. Thus $\hat{K}_{\underline{\underline{p}}}(\omega)$ is a splitting field of $\hat{B}_{\underline{\underline{p}}}$ for every finite and infinite prime of K, and thus $K(\omega)$ is a splitting field for B (cf. 6.8). #

7.18 Lemma: B contains a root ω of f(X).

Proof: This follows from (6.10). #

Let us summarize: Under the hypotheses of (7.2) there exists an element $\omega \in B$ that satisfies a monic polynomial $f(X) \in R[X]$ the constant term of which is $(-1)^m$, and such that

$$f(X) \equiv (X-1)^m \bmod (r^m R).$$

Hence, $(\omega - 1)^m / r^m = f_1(\omega) \in R$, since ω is integral over R. But then $y = (\omega - 1)/r$ is also integral over R. Replacing y resp. ω by a suitable conjugate in B, we may assume that $y \in \Omega_j$ for some $1 \leq j \leq s$ (cf. 7.12), since every integral element in B is contained in a maximal order (cf. IV, 1.3, 4.6). Hence, by (7.13,i), $r_1 y = (\omega - 1)/r_2 r_3 \in \Omega$; but $r_3 u_i^{-1} \in \Omega, 1 \leq i \leq t$, (cf. 7.13,iv) and consequently,

$$\omega_i = u_i \omega u_i^{-1} = 1 + u_i r_1 r_2 y r_3 u_i^{-1} \in \Omega.$$

Moreover, $\omega_i - 1 \in r_2 \Omega \subset \underline{\underline{a}}\Omega$ (cf. 7.13,ii) and hence $\omega_i \equiv 1 \bmod \underline{\underline{a}}\Omega$.

7.19 Claim: $\tau_i : M \longrightarrow M$,

$$m \longmapsto m\,\omega_i,$$

is an automorphism of M, with $\tau_i \equiv 1 \bmod \underline{\underline{a}}\Omega, 1 \leq i \leq t$.

Proof: Since ω_i is conjugate to ω, $f(\omega_i) = 0 = \omega_i \cdot f_2(\omega_i) + (-1)^m$, which shows that ω_1 is a unit in Ω, since $f_2(\omega_1) \in \Omega$. Thus, τ_1 is an automorphism of M. Moreover, $(y-y\omega_1)=(1-\omega_1)y \in \underline{\underline{a}}M\Omega = \underline{\underline{a}}M$, for $y \in M$. #

7.20 <u>Lemma</u>: Either Ker φ = Ker ψ or else there exists $1 \leq i \leq t$ such that $\tau_i : \operatorname{Ker}\varphi \xrightarrow{\sim} \operatorname{Ker}\psi$ is an isomorphism.

<u>Proof</u>: If there exists an automorphism α of U such that $\varphi\alpha = \psi$, then Ker φ = Ker ψ and nothing has to be proved. Thus we shall show: If this is not the case, then there exists an index i such that $\tau_i\varphi = \psi$. This automatically implies $\tau_i : \operatorname{Ker}\psi \xrightarrow{\sim} \operatorname{Ker}\varphi$. Let us denote by "-" reduction modulo $\underline{\underline{p}}_o$, where $\underline{\underline{p}}_o = \operatorname{ann}_R(U)$. Then $\tau_i\varphi = \psi$ if and only if $\overline{\tau}_i\overline{\varphi} = \overline{\psi}$ as follows from the commutative diagram:

$$\begin{array}{ccccc}
 & & \overset{\psi}{\longrightarrow} & & \\
M & \xrightarrow{\text{can}} & \overline{M} & \xrightarrow{\overline{\psi}} & U \\
\tau_i \downarrow & & \downarrow \overline{\tau}_i & & \downarrow 1_U \\
M & \xrightarrow{\text{can}} & \overline{M} & \xrightarrow{\overline{\varphi}} & U \\
 & & \underset{\varphi}{\longrightarrow} & &
\end{array}$$

where "can" denotes the canonical epimorphism $M \longrightarrow M/\underline{\underline{p}}_o M$. Thus it suffices to show $\overline{\tau}_i\overline{\varphi} = \overline{\psi}$ for some $1 \leq i \leq t$.

Since $\varphi, \psi \neq 0$, we have two non-zero elements

$$\overline{\varphi}, \overline{\psi} \in \operatorname{Hom}_{\overline{\Gamma}}(\overline{M}, U) = V.$$

But V is an $[\operatorname{End}_{\overline{\Gamma}}(\overline{M}), \operatorname{End}_{\overline{\Gamma}}(U)]$-bimodule. However, U is a simple Γ-module; i.e., it is a simple $\Gamma_{\underline{\underline{p}}_o}$-module. But A is not ramified at $\underline{\underline{p}}_o$ (cf. 7.13), and thus $\operatorname{End}_{\overline{\Gamma}}(U) = \overline{R}$. This implies that V is an $\overline{R}$-vector-space. We now distinguish two cases:

(i) $\overline{\varphi}$ and $\overline{\psi}$ are linearly dependent over $\overline{R}$; i.e., there exists $\sigma \in \operatorname{End}_{\overline{\Gamma}}(U)$ such that $\overline{\varphi}\sigma = \overline{\psi}$; but then $\varphi\sigma = \psi$, and this case we had excluded.

(ii) $\overline{\varphi}$ and $\overline{\psi}$ are linearly independent over $\overline{R}$. We recall, that $f(X)$ was choosen in such a way, that $f(X)$ has no root modulo $\underline{\underline{p}}_o$ in $\overline{R}$

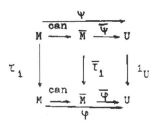

(cf. 7.15); and since ω is a root of $f(X)$, $\overline{\omega}$ can not be a multiple of $\overline{1}$ in $\overline{\Omega}$. We claim, that there exists a $v \in V$ such that $\overline{\omega}v$ and v are linearly independent. (It should be observed, that $M \in {}_{\Gamma}\underline{\underline{M}}^{o}$ is a pro-generator (cf. IV, 5.5) and thus $\mathrm{End}_{\overline{\Gamma}}(\overline{M}) = \overline{\mathrm{End}_{\Gamma}(M)}$ (cf. IV, 3.7).) Since B is unramified at $\underline{\underline{p}}_o$, $\overline{\Omega} = (\overline{R})_m$ and thus $\overline{\Omega} = \mathrm{Hom}_{\overline{R}}(V,V)$. Let $\{v_i\}_{1 \le i \le m}$ be an $\overline{R}$-basis for V. If v_i and $\overline{\omega}v_i$ are linearly independent for some i, we are done. Thus $v_i r_i = \overline{\omega} v_i$, $r_i \in \overline{R}, 1 \le i \le m$. If $r_j = r_i$ for all i then $\overline{\omega}$ is a multiple of $\overline{1} \in \overline{\Omega}$, but this was excluded. Thus, not all r_i are the same and $\overline{\omega}(\sum_{i=1}^{m} v_i)$ and $\sum_{i=1}^{m} v_i$ are linearly independent. Now we choose $\overline{R}$-bases $\overline{\varphi}, \overline{\psi}, v_3, \dots, v_m$ and $v, \overline{\omega}v, u_3, \dots, u_m$. Then there exists a linear transformation $\overline{\kappa} \in \overline{\Omega} = \mathrm{Hom}_{\overline{R}}(V,V)$ sending one basis into the other. Then $\overline{\kappa}\overline{\varphi} = v$ and $\overline{\kappa}\overline{\psi} = \overline{\omega}v$; i.e., $\overline{\omega}\overline{\kappa}\overline{\varphi} = \overline{\kappa}\overline{\psi}$. Since $\overline{\kappa}$ is invertible, we have $\overline{\kappa}^{-1}\overline{\omega}\overline{\kappa}\overline{\varphi} = \overline{\psi}$. However, there exists exactly one u_i such that $\overline{u}_i = \overline{\kappa}$ (cf. 7.13); thus there exists exactly one i such that $\overline{\tau}_i = \overline{\kappa}^{-1}\overline{\omega}\overline{\kappa}$; i.e., $\overline{\tau}_i\overline{\varphi} = \overline{\psi}$.

This completes the proof of (7.2). #

Exercises §7:

1.) Let $K = \underline{\underline{Q}}(\alpha)$ be a finite extension of $\underline{\underline{Q}}$. Let $\{\alpha_i\}_{1 \le i \le n}$ be an integral basis for K over $\underline{\underline{Q}}$, and let $\{\alpha_i^{(j)}\}_{1 \le j \le n}$ be the conjugates of $\{\alpha_i\}$. Show that the discriminant of K over $\underline{\underline{Q}}$ with respect to the basis $\{\alpha_i\}_{1 \le i \le n}$ is given by $\det(\alpha_i^{(j)})$.

2.) Minkovski's lemma: Let $\{a_{ij}\}_{1 \le i, j \le n}$ be a family of real numbers, $n > 1$, such that $\Delta = \det(a_{ij}) \ne 0$. Let $\{k_i\}_{1 \le i \le n}$ be positive real numbers such that $\prod_{i=1}^{n} k_i \ge |\Delta|$. Then there exist integers $\{x_i\}_{1 \le i \le n}$, not all zero, such that

$$\left| \sum_{i=1}^{n} a_{ji} x_i \right| < k_j, \quad 1 \le j \le n-1,$$

$$\left| \sum_{i=1}^{n} a_{ni} x_i \right| \le k_n.$$

(<u>Hint</u>: A point $(x_1,\dots,x_n) \in \mathbb{R}^{(n)}$, the n-dimensional vectorspace over $\mathbb{R}$, is called a <u>lattice point</u> if $x_i \in \mathbb{Z}, 1 \le i \le n$, and $x_i \ne 0$ for at least one i. We shall write $\mathbf{x}$ for the lattice point $(x_1,\dots,x_n)$ and $L_i(\mathbf{x}) = \sum_{j=1}^{n} a_{ij}x_j$. To prove Minkovski's lemma, we assume it to be false; i.e., every lattice point $\mathbf{x}$ satisfies at least one of the inequalities

$$|L_i(\mathbf{x})| \ge k_i, \quad 1 \le i \le n-1,$$

$$|L_n(\mathbf{x})| > k_n.$$

Now, given a set $\{\alpha_i\}_{1 \le i \le n}$ of positive real numbers, show that there are only finitely many lattice points $\mathbf{x}$ satisfying $|L_i(\mathbf{x})| < \alpha_i$. Using this, we conclude that there exists a sufficiently small positive number ε such that each lattice point $\mathbf{x}$ satisfies at least one of the inequalities

$$|L_i(\mathbf{x})| \ge k_i \quad , 1 \le i \le n-1,$$

$$|L_n(\mathbf{x})| \ge k_n + \varepsilon.$$

We put $k_i' = k_i, 1 \le i \le n-1$, and $k_n' = k_n + \varepsilon$. For every lattice point $\mathbf{y}$ we consider the parallelotope $P_{\mathbf{y}}$ defined by $\{\mathbf{z} \in \mathbb{R}^{(n)} : |L_i(\mathbf{z} - \mathbf{y})| < k_i'/2\}_{1 \le i \le n}$. By varying $\mathbf{y}$ we obtain an infinite number of geometrically congruent parallelotopes. Show that $P_{\mathbf{y}} \cap P_{\mathbf{y}'} = \emptyset$ if $\mathbf{y} \ne \mathbf{y}'$.

Next we shall compute the volume of $P_{\mathbf{o}}$ in two different ways, thus obtaining a contradiction: For a positive integer m, the sum of the volumes of $P_{\mathbf{y}}$ which lie completely in the hypercube $H_m = \{\mathbf{z} : |\mathbf{z}_i| \le m\}$ is not bigger than the volume of H_m, $(2m)^n$. Since in $P_{\mathbf{o}}$ there are only finitely many lattice points, $c = \max_{\substack{\mathbf{x} \in P_{\mathbf{o}} \\ i=1,\dots n}} (x_i)$ is bounded.*)

Thus, as soon as $\mathbf{y}$ is a lattice point satisfying $\max_{1 \le i \le n} y_i \le m$, $P_{\mathbf{y}} \subset H_{m+c}$. But there are exactly $(2m+1)^n$ such lattice points $\mathbf{x}$. Thus

---

*) By assumption, $P_{\mathbf{o}} = \emptyset$ and so $c = 1$ will do.

we have

$$(2m + 1)^n \cdot J \leq 2(m + c)^n,$$

where J is the volume of $\underline{\underline{P}}_o$. Hence

$$J \leq [(2m + 2c)/(2m + 1)]^n.$$

Now letting m tend to infinity, we obtain $J \leq 1$. On the other hand the volume J of $\underline{\underline{P}}_o$ is given by

$$J = \int_{\substack{\underline{\underline{z}} \in \underline{\underline{R}}^{(n)} \\ |L_i(\underline{\underline{z}})| < k_i'/2}} d\underline{\underline{z}}.$$

Changing to the variables $z_i' = L_i(\underline{\underline{z}})$, we obtain

$$J = (1/|\Delta|) \int_{|z_i'| < k_i'/2} dz_1' \ldots dz_n' = \prod_{i=1}^n k_i'/|\Delta|.$$

But $\prod_{i=1}^n k_i' > \prod_{i=1}^n k_i \geq |\Delta|$. Thus, $J > 1$, a contradiction.

3.) Prove <u>Minkovski's lemma for complex numbers</u>: Let $\{a_{ij}\}_{1 \leq i, j \leq n}$ be a family of complex numbers, $n > 1$ such that $\Delta = \det(a_{ij}) \neq 0$ and let $\{k_i\}_{1 \leq i \leq n}$ be positive real numbers such that $\prod_{i=1}^n k_i \geq |\Delta|$. Assume moreover, that if one of the forms $L_j = \sum_{i=1}^n a_{ji} x_i$ is complex, then there exists another one $L_k = \sum_{i=1}^n a_{ki} x_i$, which is the complex conjugate to L_j. We assume that the k_j are the same for complex conjugate forms. Then there exists a lattice point $\underline{\underline{x}}$ such that

$$|L_j(\underline{\underline{x}})| \leq k_j; \ 1 \leq j \leq n.$$

Moreover, the inequality can be replaced by a strict one except for one real form L_j or two complex conjugate forms chosen in advance (<u>Hint</u> use 2.): Replace the forms L_j by L_j', where $L_j = L_j'$ if the form L_j is real, and if the form L_j is complex and L_{j+1} is its complex conjugate, then $L_j' = (L_j + L_{j+1})/\sqrt{2}$ and $L_{j+1}' = (L_j - L_{j+1})/\sqrt{2}$.)

§8 Ideals and norms of ideals

In this section we consider one-sided ideals of maximal orders Γ in central simple algebras A over $\underline{\underline{A}}$-fields. If A is not a totally definite quaternion algebra, then a Γ-ideal is up to isomorphism uniquely determined by its norm.

In this section, K is an $\underline{\underline{A}}$-field with Dedekind domain R and A is a finite dimensional central simple K-algebra. We recall that A can not be a totally definite quaternion algebra if K is an $\underline{\underline{A}}$-field of characteristic $p > 0$.

8.1 Definition: An R-lattice M in A is called a normal ideal, if $\Lambda_1(M)$, the left order of M, is maximal. M is called a normal integral ideal, if M is normal and $M \subset \Lambda_1(M)$.

8.2 Theorem: If M is a normal ideal in A, then $\Lambda_r(M)$, the right order of M, is maximal; and if M is integral, then $M \subset \Lambda_r(M)$. Moreover, there exists a unique normal ideal M^{-1} with $\Lambda_1(M) = \Lambda_r(M^{-1})$, $\Lambda_r(M) = \Lambda_1(M^{-1})$ such that $MM^{-1} = \Lambda_1(M)$ and $M^{-1}M = \Lambda_r(M)$. M^{-1} is called the inverse of M.

Proof: Let $\Gamma_1 = \Lambda_1(M)$. Since M is a faithful Γ_1-lattice, and since Γ_1 is maximal, $M \in {}_{\Gamma_1}\underline{\underline{M}}^o$ is a progenerator (cf. IV, 5.4). Hence $\Gamma_2 = \Lambda_r(M)$ is maximal, and we have two isomorphisms

$$\mu_M : \mathrm{Hom}_{\Gamma_1}(M, \Gamma_1) \otimes_{\Gamma_1} M \longrightarrow \Gamma_2,$$

$$\tau_M : M \otimes_{\Gamma_2} \mathrm{Hom}_{\Gamma_1}(M, \Gamma_1) \longrightarrow \Gamma_1$$

(cf. III, §1). We put $M^{-1} = \mathrm{Hom}_{\Gamma_1}(M, \Gamma_1)$. Then one finds readily, that $\mathrm{Im}\,\mu_M = M^{-1}M$ and $\mathrm{Im}\,\tau_M = MM^{-1}$. Moreover $\Lambda_1(M^{-1}) = \Gamma_2$ and $\Lambda_r(M^{-1}) = \Gamma_1$, since these orders are maximal. Thus, M^{-1} is an inverse. If N were another inverse, then

$$N = N\Lambda_1(M) = NMM^{-1} = \Lambda_r(M)M^{-1} = M^{-1}.$$

If M is integral, then $MM \subset M \subset \Gamma_1$ and $M \subset \Lambda_r(M) = \Gamma_2$. #

8.3 <u>Definition</u>: (i) A normal ideal M is called a <u>maximal normal ideal</u>, if it is integral and if it is a maximal left ideal in its left order.

(ii) A product MN of normal ideals is called a <u>proper product</u>, if $\Lambda_r(M) = \Lambda_l(N)$.

8.4 <u>Lemma</u>: Let M and N be normal ideals with $\Lambda_l(M) = \Lambda_l(N)$. Then there exists a unique normal ideal M' with $\Lambda_r(M') = \Lambda_r(N)$ such that $MM' = N$ and this product is proper. Moreover, if $M \supset N$, then M' is integral.

<u>Proof</u>: We put $M' = M^{-1}N$; then M' is normal and $\Lambda_r(M') = \Lambda_r(N)$ and $\Lambda_l(M') = \Lambda_r(M)$. Thus the product $MM' = N$ is proper, and if $N \subset M$ then $M^{-1}N \subset M^{-1}M = \Lambda_r(M)$ and M' is integral by (8.2). #

8.5 <u>Lemma</u>: If M is a maximal normal ideal, then it is also a maximal right ideal in its right order.

<u>Proof</u>: Let N be a maximal right $\Gamma_2 = \Lambda_r(M)$-ideal, $M \subset N \subsetneq \Gamma_2$ and put $\Gamma_1 = \Lambda_l(M)$. Then $N^{-1} \supset \Gamma_2$ and $M = (MN^{-1})N$, and $M_1 = MN^{-1}$ is a normal integral ideal (cf. 8.4). Thus $M \subsetneq M_1 \subset \Lambda_l(M_1) = \Gamma_1$, and $M_1 = \Gamma_1$; i.e., $MN^{-1} = \Gamma_1$. This implies $\Gamma_2 N^{-1} = N^{-1} = M^{-1}$; whence $M = N$ and M is maximal right Γ_2-ideal. #

8.6 <u>Theorem</u>: Every normal integral ideal M is the proper product of s maximal normal ideals, where s is the length of a composition series of $\Lambda_l(M)/M$. Moreover, given any maximal ideal $\underline{\underline{p}}$ of R dividing $\operatorname{ann}_R(\Lambda_l(M)/M)$ we can find a proper product representation

$$M = M_1 M_2 \cdots M_s$$

of normal maximal ideals such that $\underline{\underline{p}}\Lambda_l(M) \subset M_1$.

<u>Proof</u>: Let $\Gamma = \Lambda_l(M)$. Since M is an ideal in Γ such that $KM = A$, Γ/M is a left artinian and noetherian module (cf. proof of IV, 2.2); hence

it has a composition series (cf. I, 4.7), say

$$\Gamma/M \underset{\neq}{\supset} T_1 \underset{\neq}{\supset} \cdots \underset{\neq}{\supset} T_{s-1} \underset{\neq}{\supset} T_s = 0.$$

Moreover, since Γ/M is an R-torsion module, it is the direct sum of its $\underline{\underline{p}}$-primary components (cf. I, 8.9), and we may thus arrange the composition series in such a way that $\underline{\underline{p}} \cdot (\Gamma/M) \subset T_1$, where $\underline{\underline{p}}$ is any maximal ideal dividing $\mathrm{ann}_R(\Gamma/M)$, since the annihilator of every simple Γ-module is a maximal ideal in R. Now, let M_i be the inverse image of the T_i in Γ, $1 \le i \le s$. Then we have the proper chain of Γ-submodules of Γ

$$\Gamma \underset{\neq}{\supset} M_1 \underset{\neq}{\supset} M_2 \cdots \underset{\neq}{\supset} M_{s-1} \underset{\neq}{\supset} M_s = M,$$

where M_i is a maximal Γ-submodule in M_{i-1}, $1 \le i \le s$, $(M_o = \Gamma)$. We put $\Gamma_i = \Lambda_r(M_i)$, $1 \le i \le s$, and observe that the Γ_i are all maximal, since $\Lambda_l(M_i) = \Gamma$ is maximal. Now set $N_1 = M_1$, $N_{i+1} = M_i^{-1}M_{i+1}$, $1 \le i \le s$. Then $\Lambda_r(N_i) = \Lambda_l(N_{i+1}) = \Gamma_i$ and $M = N_1N_2 \ldots N_s$ is a proper product. Moreover, all the N_i, $1 \le i \le s$, are maximal normal ideals. For, since $M_i \supset M_{i+1}$, we have $M_i^{-1}M_i = \Gamma_i \supset N_i = M_i^{-1}M_{i+1}$. Also, for any left Γ_i-lattice X, $\Gamma_i \supset X \supset N_{i+1}$ implies $M_i\Gamma_i = M_i = M_iX$ or $M_iX = M_iN_{i+1} = M_{i+1}$, since M_{i+1} is a maximal $\Gamma = M_iM_i^{-1}$ submodule of M_i. But then, $X = M_i^{-1}M_iX = M_i^{-1}M_i = \Gamma_i$ or $X = M_i^{-1}M_{i+1} = N_{i+1}$, and thus the N_i are maximal. #

8.7 <u>Definition</u>: Let M be a normal ideal in A. Then the <u>reduced norm of M</u>, $\nu(M)$ is defined as the

$$\text{R-ideal generated by } \{\mathrm{Nrd}_{A/K}(m) : m \in M\}.$$

This is in general a fractional ideal. (If it is necessary to indicate A and K, we write $\nu_{A/K}(M)$.)

8.8 <u>Lemma</u>: Let M and N be normal ideals. Then

(i) $\nu(M_{\underline{\underline{p}}}) = \nu(M)_{\underline{\underline{p}}}$ for every maximal ideal $\underline{\underline{p}}$ of R,

(ii) $\nu(\hat{M}_{\underline{\underline{p}}}) = \nu(M)\hat{}_{\underline{\underline{p}}}$ for every maximal ideal $\underline{\underline{p}}$ of R, "^" denoting the completion at $\underline{\underline{p}}$.

(iii) If the product MN is proper, then $\nu(MN) = \nu(M)\,\nu(N)$.

__Proof:__ (i) Let $(A : K) = n^2$. Then

$$\mathrm{Nrd}_{A/K}(m/s) = s^{-n}\mathrm{Nrd}_{A/K}(m),$$

for $m \in M$, $s \in R \setminus \{\underline{\underline{p}}\}$. Thus $\nu(M_{\underline{\underline{p}}}) \subset \nu(M)_{\underline{\underline{p}}}$. Conversely,

$$s^{-1}\mathrm{Nrd}_{A/K}(m) = s^{n-1}\mathrm{Nrd}_{A/K}(m/s),$$

and $\nu(M)_{\underline{\underline{p}}} \subset \nu(M_{\underline{\underline{p}}})$; thus we must have equality.

(ii) In view of (i) it suffices to show $\nu(M_{\underline{\underline{p}}})^{\wedge} = \nu(\hat{M}_{\underline{\underline{p}}})$. But $\Gamma_{\underline{\underline{p}}} = \Lambda_1(M_{\underline{\underline{p}}})$ is a principal ideal ring (cf. IV, 5.7) and $M_{\underline{\underline{p}}}$ is of the form $M_{\underline{\underline{p}}} = \Gamma_{\underline{\underline{p}}}a$ for some regular element $a \in A$. Thus

$$\nu(\Gamma_{\underline{\underline{p}}}a)^{\wedge} = \hat{R}\,\mathrm{Nrd}_{A/K}(a) = \hat{R}\,\mathrm{Nrd}_{\hat{A}_{\underline{\underline{p}}}/\hat{K}_{\underline{\underline{p}}}}(a) = \nu(\hat{\Gamma}_{\underline{\underline{p}}}a).$$

(iii) Because of (i) it suffices to show this locally. Since MN is a proper product, $M_{\underline{\underline{p}}} = \Gamma_{1_{\underline{\underline{p}}}}a$ and $N_{\underline{\underline{p}}} = \Gamma_{2_{\underline{\underline{p}}}}b$ with $\Gamma_{2_{\underline{\underline{p}}}} = a^{-1}\Gamma_{1_{\underline{\underline{p}}}}a$. Thus

$$\nu(M_{\underline{\underline{p}}}N_{\underline{\underline{p}}}) = \nu(\Gamma_{1_{\underline{\underline{p}}}}a\,a^{-1}\Gamma_{1_{\underline{\underline{p}}}}ab) = R_{\underline{\underline{p}}}\cdot\mathrm{Nrd}_{A/K}(ab) =$$

$$= R_{\underline{\underline{p}}}\,\mathrm{Nrd}_{A/K}(a)\,\mathrm{Nrd}_{A/K}(b) = \nu(M_{\underline{\underline{p}}})\,\nu(N_{\underline{\underline{p}}}). \qquad \#$$

8.9 __Lemma:__ M is a maximal normal ideal if and only if $\nu(M) = \underline{\underline{p}}$ is a maximal ideal in R.

__Proof:__ Let M be a maximal normal ideal and $\Gamma = \Lambda_1(M)$; then Γ/M is a simple left Γ-module and $\mathrm{ann}_R(\Gamma/M) = \underline{\underline{p}}_o$ is a maximal ideal in R. Thus $\Gamma_{\underline{\underline{p}}} = M_{\underline{\underline{p}}}$ for every maximal ideal $\underline{\underline{p}} \neq \underline{\underline{p}}_o$, and with (8.8) we conclude $\nu(M) = \underline{\underline{p}}_o^s$, and it suffices to show $\nu(\hat{M}) = \hat{\underline{\underline{p}}}_o$, where "$\hat{}$" denotes the completion at $\underline{\underline{p}}_o$. Since all maximal $\hat{R}$-orders are conjugate, and since conjugate ideals have the same norm, we may assume $\hat{A} = (\hat{D})_n$ and $\hat{\Gamma} = (\hat{\Omega})_n$, where $\hat{\Omega}$ is the maximal $\hat{R}$-order in the skewfield $\hat{D}$. However, every maximal ideal contains rad $\hat{\Gamma} = (\omega_o \hat{\Omega})_n$, where $\omega_o \hat{\Omega}$ = rad $\hat{\Omega}$. We have

$$\nu(\text{rad}\,\hat{\Gamma}) = \hat{R}\, \mathrm{Nrd}_{\hat{A}/\hat{K}}(\omega_o \underline{\underline{E}}),$$

where $\underline{\underline{E}}$ is the n-dimensional identity matrix. Thus

$$\nu(\text{rad}\,\hat{\Gamma}) = \hat{R}\, \mathrm{Nrd}_{D/K}(\omega_o)^n = \hat{R}\underline{\underline{p}}_o^n$$

by (IV, 6.13). On the other hand $\hat{\Gamma}/\text{rad}\,\hat{\Gamma}$ has a composition series of n terms, and since $\hat{\Gamma}/\hat{M} \overset{nat}{\cong} (\hat{\Gamma}/\text{rad}\,\hat{\Gamma})/(\hat{M}/\text{rad}\,\hat{\Gamma})$, we can find a proper product representation (cf. 8.6)

$$\text{rad}\,\hat{\Gamma} = \hat{M} \cdot \hat{M}_2 \cdots \hat{M}_n,$$

where $\hat{M}_i$ are maximal normal ideals. It thus suffices to show $\nu(\hat{M}) = \hat{R}$ implies $\hat{M} = \hat{\Gamma}$. However, $\hat{M} = \hat{\Gamma}a$ for some regular element a in A and so $\nu(\hat{M}) = \hat{R}$ if and only if $\mathrm{Nrd}_{\hat{A}/\hat{K}}(a) = u$ for some unit u in $\hat{R}$. Since $\hat{M} \subset \hat{\Gamma}$ we have $a \in \hat{\Gamma}$ and $\mathrm{Pcrd}_{\hat{A}/\hat{K}}(a,X) = X^m + \ldots + (-1)^m u \in \hat{R}[X]$, and one readily finds that a is a unit in $\hat{\Gamma}$. Thus $\hat{M} = \hat{\Gamma}$, a contradiction; and consequently $\nu(\hat{M}) = \underline{\underline{p}}_o \hat{R}$.

Conversely, if $\nu(\hat{M}) = \underline{\underline{p}}_o \hat{R}$ then $\hat{M}$ has to be maximal and thus M is maximal, since $\hat{M}_{\underline{\underline{p}}} = \hat{\Gamma}_{\underline{\underline{p}}}$ for all $\underline{\underline{p}} \neq \underline{\underline{p}}_o$, since $\nu(\hat{M}_{\underline{\underline{p}}}) = \hat{R}_{\underline{\underline{p}}}$. #

8.10 Theorem: Let Γ be a maximal R-order in A, and denote by $\underline{\underline{M}}_{\Gamma}(A)$ the set of Γ-lattices in A and by I(R) the group of non-zero fractional R-ideals in K. Then the map

$$\nu : \underline{\underline{M}}_{\Gamma}(A) \longrightarrow I(R),$$
$$M \longmapsto \nu(M)$$

is an epimorphism.

Proof: In (8.9) we have seen that the norm of every maximal normal ideal is a prime ideal. Given an ideal $\underline{\underline{a}}$ in I(R), we may assume that $\underline{\underline{a}}$ is integral, since the norm is multiplicative. We write

$\underline{\underline{a}} = \prod_{i=1}^{t} \underline{\underline{p}}_i$, $\underline{\underline{p}}_i$ maximal ideals in R and use induction on t. For t = 1, we pick a simple $\Gamma/\underline{\underline{p}}_1\Gamma$-module U, and an epimorphism

$$\varphi : \Gamma \longrightarrow U.$$

The kernel of φ, M_1 is a maximal left ideal in Γ with $\nu(M_1) = \underline{\underline{p}}_1$. Assume now that we have a left Γ-ideal M' with $\nu(M') = \prod_{i=1}^{t-1} \underline{\underline{p}}_i$. Then we pick a maximal $\Lambda_r(M')$-left ideal M" such that $\nu(M") = \underline{\underline{p}}_t$. Then it follows from (8.8,iii) that $\nu(M'M") = \underline{\underline{a}}$. #

8.11 __Lemma:__ Let M be a normal ideal with left order Γ. Then there exists an integral Γ-ideal N isomorphic to M as left Γ-module such that $N_{\underline{\underline{p}}} = \Gamma_{\underline{\underline{p}}}$ for every finite prescribed set of maximal ideals $\{\underline{\underline{p}}_i\}_{1 \le i \le s}$.

__Proof:__ After multiplication with $0 \neq r \in R$ we may assume that M is integral. Since KM = A, we have $M_{\underline{\underline{p}}} = \Gamma_{\underline{\underline{p}}}$ for almost all maximal ideals $\underline{\underline{p}}$ (cf. IV, 1.8), say $M_{\underline{\underline{q}}_i} \neq \Gamma_{\underline{\underline{q}}_i}$, $1 \le i \le t$. Since $\Gamma_{\underline{\underline{q}}_i}$ is a principal ideal ring, there exist regular elements $a_i \in A$ such that $M_{\underline{\underline{q}}_i} = \Gamma_{\underline{\underline{q}}_i} a_i$, $1 \le i \le t$. Since $M \subset \Gamma$ we have $a_i \in \Gamma_{\underline{\underline{q}}_i}$; i.e., $a_i = \gamma_i / r_i$, with $\gamma_i \in \Gamma$ and r_i a unit in $\Gamma_{\underline{\underline{q}}_i}$. Thus, we may assume $a_i \in \Gamma$, $1 \le i \le n$. According to the Chinese remainder theorem (I, 7.7), we can determine $a \in \Gamma$ such that

$$a \equiv a_{\underline{\underline{p}}} \mod (\underline{\underline{p}}\Gamma)$$

for every $\underline{\underline{p}} = \underline{\underline{p}}_i$, $1 \le i \le s$, where $a_{\underline{\underline{p}}} = a_i$ if $\underline{\underline{p}} = \underline{\underline{q}}_i$ and $a_{\underline{\underline{p}}} = 1$ for $\underline{\underline{p}} \neq \underline{\underline{q}}_i$ for $1 \le i \le t$. We now consider the left Γ-ideal

$$Ma^{-1} \cong M.$$

For $\underline{\underline{p}}_i$ we have

$$(Ma^{-1})_{\underline{\underline{p}}_i} = \Gamma_{\underline{\underline{p}}_i} a_{\underline{\underline{p}}_i} a^{-1}.$$

But for every i, $1 \le i \le s$, we have

$$a_{\underline{\underline{p}}_i} a^{-1} = 1 + \underline{\underline{p}}_i \gamma, \text{ for some } \gamma \in \Gamma.$$

Thus an application of Nakayama's lemma shows that $a_{\underline{\underline{p}}_i} a^{-1}$ is a unit in $\Gamma_{\underline{\underline{p}}_i}$; i.e., $(Ma^{-1})_{\underline{\underline{p}}_i} = \Gamma_{\underline{\underline{p}}_i}$, $1 \leq i \leq s$. Multiplying by a suitable $0 \neq r \in R$, $r \equiv 1 \bmod \underline{\underline{p}}_i$, $1 \leq i \leq s$, we may assume $Ma^{-1} \subset \Gamma$. #

8.12 Theorem (Eichler [1]): Assume that A is not a totally definite quaternion algebra, and let $\underline{\underline{a}}$ be any non-zero ideal in R. If M and N are normal ideals with the same left order Γ such that $\nu(M) = \nu(N)$, then there exists a maximal R-order Γ' in A and elements $\beta_n \in \Gamma'$, $n \in \underline{\underline{N}}$ such that

$$M\beta_n = N \text{ and } \beta_n \equiv 1 \bmod(\underline{\underline{a}}^n \Gamma').$$

Proof: We shall apply (7.2), since M and N satisfy Eichler's condition. Let $\underline{\underline{S}}(\Gamma)$ be the finite set of maximal ideals, the existence of which was established in (7.2). Let

$$\underline{\underline{S}}_o = \underline{\underline{S}}(\Gamma) \cup \{\underline{\underline{p}} : \underline{\underline{p}} \text{ a maximal ideal such that } \underline{\underline{p}} \text{ divides } \underline{\underline{a}}\}.$$

By (8.11) we can find a regular element $a \in A$ such that $M' = Ma$ and $N' = Na$ are integral ideals and $M'_{\underline{\underline{p}}} = \Gamma_{\underline{\underline{p}}}$ for every $\underline{\underline{p}} \in \underline{\underline{S}}_o$. We still have $\nu(M') = \nu(N')$ (cf. 8.8). Moreover, $\nu(M') = \nu(N')$ is coprime to every $\underline{\underline{p}} \in \underline{\underline{S}}_o$ (cf. 8.9). This implies in particular $N'_{\underline{\underline{p}}} = \Gamma_{\underline{\underline{p}}}$ for every $\underline{\underline{p}} \in \underline{\underline{S}}_o$. We write

$$M' = \prod_{i=1}^{t} M_i, \quad N' = \prod_{i=1}^{s} N_i$$

as proper products of maximal normal ideals (cf. 8.6). Since $\nu(M') = \nu(N')$, we have $s = t$. If $t = 1$, we may assume that $\nu(M_t) = \nu(N_t) = \underline{\underline{p}}_t \notin \underline{\underline{S}}_o$ (cf. 8.6, 8.9), and then M' and N' are maximal (cf. 8.9). Thus $\underline{\underline{p}}_t = \text{ann}_R(\Gamma/M') = \text{ann}_R(\Gamma/N')$, and M' and N' are maximal submodules of Γ and Γ/M' and Γ/N' are simple Γ-modules. However, A is central simple and Γ is maximal, and there exists only one class of simple left $\Gamma/\underline{\underline{p}}_t\Gamma$-modules (cf. Ex. 2,4). Thus, we have two epimorphisms

$$\varphi, \psi : \Gamma \longrightarrow U,$$

where U is the simple $\Gamma/\underline{\underline{p}}_t\Gamma$ -module. According to (7.2), there exists $\beta_n \in \Gamma$ such that $\beta_n : \mathrm{Ker}\,\varphi \xrightarrow{\sim} \mathrm{Ker}\,\psi$ and $\beta_n \equiv 1 \bmod (\underline{\underline{a}}^n\Gamma)$. This means $\beta_n : M' \xrightarrow{\sim} N'$. (Observe $(\underline{\underline{a}}^n, \underline{\underline{p}}_t) = 1$.)

Assume now, that the statement is true for ideals with less than t factors. Then we can write

$$\prod_{i=1}^{t-1} M_i\beta = \prod_{i=1}^{t-1} N_i,$$

for some integral $\beta \in A$. We put $X = \prod_{i=1}^{t-1} M_i$. Then

$$M' = X \cdot M_t, N' = X\beta \cdot N_t = X \cdot \beta N_t.$$

Since M' and N' are written as proper products, we have $\Lambda_l(M_t) = \Lambda_r(M_{t-1})$. Thus $\Lambda_r(N_{t-1}) = \beta^{-1}\Lambda_r(M_{t-1})\beta = \Lambda_l(N_t)$, and the above products are proper. From the product formula for reduced norms (8.8) we get $\nu(M_t) = \nu(\beta N_t) = \underline{\underline{p}}_t$. Let $\Gamma_1 = \Lambda_l(M_t) = \Lambda_l(\beta N_t)$. Then one shows as above - observe that $\underline{\underline{S}}(\Gamma)$ depends only on $\mathrm{End}_A(K\Gamma) = A$ - that there exists $\beta_{n,t} \in \Gamma_1$, $\beta_{n,t} \equiv 1 \bmod (\underline{\underline{a}}^n\Gamma_1)$ with $M_t\beta_{n,t} = \beta N_t$; i.e., $M'\beta_{n,t} = N'$. However $M' = Ma$, $N' = Na$, and thus $Ma\beta_{n,t}a^{-1} = N$ and since a commutes with R, we have $a\beta_{n,t}a^{-1} \equiv 1 \bmod (\underline{\underline{a}}^n a\Gamma_1 a^{-1})$. #

8.13 __Remark:__ We recall the definition of the __ray of A over K__ (cf. 6.9) $St_K(A) = \{\alpha \in A : 0 \neq \alpha$ is positive at every infinite prime at which A is ramified$\}$.

Let I(R) be the group of non-zero fractional ideals of R in K; i.e., R-lattices in K. We now consider the subgroup $St_K(A)_o = \{(\alpha) : \alpha \in St_K(A)\}$, where (α) denotes the principal ideal $R\alpha$. The factorgroup of I(R) modulo $St_K(A)_o$ is denoted by

$$V_R(A) = I(R)/St_K(A)_o.$$

8.14 __Theorem:__ Let Γ be a fixed maximal R-order in A and let $\underline{\underline{M}}_\Gamma(A)_o$ denote the set of isomorphism classes of Γ-lattices in A. If A is not a totally definite quaternion algebra, then the map

$$\overline{\nu} : \underline{\underline{M}}_{\Gamma}(A)_o \longrightarrow V_R(A),$$

$$(M) \longmapsto \nu(M) \cdot St_K(A)_o = \overline{\nu(M)}$$

is a bijection.

<u>Proof</u>: If $M \cong N$ for Γ-lattices in A, then there exists a regular element $a \in A$ such that $Ma = N$ and $\nu(Ma) = \nu(M) \cdot \nu(\Lambda_r(M)a) = \nu(M) \cdot Nrd_{A/K}(a) = \nu(N)$. In (6.9) we have seen that $R\, Nrd_{A/K}(a) \in St_K(A)_o$; i.e., $\overline{\nu(M)} = \overline{\nu(N)}$.
<u>Conversely</u>, assume that $\overline{\nu(M)} = \overline{\nu(N)}$; i.e., there exists $(\alpha) \in St_K(A)_o$ such that $\nu(M)\alpha = \nu(N)$. However, since $(\alpha) \in St_K(A)_o$, there exist a regular $a \in A$ such that $Nrd_{A/K}(a) = \alpha$; i.e., $\nu(Ma) = \nu(N)$. With (8.12) we conclude $M \cong N$. #

8.15 We <u>remark</u>, that all the statements in this section, except (8.9, 8.10, 8.12, 8.13, 8.14) remain valid for any Dedekind domain R with quotient field K.

<u>Exercises §8</u>:

1.) Show that in $A = (\underline{\underline{Q}})_2$, there exists a full $\underline{\underline{Z}}$-lattice M consisting entirely of integral elements, which is not contained in any R-order. (<u>Hint</u>: Let

$$\underline{\underline{A}} = (1/3) \begin{pmatrix} 1 & 4 \\ 5 & 2 \end{pmatrix}, \quad \underline{\underline{B}} = (1/3) \begin{pmatrix} 1 & 2 \\ 10 & 2 \end{pmatrix}, \quad \underline{\underline{C}} = \begin{pmatrix} 0 & 0 \\ 0 & 1 \end{pmatrix}.$$

Then the matrices A, B, C, A^2 are integral and so are their sums. Moreover, they are linearly independent over $\underline{\underline{Z}}$ and span a $\underline{\underline{Z}}$-lattice M. But AB is not integral, and so M can not be contained in a $\underline{\underline{Z}}$-order in A.)

2.) Let K be an $\underline{\underline{A}}$-field with Dedekind domain R and A a finite dimensional separable K-algebra.
<u>Lemma</u>: If M is a normal R-lattice in A consisting entirely of integral elements, then $M \subset \Lambda$ for some R-order Λ in A. (<u>Hint</u>: It suffices to

prove this lemma for $\hat{M}$, where "$\hat{}$" denotes the completion of some prime ideal. Moreover, we can assume A to be central simple. Then $\hat{M} = \hat{\Gamma}a$ for some integral element a, and we also may assume $\hat{\Gamma} = (\hat{\Omega})_n$, where $\hat{\Omega}$ is the maximal $\hat{R}$-order in the skewfield $\hat{D}$, $\hat{A} = (\hat{D})_n$. Use (5.12) to conclude $a \in \hat{\Gamma}$.)

CHAPTER VII

GENERA OF LATTICES [*)]

§1 Preliminaries on genera

Various equivalent conditions for lattices over orders to lie in the same genus are derived; clones are defined, and the question of local equivalence is reduced to the study of one-sided ideals in maximal orders.

Let R be a Dedekind domain with quotient field K and Λ an R-order in the finite dimensional separable K-algebra A.

1.1 Definition: $M, N \in {}_{\Lambda}\underline{\underline{M}}^{o}$ are said to lie in the same genus, notation $M \vee N$, if M is locally isomorphic to N; i.e., $M_{\underline{\underline{p}}} \cong N_{\underline{\underline{p}}}$ for every prime ideal $\underline{\underline{p}}$ of R. We write

$$\mathfrak{G}(M) = \{ N \in {}_{\Lambda}\underline{\underline{M}}^{o} : N \vee M \}$$

and $g(M)$ is the number of non-isomorphic lattices in $\mathfrak{G}(M)$.

1.2 Remark: One could also define a "genus" via completions; i.e., M and N lie in the same "genus" if $\hat{M}_{\underline{\underline{p}}} \cong \hat{N}_{\underline{\underline{p}}}$ for every prime ideal $\underline{\underline{p}}$ of R. However, because of (VI, 1.2) both definitions coincide.

1.3 Definition: Let spec R be the set of all non-zero prime ideals in R. An idèle in A is a family $(a_{\underline{\underline{p}}})_{\underline{\underline{p}} \in \operatorname{spec} R}$, where $a_{\underline{\underline{p}}} \in A$ and $a_{\underline{\underline{p}}} = 1$ for almost all $\underline{\underline{p}} \in \operatorname{spec} R$.

1.4 Lemma: There is a one-to-one correspondence between the elements in $\mathfrak{G}(\Lambda)$ and the idèles in A.

Proof: Let $M \vee \Lambda$ - for M, N modules in the same genus, we always assume $KM = KN$ - then $KM = A$ implies $M_{\underline{\underline{p}}} = \Lambda_{\underline{\underline{p}}}$ for almost all $\underline{\underline{p}} \in \operatorname{spec} R$ (cf. IV, 1.8). For all other primes we have $M_{\underline{\underline{p}}'} \cong \Lambda_{\underline{\underline{p}}'}$; i.e.,

*) In this chapter, prime ideals are assumed to be different from zero.

$M_{\underline{\underline{p}}'} = \Lambda_{\underline{\underline{p}}'} a_{\underline{\underline{p}}'}$ for some element $a_{\underline{\underline{p}}'} \in A$. Taking $a_{\underline{\underline{p}}} = 1$ if $M_{\underline{\underline{p}}} = \Lambda_{\underline{\underline{p}}}$ we define the map

$$\Psi : \mathfrak{G}(\Lambda) \longrightarrow \text{Idèles in } A,$$
$$M \longmapsto (a_{\underline{\underline{p}}})_{\underline{\underline{p}} \in \operatorname{spec} R}.$$

Then Ψ is obviously injective; but it is also surjective, for given an idèle $(a_{\underline{\underline{p}}})_{\underline{\underline{p}} \in \operatorname{spec} R}$ in A, we define $M_{\underline{\underline{p}}} = \Lambda_{\underline{\underline{p}}} a_{\underline{\underline{p}}}$. Since $a_{\underline{\underline{p}}} = 1$ for almost all $\underline{\underline{p}} \in \operatorname{spec} R$, there exists $M \in {}_{\Lambda}\underline{\underline{M}}^{o}$ such that its localizations are $\{M_{\underline{\underline{p}}}\}_{\underline{\underline{p}} \in \operatorname{spec} R}$ (cf. IV, 1.8). #

1.5 __Definition:__ Two idèles $(a_{\underline{\underline{p}}})_{\underline{\underline{p}} \in \operatorname{spec} R}$ and $(b_{\underline{\underline{p}}})_{\underline{\underline{p}} \in \operatorname{spec} R}$ are said to be __equivalent__ if $\Psi^{-1}(a_{\underline{\underline{p}}})_{\underline{\underline{p}} \in \operatorname{spec} R} \cong \Psi^{-1}(b_{\underline{\underline{p}}})_{\underline{\underline{p}} \in \operatorname{spec} R}$.

1.6 __Lemma:__ Let $\underline{\underline{p}}_1, \dots, \underline{\underline{p}}_s$ be a finite set of prime ideals in spec R. In each equivalence class of idèles there can be found an idèle $(a_{\underline{\underline{p}}})_{\underline{\underline{p}} \in \operatorname{spec} R}$ such that $a_{\underline{\underline{p}}_i} = 1, 1 \le i \le s$; and $a_{\underline{\underline{p}}} \in \Lambda$ for every $\underline{\underline{p}} \in \operatorname{spec} R$. We call $(a_{\underline{\underline{p}}})_{\underline{\underline{p}} \in \operatorname{spec} R}$ an __integral idèle coprime to__ $\{\underline{\underline{p}}_i\}_{1 \le i \le s}$.

__Proof:__ This is an immediate consequence of the proof of (VI, 8.11). #

1.7 __Theorem:__ Let $M \in {}_{\Lambda}\underline{\underline{M}}^{o}$.

(i) There exists a one-to-one correspondence between the isomorphism classes of Λ-lattices in $\mathfrak{G}(M)$ and the equivalence classes of idèles in $\operatorname{End}_A(KM)$.

(ii) $N \in \mathfrak{G}(M)$ if and only if there exists a left $\operatorname{End}_\Lambda(M)$-ideal $I \vee \operatorname{End}_\Lambda(M)$ such that $N = MI$. In particular, we can take $I = \operatorname{Hom}_\Lambda(M,N)$.

__Proof:__ (i) is obvious with (1.4, 1.5) and the first part of (ii) follows from (i) and (1.4). If $N \in \mathfrak{G}(M)$, then

$$M \circ \operatorname{Hom}_\Lambda(M,N) = \left\{ \sum_{\text{finite}} m_i \varphi_i : m_i \in M, \varphi_i \in \operatorname{Hom}_\Lambda(M,N) \right\}$$

is contained in N. However, locally we have $M_{\underline{p}} a_{\underline{p}} = N_{\underline{p}}$ for some $a_{\underline{p}} \in \operatorname{Hom}_{\Lambda_{\underline{p}}}(M_{\underline{p}}, N_{\underline{p}})$. Thus $M_{\underline{p}} \circ \operatorname{Hom}_{\Lambda_{\underline{p}}}(M_{\underline{p}}, N_{\underline{p}}) = N_{\underline{p}}$; and

$$N = \bigcap_{\underline{p} \in \underline{S}} N_{\underline{p}} = \bigcap_{\underline{p} \in \underline{S}} M_{\underline{p}} \circ \operatorname{Hom}_{\Lambda_{\underline{p}}}(M_{\underline{p}}, N_{\underline{p}}) = \bigcap_{\underline{p} \in \underline{S}} (M \circ \operatorname{Hom}_{\Lambda}(M,N))_{\underline{p}} =$$

$= M \circ \operatorname{Hom}_{\Lambda}(M,N)$ (cf. Ex. 1,3) where $\underline{S} = \operatorname{spec} R$. #

1.8 __Definition__: Let $M \in {}_{\Lambda}\underline{M}^{o}$ and let $\{\underline{p}_i\}_{1 \le i \le s}$ be a finite __non-empty__ set of prime ideals, then $N \in {}_{\Lambda}\underline{M}^{o}$ belongs to the __clone determined by M relative to__ $\{\underline{p}_i\}_{1 \le i \le s}$, if $M_{\underline{p}_i} = N_{\underline{p}_i}$, $1 \le i \le s$ and $N_{\underline{p}} \subset M_{\underline{p}}$ for all $\underline{p} \in \operatorname{spec} R$. By $\mathcal{C}_{\{p_i\}_{1 \le i \le s}}(M)$ we denote the clone determined by M relative to $\{p_i\}_{1 \le i \le s}$.

1.9 __Lemma__: Let $M \in {}_{\Lambda}\underline{M}^{o}$.

(i) If $N \vee M$, then for every non-empty finite set of prime ideals $\underline{S}_o = \{p_i\}_{1 \le i \le s}$, there exists $N' \cong N$ such that $N' \in \mathcal{C}_{\underline{S}_o}(M)$.

(ii) If $\underline{S}_o$ contains all prime ideals for which $\Lambda_{\underline{p}}$ is not maximal, then $N \in \mathcal{C}_{\underline{S}_o}(M)$ implies $N \vee M$.

__Proof__: (i) $N \in \mathcal{G}(M)$ corresponds to an idèle $(a_{\underline{p}})_{\underline{p} \in \operatorname{spec} R}$ in $\operatorname{End}_A(KM)$; and by (1.6) we can find an idèle $(b_{\underline{p}})_{\underline{p} \in \operatorname{spec} R}$ equivalent to $(a_{\underline{p}})_{\underline{p} \in \operatorname{spec} R}$ which is integral*) and coprime to $\underline{S}_o$. But $(b_{\underline{p}})_{\underline{p} \in \operatorname{spec} R}$ corresponds to $N' = \bigcap_{\underline{p} \in \operatorname{spec} R} Mb_{\underline{p}} \cong N$ and $N' \in \mathcal{C}_{\underline{S}_o}(M)$ (cf. 1.7).

(ii) If $\underline{S}_o$ contains all $\underline{p}$ for which $\Lambda_{\underline{p}}$ is not maximal and if $N \in \mathcal{C}_{\underline{S}_o}(M)$, then $N_{\underline{p}} = M_{\underline{p}}$ for all $\underline{p} \in \underline{S}_o$; and for $\underline{p} \notin \underline{S}_o$, $KM = KN$ implies $M_{\underline{p}} \cong N_{\underline{p}}$, $\Lambda_{\underline{p}}$ being maximal (cf. IV, 5.7). Thus $M \vee N$. (Observe that $\underline{S}_o \neq \emptyset$ is necessary to conclude $KM = KN$.) #

1.10 __Corollary__: If $N \in \mathcal{G}(M)$ and $N \in \mathcal{C}_{\underline{S}_o}(M)$ for some finite non-

---

*) "Integral" should be understood as "integral with respect to $\operatorname{End}_{\Lambda}(M)$; i.e., $b_{\underline{p}} \in \operatorname{End}_{\Lambda}(M)$.

empty set of prime ideals $\underline{\underline{S}}_o$, then $N = MI$ for some $I \vee \mathrm{End}_\Lambda(M)$ and I is integral [*] and coprime to $\underline{\underline{S}}_o$; i.e., $I_{\underline{\underline{p}}} = \mathrm{End}_{\Lambda_{\underline{\underline{p}}}}(M_{\underline{\underline{p}}})$ for every prime ideal $\underline{\underline{p}} \in \underline{\underline{S}}_o$.

Proof: This is an immediate consequence of the proof of (1.7, 1.9). #

1.11 Theorem (Jacobinski [3]): Let Γ be a maximal R-order in A containing Λ and let $\underline{\underline{F}}$ be a fixed two-sided Γ-ideal in A, $\underline{\underline{F}} \subset \Lambda$ and let $\underline{\underline{S}}_{\underline{\underline{F}}}$ be a finite non-empty set of prime ideals containing all prime ideals $\underline{\underline{p}}$ for which $F_{\underline{\underline{p}}} \neq \Gamma_{\underline{\underline{p}}}$. For a fixed $M \in {}_\Lambda\underline{\underline{M}}^o$, the following conditions are equivalent:

(i) $N \in \mathfrak{g}(M)$,

(ii) $N \cong N'$ for some $N' \in \mathcal{C}_{\underline{\underline{S}}_{\underline{\underline{F}}}}(M)$,

(iii) $N \cong M \cap \Gamma MI$ for some integral left $\mathrm{End}_\Gamma(\Gamma M)$-ideal I in $\mathrm{End}_A(KM)$, which is coprime to $\underline{\underline{S}}_{\underline{\underline{F}}}$.

Proof: (i) and (ii) are equivalent by (1.9) since $\underline{\underline{S}}_{\underline{\underline{F}}}$ contains all prime ideals for which $\Lambda_{\underline{\underline{p}}} \neq \Gamma_{\underline{\underline{p}}}$.

(ii) $\Longrightarrow$ (iii): We may assume $N \in \mathcal{C}_{\underline{\underline{S}}_{\underline{\underline{F}}}}(M)$; i.e., $N = MI'$, where $I' \vee \mathrm{End}_\Lambda(M)$, I' is coprime to $\underline{\underline{S}}_{\underline{\underline{F}}}$ and integral (cf. 1.10). Then we have

$$M \cap \Gamma MI' = \bigcap_{\underline{\underline{p}} \in \mathrm{spec}\, R} M_{\underline{\underline{p}}} \cap \Gamma_{\underline{\underline{p}}} M_{\underline{\underline{p}}} I'_{\underline{\underline{p}}},$$

as is easily seen. If $\underline{\underline{p}} \notin \underline{\underline{S}}_{\underline{\underline{F}}}$, then $\Gamma_{\underline{\underline{p}}} M_{\underline{\underline{p}}} = M_{\underline{\underline{p}}}$ and $M_{\underline{\underline{p}}} \cap \Gamma_{\underline{\underline{p}}} M_{\underline{\underline{p}}} I'_{\underline{\underline{p}}} = M_{\underline{\underline{p}}} I'_{\underline{\underline{p}}}$; if $\underline{\underline{p}} \in \underline{\underline{S}}_{\underline{\underline{F}}}$, then $I'_{\underline{\underline{p}}} = \mathrm{End}_{\Lambda_{\underline{\underline{p}}}}(M_{\underline{\underline{p}}})$ and thus $M_{\underline{\underline{p}}} \cap \Gamma_{\underline{\underline{p}}} M_{\underline{\underline{p}}} = M_{\underline{\underline{p}}} I'_{\underline{\underline{p}}}$. Hence $M \cap \Gamma MI' = \bigcap_{\underline{\underline{p}} \in \mathrm{spec}\, R} M_{\underline{\underline{p}}} I'_{\underline{\underline{p}}} = MI' = N$. But ΓM is a right $\mathrm{End}_\Gamma(\Gamma M)$-module, and thus, putting $I = \mathrm{End}_\Gamma(\Gamma M)I'$ we get $N = M \cap \Gamma MI$ and I is an

[*] "Integral" means "integral with respect to $\mathrm{End}_\Lambda(M)$; i.e., $I \subset \mathrm{End}_\Lambda(M)$.

integral left $\mathrm{End}_{\Gamma}(\Gamma M)$-ideal in $\mathrm{End}_A(KM)$, coprime to $\underline{\underline{S}}_{\underline{\underline{F}}}$.

(iii) $\Longrightarrow$ (ii): Let $N' = M \cap \Gamma MI$, $N' \cong N$, where I is an integral left $\mathrm{End}_{\Gamma}(\Gamma M)$-ideal in $\mathrm{End}_A(KM)$ coprime to $\underline{\underline{S}}_{\underline{\underline{F}}}$. Then for all $\underline{\underline{p}} \in \underline{\underline{S}}_{\underline{\underline{F}}}$

$N'_{\underline{\underline{p}}} = M_{\underline{\underline{p}}} \cap \Gamma_{\underline{\underline{p}}} M_{\underline{\underline{p}}} I_{\underline{\underline{p}}} = M_{\underline{\underline{p}}} \cap \Gamma_{\underline{\underline{p}}} M_{\underline{\underline{p}}} = M_{\underline{\underline{p}}}$ and $N' \in \mathcal{C}_{\underline{\underline{S}}_{\underline{\underline{F}}}}(M)$. #

1.12 Theorem (Jacobinski [3]): Under the hypotheses of (1.11) there is a one-to-one correspondence between the integral left $\mathrm{End}_{\Gamma}(\Gamma M)$-ideals coprime to $\underline{\underline{S}}_{\underline{\underline{F}}}$ and the Λ-lattices in $\mathcal{C}_{\underline{\underline{S}}_{\underline{\underline{F}}}}(M)$.

Proof: Let $N \in \mathcal{C}_{\underline{\underline{S}}_{\underline{\underline{F}}}}(M)$. Then $\Gamma N \in \mathcal{C}_{\underline{\underline{S}}_{\underline{\underline{F}}}}(\Gamma M)$ and by (1.7, 1.9) we can write $\Gamma N = \Gamma MI$, where I is a full integral left $\mathrm{End}_{\Gamma}(\Gamma M)$-ideal coprime to $\underline{\underline{S}}_{\underline{\underline{F}}}$. We now define for $N \in \mathcal{C}_{\underline{\underline{S}}_{\underline{\underline{F}}}}(M)$

$$\Phi : N \longmapsto I, \text{ where } \Gamma N = \Gamma MI.$$

If $\Gamma MI = \Gamma MI'$, then $I = I'$; in fact I and I' are normal ideals, and thus invertible (cf. VI, 8.2; it should be observed that the existence of I^{-1} does not depend on the fact that we work in a central simple algebra, as long as $\Lambda_1(I)$ is maximal.) Thus $\Gamma MII'^{-1} = \Gamma M$ and $\Gamma M = \Gamma MI'I^{-1}$; i.e., both II'^{-1} and $I'I^{-1}$ are integral normal two-sided $\mathrm{End}_{\Gamma M}$-ideals. But then it follows from (IV, 4.14) that $II'^{-1} = \mathrm{End}_{\Gamma}(\Gamma M)$ and $I = I'$. This shows that Φ is well-defined. To show that Φ is a bijection, we construct an inverse Ψ. If I is a full integral left $\mathrm{End}_{\Gamma}(\Gamma M)$-ideal in $\mathrm{End}_A(KM)$ coprime to $\underline{\underline{S}}_{\underline{\underline{F}}}$, then $N = M \cap \Gamma MI \in \mathcal{C}_{\underline{\underline{S}}_{\underline{\underline{F}}}}(M)$ by (1.11) and we define

$$\Psi : I \longmapsto N = M \cap \Gamma MI.$$

Then

$$\Psi\Phi : N \longmapsto M \cap \Gamma MI \text{ with } \Gamma N = \Gamma MI,$$

but $M \cap \Gamma MI = \bigcap_{\underline{\underline{p}} \in \mathrm{spec}\, R} M_{\underline{\underline{p}}} \cap \Gamma_{\underline{\underline{p}}} M_{\underline{\underline{p}}} I_{\underline{\underline{p}}}$; and if $\underline{\underline{p}} \in \underline{\underline{S}}_{\underline{\underline{F}}}$, then $I_{\underline{\underline{p}}} = \mathrm{End}_{\Gamma_{\underline{\underline{p}}}}(\Gamma M_{\underline{\underline{p}}})$ and $M_{\underline{\underline{p}}} \cap \Gamma_{\underline{\underline{p}}} M_{\underline{\underline{p}}} I_{\underline{\underline{p}}} = M_{\underline{\underline{p}}} = N_{\underline{\underline{p}}}$ since $N \in \mathcal{C}_{\underline{\underline{S}}_{\underline{\underline{F}}}}(M)$. If $\underline{\underline{p}} \notin \underline{\underline{S}}_{\underline{\underline{F}}}$, then $M_{\underline{\underline{p}}} = \Gamma_{\underline{\underline{p}}} M_{\underline{\underline{p}}}$

and $M_{\underline{p}} \cap \Gamma_{\underline{p}} M_{\underline{p}} I_{\underline{p}} = \Gamma_{\underline{p}} M_{\underline{p}} I_{\underline{p}} = N_{\underline{p}}$, since $I_{\underline{p}}$ is integral. Thus $M \cap \Gamma MI = N$, and $\Psi\Phi = 1$. Now,

$$\Phi\Psi : I \longmapsto N = M \cap \Gamma MI \longmapsto I' \text{ with } \Gamma N = \Gamma MI'.$$

But $\Gamma N = \Gamma(M \cap \Gamma MI) = \Gamma MI$, as is easily seen by localizing, and since Φ is well-defined, we conclude $I = I'$; i.e., $\Phi\Psi = 1$. #

Exercises §1:

We keep the notation of §1.

1.) Jacobinski [3]: Let Γ be a maximal R-order in A containing Λ, and let F be a full two-sided Γ-ideal contained in Λ such that $Fe \neq \Gamma e$ for every central idempotent e of A. For $M \in {}_{\Lambda}\underline{M}^{o}$, we have $N \in \mathfrak{G}(M)$ if and only if there exists a Γ-isomorphism

$$\tilde{\varphi} : \Gamma N/FN \xrightarrow{\sim} \Gamma M/FM \text{ such that}$$
$$\tilde{\varphi}\big|_{N/FN} : \dot{N}/FN \xrightarrow{\sim} M/FM.$$

(Hint: If $N \vee M$, use (1.11,iii) to construct $\tilde{\varphi}$. Conversely, decompose $\tilde{\varphi}$ into its $\underline{p}$-primary components

$\tilde{\varphi}_{\underline{p}} : (\Gamma N)_{\underline{p}}/(FN)_{\underline{p}} \xrightarrow{\sim} (\Gamma M)_{\underline{p}}/(FM)_{\underline{p}}$. Let e_M be the minimal central idempotent in A with $e_M M = M$. If $\tilde{\varphi}_{\underline{p}} = 0$, then $\Gamma_{\underline{p}} e_M = F_{\underline{p}} e_M$. Show that we may assume M to be faithful and that it suffices to consider only those prime ideals $\underline{p}$ for which $\tilde{\varphi}_{\underline{p}} \neq 0$ (this needs proof). Now, decompose Γ into maximal orders in simple algebras and decompose $\tilde{\varphi}$ accordingly. Use the hypothesis $\Gamma e \neq Fe$ for every primitive central idempotent of A, to show that there exists a $\Gamma_{\underline{p}}$-isomorphism $\varphi_{\underline{p}} : \Gamma_{\underline{p}} N_{\underline{p}} \longrightarrow \Gamma_{\underline{p}} M_{\underline{p}}$ such that $\varphi_{\underline{p}}\big|_{N_{\underline{p}}} : N_{\underline{p}} \xrightarrow{\sim} M_{\underline{p}}$. Then conclude $M \vee N$.)

2.) Let $M \in {}_{\Lambda}\underline{M}^{o}$ show that $N \in \mathfrak{G}(M)$ if $KM = KN$ and $M_{\underline{p}} \cong N_{\underline{p}}$ for every $\underline{p} \in \operatorname{spec} R$ for which $e_M \Lambda_{\underline{p}}$ is not maximal, where e_M is the minimal central idempotent for which $e_M M = M$. We recall that for two central idempotents e_1, e_2, $e_1 \geq e_2$ if $e_1 e_2 = e_2$.

3.) Let $M, N \in {}_{\Lambda}\underline{\underline{M}}^{o}$. Show that

$$(M \text{ o } \mathrm{Hom}_{\Lambda}(M,N))_{\underline{\underline{p}}} = M_{\underline{\underline{p}}} \text{ o } \mathrm{Hom}_{\Lambda_{\underline{\underline{p}}}}(M_{\underline{\underline{p}}}, N_{\underline{\underline{p}}})$$

(cf. proof 1.7).

§2 The number of non-isomorphic lattices in a genus

We construct a one-to-one correspondence between the isomorphism classes of lattices in the genus of M and the ideal classes in the center of $\mathrm{End}_A(KM)$, provided M satisfies Eichler's condition. This is used to show that g(M) is bounded independently of M.

In this section let us assume that K is an $\underline{\underline{A}}$-field and R a Dedekind domain with quotient field K (cf. VI, 4.10).

A.) Maximal orders

For this part we assume that Γ is a maximal R-order in the central simple K-algebra A.

2.1 **Definition:** (i) Let $M \in {}_{\Gamma}\underline{\underline{M}}^{o}$. Then we denote by I(R) the group of non-zero fractional ideals in R and by $St_K(A)_o$ the ray of A; i.e., $St_K(A)_o = \{(\alpha) : \alpha \in K,\ 0 \neq \alpha$ is positive at all infinite primes at which A is ramified$\}$ and set $V_R(A) = I(R)/St_K(A)_o$. If $L \in {}_{A}\underline{\underline{M}}^{f}$, then L is a progenerator for ${}_{A}\underline{\underline{M}}^{f}$, A being simple, and A and $\mathrm{End}_A(L)$ are ramified at exactly the same primes. Thus $St_K(A)_o = St_K(\mathrm{End}_A(L))_o$.
(ii) If $N \vee M$, then there exists a left $\mathrm{End}_{\Gamma}(M)$-ideal I such that $N = MI$ (cf. 1.7); we put $B_M = \mathrm{End}_A(KM)$ and define

$$\nu(N,M) = \nu_{B_M/K}(I) \quad \text{(cf. VI, 8.7)},$$

where $\nu_{B_M/K}(I)$ is the R-ideal generated by $\{\mathrm{Nrd}_{B_M/K}(x) : x \in I\}$.
We recall that M is said to satisfy Eichler's condition if $\mathrm{End}_A(KM) = B_M$ is not a totally definite quaternion algebra (cf. VI, 6.12, 7.1).
We write $\Omega_M = \mathrm{End}_{\Gamma}(M)$.

2.2 **Theorem:** Let $M \in {}_{\Gamma}\underline{\underline{M}}^{o}$ satisfy Eichler's condition. Then there is a one-to-one correspondence between the isomorphism classes of Γ-lattices in $\mathfrak{G}(M)$, the genus of M, and the elements in $V_R(B_M)$, the correspondence being

$$(N) \longmapsto \nu(N,M) \cdot St_K(B_M)_o = \overline{\nu(N,M)}.$$

<u>Proof</u>: The correspondence

$$\mathfrak{G}(M) \longrightarrow \text{left } \Omega_M\text{-ideals in } B_M,$$

$$N \longmapsto I \text{ if } MI = N$$

is one-to-one (cf. 1.7), and $N \cong N'$ if and only if $I \cong I'$. In fact, if $N \cong N'$, then $N = N'b$ for some regular element $b \in B_M$ and $N = MI$, $N' = MI'$ implies $MI = MI'b$, which implies $I = I'b$ (cf. proof of 1.12). <u>Conversely</u>, if $I \cong I'$ then obviously $N \cong N'$. Moreover, if B_M is not a totally definite quaternion algebra; i.e., if M satisfies Eichler's condition, then it was shown in (VI, 8.14), that the correspondence

$$I \longmapsto \overline{\nu(I)}$$

induces a bijection between the isomorphism classes of full left Ω_M-ideals and the elements in $V_R(B_M)$; i.e., $N \longmapsto \overline{\nu(N,M)}$ induces the stated bijection. #

2.3 <u>Remark</u>: If $M \in {}_{\Gamma}\underline{\underline{M}}^{o}$ satisfies Eichler's condition, (2.2) states that $g(M)$, the number of non-isomorphic Γ-lattices in $\mathfrak{G}(M)$, is equal to the order of the group $V_R(B_M)$; and this order is finite and can be computed explicitely. Moreover, since locally there exists only one isomorphism class of indecomposable Γ-lattices, $g(M)$ is at the same time the number of Γ-lattices N with $KN = KM$; observe that for modules in the same genus, we always assume that they span the <u>same</u> A-module.

We now turn to the study of maximal R-orders Γ in finite dimensional separable K-algebras. The situation here is not much different from the one above; however, one has to introduce a large amount of notation which obscures things.

2.4 <u>Notation</u>:

$A = \oplus_{i=1}^{n} Ae_i$, e_i central primitive idempotents in A, $1 \le i \le n$,

F_i = center of Ae_i, $1 \le i \le n$,

$\Gamma = \oplus_{i=1}^{n} \Gamma_i$, the decomposition of Γ into maximal R-orders in the simple algebras $A_i = Ae_i$,

$C = \oplus_{i=1}^{n} C_i$, C_i = center of Γ_i, $1 \le i \le n$,

$M \in {}_{\Gamma}\underline{\underline{M}}^{o}$,

$e_M = \sum_{e_i KM \ne 0} e_i$,

$B_M = \mathrm{End}_A(KM)$,

$\Omega_{\Gamma M} = \mathrm{End}_{\Gamma}(M)$,

$e_M(\oplus_{i=1}^{n} F_i)$ = center of B_M,

$e_M C$ = center of $\Omega_{\Gamma M}$,

$I(e_M) = \{(\underline{\underline{a}}_i)_{1 \le i \le n} : \underline{\underline{a}}_i \ne 0$ is a fractional ideal in F_i and $\underline{\underline{a}}_i = C_i$ for all i for which $e_i M = 0\}$.

$St(e_M)_o = \{((\alpha_i))_{1 \le i \le n} : \alpha_i = 1$ if $e_i M = 0$ and $0 \ne \alpha_i \in F_i$ is positive at all infinite primes of F_i at which $e_i B_M$ is ramified, for all i with $e_i M \ne 0\}$. (Observe that 1 is positive at every infinite prime.)

Obviously $I(e_M)$ is a subgroup of $I(1) = \{(\underline{\underline{a}}_i)_{1 \le i \le n} : 0 \ne \underline{\underline{a}}_i$ is a fractional ideal in $F_i, 1 \le i \le n\}$. Moreover, $St(e_M)_o$ is a subgroup of $I(e_M)$, and we put

$$V(e_M) = I(e_M)/St(e_M)_o.$$

We <u>remark</u> that $St(e_M)_o$ is a subgroup of $St(1)_o = \{((\alpha_i))_{1 \le i \le n} : 0 \ne \alpha_i \in F_i$ is positive at all infinite primes at which $e_i A$ is ramified, $1 \le i \le n\}$.

If I is a full left $\Omega_{\Gamma M}$-ideal, we put $\nu_{e_M}(I) = (\nu_{e_i B_M/F_i}(e_i I))_{1 \le i \le n}$, where we define

$$\nu(e_i I) = C_i \text{ if } e_i M = 0.$$

Then $\nu_{e_M}(I) \in I(e_M)$ and we put $\overline{\nu_{e_M}(I)} = \nu_{e_M}(I) \cdot St(e_M)_o \in V(e_M)$.

We point out that ν_{e_M} is a multiplicative function with respect to proper products of normal ideals in B_M and that it commutes with localizations and completions (cf. VI, 8.8).

2.5 Theorem (Jacobinski [3]): Let $M \in {}_{\Gamma}\underline{\underline{M}}^{o}$ satisfy Eichler's condition. Then there is a one-to-one correspondence between the isomorphism classes in $\mathfrak{g}(M)$ and the elements in $V(e_M)$:

$$(N) \longmapsto \overline{\nu_{e_M}(N,M)},\ N \in \mathfrak{g}(M).$$

Proof: The correspondence

$$N \longmapsto I,\ \text{where } N = MI \in \mathfrak{g}(M)$$

and I is a full left $\Omega_{\Gamma M}$-ideal is one-to-one and preserves isomorphisms (cf. proof of 2.2). We recall that $\nu_{e_M}(N,M) = \nu_{e_M}(I)$. It thus remains to show that $I \longmapsto \overline{\nu_{e_M}(I)}$ induces a bijection between the isomorphism classes of left $\Omega_{\Gamma M}$-ideals and the elements in $V(e_M)$.

$I_1 \cong I_2$ if and only if $I_1 = I_2 b$ for some regular element $b \in B_M$, and

$\nu_{e_M}(I_1) = \nu_{e_M}(I_2 b) = \nu_{e_M}(I_2)\,\nu_{e_M}(\Lambda_r(I_2)b) = \nu(I_2)(Nrd_{e_i B_M/F_i}(e_i b))_{1 \le i \le n}$,

where $Nrd_{e_i B_M/F_i}(e_i b) = 1$ if $e_i M = 0$. But $(Nrd_{e_i B_M/F_i}(e_i b)) \in St_K(Ae_i)_o$ (cf. VI, 5.9). Thus

$\nu_{e_M}(I_1) \equiv \nu_{e_M}(I_2) \bmod St(e_M)_o$; i.e., $\overline{\nu_{e_M}(I_1)} = \overline{\nu_{e_M}(I_2)}$.

Conversely, if $\overline{\nu_{e_M}(I_1)} = \overline{\nu_{e_M}(I_2)}$, then there exists a family

$(\alpha_i)_{1 \le i \le n} \in St(e_M)_o$, where $\alpha_i = 1$ for i with $e_i M = 0$, and otherwise α_i is positive at every infinite prime at which $e_i A$ is ramified, such that $\nu_{e_M}(I_1) = \nu_{e_M}(I_2)(\alpha_i)_{1 \le i \le n}$. According to (VI, 6.9), applied to $\{e_i B_M\}$, there exists a regular element $b \in B$ such that $Nrd_{e_i B_M/F_i}(e_i b) = \alpha_i$ for all α_i with $e_i B_M \ne 0$. Thus $\nu_{e_M}(\Omega_{\Gamma M} b) = (\alpha_i)_{1 \le i \le n}$ and we have

$$\nu_{e_M}(I_1) = \nu_{e_M}(I_2 b).$$

Applying (VI, 8.14) we conclude that $I_1 \cong I_2 b$; i.e., $I_1 \cong I_2$. #

2.6 Corollary: If $M \in {}_{\underline{\underline{\Gamma}}}M^{o}$ satisfies Eichler's condition, then $g(M)$, the number of isomorphism classes in $\mathfrak{G}(M)$, is equal to the order of $V(e_M)$.

B.) Non-maximal orders

Let Λ be an R-order in the finite dimensional separable K-algebra A and let Γ be a maximal R-order in A containing Λ. (In view of (2.5) we may assume $\Lambda \neq \Gamma$). Then $0 \neq \underline{\underline{b}} = \{r \in R : r\Gamma \subset \Lambda\}$ and we put

$$\underline{\underline{S}}_o = \{\underline{\underline{p}} \in \text{spec } R : \underline{\underline{p}} | \underline{\underline{b}}\}.$$

2.7 Notation: For a fixed $M \in {}_{\Lambda}\underline{\underline{M}}^{o}$ we use the same notation for ΓM as listed in (2.4); in addition we define

$$\Omega_M = \text{End}_{\Lambda}(M),$$

$$I_{\underline{\underline{S}}_o}(e_M) = \{(\underline{\underline{a}}_i)_{1 \le i \le n} \in I(e_M) : \underline{\underline{a}}_{i_{\underline{\underline{p}}}} = C_{i_{\underline{\underline{p}}}} \text{ for every } \underline{\underline{p}} \in \underline{\underline{S}}_o\}$$

$St_{\underline{\underline{S}}_o}(e_M)_o$ is the subgroup of $I_{\underline{\underline{S}}_o}(e_M) \cap St(e_M)_o$ generated by $((\alpha_i))_{1 \le i \le n} \in St(e_M)_o$ such that $\alpha_i \equiv 1 \mod(\underline{\underline{b}}C_i), 1 \le i \le n$.

$T_{\underline{\underline{S}}_o}(M)$ = subgroup of $I_{\underline{\underline{S}}_o}(e_M)$ generated by all elements of the form

$$\nu_{e_M}(\Omega_{\Gamma M} a),\ a \in \Omega_M.$$

We point out, that $T_{\underline{\underline{S}}_o}(M)$ is in general not contained in $St_{\underline{\underline{S}}_o}(e_M)_o$; but $T_{\underline{\underline{S}}_o}(M) \subset St(e_M)_o \cap I_{\underline{\underline{S}}_o}(e_M)$.

Now we put

$$W_{\underline{\underline{S}}_o}(e_M) = I_{\underline{\underline{S}}_o}(e_M)/T_{\underline{\underline{S}}_o}(M).$$

We recall that $\mathfrak{C}_{\underline{\underline{S}}_o}(M)$ consists of all Λ-lattic in $\mathfrak{G}(M)$, $N \subset M$ with $M_{\underline{\underline{p}}} = N_{\underline{\underline{p}}}$ for all $\underline{\underline{p}} \in \underline{\underline{S}}_o$ (cf. 1.8).

If I is a full left $\Omega_{\Gamma M}$-ideal which is coprime to $\underline{\underline{S}}_o$, then

$$w_{e_M}(I) = \nu_{e_M}(I) \cdot T_{\underline{\underline{S}}_o}(M) \in W_{\underline{\underline{S}}_o}(e_M).$$

Since the norm localizes properly, we have $\nu_{e_M}(I) \in I_{\underline{\underline{S}}_o}(e_M)$.

2.8 <u>Theorem</u> (Jacobinski [3]): Let $N \in \mathcal{C}_{\underline{\underline{S}}_o}(M)$, with $M \in {}_\Lambda\underline{\underline{M}}^o$ satisfying Eichler's condition. Then $\Gamma N = \Gamma MI$, where I is a full integral $\Omega_{\Gamma M}$-ideal coprime to $\underline{\underline{S}}_o$, and the map

$$w : \mathcal{C}_{\underline{\underline{S}}_o}(M) \longrightarrow W_{\underline{\underline{S}}_o}(e_M),$$

$$N \longmapsto w_{e_M}(I) = w(N)$$

induces a bijection between the isomorphism classes of lattices in $\mathcal{C}_{\underline{\underline{S}}_o}(M)$ and the elements of $W_{\underline{\underline{S}}_o}(e_M)$.

Before we turn to the <u>proof of (2.8)</u>, we shall establish some lemmata:

2.9 <u>Proposition</u>: If $N \in \mathcal{C}_{\underline{\underline{S}}_o}(M)$, then $M \cong N$ if and only if $N = M \cap \Gamma Mb$, for some $b \in \Omega_M$ such that $\Omega_{\Gamma M}b$ is coprime to $\underline{\underline{S}}_o$.

<u>Proof</u>: Since $N \in \mathcal{C}_{\underline{\underline{S}}_o}(M)$, $N \subset M$ (cf. 1.10) and if $N \cong M$ then there exists $b \in \Omega_M$ such that $N = Mb$ and $\Omega_{\Gamma M}b$ is coprime to $\underline{\underline{S}}_o$. The usual localization argument shows $N = M \cap \Gamma Mb$. If conversely, $N = M \cap \Gamma Mb$ where $b \in \Omega_M$ and $\Omega_{\Gamma M}b$ is coprime to $\underline{\underline{S}}_o$, then $N \subset M$ and $N \in \mathcal{C}_{\underline{\underline{S}}_o}(M)$, and one finds readily that $Mb = M \cap \Gamma Mb$. However, we have shown in (1.12), that there is a one-to-one correspondence between the modules in $\mathcal{C}_{\underline{\underline{S}}_o}(M)$ and the full left $\Omega_{\Gamma M}$-ideals coprime to $\underline{\underline{S}}_o$. Thus $N = Mb$. #

2.10 <u>Lemma</u>: With the notation of (2.8), $w(N) = 1$ implies $N = M$, for $N \in \mathcal{C}_{\underline{\underline{S}}_o}(M)$.

<u>Proof</u>: $w(N) = 1$ means $N = M \cap \Gamma MI$, where I is a full left $\Omega_{\Gamma M}$-ideal coprime to $\underline{\underline{S}}_o$. Moreover, since $\nu(I) \in T_{\underline{\underline{S}}_o}(e_M) \subset St(e_M)_o$ we apply (2.5) to conclude $I = \Omega_{\Gamma M}b$ for some regular element $b \in \Omega_{\Gamma M}$. On the other hand, $\nu(I) \in T_{\underline{\underline{S}}_o}(e_M)$ implies that there exist regular elements $a, a' \in \Omega_M$ such that $\nu(\Omega_{\Gamma M}b) = \nu(\Omega_{\Gamma M}aa'^{-1})$; and consequently,

$$\nu(\Omega_{\Gamma M}ba') = \nu(\Omega_{\Gamma M}a),$$

as follows from the multiplicativity of the norm function. However, $\nu_{e_M}(Na',M) = \nu(\Omega_{\Gamma M}a)$. In fact, $Na' = Ma' \cap \Gamma M\Omega_{\Gamma M}ba' = M \cap \Gamma M\Omega_{\Gamma M}a' \cap \Gamma M\Omega_{\Gamma M}ba'$ (cf. 2.9). But $a', b \in \Omega_{\Gamma M}$ implies $\Gamma M\Omega_{\Gamma M}a' \cap \Gamma M\Omega_{\Gamma M}ba' = \Gamma M\Omega_{\Gamma M}ba'$. Thus $Na' = M \cap \Gamma M\Omega_{\Gamma M}ba'$ and hence $\nu_{e_M}(Na',M) = \nu(\Omega_{\Gamma M}ba')$. If we can show $Na' \cong M$, then obviously $N \cong M$. Therefore we may assume

$$\nu(\Omega_{\Gamma M}b) = \nu(\Omega_{\Gamma M}a),$$

where $b \in \Omega_{\Gamma M}$ and $a \in \Omega_M$. By (VI, 8.12), there exists an R-order Ω_1 in B_M such that for every $n \in \underline{\underline{N}}$ we have an element $\beta_n \in \Omega_1$ such that

$$\Omega_{\Gamma M}a\beta_n = \Omega_{\Gamma M}b$$

and $\beta_n \equiv 1 \mod \underline{\underline{a}}^n\Omega_1$, where $\underline{\underline{a}} = \prod_{\underline{\underline{p}} \in \underline{\underline{S}}_o} \underline{\underline{p}}$. Since $b \in \Omega_{\Gamma M}$, we conclude $a\beta_n \in \Omega_{\Gamma M}$. Moreover,

$$a\beta_n = a + \alpha_n a\gamma_n \in \Omega_{\Gamma M},$$

where $\alpha_n \in \underline{\underline{a}}^n$ and $\gamma_n \in \Omega_1$, implies $\alpha_n a\gamma_n \in \Omega_{\Gamma M}$. By Ex. 2,1 there exists $n_o \in \underline{\underline{N}}$ such that, $\alpha_{n_o} a\gamma_{n_o} \in \underline{\underline{a}}^s\Omega_{\Gamma M}$, where s is such that $\underline{\underline{a}}^s\Omega_{\Gamma M} \subset O_M$. This is possible because of the choice of $\underline{\underline{S}}_o$. Then

$$\Omega_{\Gamma M}b = \Omega_{\Gamma M}a\beta_{n_o}$$

and $a\beta_{n_o} = a + \alpha_{n_o}a\gamma_{n_o}$, with $\alpha_{n_o}a\gamma_{n_o} \in \Omega_M$. Thus

$$Ma\beta_{n_o} \subset Ma + M\alpha_{n_o}a\gamma_{n_o} \subset M,$$

since $a \in \Omega_M$, and we conclude $a\beta_{n_o} \in \Omega_M$. Hence we have shown that $\Omega_{\Gamma M}b = \Omega_{\Gamma M}c$ where $c \in \Omega_M$ is regular, and consequently

$$N = M \cap \Gamma M\Omega_{\Gamma M}b = M \cap \Gamma M\Omega_{\Gamma M}c$$

is isomorphic to M (cf. 2.9). #

Now we turn to the <u>proof of (2.8)</u>:

(i) Since $N \in \mathcal{C}_{\underline{\underline{S}}_o}(M)$, $w_{e_M}(I) \in W_{\underline{\underline{S}}_o}(e_M)$, and the map w is well-defined. We shall write

$$w_{e_M}(I) = w_{e_M}(N,M).$$

(ii) To show that w is an epimorphism, let $(\underline{\underline{a}}_i)_{1\leq i\leq n}$ be an element in $I_{\underline{\underline{S}}_o}(e_M) \subset I(e_M)$. By (VI, 7.10), we can find a full left $\Omega_{\Gamma M}$-ideal I' such that $\nu_{e_M}(I') = (\underline{\underline{a}}_i)_{1\leq i\leq n}$. However, since the norm localizes properly, one sees that I' is coprime to $\underline{\underline{S}}_o$ (but not necessarily integral). We choose $0 \neq r \in R, r \equiv 1 \mod \underline{\underline{p}}$ for all $\underline{\underline{p}} \in \underline{\underline{S}}_o$ such that $I = rI'$ is integral. Then I is coprime to $\underline{\underline{S}}_o$, and $\nu_{e_M}(I) = \nu_{e_M}(I')\, \nu_{e_M}(\Omega_{\Gamma M} r)$. Since $\nu_{e_M}(\Omega_{\Gamma M} r) \in T_{\underline{\underline{S}}_o}(M)$, we find $w_{e_M}(I) = w_{e_M}(I')$. It now follows from (1.12) that, for $N = M \cap \Gamma MI \in \mathcal{C}_{\underline{\underline{S}}_o}(M)$, $w_{e_M}(N,M) = w_{e_M}(I)$ and w is an epimorphism.

(iii) If $N,N' \in \mathcal{C}_{\underline{\underline{S}}_o}(M)$ are isomorphic, then $N = N'b$ for some regular element $b \in B_M$; however, $N_{\underline{\underline{p}}} = M_{\underline{\underline{p}}} = N'_{\underline{\underline{p}}}$ for every $\underline{\underline{p}} \in \underline{\underline{S}}_o$, and thus $(\Omega_{\Gamma M})_{\underline{\underline{p}}} b = (\Omega_{\Gamma M})_{\underline{\underline{p}}}$ for $\underline{\underline{p}} \in \underline{\underline{S}}_o$. Moreover, if we mutiply b by a suitable $0 \neq r \in R$, $r \equiv 1 \mod(\underline{\underline{p}})$ for every $\underline{\underline{p}} \in \underline{\underline{S}}_o$, then we may assume (cf. definition of $\underline{\underline{S}}_o$)

(α) $r\, b \in \Omega_M$,

(β) $\Omega_{\Gamma M} b\, r$ is integral and coprime to $\underline{\underline{S}}_o$,

(γ) $w_{e_M}(N,M) = w_{e_M}(Nr,M)$ (cf. proof of (ii)).

Thus we may assume $N = N'b$ where $b \in \Omega_M$ and $\Omega_{\Gamma M} b$ is integral and coprime to $\underline{\underline{S}}_o$. Thus

$$\nu_{e_M}(N,M) = \nu_{e_M}(N'b,M) = \nu_{e_M}(N',M)\, \nu_{e_M}(\Omega_{\Gamma M} b).$$

But $\nu_{e_M}(\Omega_{\Gamma M} b) \in T_{\underline{\underline{S}}_o}(e_M)$ and thus $w_{e_M}(N,M) = w_{e_M}(N',M)$; i.e., isomorphic lattices have the same image under w.

(iv) It remains to show that $w_{e_M}(N,M) = w_{e_M}(N',M)$ implies $N \cong N'$, where $N,N' \in \mathcal{C}_{\underline{\underline{S}}_o}(M)$. Let $N = M \cap \Gamma MI$, $N' = M \cap \Gamma MI'$, where I and I' are full integral ideals in $\Omega_{\Gamma M}$ coprime to $\underline{\underline{S}}_o$, $N, N' \in \mathcal{C}_{\underline{\underline{S}}_o}(M)$. Since I, I' are coprime to $\underline{\underline{S}}_o$, we can find $0 \neq r \in \Omega_M$, $r \equiv 1 \bmod \underline{\underline{p}}$ for every $\underline{\underline{p}} \in \underline{\underline{S}}_o$ such that $rI \subset I'$. Then

$$Nr = Mr \cap \Gamma MIr = M \cap \Gamma Mr \cap \Gamma MIr;$$

however, $\Gamma Mr \supset \Gamma MIr$ and so $Nr = M \cap \Gamma MIr$, and

$$w_{e_M}(Nr, M) = w_{e_M}(N, M).$$

Thus we may assume $I \subset I'$. Then $I'^{-1}I \subset \Lambda_r(I')$ and this product is proper. Thus $\nu_{e_M}(I'^{-1}I) = \nu_{e_M}(I'^{-1})\,\nu_{e_M}(I) \equiv 1 \bmod T_{\underline{\underline{S}}_o}(M)$. In particular, $I'^{-1}I = \Lambda_r(I') \cdot b$ is principal and $b \in \Lambda_r(I')$. However, $\Lambda_r(I')$ is maximal, and its localizations coincide with those of Ω_M for all primes not in $\underline{\underline{S}}_o$. A similar technique as employed in the proof of (2.10) shows that we can find an element $c \in \Omega_M$ such that

$$\Lambda_r(I')b = \Lambda_r(I')c; \text{ i.e.,}$$

$I'^{-1}I = \Lambda_r(I')c$ and thus $I = I'c$ with $c \in \Omega_M$, a regular element. Now

$$N'c = Mc \cap \Gamma MI'c = M \cap \Gamma Mc \cap \Gamma MI'c.$$

Since $\Gamma Mc \supset \Gamma MI'c$, we get

$$N'c = M \cap \Gamma MI = N. \qquad \#$$

2.11 <u>Theorem</u> (Eichler [4]): Let A_1 be a central simple K-algebra, where K is an algebraic number field with ring of integers R, which is not a totally definite quaternion algebra, and let $\underline{\underline{a}}$ be an ideal in R. If $x \in \Gamma_1$, where Γ_1 is a maximal R-order in A, is regular and such that $\mathrm{Nrd}_{A_1/K}(x) \equiv u \bmod (\underline{\underline{a}})$, where u is a unit in R, then there exists a unit γ_o in Γ_1 such that $\gamma_o \equiv x \bmod (\underline{\underline{a}}\,\Gamma_1)$ and $\mathrm{Nrd}_{A_1/K}(x) \equiv \mathrm{Nrd}_{A_1/K}(\gamma_o)$ mod $\underline{\underline{a}}$.

The <u>proof</u> may be found in Eichler [4], Hilfssatz 5, p. 235.

We now restrict K to be an algebraic number field.

2.12 <u>Corollary</u>: Let $M \in {}_{\Lambda}\underline{\underline{M}}^o$ satisfy Eichler's condition. Then

$St_{\underline{\underline{S}}_o}(e_M)_o \subset T_{\underline{\underline{S}}_o}(M)$.

Proof: Given $((\alpha_i))_{1\le i\le n} \in St_{\underline{\underline{S}}_o}(e_M)_o$; then $((\alpha_i))_{1\le i\le n} \in St(e_M)_o$ and we may assume $\alpha_i \equiv 1 \bmod(\underline{\underline{b}}C_i)$ (cf. 2.7). Then we can find a regular element $b \in B_M$ such that $Nrd_{e_iB_M/F_i}(b) = u\alpha_i$ for all i with $e_iM \ne 0$, where u is a unit in e_MC. And we might as well assume that $b \in \Omega_{\Gamma M}$, after multiplication with a suitable $0 \ne r \in R$. Now we apply (2.11) to obtain a unit $\omega_o \in \Omega_{\Gamma M}$ such that $\omega_o \equiv b \bmod(\underline{\underline{b}}\Omega_{\Gamma M})$ Observe that $\underline{\underline{b}}\Omega_{\Gamma M} \subset \Omega_M$. Thus we can write $b = \omega_o(1 + \gamma_o)$, where $\gamma_o \in \Omega_M$. Thus $\nu_{e_M}(\Omega_{\Gamma M}) = \nu_{e_M}(\Omega_{\Gamma M}(1+\gamma_o))$. But $1+\gamma_o \in \Omega_M$, and hence $\nu_{e_M}(\Omega_{\Gamma M}b) \in T_{\underline{\underline{S}}_o}(M)$. #

2.13 Theorem (Jacobinski [4]): There exists a natural number n_o depending only on Λ such that for every $M \in {}_{\Lambda}\underline{\underline{M}}^o$ which satisfies Eichler's condition, there are at most n_o non-isomorphic Λ-lattices in the genus of M.

This theorem has been proved in more generality by Roiter [4] cf. Ex. 3,1.

Proof: If $M \in {}_{\Lambda}\underline{\underline{M}}^o$ satisfies Eichler's condition, then $g(M) = |W_{\underline{\underline{S}}_o}(e_M)|$ (cf. 2.8) But in (2.12) we have shown

$$|W_{\underline{\underline{S}}_o}(e_M)| = |I_{\underline{\underline{S}}_o}(e_M)|/|T_{\underline{\underline{S}}_o}(M)| \le |I_{\underline{\underline{S}}_o}(e_M)|/|St_{\underline{\underline{S}}_o}(e_M)_o| \le$$
$$\le |I_{\underline{\underline{S}}_o}(1)|/|St_{\underline{\underline{S}}_o}(1)_o| = n_o.$$

Thus $g(M) \le n_o$ for every $M \in {}_{\Lambda}\underline{\underline{M}}^o$ which satisfies Eichler's condition.#

Exercises §2:

We keep the notation of §2.

1.) Let Λ and Λ_1 be R-orders in A and let $\underline{\underline{a}}$ be an ideal in R. Show that there exists $n_o \in \underline{\underline{N}}$ such that for every $s \in \underline{\underline{N}}$,

$$\Lambda_1\underline{\underline{a}}^{n_o+s} \cap \Lambda \subset \underline{\underline{a}}^s\Lambda.$$

2.) Heller-Reiner [3]: Let S be a ring and $M_i, N_i \in {}_{\underline{\underline{S}}}\underline{\underline{M}}^f$, $1 \leq i \leq 3$. Given two exact sequences

$$0 \longrightarrow M_1 \xrightarrow{\alpha} M_2 \longrightarrow M_3 \longrightarrow 0$$

$$0 \longrightarrow N_1 \xrightarrow{\beta} N_2 \longrightarrow N_3 \longrightarrow 0$$

and an S-homomorphism $\varphi: M_2 \longrightarrow N_2$. Assume that $\varphi|_{M_1} \in \mathrm{Hom}_S(\alpha M_1, \beta N_1)$ - this is satisfied e.g. if $\mathrm{Hom}_S(M_1, N_3) = 0$. Then one can complete the diagram

$$\begin{array}{ccccccccc} 0 & \to & M_1 & \to & M_2 & \to & M_3 & \to & 0 \\ & & {\scriptstyle\varphi_1}\downarrow & & {\scriptstyle\varphi}\downarrow & & \downarrow{\scriptstyle\varphi_3} & & \\ 0 & \to & N_1 & \to & N_2 & \to & N_3 & \to & 0 . \end{array}$$

We assume now that $\mathrm{Hom}_S(N_1, M_3) = 0$. Show that φ_1 and φ_3 are isomorphisms if φ is an isomorphism. (<u>Hint</u>: Use II, Ex. 2,1.)

3.) Use (2.11) to give a short proof of (2.8).

4.) Jacobinski [4]: Let K be an algebraic number field with ring of integers R, Λ an R-order in A. Then there exists an extension field K' of K, such that for every $M, N \in {}_{\Lambda}\underline{\underline{M}}^o$,

$$M \vee N \text{ if and only if } R' \otimes_R M \cong R' \otimes_R N,$$

where R' is the ring of integers in K'.

(<u>Hint</u>: $\Longleftarrow$ This follows from the integral version of the Noether-Deuring theorem (cf. VI, 3.8). $\Longrightarrow$ Extend K to a finite splitting field K_1. We may assume $N \in \mathcal{C}_{\underline{\underline{S}}_o}(M)$, and we extend K_1 to K_2 such that in R_2, the ring of integers in K_2,

$$\nu_{e_M}(N, M) = (a) \text{ is principal.}$$

Now we choose an extension K_3 such that in the ring of integers R_3 of K_3, $a \equiv$ unit mod$(\underline{\underline{S}}_o)$. This is possible (cf. Jacobinski [3,4]). Now take $R' = R_3$. Then $\nu_{e_M}(R' \otimes_R M, R' \otimes_R N) \in St_{\underline{\underline{S}}_o}(e_M) \subset T_{\underline{\underline{S}}_o}(e_M)$ and

$R' \otimes_R N \cong R' \otimes_R N$.)

5.) Show that $V(e_M)$ and $W_{\underline{\underline{S}}_o}(e_M)$ are finite groups.

§3 Embedding theorems for modules in the same genus

Lattices in the same genus can be embedded into each other in a very special way; this is used to give criteria, for a local direct summand to be a global direct summand.

Let R be Dedekind domain, the quotient field K of which is an $\underline{\underline{A}}$-field, and Λ an R-order in the finite dimensional separable K-algebra A.

3.1 Theorem (Roiter [4]): For $M, N \in {}_{\Lambda}\underline{\underline{M}}^{o}$, $M \vee N$ if and only if, for every non-zero ideal $\underline{\underline{a}}$ of R, we have an embedding

$$\varphi : N \longrightarrow M$$

such that

(i) $M/N\varphi = \oplus_{i=1}^{s} U_i$, where $\{U_i\}_{1\le i\le s}$ are simple Λ-modules,

(ii) $(\mathrm{ann}_R(U_i), \underline{\underline{a}}) = 1, 1\le i\le s$,

(iii) $(\mathrm{ann}_R U_i, \mathrm{ann}_R U_j) = 1, 1\le i,j\le s, i\ne j$.

Proof: If (i,ii,iii) can be satisfied for every ideal $\underline{\underline{a}}$ of R, we choose $\underline{\underline{a}}$ such that every maximal ideal $\underline{\underline{p}}$ for which $\Lambda_{\underline{\underline{p}}}$ is not maximal, divides $\underline{\underline{a}}$. Then $M_{\underline{\underline{p}}} \cong N_{\underline{\underline{p}}}$ for every $\underline{\underline{p}}$ dividing $\underline{\underline{a}}$, because of (ii), for the other prime ideals $\Lambda_{\underline{\underline{p}}}$ is maximal and (i) implies $KM \cong KN$; i.e., $M_{\underline{\underline{p}}} \cong N_{\underline{\underline{p}}}$.

Conversely, let $\{M_i\}_{1\le i\le t}$ be representatives of the isomorphism classes of Λ-lattices N with KM = KN. This number is finite by the Jordan-Zassenhaus theorem (cf. VI, 4.5, 4.7). For a fixed pair $(i,j), 1\le i,j\le t$ we look at all possible embeddings of M_i into M_j as maximal submodule; i.e., at short exact sequences

$$E(i,j) : 0 \longrightarrow M_i \longrightarrow M_j \longrightarrow U \longrightarrow 0,$$

where U is a simple Λ-module. Let $\underline{\underline{S}}_{ij}$ denote the set of prime ideals, which occur for a fixed pair (i,j) as $\mathrm{ann}_R(U)$ in all possible sequences E(i,j). There are two possibilities

α.) $\underline{\underline{S}}_{i,j}$ is finite for a pair (i,j),

β.) $\underline{\underline{S}}_{i,j}$ is infinite.

We put $\underline{\underline{b}} = \prod \underline{\underline{p}}$, where the product is taken over all $\underline{\underline{p}} \in \underline{\underline{S}}_{ij}$ for all pairs (i,j) for which α.) occurs or $\underline{\underline{b}} = R$ if α.) does not occur at all. Then $\underline{\underline{b}}$ is a non-zero ideal in R. By (1.9) we can embed N into M - φ: $N \longrightarrow M$ a monomorphism - such that $(\mathrm{ann}_R(M/N\varphi), \underline{\underline{b}}) = 1$. We identify N and $\mathrm{Im}\,\varphi$. Then M/N is a $\Lambda/\mathrm{ann}_R(M/N)\Lambda$ -module; i.e., a module over an artinian and noetherian ring, and it has a composition series, which we can lift to a "composition series between M and N":

$$M = X_0 \supsetneqq X_1 \supsetneqq \dots \supsetneqq X_s = N$$

such that X_i/X_{i+1}, $0 \le i \le s-1$, are simple Λ-modules. Since $KM = KN$, $X_i \in \{M_j\}_{1 \le j \le t}$ for every $0 \le i \le s$. Moreover, $(\mathrm{ann}_R(X_i/X_{i+1}), \underline{\underline{b}}) = 1$, and we can embed X_{i+1} into X_i as maximal submodules in infinitely many ways. We embed X_1 into X_0 as maximal submodule such that

$$(\mathrm{ann}_R(M/X_1\varphi_1), \underline{\underline{a}}) = 1,$$

where $\varphi_1 : X_1 \longrightarrow M$ is the embedding. Then we embed recursively X_{i+1} into $X_i\varphi_i$, - $\varphi_{i+1} : X_{i+1} \longrightarrow X_i\varphi_i$ - as maximal submodule such that

$$(\mathrm{ann}_R(X_i\varphi_i/X_{i+1}\varphi_{i+1}), \underline{\underline{a}}\ \mathrm{ann}_R(M/X_i\varphi_i)) = 1, 1 \le i \le s.$$

Putting $\varphi_0 = 1_M$, and $U_i = X_i\varphi_i/X_{i+1}\varphi_{i+1}$, $0 \le i \le s-1$, we see that the U_i are simple left Λ-modules such that $(\mathrm{ann}_R(U_i), \underline{\underline{a}}) = 1$, $0 \le i \le s-1$ and

$$(\mathrm{ann}_R(U_i), \mathrm{ann}_R(U_j)) = 1 \text{ for } 0 \le i \ne j \le s-1.$$

Moreover,

$$M/N\varphi_s \supsetneqq X_1\varphi_1/N\varphi_s \supsetneqq \dots \supsetneqq X_{s-1}\varphi_{s-1}/N\varphi_s \supsetneqq 0$$

is a composition series of $M/N\varphi_s$, and the composition factors are $X_i\varphi_i/X_{i+1}\varphi_{i+1} = U_i$, $0 \le i \le s-1$. Since the annihilators of the composition factors are relatively prime

$$M/N\varphi_s = \oplus_{i=0}^{s-1} U_i. \qquad \#$$

3.2 Notation: Let $A = \oplus_{i=1}^{s} Ae_i$ be the decomposition of A into simple K-algebras. If Λ is an R-order in A and $M \in {}_{\Lambda}\underline{\underline{M}}^{o}$, then we put $e_M = \sum_{e_i M \neq 0} e_i$, and if $U \in {}_{\Lambda}\underline{\underline{M}}^{f}$ is an R-torsion Λ-module, we put $e_U = \sum_{e_i U \neq 0} e_i$.

3.3 Theorem (Roiter [4]): Let Λ be an R-order in A and denote by $\underline{\underline{H}}(\Lambda)$ the Higman ideal of Λ. Let $\underline{\underline{S}}_o$ be a finite non-empty set of prime ideals of R containing all $\underline{\underline{p}}$ that divide $\underline{\underline{H}}(\Lambda)$. If $M \in {}_{\Lambda}\underline{\underline{M}}^{o}$ and if $U \in {}_{\Lambda}\underline{\underline{M}}^{f}$ is an R-torsion module such that

(i) $(\mathrm{ann}_R(U), \underline{\underline{S}}_o) = 1$ (This means $(\mathrm{ann}_R(U), \underline{\underline{p}}) = 1$ for every $\underline{\underline{p}} \in \underline{\underline{S}}_o$),

(ii) $U = \oplus_{i=1}^{s} U_i^{(s_i)}$, where $\{U_i\}_{1 \leq i \leq s}$ are non-isomorphic simple Λ-modules,

(iii) $e_M e_U = e_U$,

then there is an epimorphism

$$\varphi : M^{(n)} \longrightarrow U$$

for some $n \leq \max_i s_i$. Moreover, $\mathrm{Ker}\,\varphi \vee M^{(n)}$.

Proof: Since $e_M e_U = e_U$, we may assume that M is a faithful Λ-lattice. We decompose U into its $\underline{\underline{p}}$-primary components

$$U = \oplus_{j=1}^{t} V_j$$

where $\mathrm{ann}_R(V_j) = \underline{\underline{p}}_j$, since U is a direct sum of simple Λ-modules. Assume now that for each $j, 1 \leq j \leq t$, we have an epimorphism

$$\varphi_j' : M^{(n_j)} \longrightarrow V_j,$$

where n_j is not larger then the number of non-isomorphic simple Λ-modules in V_j (observe that the Krull-Schmidt theorem is valid for the summands of U). Then we have epimorphisms

$$\varphi_j : M^{(n)} \longrightarrow V_j$$

where $n = \max_j n_j \leq \max_i s_i$.

(Observe that simple modules can only be isomorphic if their annihilators are the same.) We define

$$\varphi : M^{(n)} \longrightarrow U = \bigoplus_{j=1}^{t} V_j,$$

$$m' \longmapsto (m'\varphi_j)_{1 \le j \le t}, m' \in M^{(n)}.$$

By the Chinese remainder theorem (I, 7.7) we can choose for every $1 \le k \le t$ an element $r_k \in R$ satisfying $r_k \equiv 1 \bmod \underline{\underline{p}}_k$ and $r_k \equiv 0 \bmod \underline{\underline{p}}_j$ for $j \ne k$. Then

$$\varphi : r_k m' \longrightarrow (0,\dots,0,m'\varphi_k,0,\dots,0)$$

and φ is an epimorphism since each φ_k is one. Because of $(\mathrm{ann}_R(U), \underline{\underline{H}}(\Lambda)) = 1$, we have $\mathrm{Ker}\,\varphi \vee M^{(n)}$ (cf. 1.9). Thus, it remains to prove the theorem in case $\mathrm{ann}_R(U) = \underline{\underline{p}}_o$ with $(\underline{\underline{p}}_o, \underline{\underline{H}}(\Lambda)) = 1$. But then

$$\Lambda/\underline{\underline{p}}_o\Lambda \cong \hat{\Lambda}_{\underline{\underline{p}}_o}/\underline{\underline{p}}_o\hat{\Lambda}_{\underline{\underline{p}}_o}$$

is semi-simple (cf. VI, 2.2). Since M is faithful, so is $\hat{M}_{\underline{\underline{p}}_o}$ but $\hat{\Lambda}_{\underline{\underline{p}}_o}$ is separable and thus maximal (cf. VI, 2.5) and $\hat{M}_{\underline{\underline{p}}_o}$ is a progenerator (cf. IV, 5.5). But then $\overline{M} = M/\underline{\underline{p}}_o M \cong \hat{M}_{\underline{\underline{p}}_o}/\underline{\underline{p}}_o\hat{M}_{\underline{\underline{p}}_o}$ is a progenerator (cf. IV, 3.7), and we have an epimorphism

$$\overline{\varphi} : \overline{M}^{(n)} \longrightarrow U,$$

since U is a $\Lambda/\underline{\underline{p}}_o\Lambda$ -module (cf. III, 1.10). Moreover, one sees easily that n can be choosen so that n = s, where $U = U_1^{(s)}$ and U_1 is a simple $\overline{\Lambda}$-module. But then we also have an epimorphism

$$\varphi : M^{(n)} \longrightarrow \overline{M}^{(n)} \longrightarrow U.$$

Since $(\underline{\underline{p}}_o, \underline{\underline{H}}(\Lambda)) = 1$, we have $M^{(n)} \vee \mathrm{Ker}\,\varphi$. #

3.4 <u>Theorem</u> (Roiter [2], Jacobinski [4]): Let $M_1, M_2 \in {}_{\Lambda}\underline{\underline{M}}^o$ lie in the same genus, and let $N_1 \in {}_{\Lambda}\underline{\underline{M}}^o$ be such that $e_{N_1}e_{M_2} = e_{M_2}$. Then there exists $N_2 \vee N_1$ such that

$$M_1 \oplus N_1 \cong M_2 \oplus N_2.$$

We <u>remark</u> that this result has been obtained for maximal orders in skewfields by Chevalley [1], using some work of Steinitz [1]; and it has been generalized to maximal R-orders by Swan [3].

__Proof:__ By (3.1) we have an embedding $\varphi : M_1 \longrightarrow M_2$ such that $U = M_2/M_1\varphi$ satisfies the hypotheses of (3.3) with $s_i = 1, 1 \le i \le s$. Moreover, we surely have $e_{M_2} e_U = e_U$, and thus $e_{N_1} e_U = e_U$. Hence we may apply (3.3) with $n = 1$, and we obtain an epimorphism

$$\psi : N_1 \longrightarrow U$$

where $N_2 = \operatorname{Ker}\psi$ lies in the same genus as N_1; i.e., we have two exact sequences

$$0 \longrightarrow N_2 \longrightarrow N_1 \xrightarrow{\alpha} U \longrightarrow 0$$

$$0 \longrightarrow M_1 \longrightarrow M_2 \xrightarrow{\beta} U \longrightarrow 0.$$

But $(\operatorname{ann}_R(U), \underline{\underline{H}}(\Lambda)) = 1$, where $\underline{\underline{H}}(\Lambda)$ is the Higman ideal of Λ. Thus, α and β are projective homomorphisms (cf. V, 2.1). In fact, we can choose $0 \ne r \in \underline{\underline{H}}(\Lambda)$ such that $rU = U$ and thus $\alpha = r \cdot (\alpha/r)$, where $\alpha' = \alpha/r \in \operatorname{End}_\Lambda(U)$, and α is projective. We now apply Schanuel's lemma (V, 2.6) to conclude

$$M_1 \oplus N_1 \cong M_2 \oplus N_2. \qquad \#$$

From the proof of this we obtain

3.5 __Corollary:__ If $U \in {}_{\Lambda}\underline{\underline{M}}^f$ is an R-torsion module with $(\operatorname{ann}_R(U), \underline{\underline{H}}(\Lambda)) = 1$, and if we have an exact sequence

$$0 \longrightarrow N \longrightarrow M \xrightarrow{\varphi} U \longrightarrow 0,$$

then φ is a projective homomorphism.

3.6 __Definition:__ Let $M, N \in {}_{\Lambda}\underline{\underline{M}}^o$. We say that N is a __local direct summand of M__, $N \,_{\mathrm{loc}}| \, M$, if $N_{\underline{\underline{p}}}$ is isomorphic to a direct summand of $M_{\underline{\underline{p}}}$ for every maximal ideal $\underline{\underline{p}}$ of R.

3.7 __Lemma:__ Let $M, N \in {}_{\Lambda}\underline{\underline{M}}^o$ and let $\underline{\underline{S}}_o$ be a non-empty finite set of prime ideals containing all prime ideals for which $\Lambda_{\underline{\underline{p}}}$ is not maximal. Then $N \underset{loc}{|} M$ if and only if $N_{\underline{\underline{p}}} \underset{\cong}{|} M_{\underline{\underline{p}}}$ for all $\underline{\underline{p}} \in \underline{\underline{S}}_o$. ($N \underset{\cong}{|} M$ indicates that N is isomorphic to a direct summand of M.)

<u>Proof</u>: It suffices to show one direction. Since $\underline{\underline{S}}_o \neq \emptyset$, $KN \underset{\cong}{|} KM$, and for all $\underline{\underline{p}} \in \underline{\underline{S}}_o$ we have $N_{\underline{\underline{p}}} \underset{\cong}{|} M_{\underline{\underline{p}}}$. But if $\underline{\underline{p}} \notin \underline{\underline{S}}_o$, then $\Lambda_{\underline{\underline{p}}}$ is maximal and $KN \underset{\cong}{|} KM$ implies $N_{\underline{\underline{p}}} \underset{\cong}{|} M_{\underline{\underline{p}}}$, since then $X_{\underline{\underline{p}}} \cong Y_{\underline{\underline{p}}}$ if and only if $KX_{\underline{\underline{p}}} \cong KY_{\underline{\underline{p}}}$ (cf. IV, 5.7). Thus $N \underset{loc}{|} M$. #

3.8 <u>Theorem</u>: Let $M, N \in {}_{\Lambda}\underline{\underline{M}}^o$ and assume $N \underset{loc}{|} M$. Then there exists $N' \vee N$ such that $N' \underset{\cong}{|} M$.

<u>Proof</u>: Let

$$M_{\underline{\underline{p}}} \cong N_{\underline{\underline{p}}} \oplus X_{\underline{\underline{p}}}.$$

By $N(\underline{\underline{p}})$ we denote the embedding of $N_{\underline{\underline{p}}}$ into $M_{\underline{\underline{p}}}$ as a direct summand. For almost all prime ideals $\underline{\underline{p}}$ of R, $N_{\underline{\underline{p}}} = N(\underline{\underline{p}})$ since $KN(\underline{\underline{p}}) = KN_{\underline{\underline{p}}}$ (this we always can assume). But then there exists a Λ-lattice N_1 such that $N_{1_{\underline{\underline{p}}}} = N(\underline{\underline{p}})$ for all prime ideals $\underline{\underline{p}}$ of R (cf. IV, 1.8). Obviously $N_1 \vee N$ since $N(\underline{\underline{p}}) \cong N_{\underline{\underline{p}}}$. Furthermore, since $N_{1_{\underline{\underline{p}}}} \subset M_{\underline{\underline{p}}}$ for all $\underline{\underline{p}}$, $N_1 \subset M$ and it even is an R-pure submodule, since this is true locally. We thus have an exact sequence of Λ-lattices

$$E : 0 \longrightarrow N_1 \xrightarrow{\varphi} M \xrightarrow{\psi} M/N_1 \longrightarrow 0$$

with canonical homomorphisms.

In (V, 3.7) we have shown that

$$\mathrm{Ext}^1_{\Lambda}(M/N_1, N_1) \overset{nat}{\cong} \bigoplus_{\underline{\underline{p}} | \underline{\underline{H}}(\Lambda)} \mathrm{Ext}^1_{\Lambda_{\underline{\underline{p}}}}(M_{\underline{\underline{p}}}/N_{1_{\underline{\underline{p}}}}, N_{1_{\underline{\underline{p}}}}).$$

However, for every prime $\underline{\underline{p}}$, the sequence

$$E_{\underline{\underline{p}}} : 0 \longrightarrow N_{1_{\underline{\underline{p}}}} \xrightarrow{\varphi_{\underline{\underline{p}}}} M_{\underline{\underline{p}}} \xrightarrow{\psi_{\underline{\underline{p}}}} M_{\underline{\underline{p}}}/N_{1_{\underline{\underline{p}}}} \longrightarrow 0$$

is split exact, where $\varphi_{\underline{\underline{p}}} = 1_{R_{\underline{\underline{p}}}} \otimes \varphi$ and $\psi_{\underline{\underline{p}}} = 1_{R_{\underline{\underline{p}}}} \otimes \psi$. This shows that under the above isomorphism, $[E] \longmapsto 0$. Hence $[E] = 0$, and E is split exact; i.e., $N_1 \underset{\cong}{|} M$. #

3.9 <u>Theorem</u> (Jacobinski [4], Roiter [2]): Let $M, N \in {}_{\Lambda}\underline{\underline{M}}^{o}$ such that $N \underset{loc}{|} M$. If $e_{KM/KN} e_{KN} = e_{KN}$, then $N \underset{\cong}{|} M$.

<u>Proof</u>: In view of (3.8) we have $N_1 \oplus X \cong M$ for some $N_1 \vee N$, and the condition on the idempotents means $e_X e_N = e_N$. We therefore can apply (3.4) to conclude that $M \cong N_1 \oplus X \cong X_1 \oplus N$, where $X_1 \vee X$. Thus $N \underset{\cong}{|} M$. #

<u>Remark</u>: This generalizes a theorem of Serre (cf. Bass [5]) to lattices over orders.

3.10 <u>Corollary</u>: If $M \cong M_1 \oplus M_2$, $M \in {}_{\Lambda}\underline{\underline{M}}^{o}$ and if $N \vee M$, then $N \cong N_1 \oplus N_2$, with $N_1 \vee M_1$, and $N_2 \vee M_2$. Thus one can say that a <u>genus decomposes</u>.

The <u>proof</u> follows immediately from (3.8). #

3.11 <u>Corollary</u>: Let $M = M_1 \oplus M_2$, $M \in {}_{\Lambda}\underline{\underline{M}}^{o}$ be such that $e_{M_1} e_{M_2} = e_{M_2}$. Then

$$g(M) \leq g(M_1),$$

where $g(X)$ denotes the number of non-isomorphic lattices in the same genus as X. We have equality if and only if

$$N_i \oplus M_2 \cong N_j \oplus M_2 \text{ implies } N_i \cong N_j$$

for $N_i, N_j \in \mathfrak{G}(M_1)$, the genus of M_1. In particular, if $e_{M_1} = e_{M_2}$, then

$$g(M) \leq \min(g(M_1), g(M_2)).$$

<u>Proof</u>: Given $N \vee M$, then $N = N_1 \oplus N_2$, $N_i \vee M_i$, $i=1,2$, by (3.10), and with (3.4) we can find $N_3 \vee M_1$ such that

$$N \cong N_3 \oplus M_2; \text{ i.e., } g(M) \leq g(M_1).$$

(Observe that $e_{M_1} = e_{N_1}$.)

If we have equality, then $N_i \oplus M_2 \cong N_j \oplus M_2$ for $N_i, N_j \in \mathcal{G}(M_1)$ implies $N_i \cong N_j$.

Conversely, assume that we can cancel M_2; i.e., $N_i \oplus M_2 \cong N_j \oplus M_2$ implies $N_i \cong N_j$, $N_i, N_j \in \mathcal{G}(M_1)$. Let $\{N_i\}_{1 \le i \le t}$ be representatives of the non-isomorphic Λ-lattices in $\mathcal{G}(M_1)$. Then all the lattices $N_i \oplus M_2$, $1 \le i \le t$ are non-isomorphic; i.e., $g(M) \geq g(M_1)$. Thus $g(M) = g(M_1)$. #

3.12 Theorem (Jacobinski [5]): Let K be an algebraic number field and R the ring of integers in K. Assume that $M \in {}_{\Lambda}\underline{\underline{M}}^{o}$ satisfies Eichler's condition and that $KM = \oplus_{i=1}^{t} L_i^{(\alpha_i)}$, where $\{L_i\}_{1 \le i \le t}$ are non-isomorphic simple A-modules, $\alpha_i > 0, 1 \le i \le t$. If $N \vee M$, then there exists an embedding

$$\varphi : N \longrightarrow M \text{ such that } l(M/N\varphi) \le t,$$

where $l(M/N\varphi)$ denotes the length of a composition series of $M/N\varphi$.

We remark that this answers a question of Roiter [4].

We need a deep result from algebraic number theory, which we quote without proof (cf. e.g. Hecke [1], Weil [1, Ch. VII, § 8]).

3.13 Theorem: Let $\underline{\underline{S}}_o$ be a finite non-empty set of prime ideals, then every ideal class modulo $St_{\underline{\underline{S}}_o}(1)_o$ contains infinitely many prime ideals (for the notation cf. 2.8).

Now we turn to the proof of (3.12): We may assume that $N \in \mathcal{C}_{\underline{\underline{S}}_o}(M)$ (cf. 1.8), where $\underline{\underline{S}}_o$ is a finite non-empty set of prime ideals containing all prime ideals for which $\Lambda_{\underline{\underline{p}}} \neq \Gamma_{\underline{\underline{p}}}$, Γ a maximal R-order in A containing Λ. We then write

$$N = M \cap \Gamma MI \text{ (cf. 1.12)}$$

where I is a full left $\Omega_{\Gamma M} = \text{End}_{\Gamma}(\Gamma M)$-ideal coprime to $\underline{\underline{S}}_o$. It is easily seen, that the correspondence set up in (1.12) preserves inclusions. Since the norm function too preserves inclusions, N is a maximal submodule of M if and only if $\nu_{e_M}(N,M)$ is a maximal ideal in

$e_M C$. (For the notation cf. 2.4, 2.8.) This implies that a composition series between M and N

$$M \supsetneqq M_1 \supsetneqq \cdots \supsetneqq N$$

gives rise to a composition series

$$\Omega\Gamma M \supsetneqq I_1 \supsetneqq \cdots \supsetneqq I$$

between $\Omega\Gamma M$ and I, which gives rise to a composition series

$$e_M C \supsetneqq \nu_{e_M}(I_1) \supsetneqq \cdots \supsetneqq \nu_{e_M}(I) = \nu_{e_M}(N,M)$$

between $e_M C$ and $\nu_{e_M}(N,M)$. According to (2.8) we do not change the isomorphism class of N if we replace I by an integral ideal I_o, coprime to $\underline{\underline{S}}_o$ such that

$$\nu_{e_M}(I_o) \in \nu_{e_M}(I) T_{\underline{\underline{S}}_o}(M).$$

However, $T_{\underline{\underline{S}}_o}(M) \supset St_{\underline{\underline{S}}_o}(e_M)_o$ (cf. 2.12). Thus we can replace I by I_o provided that

$$\nu_{e_M}(I_o) \in \nu_{e_M}(I) \prod_{e_i M \neq 0} St_{\underline{\underline{S}}_o}(e_i)_o.$$

By the generalized theorem on arithmetic progressions (3.13), every class $\nu_{e_i B_M/F_i}(e_i I) St_{\underline{\underline{S}}_o}(e_i)_o$ contains a prime ideal $\underline{\underline{p}}_i$ of $e_i C$. According to (VI, 8.10, 8.11) we can find maximal left $\Omega\Gamma M e_i$-ideals P_i in $e_i B_M$ such that $\nu_{e_i B_M/F_i}(P_i) = \underline{\underline{p}}_i$ for all i for which $e_i M \neq 0$. We put $P = \bigoplus_{e_i M \neq 0} P_i$; then P is a full left $\Omega\Gamma M$-ideal coprime to $\underline{\underline{S}}_o$, such that

$$\nu_{e_M}(P) \in \nu_{e_M}(I) St_{\underline{\underline{S}}_o}(e_M)_o.$$

Hence $N_1 = M \cap \Gamma MP$ is isomorphic to N and $N_1 \in \mathcal{C}_{\underline{\underline{S}}_o}(M)$. Moreover, $l(M/N_1) \leq t$ as follows from the above argument, since t is the number of e_i with $e_i M \neq 0$. #

<u>Exercises</u> §3:

In these exercises K with Dedekind domain R is an $\underline{\underline{A}}$-field, and we keep the notation of §3.

1.) Use (3.1) and (3.3) to prove the following theorem of Roiter:

Theorem (Roiter [4]): There exists a number n depending only on Λ such that $g(M) \leq n$ for every $M \in {}_{\Lambda}\underline{\underline{M}}^{o}$. *)

(<u>Hint</u>: Given $M \in {}_{\Lambda}\underline{\underline{M}}^{o}$, let $\{M_i\}_{1 \leq i \leq t}$ be representatives of the different isomorphism classes of Λ-lattices in $\mathfrak{g}(M)$. Put $M = M_1$, let Γ be a maximal R-order in A containing Λ and let $\underline{\underline{H}}(\Lambda)$ be the Higman ideal of Λ. We embed M_i into M, $2 \leq i \leq t$ such that

(i) $M/M_i = \oplus_{j=1}^{s_i} U_{ij}, 2 \leq i \leq t, U_{ij}$ simple Λ-modules,

(ii) $(\mathrm{ann}_R U_{ij}, \mathrm{ann}_R U_{ij'}) = 1$ for $j \neq j'$,

(iii) $(\mathrm{ann}_R U_{ij}, \underline{\underline{H}}(\Lambda) \prod_{l=1}^{s_k} \mathrm{ann}_R U_{kl}) = 1, i \neq k$.

This can be done by (3.1); to satisfy (iii), the embedding has to be defined recuresively. Then we have an exact sequence

$$0 \longrightarrow \bigcap_{i=2}^{t} M_i \longrightarrow M \longrightarrow \oplus_{i=2}^{t} M/M_i \longrightarrow 0.$$

If now $N \in {}_{\Lambda}\underline{\underline{M}}^{o}$ is faithful, we get an epimorphism (cf. 3.3)

$$\varphi : N \longrightarrow \oplus_{i=2}^{t} M/M_i .$$

However, φ is projective (cf. 3.5) and we can complete the following diagram

$$\begin{array}{ccccccccc} & & & & & & 0 & & \\ & & & & & & \uparrow & & \\ 0 & \longrightarrow & \bigcap_{i=2}^{t} M_i & \longrightarrow & M & \longrightarrow & \oplus_{i=2}^{t} M/M_i & \longrightarrow & 0 \\ & & & & & & \uparrow \varphi & & \\ & & & & & & N & & . \end{array}$$

($\psi : N \dashrightarrow M$)

Putting $N_i = KN \cap M$, we get the following commutative diagrams, $2 \leq i \leq t$:

$$\begin{array}{lccccccccc} E_i : & 0 & \longrightarrow & M_i \cap N & \longrightarrow & M_i & \longrightarrow & M/N_i & \longrightarrow & 0 \\ & & & \uparrow & & \uparrow & & \uparrow 1_{M/N_i} & & \\ E : & 0 & \longrightarrow & N \cap N_i & \longrightarrow & N & \longrightarrow & M/N_i & \longrightarrow & 0 \end{array} .$$

Use this to prove the theorem.)

---

*) Recently, Drozd (Izv. Akad. Nauk SSSR <u>33</u> (69), 1080) has given an independent proof of this statement.

2.) Let Γ be a maximal R-order in the central simple K-algebra A. If I is a full integral left Γ-ideal, show $l(\Gamma/I) = l(R/\nu(I))$.

§4 Genera of special types of lattices

The genera of lattices over maximal orders are listed, and for maximal orders a necessary condition for cancellation is given. The genera of absolutely irreducible lattices are computed, and the concept of restricted genera is introduced.

For this section we assume that K with Dedekind domain R is an $\underline{\underline{A}}$-field. As one sees from the previous sections, the theory of genera is closely related to the difficult concept of arithmetic in orders (cf. Jacobinski [3,4], Takahashi [1]). It seems to be very difficult in general, to give an explicit description of the non-isomorphic lattices in one genus (cf. Drozd-Turchin [1]). However, in some special cases this is possible.

4.1 <u>Lemma</u> (Swan [4]): Let A be a central simple K-algebra and Γ a maximal R-order in A. If $M \in {}_{\Gamma}\underline{\underline{M}}^{o}$ satisfies Eichler's condition, then

$$M \oplus X \cong N \oplus X \text{ implies } M \cong N,$$

for $X, N \in {}_{\Gamma}\underline{\underline{M}}^{o}$.

<u>Proof</u>: Since A is simple, $e_M e_X = e_X = e_M$, and thus we can cancel if and only if $g(M \oplus X) = g(M)$ (cf. 3.11). But with M also $M \oplus X$ satisfies Eichler's condition, and with (2.2) we conclude $g(M) = g(M \oplus X)$, where $g(Y)$ denotes the number of non-isomorphic lattices in the genus of Y, $\mathfrak{g}(Y)$. #

4.2 <u>Remark</u>: This shows that $g(M) = g(N)$ if M and N both satisfy Eichler's condition and $g(M) \leq g(N)$ if N does not satisfy Eichler's condition, as follows by applying (3.11) to $M \oplus N$.

4.3 <u>Theorem</u> (Jacobinski [4]): Let A be a central simple K-algebra and let Γ be a maximal R-order in A. For a fixed irreducible Γ-lattice M_o we put $\Omega = \mathrm{End}_{\Gamma}(M_o)$. Let $\{I_j\}_{1 \leq j \leq n}$ be representatives of the different classes of left ideals in Ω. If $M \in {}_{\Gamma}\underline{\underline{M}}^{o}$ is such that $KM \cong KM_o^{(s)}$, then every $N \vee M$ has the form

$$N \cong M_o \otimes_{\Omega} (\Omega^{(s-1)} \oplus I_j) \cong M_o^{(s-1)} \oplus M_o I_j \text{ for some } 1 \leq j \leq n.$$

Moreover, if A is not a full matrix algebra over a totally definite quaternion-algebra, then no two of the above modules are isomorphic; i.e., $g(N) = n$.

<u>Proof</u>: By (IV, 5.7 and VII 3.7) we have

$$M \cong \oplus_{i=1}^{s} M_i',$$

where $M_i' \vee M_o$. However, from (3.4) it follows that $M \cong M_o^{(s-1)} \oplus M_s$, $M_s \vee M_o$. Since locally there is only one isomorphism class of irreducible Γ_p-lattices, the representatives of Γ-lattices in the same genus as $\underline{\underline{M}}_o$ are the lattices $M_o \otimes_{\Omega} I_j, 1 \leq j \leq t$. (Observe that we have a Morita equivalence between ${}_{\Gamma}\underline{\underline{M}}^o$ and ${}_{\Omega}\underline{\underline{M}}^o$.) Thus $M_s \cong M_o \otimes_{\Omega} I_j$ for some $1 \leq j \leq n$, and $M \cong M_o \otimes_{\Omega} (\Omega^{(s-1)} \oplus I_j)$. Similarly for every $N \vee M$. We remark, that obviously all the lattices $M_o \otimes_{\Omega} (\Omega^{(s-1)} \oplus I_j)$ lie in the same genus as M (cf. IV, 5.7). (Since Ω is maximal $M_o \otimes_{\Omega} I_j \cong M_o I_j$.) Now we assume that A is not a full matrix algebra over a totally definite quaternion algebra. Then

$$M_o \otimes_{\Omega} (\Omega^{(n-1)} \oplus I_j) \cong M_o \otimes_{\Omega} (\Omega^{(n-1)} \oplus I_k)$$

implies (cf. 4.1) $M_o \otimes_{\Omega} I_j \cong M_o \otimes_{\Omega} I_k$; but then $I_j \cong I_k$ (cf. IV, 5.5; III 2.1, 2.3); i.e., $g(M) = n$. #

4.4 <u>Remark</u>: We shall state some immediate consequences of (3.4) and the proof of (4.3). Let Λ be an R-order in the separable finite dimensional K-algebra A. Let $M \in {}_{\Lambda}\underline{\underline{M}}^o$ be given. For a given $N \vee M^{(s)}$ and given $M_i, 1 \leq i \leq s-1$, $M_i \vee M$, there exists - not necessarily only one - $M_s \vee M$ such that

$$N \cong \oplus_{i=1}^{s} M_i.$$

In particular, if we choose $M_i = M, 1 \leq i \leq s-1$ then $N \cong M^{(s-1)} \oplus M_s$, $M_s \vee M$.

We shall now apply this to various situations:

(i) $\Lambda = R$, $M = R$. Given any R-lattice N with $KN \cong K^{(s)}$ and $\{\underline{\underline{a}}_i\}_{1 \le i \le s-1}$ R-ideals in K, then there exists an R-ideal $\underline{\underline{a}}_s$ in K such that

$$N \cong \oplus_{i=1}^{s} \underline{\underline{a}}_i,$$

in particular,

$$N \cong R^{(s-1)} \oplus \underline{\underline{a}}_s',$$

and the ideal class of $\underline{\underline{a}}_s'$ is uniquely determined (cf. 4.3), since K is not a totally definite quaternion algebra.

(ii) If A is a skewfield and Γ a maximal R-order in A, then every $N \in {}_{\Gamma}\underline{\underline{M}}^{o}$ with $KN \cong A^{(s)}$ has a decomposition

$$N \cong \oplus_{i=1}^{s} I_i,$$

where $I_i, 1 \le i \le s-1$ are preassigned ideals and I_s is an ideal. If A is not a totally definite quaternion algebra, then I_s is uniquely determined (up to isomorphism).

(iii) If Λ is any R-order in A, then every Λ-lattice N in the same genus as ${}_{\Lambda}\Lambda^{(s)}$ has a decomposition

$$N \cong {}_{\Lambda}\Lambda^{(s-1)} \oplus I,$$

where $I \vee {}_{\Lambda}\Lambda$.

If G is a finite group such that the characteristic of K does not divide the order of G, and if no rational prime dividing the order of G is a unit in R, then it will be shown in (IX, 1.9) that every projective Λ-lattice is locally free, for $\Lambda = RG$. Thus we obtain a result of Swan [1,2]: Every projective Λ-lattice P has the form

$$P = {}_{\Lambda}\Lambda^{(s-1)} \oplus I,$$

where $I \vee {}_{\Lambda}\Lambda$.

4.5 Remark: The results of (4.1) and (4.3) can readily be generalized - with the appropriate changes - to maximal orders in separable algebras.

We now assume that A is a finite dimensional separable K-algebra, Λ an

R-order in A and Γ a maximal R-order in A containing Λ.

4.6 <u>Lemma</u>: Let $M \in {}_{\Lambda}\underline{\underline{M}}^{o}$ be a Γ-lattice. Then every Λ-lattice in $\mathfrak{G}(M)$ is a Γ-lattice, and, in view of (4.3, 4.4), $\mathfrak{G}(M)$ can assumed to be known.

<u>Proof</u>: If $N \vee M$, then $N_{\underline{\underline{p}}} \in {}_{\Gamma_{\underline{\underline{p}}}}\underline{\underline{M}}^{o}$ for every prime ideal. Thus $N \in {}_{\Gamma}\underline{\underline{M}}^{o}$, and for $N_1, N_2 \in \mathfrak{G}(M)$ we have $N_1 \cong_{\Lambda} N_2$ if and only if $N_1 \cong_{\Gamma} N_2$ (cf. IV, 1.14). #

4.7 <u>Theorem</u> (Maranda [2], Roggenkamp [6]): Let M be an absolutely irreducible Λ-lattice, and let $\{\Gamma_i\}_{1 \le i \le s}$ be the different maximal R-orders in Ae_M containing Λe_M. If M_i is an irreducible Γ_i-lattice and if $\{I_j\}_{1 \le i \le h}$ are representatives of the different classes of ideals in K, then

(i) no two of the Λ-lattices $\{M_i\}_{1 \le i \le s}$ lie in the same genus, and there are exactly s different genera of Λ-lattices N with $KN \cong KM$,

(ii) each genus $\mathfrak{G}(M_i)$, $1 \le i \le s$, contains exactly h non-isomorphic Λ-lattices,

$$\mathfrak{G}(M_i) = \{(M_i \otimes_R I_j), 1 \le j \le h\} = \{M_i I_j,\ 1 \le j \le h\}.$$

<u>Proof</u>: This is an immediate consequence of (VI, §5 and VII, 4.6). #

4.8 <u>Definition</u>: Let $M \in {}_{\Lambda}\underline{\underline{M}}^{o}$. We say that $N \in {}_{\Lambda}\underline{\underline{M}}^{o}$ lies in the same <u>restricted genus</u>, notation $N \curlyvee M$, if there exists $T \in {}_{\Lambda}\underline{\underline{M}}^{o}$ such that $M \oplus T \cong N \oplus T$.

We remark that we always may assume $e_T = e_M = e_N$ and that $N \curlyvee M$ implies $M \vee N$ (cf. VI, 3.5).

4.9 <u>Lemma</u> (Jacobinski [4]): Assume that $M \in {}_{\Lambda}\underline{\underline{M}}^{o}$ satisfies Eichler's condition. Then $N \curlyvee M$ if and only if $N \vee M$ and $\Gamma M \cong \Gamma N$ where Γ is any maximal R-order in A containing Λ.

We <u>remark</u>, that Jacobinski [3] has introduced restricted genera differently. He defined two Λ-lattices M and N to lie in the same restricted genus relative to some maximal R-order Γ containing Λ, if $M \vee N$ and $\Gamma M \cong \Gamma N$. This classification depends on Γ. However, (4.9) shows that in case M satisfies Eichler's condition, the restricted genera are independent of a particular maximal order. In this case, our definition (4.8) coincides with Jacobinski's. However, this need not be so if M does not satisfy Eichler's condition. We have chosen (4.8) as the definition, since this depends only on Λ.

Proof: If $N \overset{\vee}{\vee} M$, then $M \oplus T \cong N \oplus T$ with $e_M = e_T$; hence $\Gamma M \oplus \Gamma T = \cong \Gamma N \oplus \Gamma T$. However, with M also ΓM satisfies Eichler's condition, and since $e_{\Gamma M} = e_{\Gamma T}$ we can apply (4.1) to conclude $\Gamma M \cong \Gamma N$.
Conversely, if $M \vee N$ and $\Gamma M \cong \Gamma N$, we can find $X \vee \Gamma M$ such that $M \oplus X \cong N \oplus \Gamma M$ (cf. 3.4), since $e_M = e_{\Gamma M}$. By (4.6), $X \in {}_{\Gamma}\underline{\underline{M}}^{o}$ and $\Gamma M \oplus X = \Gamma N \oplus \Gamma M$ implies $X \cong \Gamma M$, since M satisfies Eichler's condition (cf. 4.1); i.e., $M \oplus \Gamma M \cong N \oplus \Gamma M$. #

4.11 Theorem (Jacobinski [4]): Assume that $M \in {}_{\Lambda}\underline{\underline{M}}^{o}$ satisfies Eichler's condition. Then the genus of M, $\mathfrak{G}(M)$ is partitioned into restricted genera; their number is equal to the number $g(\Gamma M)$ of non-isomorphic Γ-lattices in $\mathfrak{G}(\Gamma M)$, where Γ is any fixed maximal R-order in A containing Λ. Moreover, each restricted genus of $\mathfrak{G}(M)$ contains the same number of non-isomorphic Λ-lattices.

Proof: Lying in the same restricted genus is obviously an equivalence relation. To prove the remaining statements, we return to the notation of (2.8). In (1.12) we had shown that there is a one-to-one correspondence between the lattices in $\mathfrak{C}_{\underline{\underline{S}}_o}(M)$ and the full left integral $\Omega_{\Gamma M}$-ideals coprime to $\underline{\underline{S}}_o$:

$$N \longmapsto I \text{ where } N = M \cap \Gamma M I.$$

Moreover, this way we obtain - up to isomorphism - all Λ-lattices in $\mathfrak{G}(M)$. Obviously, the number of district restricted genera in $\mathfrak{G}(M)$

can not be larger than $g(\Gamma M)$. On the other hand, by varying I we can obtain $N \vee M$ such that ΓN is isomorphic to any $X \vee \Gamma M$; in fact $X \cong \Gamma MI$ for some I (cf. 1.7, 1.10). Then $N = M \cap \Gamma MI$ has the desired property (cf. proof of 1.12).

It remains to show that the number of non-isomorphic Λ-lattices in each restricted genus of $\mathfrak{G}(M)$ is the same. We have shown in (2.8), that

$$g(M) = |I_{\underline{\underline{S}}_o}(e_M)/T_{\underline{\underline{S}}_o}(e_M)|,$$

and from the proof of (2.8) one sees easily that $N \veebar M$ if and only if

$$w_{e_M}(N,M) \in (I_{\underline{\underline{S}}_o}(e_M) \cap St(e_M)_o)/T_{\underline{\underline{S}}_o}(e_M).$$

However $I_{\underline{\underline{S}}_o}(e_N) \cap St(e_N)_o = I_{\underline{\underline{S}}_o}(e_M) \cap St(e_M)_o$ if $N \vee M$, and it remains to show that $T_{\underline{\underline{S}}_o}(e_M) = T_{\underline{\underline{S}}_o}(e_N)$ if $N \vee M$.

We have shown in (2.12), that $T_{\underline{\underline{S}}_o}(e_M) \supset St_{\underline{\underline{S}}_o}(e_M)_o$. But $St_{\underline{\underline{S}}_o}(e_M)_o =$ $= St_{\underline{\underline{S}}_o}(e_N)_o$, and thus it suffices to show $T_{\underline{\underline{S}}_o}(e_M)/St_{\underline{\underline{S}}_o}(e_M)_o =$ $T_{\underline{\underline{S}}_o}(e_N)/St_{\underline{\underline{S}}_o}(e_M)_o$. We may assume $N \subset M$, $N \in \mathcal{C}_{\underline{\underline{S}}_o}(M)$. Then there exists $0 \neq r \in R$, $r \equiv 1 \bmod(\underline{\underline{S}}_o)$ such that $rM \subset N \subset M$. Let $a \in \Omega_M$, then $r\,a \in \Omega_N$ and $\nu_{e_M}(\Omega_{\Gamma M} a) \in T_{\underline{\underline{S}}_o}(e_M)$, and $\nu_{e_M}(\Omega_{\Gamma M} r\,a) =$ $R\,\nu_{e_M}(r)\,\nu_{e_M}(a) \in T_{\underline{\underline{S}}_o}(e_N)$. But $R\,\nu_{e_M}(r) \in St_{\underline{\underline{S}}_o}(e_M)_o$ and $\nu_{e_M}(\Omega_{\Gamma M} a)$ $\in T_{\underline{\underline{S}}_o}(e_N)$. Thus $T_{\underline{\underline{S}}_o}(e_N) \subset T_{\underline{\underline{S}}_o}(e_M)$. Similarly one shows the other inclusion. Hence each restricted genus in $\mathfrak{G}(M)$ contains the same number of non-isomorphic lattices. #

<u>Exercise</u> §4:

We keep the <u>notation</u> of §4.

1.) Let $M \in {}_{\Lambda}\underline{\underline{M}}^o$ satisfy Eichler's condition. Give a necessary and sufficient condition for the restricted genus of M to contain only one isomorphism class of lattices.

CHAPTER VIII

GROTHENDIECK GROUPS

§1 Grothendieck groups and other groups associated with modules

The groups $\underline{\underline{K}}_i(\underline{\underline{C}})$, $i = 0,1$ and $\underline{\underline{G}}_i(\underline{\underline{C}})$ are defined for an admissible subcategory $\underline{\underline{C}}$ of a category of modules. Some examples are given and some elementary properties are derived.

1.1 Definition: Let S be a ring. A subcategory $\underline{\underline{C}}$ of ${}_S\underline{\underline{M}}^f$ is called an admissible subcategory if it satisfies the following conditions:

(i) For $C_1, C_2 \in ob(\underline{\underline{C}})$, $morph_{\underline{\underline{C}}}(C_1,C_2) = Hom_S(C_1,C_2)$.

(ii) $0 \in ob(\underline{\underline{C}})$.

(iii) Finite direct sums of objects in $\underline{\underline{C}}$ lie in $\underline{\underline{C}}$.

(iv) If $0 \longrightarrow C' \longrightarrow M \longrightarrow C'' \longrightarrow 0$ is an exact sequence in ${}_S\underline{\underline{M}}^f$, and if $C', C'' \in ob(\underline{\underline{C}})$, then $M \in ob(\underline{\underline{C}})$.

1.2 Definition: Let S be a ring and $\underline{\underline{C}}$ an admissible subcategory of ${}_S\underline{\underline{M}}^f$. Then $\underline{\underline{G}}_o(\underline{\underline{C}})$ is the abelian group generated by symbols $[C]$, where C runs through the objects in $\underline{\underline{C}}$, and subject to the relations $[C] = [C'] + [C'']$ for every short exact sequence $0 \longrightarrow C' \longrightarrow C \longrightarrow C'' \longrightarrow 0$ in $\underline{\underline{C}}$. $\underline{\underline{G}}_o(\underline{\underline{C}})$ is called the Grothendieck group of $\underline{\underline{C}}$ relative to short exact sequences.

1.3 Remark: Let A be an abelian group and $f : ob(\underline{\underline{C}}) \longrightarrow A$ a map such that $f(C) = f(C') + f(C'')$ for each exact sequence

$$0 \longrightarrow C' \longrightarrow C \longrightarrow C'' \longrightarrow 0 \text{ in } \underline{\underline{C}},$$

then f factors uniquely through the map $ob(\underline{\underline{C}}) \longrightarrow \underline{\underline{G}}_o(\underline{\underline{C}})$; i.e., $\underline{\underline{G}}_o(\underline{\underline{C}})$ is universal with the above property.

1.4 Lemma:

(i) $[0] = 0$,

(ii) $C_1 \cong C_2$ implies $[C_1] = [C_2]$,

(iii) $[C_1 \oplus C_2] = [C_1] + [C_2]$.

Proof:

(i) $0 \longrightarrow 0 \longrightarrow 0 \longrightarrow 0 \longrightarrow 0$ is exact;

(ii) $0 \longrightarrow C_1 \longrightarrow C_2 \longrightarrow 0 \longrightarrow 0$ is exact;

(iii) $0 \longrightarrow C_1 \longrightarrow C_1 \oplus C_2 \longrightarrow C_2 \longrightarrow 0$ is exact.

1.5 <u>Theorem</u> (Heller-Reiner [5]): Let $C_1, C_2 \in ob(\underline{\underline{C}})$. Then $[C_1] = [C_2]$ in $\underline{\underline{G}}_o(\underline{\underline{C}})$ if and only if there exist two exact sequences

$$0 \longrightarrow X' \longrightarrow C_1 \oplus X \longrightarrow X'' \longrightarrow 0$$

$$0 \longrightarrow X' \longrightarrow C_2 \oplus X \longrightarrow X'' \longrightarrow 0 \text{ in } \underline{\underline{C}}:$$

i.e., $X, X', X'' \in ob(\underline{\underline{C}})$.

<u>Proof</u>: If those two exact sequences do exist, then obviously $[C_1] = [C_2]$ in $\underline{\underline{G}}_o(\underline{\underline{C}})$. To show the converse, we represent $\underline{\underline{G}}_o(\underline{\underline{C}})$ as A/B, where A is the free abelian group generated by the isomorphism classes (C) of objects in $\underline{\underline{C}}$, and B is the subgroup of A generated by all elements of the form (C) - (C') - (C") if there is an exact sequence $0 \longrightarrow C' \longrightarrow C \longrightarrow C'' \longrightarrow 0$ in $\underline{\underline{C}}$. Then $[C_1] = [C_2]$ in $\underline{\underline{G}}_o(\underline{\underline{C}})$ implies

$$(C_1) - (C_2) = \sum_i \{(M_i) - (M_i') - (M_i'')\} + \sum_i \{-(N_i) + (N_i') + (N_i'')\},$$

where $0 \longrightarrow M_i' \longrightarrow M_i \longrightarrow M_i'' \longrightarrow 0$ and $0 \longrightarrow N_i' \longrightarrow N_i \longrightarrow N_i'' \longrightarrow 0$ are exact sequences in $\underline{\underline{C}}$. Thus

$$(C_1) + \sum_i (N_i) + \sum_i (M_i') + \sum_i (M_i'') = (C_2) + \sum_i (M_i) + \sum_i (N_i') + \sum_i (N_i'').$$

If we put $N = \oplus_i N_i$, $N' = \oplus_i N_i'$, $N'' = \oplus_i N_i''$, $M = \oplus_i M_i$, $M' = \oplus_i M_i'$ and $M'' = \oplus_i M_i''$, then

$$C_1 \oplus N \oplus M' \oplus M'' \cong C_2 \oplus M \oplus N' \oplus N''.$$

Now, for $Y = C_2 \oplus M \oplus N' \oplus N''$, the sequences

$$0 \longrightarrow M' \oplus N' \longrightarrow C_1 \oplus Y \longrightarrow C_1 \oplus C_2 \oplus M'' \oplus N'' \longrightarrow 0$$

and

$$0 \longrightarrow M' \oplus N' \longrightarrow C_2 \oplus Y \longrightarrow C_1 \oplus C_2 \oplus M'' \oplus N'' \longrightarrow 0$$

are exact sequences in $\underline{\underline{C}}$,

$$0 \longrightarrow N' \longrightarrow N \longrightarrow N'' \longrightarrow 0$$

and

$$0 \longrightarrow M' \longrightarrow M \longrightarrow M'' \longrightarrow 0$$

being exact. #

1.6 <u>Definition</u>: Let $\underline{\underline{E}}$ be a subcategory of ${}_{\underline{\underline{S}}}M^f$, closed under isomorphisms, which satisfies (1.1, (ii), (iii)). The <u>Grothendieck group</u> $\underline{\underline{K}}_o(\underline{\underline{E}})$ of $\underline{\underline{E}}$ <u>relative to direct sums</u> is the abelian group generated by symbols $\langle E\rangle$, where E runs through the objects in $\underline{\underline{E}}$, and subject to the relations $\langle E\rangle = \langle E_1\rangle + \langle E_2\rangle$, if $E \cong E_1 \oplus E_2$. Then $\underline{\underline{K}}_o(\underline{\underline{E}})$ is universal in an obvious sense.

1.7 <u>Lemma</u>: $\langle X\rangle = \langle Y\rangle$ in $\underline{\underline{K}}_o(\underline{\underline{E}})$ if and only if $X \oplus Z \cong Y \oplus Z$ for some $Z \in \underline{\underline{E}}$.

<u>Proof</u>: If the condition is satisfied, then trivially $\langle X\rangle = \langle Y\rangle$. Conversely, we represent $\underline{\underline{K}}_o(\underline{\underline{E}})$ as A/B, where A is the free abelian group generated by the isomorphism classes (E) of objects in $\underline{\underline{E}}$ and B is the subgroup generated by $(E \oplus E') - (E) - (E')$. Thus in A, we have

$$(X) - (Y) = \sum_i \{(E_i \oplus E_i') - (E_i) - (E_i')\} + \sum_i \{(F_i) + (F_i') - (F_i \oplus F_i')\};$$

i.e.,

$$(X) + \sum_i (E_i) + \sum_i (E_i') + \sum_i (F_i \oplus F_i') = (Y) + \sum_i (E_i \oplus E_i') + \sum_i (F_i) + \sum_i (F_i').$$

Since A is free abelian on $\{(E) : E \in ob(\underline{\underline{E}})\}$, this means

$$X \oplus Z \cong Y \oplus Z,$$

where $Z = (\bigoplus_i E_i) \oplus (\bigoplus_i E_i') \oplus (\bigoplus_i F_i) \oplus (\bigoplus_i F_i') \in ob(\underline{\underline{E}})$. #

1.8 <u>Remarks</u>: (i) Let $\underline{\underline{C}}$ be an admissible subcategory of ${}_S\underline{\underline{M}}^f$. Then we have an epimorphism of abelian groups

$$\sigma : \underline{\underline{K}}_o(\underline{\underline{C}}) \longrightarrow \underline{\underline{G}}_o(\underline{\underline{C}})$$

$$\sigma : \langle C\rangle \longmapsto [C]$$

which is a group homomorphism, because of (1.4); and since every $x \in \underline{\underline{G}}_o(\underline{\underline{C}})$ has the form $x = [C_1] - [C_2]$, $C_i \in ob(\underline{\underline{C}})$, $i = 1,2$, σ is an epimorphism.

(ii) Let S_1 and S_2 be rings and assume that $\underline{\underline{E}}_1$ and $\underline{\underline{E}}_2$ are subcategories of ${}_{S_1}\underline{\underline{M}}^f$ and ${}_{S_2}\underline{\underline{M}}^f$ resp. closed under isomorphisms, which satisfy (1.1, (ii), (iii)). An additive functor $\underline{\underline{F}} : \underline{\underline{E}}_1 \longrightarrow \underline{\underline{E}}_2$ then induces a homomorphism of abelian groups $\underline{\underline{K}}_o(\underline{\underline{E}}_1) \longrightarrow \underline{\underline{K}}_o(\underline{\underline{E}}_2)$. If $\underline{\underline{C}}_1$ and $\underline{\underline{C}}_2$ are admissible subcategories and if $\underline{\underline{F}} : \underline{\underline{C}}_1 \longrightarrow \underline{\underline{C}}_2$ is an exact functor, then $\underline{\underline{F}}$ induces a homomorphism of abelian groups $\underline{\underline{G}}_o(\underline{\underline{C}}_1) \longrightarrow \underline{\underline{G}}_o(\underline{\underline{C}}_2)$.

1.9 Examples:

I. Let K be a field and B a finite dimensional K-algebra. Then $\underline{\underline{G}}_o({}_B\underline{\underline{M}}^f)$ is a free abelian group on symbols $[L_i]$, $1 \le i \le n$, where $\{L_i\}_{1 \le i \le n}$ are representatives of the isomorphism classes of simple left B-modules. This follows immediately from the validity of the Jordan-Hölder theorem (cf. I, 4.6).

II. Let A be a semi-simple finite dimensional algebra over a field K. Then $\underline{\underline{K}}_o({}_A\underline{\underline{M}}^f) = \underline{\underline{G}}_o({}_A\underline{\underline{M}}^f)$, since every exact sequence of finitely generated left A-modules splits. Moreover, in $\underline{\underline{G}}_o({}_A\underline{\underline{M}}^f)$, $[L_1] = [L_2]$ if and only if $L_1 \cong L_2$, since the Krull-Schmidt theorem is valid for ${}_A\underline{\underline{M}}^f$ (cf. 1.7). For any ring S we shall write

$$\underline{\underline{G}}_o^f(S) = \underline{\underline{G}}_o({}_S\underline{\underline{M}}^f) \text{ and } \underline{\underline{K}}_o(S) = \underline{\underline{K}}_o({}_S\underline{\underline{P}}^f).$$

III. Let R be a Dedekind domain with quotient field K, A a finite dimensional separable K-algebra and Λ an R-order in A. Then ${}_\Lambda\underline{\underline{M}}^o$, the category of Λ-lattices, is an admissible subcategory of ${}_\Lambda\underline{\underline{M}}^f$, and we define

$$\underline{\underline{G}}_o^f(\Lambda) = \underline{\underline{G}}_o({}_\Lambda\underline{\underline{M}}^f)$$

$$\underline{\underline{G}}_o(\Lambda) = \underline{\underline{G}}_o({}_\Lambda\underline{\underline{M}}^o).$$

${}_\Lambda\underline{\underline{P}}^f$ is an admissible subcategory of ${}_\Lambda\underline{\underline{M}}^f$, (observe $0 \in {}_\Lambda\underline{\underline{P}}^f$), and

$$\underline{\underline{K}}_o(\Lambda) = \underline{\underline{K}}_o({}_\Lambda\underline{\underline{P}}^f).$$

$\underline{\underline{K}}_o(\Lambda)$ is called the Grothendieck group of projective Λ-lattices. We would like to remind that in ${}_\Lambda\underline{\underline{P}}^f$ we only admit Λ-isomorphisms as morphisms.

If $M \in {}_\Lambda\underline{\underline{M}}^o$ is fixed, then the set $\underline{\underline{E}}_M = \{N \in {}_\Lambda\underline{\underline{M}}^o$: N is isomorphic to a direct summand of $M^{(s)}$ for some positive integer s$\}$ is a subcategory of ${}_\Lambda\underline{\underline{M}}^f$ closed under isomorphisms; which satisfies (1.1, ii,iii), and we write $\underline{\underline{D}}_M = \underline{\underline{K}}_o(\underline{\underline{E}})$.

1.10 Definition (Bass' version of the Whitehead group, Bass [5]): Let S be a ring and $\underline{\underline{C}}$ an admissible subcategory of ${}_S\underline{\underline{M}}^f$. We consider a new category $\Sigma\underline{\underline{C}}$ whose objects are pairs (C,α), where $C \in ob(\underline{\underline{C}})$ and $\alpha \in End_S(C)$ is an automorphism of C. A morphism in $\Sigma\underline{\underline{C}}$, $\varphi : (C,\alpha) \longrightarrow (D,\beta)$ is an element $\varphi \in Hom_S(C,D)$ such that

$$\begin{array}{ccc} C & \xrightarrow{\varphi} & D \\ \alpha\downarrow & & \downarrow\beta \\ C & \xrightarrow{\varphi} & D \end{array}$$

is a commutative diagram. We say that a sequence of objects and morphisms in $\Sigma\underline{\underline{C}}$,

$$0 \longrightarrow (C,\alpha) \xrightarrow{\varphi} (D,\beta) \xrightarrow{\psi} (E,\gamma) \longrightarrow 0$$

is exact if we have the commutative diagram with exact rows

$$\begin{array}{ccccccccc} 0 & \longrightarrow & C & \xrightarrow{\varphi} & D & \xrightarrow{\psi} & E & \longrightarrow & 0 \\ & & \alpha\downarrow & & \beta\downarrow & & \gamma\downarrow & & \\ 0 & \longrightarrow & C & \xrightarrow{\varphi} & D & \xrightarrow{\psi} & E & \longrightarrow & 0, \end{array}$$

where the vertical maps are automorphisms. Then $\Sigma\underline{\underline{C}}$ is an admissible subcategory of the abelian category $\Sigma_S\underline{\underline{M}}^f$ (cf. II, Ex. 1,4); observe that $_S\underline{\underline{M}}^f$ is an abelian category, and that one can define equally well Grothendieck groups for admissible subcategories of abelian categories (cf. Ex. 1,1). Now, we define the <u>Whiteheadgroup of</u> $\underline{\underline{C}}$, $\underline{\underline{G}}_1(\underline{\underline{C}})$, as the abelian group generated by symbols $[C,\alpha]$ for $(C,\alpha) \in ob\,\Sigma\underline{\underline{C}}$, where α runs through all S-automorphisms of C and C runs through all objects in $\underline{\underline{C}}$ subject to the relations

(i) $[D,\beta] = [C,\alpha] + [E,\gamma]$ if
$0 \longrightarrow (C,\alpha) \longrightarrow (D,\beta) \longrightarrow (E,\gamma) \longrightarrow 0$ is an exact sequence in $\Sigma\underline{\underline{C}}$,

(ii) $[C,\alpha\beta] = [C,\alpha] + [C,\beta]$ where α and β are automorphisms of C.

It is clear that $\underline{\underline{G}}_1(\underline{\underline{C}})$ is universal with these properties.

1.11 <u>Lemma</u>: Every element in $\underline{\underline{G}}_1(\underline{\underline{C}})$ is of the form $[C,\alpha]$ for an automorphism α of $C \in ob(\underline{\underline{C}})$; moreover $[C,1_C] = 0$ and $[C,\alpha^{-1}] = -[C,\alpha]$.

<u>Proof</u>: Every element in $\underline{\underline{G}}_1(\underline{\underline{C}})$ has the form $x = [C,\alpha] - [D,\beta]$, but because of (1.10,ii), $[D,\beta] + [D,\beta^{-1}] = [D,1_D]$.
Since $[D,\beta'] + [D,1_D] = [D,\beta']$ we conclude $[D,1_D] = 0$, and $[D,\beta] = -[D,\beta^{-1}]$. Hence $x = [C,\alpha] + [D,\beta^{-1}] = [C \oplus D, \alpha \oplus \beta^{-1}]$. #

1.12 Theorem: Let S be a ring and $M \in {}_{\underline{\underline{S}}}\underline{\underline{M}}^f$ a progenerator, and put $T = \mathrm{End}_S(M)$. Then we have a Morita equivalence

$$h : {}_{\underline{\underline{S}}}\underline{\underline{M}}^f \longrightarrow {}_{\underline{\underline{T}}}\underline{\underline{M}}^f .$$

Let $\underline{\underline{C}}$ be an admissible subcategory of ${}_{\underline{\underline{S}}}\underline{\underline{M}}^f$. Then $h(\underline{\underline{C}})$ is an admissible subcategory of ${}_{\underline{\underline{T}}}\underline{\underline{M}}^f$, and we have isomorphisms of abelian groups

$$\underline{\underline{G}}_0(\underline{\underline{C}}) \cong \underline{\underline{G}}_0(h(\underline{\underline{C}})),$$

$$\underline{\underline{K}}_0(\underline{\underline{C}}) \cong \underline{\underline{K}}_0(h(\underline{\underline{C}})),$$

$$\underline{\underline{G}}_1(\underline{\underline{C}}) \cong \underline{\underline{G}}_1(h(\underline{\underline{C}})).$$

The proof is straightforward and is left as an exercise. #

Exercises §1:

1.) Let $\underline{\underline{A}}$ be an abelian category. If $\underline{\underline{C}}$ is an admissible subcategory, we can define $\underline{\underline{G}}_0(\underline{\underline{C}})$ and $\underline{\underline{G}}_1(\underline{\underline{C}})$. Show that $\Sigma\underline{\underline{C}}$ is an admissible subcategory of the abelian category $\Sigma\underline{\underline{A}}$.

2.) Let $\underline{\underline{A}}$ be an abelian category which has countable direct sums. Show $\underline{\underline{G}}_0(\underline{\underline{A}}) = 0$. In particular, if S is a ring, then $\underline{\underline{G}}_0({}_{\underline{\underline{S}}}\underline{\underline{M}}) = 0$, where ${}_{\underline{\underline{S}}}\underline{\underline{M}}$ is the category of all left S-modules. Find a subcategory $\underline{\underline{C}}$ of ${}_{\underline{\underline{S}}}\underline{\underline{M}}$ such that $\underline{\underline{G}}_0(\underline{\underline{C}}) \neq 0$.

3.) Let $\underline{\underline{Ab}}$ be the abelian category of finitely generated abelian groups. Show $\underline{\underline{G}}_0(\underline{\underline{Ab}}) \cong \underline{\underline{Z}}$. (Hint: $A \in \mathrm{ob}(\underline{\underline{Ab}}) \Longrightarrow A \cong \underline{\underline{Z}}^{(n)} \oplus \bigoplus_{i=1}^{t} \underline{\underline{Z}}/d_i\underline{\underline{Z}}$. Show that $[\underline{\underline{Z}}/d\underline{\underline{Z}}] = 0$ in $\underline{\underline{G}}_0(\underline{\underline{Ab}})$, and then show that $\underline{\underline{G}}_0(\underline{\underline{Ab}}) \longrightarrow \underline{\underline{Z}}, A \longmapsto n$ is an isomorphism of abelian groups.)

4.) Let $\underline{\underline{Ab}}^f$ be the abelian category of finite abelian groups. Show that $\underline{\underline{G}}_0(\underline{\underline{Ab}}^f)$ is free abelian with basis $\{[\underline{\underline{Z}}/p\underline{\underline{Z}}]\}$, p running over the rational primes.

5.) Prove 1.12.

6.) Let $\underline{\underline{C}}$ be an admissible subcategory of the abelian category $\underline{\underline{A}}$.

a.) We shall sketch a different proof of (1.5). Write $C \approx D$ for $C, D \in \mathrm{ob}(\underline{\underline{C}})$, whenever there exist exact sequences in $\underline{\underline{C}}$

$$0 \longrightarrow X \longrightarrow C \longrightarrow Y \longrightarrow 0 \text{ and } 0 \longrightarrow X \longrightarrow D \longrightarrow Y \longrightarrow 0.$$

Then $C \sim D$ if $C \oplus Z \approx D \oplus Z$ for some $Z \in ob(\underline{\underline{C}})$ is an equivalence relation on $ob(\underline{\underline{C}})$, compatible with direct sums; i.e., $C \sim D$ implies $C \oplus Z \sim D \oplus Z$ for every $Z \in ob(\underline{\underline{C}})$. Thus the set of equivalence classes $|C| = \{D \in ob(\underline{\underline{C}}) : D \sim C\}$ is an abelian semi-group S under the operation "+" defined by $|C| + |D| = |C \oplus D|$. Moreover, the cancellation law holds in S, since $|C| + |D| = |C'| + |D|$ implies $|C| = |C'|$. Therefore S can be embedded in a "universal" abelian group G (cf. the construction 1.2), and the map $ob(\underline{\underline{C}}) \longrightarrow G$, $C \longmapsto |C|$ is such that

$$|C| = |X| + |Y| \text{ if } 0 \longrightarrow X \longrightarrow C \longrightarrow Y \longrightarrow 0$$

is an exact sequence, since $C \sim X \oplus Y$. The universality of $\underline{\underline{G}}_0(\underline{\underline{C}})$ implies the existence of a homomorphism of abelian groups

$$\underline{\underline{G}}_0(\underline{\underline{C}}) \longrightarrow G, \ [C] \longmapsto |C|,$$

and hence $[C_1] = [C_2]$ implies $|C_1| = |C_2|$; i.e., the statement of (1.5). Together with the first part of (1.5) one obtains an isomorphism $\underline{\underline{G}}_0(\underline{\underline{C}}) \cong G$.

b.) Use a similar technique to prove (1.7). ($E_1 \sim E_2$ if $E_1 \oplus X \cong$ $\cong E_2 \oplus X$ for some $X \in ob(\underline{\underline{E}})$.)

c.) For $C \in ob(\underline{\underline{C}})$ write $Com(C)$ for the commutator subgroup of $Aut_{\underline{\underline{C}}}(C)$, the group of $\underline{\underline{C}}$-automorphisms of C. As in (1.10, Ex. 1,1) we form the category $\Sigma\underline{\underline{C}}$, and define $(C,\alpha) \approx (D,\beta)$ for $(C,\alpha),(D,\beta)$ $\in ob(\Sigma\underline{\underline{C}})$, if there exist two exact sequences in $\Sigma\underline{\underline{C}}$

$$0 \longrightarrow (X,\varphi) \longrightarrow (C,\alpha) \longrightarrow (Y,\psi) \longrightarrow 0$$

and

$$0 \longrightarrow (X,\varphi) \longrightarrow (D,\beta) \longrightarrow (Y,\psi) \longrightarrow 0.$$

We now define the equivalence relation $(C,\alpha) \sim (D,\beta)$ as follows: There exist $E,E' \in ob(\underline{\underline{C}})$, $\gamma \in Com(C \oplus E)$, $\delta \in Com(D \oplus E')$ such that $(C \oplus E, (\alpha \oplus 1_E)\gamma) \approx (D \oplus E', (\beta \oplus 1_{E'})\delta)$. Show that $[C,\alpha] =$ $= [D,\beta]$ in $\underline{\underline{G}}_1(\underline{\underline{C}})$ if and only if $(C,\alpha) \sim (D,\beta)$. (<u>Hint</u>: Use a similar technique as in (a). Note that

$$(\alpha^{-1} \oplus \alpha) = (\alpha^{-1} \oplus 1)(\iota_2\pi_1 + \iota_1\pi_2)(\alpha \oplus 1)(\iota_2\pi_1 + \iota_1\pi_2)$$

is a commutator, and that $(\alpha \oplus \beta) = (\alpha\beta \oplus 1)(\beta^{-1} \oplus \beta)$.)

§2 The Whitehead group of a ring

The Whitehead group $\underline{\underline{K}}_1(S) = GL(S)/E(S)$ is defined; for a semi-primary ring S we have an epimorphism $GL(1,S) \longrightarrow \underline{\underline{K}}_1(S)$. Moreover, $\underline{\underline{K}}_1(S) \cong \underline{\underline{G}}_1({}_{S}\underline{\underline{F}}) \cong \underline{\underline{G}}_1({}_{S}\underline{\underline{P}}^f)$. The Whitehead group of a simple algebra is computed via the Dieudonné-determinant.

2.1 Definitions: Let S be a ring, GL(n,S) is the group of invertible $(n \times n)$-matrices with entries in S, the general linear group. We embed GL(n,S) into GL(n+1,S)

$$\iota_{n,n+1} : (s_{ij})^{n\times n} \longmapsto \begin{vmatrix} (s_{ij}) & 0 \\ 0 & 1 \end{vmatrix}^{(n+1)\times(n+1)}$$

This way GL(n,S) becomes a subgroup of GL(n+1,S), and we define

$$GL(S) = \varinjlim_{n} GL(n,S)$$

to be the injective limit of $\{GL(n,S), \iota_{n,n+1}\}_{n\in\underline{\underline{N}}}$ (cf. I, Ex. 9,3). Let $\underline{\underline{E}}_{ij}$ be the matrix with 1 at the (i,j)-position and zeros elsewhere. An elementary matrix $\underline{\underline{M}}$ is of the form $\underline{\underline{M}} = \underline{\underline{E}} + s\underline{\underline{E}}_{ij}$, $i \neq j, s \in S$, where $\underline{\underline{E}}$ is the identity matrix of the proper size. Then $\underline{\underline{M}} \in GL(S)$, since $(\underline{\underline{E}} + s\underline{\underline{E}}_{ij})(\underline{\underline{E}} - s\underline{\underline{E}}_{ij}) = \underline{\underline{E}}$.

E(S) and E(n,S) are the subgroups of GL(S) and GL(n,S) resp. generated by the elementary matrices. By [GL(S), GL(S)] we denote the commutator subgroup of GL(S); i.e., the subgroup generated by all commutators $[A,B] = ABA^{-1}B^{-1}$, $A,B \in GL(S)$.

2.2 Theorem (Whitehead): [GL(S), GL(S)] = E(S).

Proof: We have $\underline{\underline{E}}_{ij}\,\underline{\underline{E}}_{kl} = \begin{cases} 0 \text{ if } j\neq k \\ \underline{\underline{E}}_{il} \text{ if } j=k \end{cases}$.

(i) $E(S) \subset [GL(S), GL(S)]$.

If i,j,k are distinct, then

$$[(\underline{\underline{E}} + s_1\underline{\underline{E}}_{ik}),(\underline{\underline{E}} + s_2\underline{\underline{E}}_{kj})] = (\underline{\underline{E}} + s_1s_2\underline{\underline{E}}_{ij}),$$

as is easily seen. Thus for $n \geq 3$, we have

$$E(n,S) \subset [GL(n,S), GL(n,S)].$$

Hence $E(S) \subset [GL(S), GL(S)]$.

(ii) $E(S) \supset [GL(S), GL(S)]$.

To show this, we first demonstrate the following assertion:

2.3 __Lemma__: Let $\underline{\underline{A}} \in GL(n,S)$. Then

$$\begin{pmatrix} \underline{\underline{A}} & 0 \\ 0 & \underline{\underline{A}}^{-1} \end{pmatrix} \in E(2n,S).$$

__Proof__: First of all we observe that, if $\underline{\underline{X}}$ is any $(n \times n)$-matrix over S and $\underline{\underline{E}}$ is the $(n \times n)$-identity matrix, then

$$\begin{pmatrix} \underline{\underline{E}} & \underline{\underline{X}} \\ 0 & \underline{\underline{E}} \end{pmatrix} \in E(2n,S) \text{ and } \begin{pmatrix} \underline{\underline{E}} & 0 \\ \underline{\underline{X}} & \underline{\underline{E}} \end{pmatrix} \in E(2n,S).$$

In fact, if

$$\underline{\underline{X}} = \sum_{i,j=1}^{n} x_{ij}\underline{\underline{E}}_{ij},$$

then

$$\begin{pmatrix} \underline{\underline{E}} & \underline{\underline{X}} \\ 0 & \underline{\underline{E}} \end{pmatrix} = \prod_{i,j=1}^{n} \left[\begin{pmatrix} \underline{\underline{E}} & 0 \\ 0 & \underline{\underline{E}} \end{pmatrix} + \begin{pmatrix} 0 & x_{ij}\underline{\underline{E}}_{ij} \\ 0 & 0 \end{pmatrix}\right].$$

Similarly for $\begin{pmatrix} \underline{\underline{E}} & 0 \\ \underline{\underline{X}} & \underline{\underline{E}} \end{pmatrix}$.

Moreover, using block multiplication, it is easily verifyed, that

$$\begin{pmatrix} 0 & \underline{\underline{E}} \\ -\underline{\underline{E}} & 0 \end{pmatrix} = \begin{pmatrix} \underline{\underline{E}} & \underline{\underline{E}} \\ 0 & \underline{\underline{E}} \end{pmatrix}\begin{pmatrix} \underline{\underline{E}} & 0 \\ -\underline{\underline{E}} & \underline{\underline{E}} \end{pmatrix}\begin{pmatrix} \underline{\underline{E}} & \underline{\underline{E}} \\ 0 & \underline{\underline{E}} \end{pmatrix} \in E(2n,S)$$

and

$$\begin{pmatrix} \underline{\underline{A}} & 0 \\ 0 & \underline{\underline{A}}^{-1} \end{pmatrix} = \begin{pmatrix} \underline{\underline{E}} & -\underline{\underline{A}} \\ 0 & \underline{\underline{E}} \end{pmatrix}\begin{pmatrix} \underline{\underline{E}} & 0 \\ \underline{\underline{A}}^{-1} & \underline{\underline{E}} \end{pmatrix}\begin{pmatrix} \underline{\underline{E}} & -\underline{\underline{A}} \\ 0 & \underline{\underline{E}} \end{pmatrix}\begin{pmatrix} 0 & \underline{\underline{E}} \\ -\underline{\underline{E}} & 0 \end{pmatrix} \in E(2n,S)$$

and hence the desired result is established. #

We now continue with the <u>proof of (2.2.11)</u>. Given $\underline{\underline{A}},\underline{\underline{B}} \in GL(S)$; we may assume that $\underline{\underline{A}}$ and $\underline{\underline{B}}$ have the same size, say $(n \times n)$. Then

$$\begin{pmatrix} \underline{\underline{ABA}}^{-1}\underline{\underline{B}}^{-1} & 0 \\ 0 & \underline{\underline{E}} \end{pmatrix} = \begin{pmatrix} \underline{\underline{AB}} & 0 \\ 0 & (\underline{\underline{AB}})^{-1} \end{pmatrix} \begin{pmatrix} \underline{\underline{A}}^{-1} & 0 \\ 0 & \underline{\underline{A}} \end{pmatrix} \begin{pmatrix} \underline{\underline{B}}^{-1} & 0 \\ 0 & \underline{\underline{B}} \end{pmatrix},$$

and the right hand side of this equation lies in $E(2n,S)$, by (2.3). Thus $\underline{\underline{A}},\underline{\underline{B}} \in E(S)$. Since $[GL(S), GL(S)]$ is generated by $\{[\underline{\underline{A}},\underline{\underline{B}}]\}$ we conclude $[GL(S), GL(S)] \subset E(S)$. #

2.4 <u>Definition</u>: $\underline{\underline{K}}_1(S) = GL(S)/E(S)$ is called the <u>Whitehead group of S</u>.

We <u>remark</u> that in view of (2.2), $\underline{\underline{K}}_1(S)$ is the largest abelian factor group of $GL(S)$.

<u>Remark</u>: Algebraic K-theory originates from topological K-theory, which was introduced and developed by Atiyah and Hirzebruch. Topological K-thoery deals with Grothendieck groups of the category of vector-bundles $\underline{\underline{B}}(X)$ over a space X, and $\underline{\underline{K}}_o(\underline{\underline{B}}(X))$ is the Grothendieck group of $\underline{\underline{B}}(X)$ as a category with coproducts. However, in topology, one can define higher Grothendieck groups, $\underline{\underline{K}}_1(\underline{\underline{B}}(X)) = \underline{\underline{K}}_o(\underline{\underline{B}}(S^1(X)))$, where S is the suspension on X. The possibility to translate topological statements into algebraic ones originates from a theorem of Serre-Swan: If $\underline{\underline{B}}(X)$ are K-vectorbundles and if X is compact, then $\underline{\underline{B}}(X) \cong {}_{K(X)}\underline{\underline{P}}^f$, where K(X) is the ring of continuous functions from X to K. Moreover, X is homeomorphic to specK(X), the prime spectrum of K(X). Hence $\underline{\underline{K}}_o({}_{K(X)}\underline{\underline{P}}^f) \cong \underline{\underline{K}}_o(\underline{\underline{B}}(X))$. However, there is no algebraic analogue to the suspension; but the topological K-theory shows that a proper algebraic analogue for $\underline{\underline{K}}_1(B(X))$ is $GL(K(X))/E(K(X)) = \underline{\underline{K}}_1(K(X))$; this group had already been introduced earlier, in 1951 by Whitehead in his studies of simple homotopy types of $\underline{\underline{Z}}G$, the group ring of a finite group G. Recently there have been introduced definitions for $\underline{\underline{K}}_2(R)$ (Milnor) and $K_1(R)$ (Villamayor-Nobile) where R is a ring. (For more details on these facts we refer to Bass [8], Swan [7].)

2.5 <u>Theorem</u> (Bass [5]): If S is a semi-primary ring, then we have an epimorphism of groups

$$GL(1,S) \longrightarrow \underline{\underline{K}}_1(S).$$

(It should be noted that GL(1,S) is the group of units in S.)

<u>Proof</u>: We recall that S is semi-primary if S/radS is left noetherian and left artinian (cf. III,(7.5)). The proof of (2.5) uses the following statement.

2.6 <u>Lemma</u>: If S is a semi-primary ring and if I is a left ideal in S and if $s_o \in S$ is such that $Ss_o + I = S$, then there exists $\alpha \in I$ such that $\alpha + s_o$ is a unit in S.

<u>Proof</u>: (i) <u>Reduction to the case of an artinian ring</u>. Let $\bar{S} = S/\mathrm{rad}S$, $\bar{I} = (I + \mathrm{rad}S)/\mathrm{rad}S$ and $\bar{s}_o = s_o + \mathrm{rad}S$. Then $\bar{S}\bar{s}_o + \bar{I} = \bar{S}$. Assume there exists $\bar{\alpha} \in \bar{I}$ such that $\bar{\alpha} + \bar{s}_o$ is a unit in $\bar{S}$. We lift $\bar{\alpha}$ to α in S. Then $S = S(\alpha + s_o) + \mathrm{rad}S$ and by Nakayama's lemma $\alpha + s_o$ is a unit in S.

(ii) We may thus assume that S is semi-simple, artinian and noetherian. By (III, 5.3) there exists a left ideal $I' \subset I$ such that $S = I' \oplus Ss_o$. If $\varphi : S \longrightarrow Ss_o$ is right multiplication by s_o, then the exact sequence

$$0 \longrightarrow \mathrm{Ker}\,\varphi \longrightarrow S \xrightarrow{\varphi} Ss_o \longrightarrow 0$$

splits. Let $\psi : S \longrightarrow \mathrm{Ker}\,\varphi$ be the projection onto $\mathrm{Ker}\,\varphi$. Because of the Krull-Schmidt theorem (I, 4.10) there is an S-isomorphism $\vartheta : \mathrm{Ker}\,\varphi \longrightarrow I'$. Since

$$(\psi,\varphi) : S \longrightarrow \mathrm{Ker}\,\varphi \oplus Ss_o,$$
$$s \longmapsto (s\psi,\ ss_o)$$

is an epimorphism, the composite

$$(\psi,\varphi)(\vartheta \oplus 1_{Ss_o})(1_{I'} + 1_{Ss_o})$$

is an automorphism of S; whence it is given by right multiplication with a unit $u_o \in S$; i.e., $s\psi\vartheta + ss_o = su_o$ for every $s \in S$. In particular $1\psi\vartheta + s_o = u_o$. Obviously $\alpha = 1\psi\vartheta \in I' \subset I$, and $\alpha + s_o$ is a unit in S. #

We now come to the <u>proof of (2.5)</u>.

Let $\underline{\underline{A}} \in GL(n,S)$, $\underline{\underline{A}} = (a_{ij})$, $\underline{\underline{A}}^{-1} = (a^*_{ij})$. Then $\underline{\underline{A}}^{-1}\underline{\underline{A}} = \underline{\underline{E}}$ implies $\sum_{i=1}^{n} a^*_{1i}\, a_{i1} = 1$. Hence

$$Sa_{11} + \sum_{i=2}^{n} Sa_{i1} = S.$$

By (2.6) there exists a unit $u \in S$ such that $u = a_{11} + \sum_{i=2}^{n} s_i\, a_{i1}$.

Hence

$$\begin{pmatrix} 1 & s_2 & \cdots & s_n \\ 0 & 1 & & 0 \\ & & \ddots & \\ 0 & & & 1 \end{pmatrix} A = \left(\begin{array}{c|c} u & * \\ \hline * & * \end{array}\right).$$

Since the first matrix is in $E(n,S)$ (cf. proof of (2.3)),

$$\underline{\underline{A}} \equiv \begin{vmatrix} u & * \\ * & * \end{vmatrix} \mod E(n,S).$$

By elementary transformations; i.e., multiplication from the left and right by elementary matrices we obtain (observe that u is a unit)

$$\underline{\underline{A}} \equiv \left|\begin{array}{c|c} u & 0 \\ \hline 0 & * \end{array}\right| \mod E(n,S).$$

Since $(*) \in GL(n-1,S)$ we may use induction on n to conclude

$$\underline{\underline{A}} \equiv \begin{pmatrix} u_1 & & 0 \\ & \ddots & \\ 0 & & u_n \end{pmatrix} \qquad \mod E(n,S),$$

where u_i, $1 \leq i \leq n$, are units in S. Because of our embedding $GL(n,S) \hookrightarrow GL(n+1,S)$ and from (2.3) it follows that all matrices

$$\begin{pmatrix} u_1 & & & & & & & 0 \\ & 1 & & & & & & \\ & & \ddots & & & & & \\ & & & 1 & & & & \\ & & & & u_1^{-1} & & & \\ & & & & & 1 & & \\ & & & & & & \ddots & \\ 0 & & & & & & & 1 \end{pmatrix} \quad n \times n$$

lie in $E(S)$.

Thus

$$\begin{pmatrix} u_1 & & & 0 \\ & \cdot & & \\ & & \cdot & \\ 0 & & & u_n \end{pmatrix} \begin{pmatrix} u_n & & & 0 \\ & 1 & & \\ & & 1 & \\ 0 & & & u_n^{-1} \end{pmatrix} = \begin{pmatrix} u_1u_n & & & & 0 \\ & u_2 & & & \\ & & \cdot & & \\ 0 & & & u_{n-1} & \\ & & & & 1 \end{pmatrix},$$

and by induction on n we conclude

$$\begin{pmatrix} u_1 & & 0 \\ & \cdot & \\ 0 & & u_n \end{pmatrix} \equiv \begin{pmatrix} u_o & & & 0 \\ & 1 & & \\ & & \cdot & \\ 0 & & & 1 \end{pmatrix} \qquad \mathrm{mod}\, E(S),$$

and the map

$$GL(1,S) \longrightarrow \underline{\underline{K}}_1(S)$$

$$u \longmapsto \begin{pmatrix} u & & & 0 \\ & 1 & & \\ & & \cdot & \\ 0 & & & 1 \end{pmatrix} + E(S)$$

is an epimorphism of groups. #

2.7 __Lemma:__ Let S be a ring and ${}_S\underline{\underline{F}}$ the category of free left S-modules on a finite basis. Then ${}_S\underline{\underline{F}}$ is an admissible subcategory of ${}_S\underline{\underline{M}}^f$, and we have an isomorphism of abelian groups

$$\Phi : \underline{\underline{K}}_1(S) \longrightarrow \underline{\underline{G}}_1({}_S\underline{\underline{F}}).$$

__Proof:__ We can form the category $\Sigma_S\underline{\underline{F}}$ (cf. 1.10, Ex. 1,1) and we define a map

$$\Phi'_n : GL(n,S) \longrightarrow \underline{\underline{G}}_1({}_S\underline{\underline{F}}),$$

$$\underline{\underline{A}} \longmapsto [{}_SS^{(n)}, \alpha] ;$$

considering the elements in ${}_SS^{(n)}$ as row vectors, we may view $\underline{\underline{A}}$ as matrix of an S-isomorphism α of ${}_SS^{(n)}$; $\underline{\underline{A}}$ is invertible.

But Φ'_n commutes with the embedding

$$\iota_{n,n+1} : GL(n,S) \longrightarrow GL(n+1,S); \text{ i.e.,}$$

$\Phi'_n = \iota_{n,n+1}\Phi'_{n+1}$. In fact,

$$\Phi'_n : \underline{\underline{A}} \longmapsto [{}_S S^{(n)}, \alpha] \quad \text{and}$$

$$\iota_{n,n+1}\Phi'_{n+1} : \underline{\underline{A}} \longmapsto [{}_S S^{(n+1)}, \alpha \oplus 1] \quad .$$

But in $\underline{\underline{G}}_1({}_S\underline{\underline{F}})$ we have

$$[{}_S S^{(n+1)}, \alpha \oplus 1_S] = [{}_S S^{(n)}, \alpha] + [{}_S S, 1_S] = [{}_S S^{(n)}, \alpha]$$

(cf. 1.11).

Thus the set $\{\Phi'_n\}_{n \in \underline{\underline{N}}}$ induces a map

$$\Phi' : GL(S) = \varinjlim GL(n,S) \longrightarrow \underline{\underline{G}}_1({}_S\underline{\underline{F}}),$$

which is a group homomorphism, each Φ'_n being one. Moreover, since $\underline{\underline{G}}_1({}_S\underline{\underline{F}})$ is commutative, $[GL(S), GL(S)] \subset \operatorname{Ker} \Phi'$. Using (2.2), we conclude that Φ' induces a group homomorphism

$$\Phi : \underline{\underline{K}}_1(S) \longrightarrow \underline{\underline{G}}_1({}_S\underline{\underline{F}}).$$

To show that Φ is an isomorphism, we construct its inverse:

$$\Psi : \underline{\underline{G}}_1({}_S\underline{\underline{F}}) \longrightarrow \underline{\underline{K}}_1(S).$$

Given $[F, \alpha] \in \underline{\underline{G}}_1({}_S\underline{\underline{F}})$, we pick a basis $f_1,\ldots,f_n$ of F, and relative to this basis, α is represented as a matrix $\underline{\underline{A}} \in GL(n,S)$. Let $(\underline{\underline{A}})$ be the image of $\underline{\underline{A}}$ in $\underline{\underline{K}}_1(S)$. Then we define

$$\Psi : [F,\alpha] \longmapsto (\underline{\underline{A}}).$$

Observe that every element in $\underline{\underline{G}}_1({}_S\underline{\underline{F}})$ has the form $[F,\alpha]$ (cf. 1.11). We have to show that Ψ is well-defined. We first show that $(\underline{\underline{A}})$ is independent of the chosen basis. Let $f'_1,\ldots,f'_m$ be another S-basis of F (observe that not necessarily $m = n$), and let α be represented by $\underline{\underline{A}}'$ relative to the basis $f'_1,\ldots,f'_m$. We have the commutative diagram

$$D : \quad \begin{array}{ccc} F \oplus F & \xrightarrow{\tau} & F \oplus F \\ \Big\downarrow 1_F \oplus \alpha & & \Big\downarrow \alpha \oplus 1_F \\ F \oplus F & \xrightarrow{\tau} & F \oplus F \end{array} \quad ,$$

where τ is the transposition: $(a,b) \longmapsto (b,a)$; then τ is an isomorphism in $\Sigma_S\underline{\underline{F}}$. In the diagram D we represent α on the first summand with respect to the basis $\{f_i\}_{1 \le i \le n}$ and on the second summand with respect to the basis $\{f'_i\}_{1 \le i \le m}$. Then $1_F \oplus \alpha$ and $\alpha \oplus 1_F$ are represented by the invertible $(n+m) \times (n+m)$ matrices.

$$\begin{pmatrix} \underline{\underline{E}}_n & 0 \\ 0 & \underline{\underline{A}}' \end{pmatrix} \quad \text{and} \quad \begin{pmatrix} \underline{\underline{A}} & 0 \\ 0 & \underline{\underline{E}}_m \end{pmatrix} \quad \text{resp.}$$

Because of the commutativity of the diagram D and since τ is invertible, there exists $\underline{\underline{C}} \in GL(n+m,S)$ such that

$$\underline{\underline{C}}^{-1} \begin{pmatrix} \underline{\underline{E}}_n & 0 \\ 0 & \underline{\underline{A}}' \end{pmatrix} \underline{\underline{C}} = \begin{pmatrix} \underline{\underline{A}} & 0 \\ 0 & \underline{\underline{E}}_m \end{pmatrix} .$$

But $\underline{\underline{K}}_1(S)$ is commutative and thus

$$(\underline{\underline{A}}) = (\underline{\underline{A}}') \text{ in } \underline{\underline{K}}_1(S),$$

and Ψ is independent of the chosen basis of F.

Now $\Phi\Psi : [F, \alpha] \overset{\Psi}{\longmapsto} (\underline{\underline{A}}) \overset{\Phi}{\longmapsto} [F, \alpha]$,

and $\Psi\Phi : (\underline{\underline{A}}) \overset{\Phi}{\longmapsto} [{}_S S^{(n)}, \alpha] \overset{\Psi}{\longmapsto} (\underline{\underline{A}})$.

Thus $\Phi\Psi = 1_{\underline{\underline{G}}_1({}_{\underline{\underline{S}}}\underline{\underline{F}})}$ and $\Psi\Phi = 1_{\underline{\underline{K}}_1(S)}$; by (Ex. 2,4) Ψ is a well-defined group homomorphism and Φ is an isomorphism. #

2.8 __Theorem__: Let ${}_{\underline{\underline{S}}}\underline{\underline{F}}$ be the category of finitely generated free left S-modules, and ${}_{\underline{\underline{S}}}\underline{\underline{P}}^f$ the category of finitely generated projective left S-modules. Then ${}_{\underline{\underline{S}}}\underline{\underline{P}}^f$ and ${}_{\underline{\underline{S}}}\underline{\underline{F}}$ are admissible subcategories of ${}_{\underline{\underline{S}}}\underline{\underline{M}}^f$ (cf. 1.1), and we have an isomorphism of abelian groups:

we have an isomorphism of abelian groups:

$$\Phi : \underline{\underline{G}}_1({}_{\underline{\underline{S}}}\underline{\underline{F}}) \cong \underline{\underline{G}}_1({}_{\underline{\underline{S}}}\underline{\underline{P}}^f).$$

__Proof__: We define

$$\Phi : \underline{\underline{G}}_1({}_{\underline{\underline{S}}}\underline{\underline{F}}) \longrightarrow \underline{\underline{G}}_1({}_{\underline{\underline{S}}}\underline{\underline{P}}^f),$$

$$[F, \alpha]_F \longmapsto [F, \alpha]_P.$$

Then Φ is a group homomorphism, and to show that it is an isomorphism, we construct an inverse

$$\Psi : \underline{\underline{G}}_1({}_{\underline{\underline{S}}}\underline{\underline{P}}^f) \longrightarrow \underline{\underline{G}}_1({}_{\underline{\underline{S}}}\underline{\underline{F}}),$$

$$[P, \alpha]_P \longmapsto [P \oplus Q, \alpha \oplus 1_Q]_F,$$

where $Q \in {}_{\underline{\underline{S}}}\underline{\underline{P}}^f$ is such that $P \oplus Q$ is S-free on a finite number of basis elements. Observe that every element in $\underline{\underline{G}}_1({}_S\underline{\underline{P}}^f)$ has the form $[P, \alpha]$ (cf. 1.11). If also $P \oplus Q'$ is S-free, then

$$[P \oplus Q \oplus P \oplus Q', \ \alpha \oplus 1_Q \oplus 1_P \oplus 1_{Q'}]$$
$$= [P \oplus Q \oplus P \oplus Q', \ 1_P \oplus 1_Q \oplus \alpha \oplus 1_{Q'}] ,$$

since we have the following commutative diagram

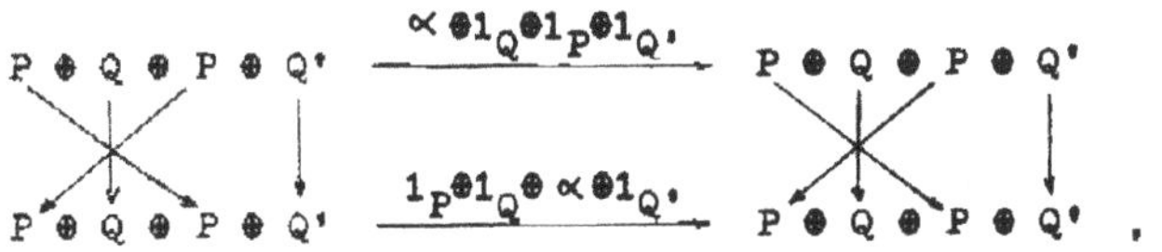

where the vertical maps are isomorphisms and the arrows indicate the corresponding permutations. But then

$$[P \oplus Q, \alpha \oplus 1_Q]_F + 0 = [P \oplus Q', \alpha \oplus 1_{Q'}]_F + 0$$

(cf. 1.11). Thus Ψ is well-defined. To show that Φ is an isomorphism it suffices to prove $\Phi\Psi = 1_{\underline{\underline{G}}_1({}_S\underline{\underline{P}}^f)}$ and $\Psi\Phi = 1_{\underline{\underline{G}}_1({}_S\underline{\underline{F}})}$, since Φ is a group homomorphism (cf. Ex. 2,4). But

$$\Phi\Psi : [P,\alpha]_P \overset{\Psi}{\longmapsto} [P \oplus Q, \alpha \oplus 1]_F \overset{\Phi}{\longmapsto} [P \oplus Q, \alpha \oplus 1]_P = [P,\alpha]_P$$

and

$$\Psi\Phi : [F,\alpha]_F \overset{\Phi}{\longmapsto} [F,\alpha]_P \overset{\Psi}{\longmapsto} [F,\alpha]_F . \qquad \#$$

Combining (2.7) and (2.8) we obtain:

2.9 __Corollary__: There is an isomorphism of abelian groups

$$\underline{\underline{K}}_1(S) \xrightarrow{\sim} \underline{\underline{G}}_1({}_S\underline{\underline{F}}) \xrightarrow{\sim} \underline{\underline{G}}_1({}_S\underline{\underline{P}}^f).$$

2.10 __Theorem__ (Dieudonné [1]): Let D be a skewfield, and denote by $D^* = D \setminus \{0\}$ its multiplicative group and by $[D^*,D^*]$ the commutator subgroup. Then there is an isomorphism

$$\underline{\underline{K}}_1(D) \cong D^*/[D^*,D^*] ;$$

the isomorphism is given via the __Dieudonné determinant__.

__Proof__: Since D is semi-primary, we have an epimorphism

$$\tilde{\sigma} : D^* \longrightarrow \underline{\underline{K}}_1(D),$$
$$d \longmapsto \begin{pmatrix} d & & & 0 \\ & 1 & & \\ & & \ddots & \\ 0 & & & 1 \end{pmatrix} + \underline{\underline{E}}(D)$$

(cf. 2.5), since $D^* = GL(1,D)$. But $\underline{\underline{K}}_1(D)$ is commutative (cf. 2.2, 2.4), and so $\tilde{\sigma}$ induces a homomorphism

$$\sigma : D^*/[D^*,D^*] \longrightarrow \underline{\underline{K}}_1(D)$$

$$d + [D^*,D^*] \longmapsto \begin{pmatrix} d & & & \\ & 1 & & \\ & & \ddots & \\ & & & 1 \end{pmatrix} + \underline{\underline{E}}(D).$$

Dieudonné has developed a determinant theory for matrices in $(D)_n$; i.e., he has constructed a multiplicative function

$$\det : (D)_n \longrightarrow D^*/[D^*,D^*] \cup \{0\}.$$

This means that we adjoin a zero to $D^*/[D^*,D^*]$ with the obvious multiplication. By "$\bar{d}$" we denote the image of $d \in D$ under the map

$$D \longrightarrow D^*/[D^*,D^*] \cup \{0\}.$$

(One word of caution; it may happen that $\overline{-1} = \bar{1}$, e.g. quaternions cf. Ex. 2,2.) The function det satisfies:

(i) Let $\underline{\underline{B}} \in (D)_n$ be obtained from $\underline{\underline{A}} \in (D)_n$ by multiplying one row of $\underline{\underline{A}}$ on the left by $d \in D$. Then

$$\det \underline{\underline{B}} = \bar{d} \det \underline{\underline{A}}.$$

(ii) If $\underline{\underline{B}}$ is obtained from $\underline{\underline{A}}$ by adding one row to another, then

$$\det \underline{\underline{B}} = \det \underline{\underline{A}}.$$

(iii) The unit matrix has determinant $\bar{1}$.

The existence of such a function is established in (Ex. 2,1). We now define a map

$$\tau_n : GL(n,D) \longrightarrow D^*/[D^*,D^*],$$

$$\underline{\underline{A}} \longmapsto \det \underline{\underline{A}}.$$

Since the determinant map is multiplicative (cf. Ex. 2,1), this is a group homomorphism; observe that $\det \underline{\underline{A}} \neq 0$ for all $\underline{\underline{A}} \in GL(n,D)$. Let $\iota_{n,n+1} : GL(n,D) \hookrightarrow GL(n+1,D)$ be the canonical embedding. Then this embedding commutes with the determinant function (cf. Ex. 2,1). Thus τ_n induces a group homomorphism

$$\tau : GL(D) \longrightarrow D^*/[D^*,D^*] .$$

Since $D^*/[D^*,D^*]$ is commutative, we obtain a map

$$\tau : \underline{\underline{K}}_1(D) \longrightarrow D^*/[D^*,D^*] ,$$

$$\underline{\underline{A}} + \underline{\underline{E}}(D) \longmapsto \det \underline{\underline{A}},$$

and

$$\sigma\tau : D^*/[D^*,D^*] \longrightarrow D^*/[D^*,D^*]$$

is the identity, since

$$\det \begin{vmatrix} d & & & \\ & 1 & & \\ & & \ddots & \\ & & & 1 \end{vmatrix} = \bar{d}.$$

Thus σ is monic. Since it is an epimorphism, (cf. 2.5), $\underline{\underline{K}}_1(D) \cong$ $= D^*/[D^*,D^*]$. #

2.11 Corollary: Let A be a simple algebra over a field K. Then $A = (D)_n$, where D is a skewfield over K and

$$\underline{\underline{K}}_1(A) \cong D^*/[D^*,D^*] .$$

The proof is left as an exercise. #

We quote without proof a theorem of Wang [1, Theorem p. 329] which gives a more explicit description of $\underline{\underline{K}}_1(A)$.

2.12 Theorem (Wang [1]): Let K be an algebraic number field and A a central simple K-algebra. By $St_K(A)$ we denote the ray (cf. VI, 5.4); i.e., $St_K(A) = \{ \alpha \in K : 0 \neq \alpha$ is positive at all infinite primes at which A is ramified$\}$. Then

$$\underline{\underline{K}}_1(A) \cong St_K(A).$$

This isomorphism is given via the reduced norm: We have the epimorphism $\sigma : GL(1,A) \longrightarrow \underline{\underline{K}}_1(A)$, and the reduced norm

$$GL(1,A) \longrightarrow St_K(A) ,$$

$$a \longmapsto Nrd_{A/K}(a),$$

and it is shown that $\operatorname{Ker} \sigma = \{a \in A : Nrd_{A/K}(a) = 1\}$. (Observe that $\operatorname{Ker} \sigma = [GL(1,A),GL(1,A)]$.)

Compare this result with (IX, 3.27)!

<u>Exercises §2:</u>

1.) Let D be a skewfield. Show the existence and uniqueness of the determinant function

$$\det : (D)_n \longrightarrow D^*/[D^*,D^*] \cup \{0\}.$$

(<u>Hint</u>: In the proof of (2.5) we have shown that every $\underline{\underline{A}}$ in GL(n,D) can be written as

$$\underline{\underline{A}} = \underline{\underline{C}} \cdot \underline{\underline{\Delta}}(d),$$

where $\underline{\underline{C}} \in E(n,D)$ and $\underline{\underline{\Delta}}(d) = \begin{pmatrix} d & & & & 0 \\ & 1 & & & \\ & & \cdot & & \\ & & & \cdot & \\ 0 & & & & 1 \end{pmatrix}$. From the axioms (i),(ii) and (iii) of the determinant function in the proof of (2.10) derive the following properties:

a.) If one adds to the row A_i of $\underline{\underline{A}}$ a left multiple λA_j of another row, the determinant does not change if $i \neq j$.

b.) If $\underline{\underline{A}}$ is singular, then $\det \underline{\underline{A}} = 0$.

c.) If the rows A_i and A_j $(i \neq j)$ are interchanged, then $\det \underline{\underline{A}}$ is multiplied by $\overline{(-1)}$.

d.) $\det(\underline{\underline{\Delta}}(d)) = \overline{d}$.

e.) If $\underline{\underline{A}}$ is non-singular and $\underline{\underline{A}} = \underline{\underline{C}} \cdot \underline{\underline{\Delta}}(d)$ where $\underline{\underline{C}} \in E(n,D)$, then $\det \underline{\underline{A}} = \overline{d}$.

f.) $\det \underline{\underline{A}}\,\underline{\underline{B}} = \det \underline{\underline{A}} \cdot \det \underline{\underline{B}}$.

h.) $\det \underline{\underline{A}}$ is not changed if a right multiple of a column is added to another.

i.) If a column of $\underline{\underline{A}}$ is multiplied by d on the right, then the determinant is multiplied by $\overline{d}$.

<u>Existence of determinants:</u>

1.) For a (1×1) matrix $\underline{\underline{A}} = (d)$ we put $\det \underline{\underline{A}} = \overline{d}$.

2.) If $\underline{\underline{A}}$ is singular, we put $\det \underline{\underline{A}} = 0$.

3.) If $\underline{\underline{A}}$ is not singular, then we know (cf. proof of 2.6)

$$\underline{\underline{A}} \equiv \left|\begin{array}{c|c} d & 0 \\ \hline 0 & \underline{\underline{B}} \end{array}\right| \mod \underline{\underline{E}}(n,D), \text{ where } 0 \neq d \in D,$$

and $\underline{\underline{B}}$ is an (n-1) × (n-1) matrix, and we define inductively

$$\det \underline{\underline{A}} = \bar{d} \cdot \det \underline{\underline{B}}.$$

Now check that det is a well-defined function

$$GL(n,D) \longrightarrow D^*/[D^*,D^*] \cup \{0\},$$

which satisfies axioms (i),(ii),(iii).

<u>Uniqueness</u>: Show that det satisfies (1,2,3) whenever it satisfies (i,ii,iii).

2.) Let $\underline{\underline{H}}(Q)$ be the quaternion skewfield. Show that $\overline{(-1)} = \bar{1}$ in $\underline{\underline{H}}(Q)^*/[\underline{\underline{H}}(Q)^*,\underline{\underline{H}}(Q)^*]$.

3.) Let A be a simple K-algebra and $A = (D)_n$, where D is a skewfield. Show that $\underline{\underline{K}}_1(A) \cong D^*/[D^*,D^*]$.

4.) Let G_1 and G_2 be groups. If there exists a group homomorphism $\varphi: G_1 \longrightarrow G_2$ and a set map $\psi: G_2 \longrightarrow G_1$ such that $\varphi\psi = 1_{G_1}$ and $\psi\varphi = 1_{G_2}$, show that ψ is a group-isomorphism.

5.) Prove 2.11!

6.) (i) Let $\underline{\underline{C}}$ and $\underline{\underline{C}}'$ be admissible subcategories of an abelian category $\underline{\underline{A}}$, such that $\underline{\underline{C}}'$ is a subcategory of $\underline{\underline{C}}$. Then we have a group homomorphism

$$\varphi: \underline{\underline{G}}_0(\underline{\underline{C}}') \longrightarrow \underline{\underline{G}}_0(\underline{\underline{C}}),$$
$$[C'] \longmapsto [C'].$$

Give necessary and sufficient conditions for φ to be a monomorphism resp. epimorphism.

(ii) Use Ex. 1,6 and (i) to prove (2.7) and (2.8).

§3 Grothendieck groups of orders

The Grothendieck group of all Λ-modules of finite type is the same as the Grothendieck group of all Λ-lattices. Various exact sequences and commutative diagrams are derived.*)

Let R be a Dedekind domain with quotient field K; A a finite dimensional separable K-algebra and Λ an R-order in A. We recall that $\underline{\underline{G}}_o^f(\Lambda) = \underline{\underline{G}}_o({}_\Lambda\underline{\underline{M}}^f)$ is the Grothendieck group of all finitely generated Λ-modules and $\underline{\underline{G}}_o(\Lambda) = \underline{\underline{G}}_o({}_\Lambda\underline{\underline{M}}^o)$ is the Grothendieck group of all Λ-lattices. By

$$\underline{\underline{G}}_o^T(\Lambda)$$

we denote the Grothendieck group of the category of all finitely generated Λ-modules that are R-torsion modules.

3.1 Theorem (Swan [2]): The map

$$\varphi: \underline{\underline{G}}_o(\Lambda) \longrightarrow \underline{\underline{G}}_o^f(\Lambda), \quad [M]_o \longmapsto [M]$$

is an isomorphism.

Proof: φ is obviously a group homomorphism. To show that it is an isomorphism, we construct its inverse:

$$\psi: \underline{\underline{G}}_o^f(\Lambda) \longrightarrow \underline{\underline{G}}_o(\Lambda).$$

It suffices to define ψ on [T] where $T \in {}_\Lambda\underline{\underline{M}}^f$. We take a presentation

$$0 \longrightarrow M \longrightarrow P \longrightarrow T \longrightarrow 0$$

in ${}_\Lambda\underline{\underline{M}}^f$, where $P \in {}_\Lambda\underline{\underline{P}}^f$ and consequently $M \in {}_\Lambda\underline{\underline{M}}^o$. Now we define

$$\psi: [T] \longmapsto [P]_o - [M]_o.$$

(i) ψ is independent of the presentation: Let

$$0 \longrightarrow M' \longrightarrow P' \longrightarrow T \longrightarrow 0$$

be another presentation of T with $P' \in {}_\Lambda\underline{\underline{P}}^f$ and $M' \in {}_\Lambda\underline{\underline{M}}^o$. Since P and P' are projective, we can apply Schanuel's lemma (V, 2.6), to conclude

$$M \oplus P' \cong M' \oplus P,$$

and ψ is independent of the presentation.

(ii) ψ preserves relations: Given an exact sequence

---

*) In this section a prime ideal can be the zero ideal.

$$0 \longrightarrow T' \longrightarrow T \longrightarrow T'' \longrightarrow 0$$

in ${}_{\Lambda}\underline{\underline{M}}^{f}$.

As in the proof of (II, 3.9) we construct a commutative diagram with exact rows and columns

$$\begin{array}{ccccccccc}
 & & 0 & & 0 & & 0 & & \\
 & & \uparrow & & \uparrow & & \uparrow & & \\
0 & \longrightarrow & T' & \longrightarrow & T & \longrightarrow & T'' & \longrightarrow & 0 \\
 & & \uparrow & & \uparrow & & \uparrow & & \\
0 & \longrightarrow & P_1' & \xrightarrow{\alpha_1} & P_1' \oplus P_1'' & \xrightarrow{\beta_1} & P_1'' & \longrightarrow & 0 \\
 & & \kappa \uparrow & & \lambda \uparrow & & \mu \uparrow & & \\
0 & \longrightarrow & P_2' & \xrightarrow{\alpha_2} & P_2' \oplus P_2'' & \xrightarrow{\beta_2} & P_2'' & \longrightarrow & 0
\end{array} ,$$

where $P_2' \longrightarrow P_1' \longrightarrow T' \longrightarrow 0$ and $P_2'' \longrightarrow P_1'' \longrightarrow T'' \longrightarrow 0$ are the tails of projective resolutions of T' and T'' resp. We put $M' = P_2'/\operatorname{Ker}\kappa$, $X = (P_2' \oplus P_2'')/\operatorname{Ker}\lambda$ and $M'' = P_2''/\operatorname{Ker}\mu$; then there are induced maps $\bar\kappa$, $\bar\lambda$ and $\bar\mu$: $\bar\kappa : M' \longrightarrow P_1'$, $\bar\lambda : X \longrightarrow P_1' \oplus P_2''$; $\bar\mu : M'' \longrightarrow P_1''$. Moreover, we obtain induced maps $\bar\alpha_2 : M' \longrightarrow X$, since $(\operatorname{Ker}\kappa)\alpha_2 \subset \operatorname{Ker}\lambda$, and $\bar\beta_2 : X \longrightarrow M''$, since $(\operatorname{Ker}\lambda)\beta_2 \subset \operatorname{Ker}\mu$. Diagram chasing shows that the following diagram is commutative with exact rows and columns

$$\begin{array}{ccccccccc}
 & & 0 & & 0 & & 0 & & \\
 & & \uparrow & & \uparrow & & \uparrow & & \\
0 & \longrightarrow & T' & \longrightarrow & T & \longrightarrow & T'' & \longrightarrow & 0 \\
 & & \uparrow & & \uparrow & & \uparrow & & \\
0 & \longrightarrow & P_1' & \xrightarrow{\alpha_1} & P_1' \oplus P_1'' & \xrightarrow{\beta_1} & P_1'' & \longrightarrow & 0 \\
 & & \bar\kappa \uparrow & & \bar\lambda \uparrow & & \bar\mu \uparrow & & \\
0 & \longrightarrow & M' & \xrightarrow{\bar\alpha_2} & X & \xrightarrow{\bar\beta_2} & M'' & \longrightarrow & 0 \\
 & & \uparrow & & \uparrow & & \uparrow & & \\
 & & 0 & & 0 & & 0 & &
\end{array}$$

(cf. Ex. 3,6).

Thus

$$\psi([T]) = \psi([T']) + \psi([T'']),$$

and ψ preserves relations; i.e., ψ is a group homomorphism.

(iii) $\psi\varphi = 1_{\underline{\underline{G}}_o(\Lambda)}$ and $\varphi\psi = 1_{\underline{\underline{G}}_o^f(\Lambda)}$.

$$\psi\varphi : [M]_o \longmapsto^{\varphi} [M]_f \longmapsto^{\psi} [P]_o - [N]_o,$$

where

$$E : 0 \longrightarrow N \longrightarrow P \longrightarrow M \longrightarrow 0$$

is an exact sequence. But $M \in {}_{\Lambda}\underline{\underline{M}}^o$, and thus E gives rise to the relation $[P]_o = [M]_o + [N]_o$ in $\underline{\underline{G}}_o(\Lambda)$, and $\psi\varphi = 1_{\underline{\underline{G}}_o(\Lambda)}$.

$$\varphi\psi : [T]_f \longmapsto^{\psi} [P]_o - [M]_o \longmapsto^{\varphi} [P]_f - [M]_f,$$

where

$$E' : 0 \longrightarrow M \longrightarrow P \longrightarrow T \longrightarrow 0$$

is an exact sequence of finitely generated Λ-modules. Hence, in $\underline{\underline{G}}_o^f(\Lambda)$ we have the relation $[P]_f = [M]_f + [T]_f$ and $\varphi\psi = 1_{\underline{\underline{G}}_o^f(\Lambda)}$. #

3.2 <u>Theorem</u> (Swan [2,4]): Let $\underline{\underline{S}}_o$ be a finite non empty set of prime ideals of R containing $\{0\}$. By $\underline{\underline{G}}_o{}^T_{\underline{\underline{S}}_o}(\Lambda)$ we denote the Grothendieck group of the Λ-modules T which are finitely generated R-torsion modules with $(\mathrm{ann}_R(T), \underline{\underline{S}}_o) = 1$ (i.e., $(\mathrm{ann}_R(T), \underline{\underline{q}}) = 1$ for every $0 \neq \underline{\underline{q}} \in \underline{\underline{S}}_o$). The category of these modules is denoted by ${}_{\Lambda}\underline{\underline{M}}^T_{\underline{\underline{S}}_o}$, and by $\underline{\underline{G}}_o(\Lambda/\underline{\underline{p}}\Lambda)$ we denote the Grothendieck group $\underline{\underline{G}}_o({}_{\Lambda/\underline{\underline{p}}\Lambda}\underline{\underline{M}}^f)$ for the maximal ideal $\underline{\underline{p}}$ of R. Let $S = R \setminus \{\bigcup_{\underline{\underline{q}} \in \underline{\underline{S}}_o} \underline{\underline{q}}\}$. Then we have a natural isomorphism

$$\sigma_{\underline{\underline{S}}_o} : \bigoplus_{\underline{\underline{p}} \notin \underline{\underline{S}}_o} \underline{\underline{G}}_o(\Lambda/\underline{\underline{p}}\Lambda) \longrightarrow \underline{\underline{G}}_o{}^T_{\underline{\underline{S}}_o}(\Lambda),$$

and a homomorphism

$$\tau_{\underline{\underline{S}}_o} : \underline{\underline{G}}_o(\Lambda_S) \longrightarrow \bigoplus_{\underline{\underline{q}} \in \underline{\underline{S}}_o} \underline{\underline{G}}_o(\Lambda_{\underline{\underline{q}}}),$$

where Λ_S is the localization of Λ at S. Moreover, the following diagram is commutative with exact rows

$$\begin{array}{ccccccc}
\mathbf{G}_0{}^{T}_{\mathbf{S}_0}(\Lambda) & \xrightarrow{\delta_{\mathbf{S}_0}} & \mathbf{G}_0(\Lambda) & \xrightarrow{\iota_{\mathbf{S}_0}} & \mathbf{G}_0(\Lambda_S) & \longrightarrow & 0 \\
\Big\downarrow{\scriptstyle \sigma^{-1}_{\mathbf{S}_0}} & & \Big\downarrow{\scriptstyle 1_{\mathbf{G}_0(\Lambda)}} & & \Big\downarrow{\scriptstyle \tau_{\mathbf{S}_0}} & & \\
\bigoplus_{\mathbf{p}\notin \mathbf{S}_0} \mathbf{G}_0(\Lambda/\mathbf{p}\Lambda) & \xrightarrow{\bigoplus_{\mathbf{p}\notin \mathbf{S}_0} \delta_{\mathbf{p}}} & \mathbf{G}_0(\Lambda) & \xrightarrow{\bigoplus_{\mathbf{q}\in \mathbf{S}_0} \iota_{\mathbf{p}}} & \bigoplus_{\mathbf{q}\in \mathbf{S}_0} \mathbf{G}_0(\Lambda_{\mathbf{q}}) & &
\end{array}$$

__Proof:__ We first observe that the category of Λ-modules T which are R-torsion modules with $(\mathrm{ann}_R T, \mathbf{S}_0) = 1$ is an admissible subcategory of ${}_{\Lambda}\mathbf{M}^f$. (i,ii,iii) of (1.1) are obviously satisfied. As for (iv), let

$$E : 0 \longrightarrow T' \longrightarrow T \longrightarrow T'' \longrightarrow 0$$

be an exact sequence of finitely generated Λ-modules, with $T', T'' \in {}_{\Lambda}\mathbf{M}^T_{\mathbf{S}_0}$.

If $\mathbf{q} \in \mathbf{S}_0$, then tensoring the sequence E with $R_{\mathbf{q}} \otimes_R -$ implies $R_{\mathbf{q}} \otimes_R T = 0$; since this is true for every $\mathbf{q} \in \mathbf{S}_0$, we have $(\mathrm{ann}_R T, \mathbf{S}_0) = 1$ and ${}_{\Lambda}\mathbf{M}^T_{\mathbf{S}_0}$ is an admissible subcategory; thus the expression $\mathbf{G}_0{}^{T}_{\mathbf{S}_0}(\Lambda)$ makes sense (cf. 1.2). If $\mathbf{p} \notin \mathbf{S}_0$, we define a map

$$\sigma_{\mathbf{p}} : \mathbf{G}_0(\Lambda/\mathbf{p}\Lambda) \longrightarrow \mathbf{G}_0{}^{T}_{\mathbf{S}_0}(\Lambda),$$

$$[T]_{\mathbf{p}} \longmapsto [T]_{\mathbf{S}_0} .$$

This obviously is a group homomorphism; and thus the family $\{\sigma_{\mathbf{p}}\}_{\mathbf{p}\notin \mathbf{S}_0}$ induces a group homomorphism (cf. I, Ex. 1,2)

$$\sigma_{\mathbf{S}_0} : \bigoplus_{\mathbf{p}\notin \mathbf{S}_0} \mathbf{G}_0(\Lambda/\mathbf{p}\Lambda) \longrightarrow \mathbf{G}_0{}^{T}_{\mathbf{S}_0}(\Lambda).$$

To show that $\sigma_{\mathbf{S}_0}$ is an isomorphism, we set up its inverse

$$\tilde{\sigma}_{\mathbf{S}_0} : \mathbf{G}_0{}^{T}_{\mathbf{S}_0}(\Lambda) \longrightarrow \bigoplus_{\mathbf{p}\notin \mathbf{S}_0} \mathbf{G}_0(\Lambda/\mathbf{p}\Lambda);$$

Given $[T]_{\mathbf{S}_0} \in \mathbf{G}_0{}^{T}_{\mathbf{S}_0}(\Lambda)$, T is the direct sum of its $\mathbf{p}$-primary components (cf. I, 8.9), $T = \oplus_{i=1}^{n} T_{\underline{\underline{p}}_i}$, where

$$\mathrm{ann}_R(T_{\underline{\underline{p}}_i}) = \underline{\underline{p}}_i^{s_i};\ \underline{\underline{p}}_i \notin \underline{\underline{S}}_o,\ \text{and}\ T_{\underline{\underline{p}}_i} \cong R_{\underline{\underline{p}}_i} \otimes_R T.$$

Each $T_{\underline{\underline{p}}_i}$ has a composition series

$$T_{\underline{\underline{p}}_i} = X_{i_o} \supsetneqq X_{i_1} \supsetneqq \cdots \supsetneqq X_{i_{s_i-1}} \supsetneqq 0 = X_{i_{s_i}}.$$

We now put

$$\tilde{\sigma}_{\underline{\underline{S}}_o} : [T]_T \longmapsto \left(\sum_{j=0}^{s_i-1} [X_{i_j}/X_{i_{j+1}}]_{\underline{\underline{p}}_i}\right)_{1 \le i \le n},$$

then $\tilde{\sigma}_{\underline{\underline{S}}_o}$ is well-defined since $R_{\underline{\underline{p}}_i} \otimes_R -$ is an additive exact functor (cf. I, 6.5). Obviously, $\sigma_{\underline{\underline{S}}_o}$ and $\tilde{\sigma}_{\underline{\underline{S}}_o}$ are inverse to each other, and $\sigma_{\underline{\underline{S}}_o}$ is an isomorphism. We define,

$$\tau_{\underline{\underline{S}}_o} : \underline{\underline{G}}_o(\Lambda_S) \longrightarrow \bigoplus_{\underline{\underline{q}} \in \underline{\underline{S}}_o} \underline{\underline{G}}_o(\Lambda_{\underline{\underline{q}}}),$$

$$[M]_S \longmapsto ([R_{\underline{\underline{q}}} \otimes_{R_S} M])_{\underline{\underline{q}} \in \underline{\underline{S}}_o}$$

(cf. I, 8.4). This is well-defined since $\underline{\underline{S}}_o$ is a finite set and since $R_{\underline{\underline{p}}} \otimes_{R_S} -$ is an exact additive functor (cf. 1.8).

We put

$$\iota_{\underline{\underline{S}}_o} : \underline{\underline{G}}_o(\Lambda) \longrightarrow \underline{\underline{G}}_o(\Lambda_S),$$

$$[M] \longmapsto [M_S],$$

$$\iota_{\underline{\underline{p}}} : \underline{\underline{G}}_o(\Lambda) \longrightarrow \underline{\underline{G}}_o(\Lambda_{\underline{\underline{p}}}),$$

$$[M] \longmapsto [M_{\underline{\underline{p}}}].$$

Since $R_S \otimes_R -$ and $R_{\underline{\underline{p}}} \otimes_R -$ are additive exact functors, these maps are group homomorphisms, and for every $\underline{\underline{q}} \in \underline{\underline{S}}_o$, we have

$R_{\mathfrak{q}} \otimes_R - \sim R_{\mathfrak{q}} \otimes_{R_S} R_S \otimes_R -$, so that $\bigoplus_{\mathfrak{q} \in \mathfrak{S}_0} \iota_{\mathfrak{q}}$ factors through $\tau_{\mathfrak{S}_0}$, and we get the commutative diagram

$$(1) \qquad \begin{array}{ccc} & & \mathfrak{G}_0(\Lambda_S) \\ & \overset{\iota_{\mathfrak{S}_0}}{\nearrow} & \\ \mathfrak{G}_0(\Lambda) & & \Big\downarrow \tau_{\mathfrak{S}_0} \\ & \underset{\bigoplus_{\mathfrak{q} \in \mathfrak{S}_0} \iota_{\mathfrak{q}}}{\searrow} & \\ & & \bigoplus_{\mathfrak{q} \in \mathfrak{S}_0} \mathfrak{G}_0(\Lambda_{\mathfrak{q}}) . \end{array}$$

Finally, to define the maps $\varrho_{\mathfrak{p}}$ and $\varrho_{\mathfrak{S}_0}$:

$$\varrho_{\mathfrak{p}} : \mathfrak{G}_0(\Lambda/\mathfrak{p}\Lambda) \longrightarrow \mathfrak{G}_0(\Lambda).$$

Given $T \in {}_{\Lambda/\mathfrak{p}\Lambda}\mathfrak{M}^0$, we take a presentation

$$0 \longrightarrow M \longrightarrow P \longrightarrow T \longrightarrow 0,$$

where $P \in {}_{\Lambda}\mathfrak{P}^f$ and $M \in {}_{\Lambda}\mathfrak{M}^0$, and define

$$\varrho_{\mathfrak{p}} : [T]_{\mathfrak{p}} \longmapsto [P]_0 - [M]_0 .$$

As in the proof of (3.1) one shows that $\varrho_{\mathfrak{p}}$ is a group homomorphism. In addition, $\bigoplus_{\mathfrak{q} \in \mathfrak{S}_0} \iota_{\mathfrak{q}} \varrho_{\mathfrak{p}} = 0$ and $\iota_{\mathfrak{S}_0} \varrho_{\mathfrak{p}} = 0$ if $\mathfrak{p} \notin \mathfrak{S}_0$ and for $\mathfrak{q} \in \mathfrak{S}_0$, since $R_{\mathfrak{q}} \otimes_R T = 0$ for $\mathfrak{q} \in \mathfrak{S}_0$ and $R_S \otimes_R T = 0$. In particular $\operatorname{Im}(\bigoplus_{\mathfrak{p} \notin \mathfrak{S}_0} \varrho_{\mathfrak{p}}) \subset \operatorname{Ker} \iota_{\mathfrak{S}_0} \cap \operatorname{Ker}(\bigoplus_{\mathfrak{q} \in \mathfrak{S}_0} \iota_{\mathfrak{q}})$. If we define

$$\begin{aligned} \varrho_{\mathfrak{S}_0} : \mathfrak{G}_0\,{}^{T}_{\mathfrak{S}_0}(\Lambda) &\longrightarrow \mathfrak{G}_0(\Lambda), \\ [T]_{\mathfrak{S}_0} &\longmapsto [P]_0 - [M]_0 , \end{aligned}$$

where $0 \longrightarrow M \longrightarrow P \longrightarrow T \longrightarrow 0$ is an exact sequence with $P \in {}_{\Lambda}\mathfrak{P}^f$, $M \in {}_{\Lambda}\mathfrak{M}^0$, then $\varrho_{\mathfrak{S}_0}$ is a group homomorphism. Because of the isomorphism (3.1), we find that the following diagram is commutative:

$$(2)\qquad \begin{array}{ccc} G_0{}^{T}_{S_0}(\Lambda) & \xrightarrow{\ \rho_{S_0}\ } & G_0(\Lambda) \\ \Big\uparrow \sigma_{S_0} & \nearrow\ \bigoplus\limits_{p \notin S_0} \rho_p & \\ \bigoplus\limits_{p \notin S_0} G_0(\Lambda/p\Lambda) & & \end{array} .$$

In particular, $\operatorname{Im} \rho_{S_0} \subset \operatorname{Ker} \iota_{S_0} \cap \operatorname{Ker} \bigoplus_{q \in S_0} \iota_q$. However, the commutative diagram (1) shows $\operatorname{Ker} \bigoplus_{q \in S_0} \iota_q \supset \operatorname{Ker} \iota_{S_0}$.

Since $\operatorname{Im} \rho_{S_0} \subset \operatorname{Ker} \iota_{S_0}$, we obtain an induced group homomorphism

$$\iota' : \operatorname{Coker}(\rho_{S_0}) = G_0(\Lambda)/\operatorname{Im} \rho_{S_0} \longrightarrow G_0(\Lambda_S).$$

To show that the sequence

$$(3)\qquad G_0{}^{T}_{S_0}(\Lambda) \xrightarrow{\ \rho_{S_0}\ } G_0(\Lambda) \xrightarrow{\ \iota_{S_0}\ } G_0(\Lambda_S) \longrightarrow 0$$

is exact, it suffices to show that ι' is an isomorphism. To do so, we set up its inverse.

$$\kappa : G_0(\Lambda_S) \longrightarrow \operatorname{Coker} \rho_{S_0}.$$

Given $M_S \in {}_{\Lambda_S}M^o$, we take a presentation; i.e., an exact sequence

$$F_{1_S} \xrightarrow{\ \alpha\ } F_{2_S} \longrightarrow M_S \longrightarrow 0,$$

where F_{1_S} and F_{2_S} are free Λ_S-lattices. Then we can find free Λ-lattices F_1 and F_2 such that $R_S \otimes_R F_1 = F_{1_S}$ and $R_S \otimes_R F_2 = F_{2_S}$. Moreover, there exists $\beta \in \operatorname{Hom}_\Lambda(F_1, F_2)$ such that $\beta = s\alpha|_{F_1}$ for some $0 \neq s \in S$. Since s is a unit in R_S, $\operatorname{Im}\alpha = \operatorname{Im} s\alpha$. Hence, we may assume $s\alpha = \alpha$. Let $(\operatorname{Coker}\beta)^T$ be the R-torsion submodule of $\operatorname{Coker}\beta$; then $(\operatorname{Coker}\beta)^T \in {}_\Lambda M^f$ and $(\operatorname{Coker}\beta)^o = \operatorname{Coker}\beta/(\operatorname{Coker}\beta)^T \in {}_\Lambda M^o$. We now define

$$\kappa : G_0(\Lambda_S) \longrightarrow \operatorname{Coker} \rho_{S_0},$$

$$[M_S] \longmapsto [(\operatorname{Coker}\beta)^o] + \operatorname{Im} \rho_{S_0}.$$

(i) κ <u>is independent of the presentation</u>: Let

$$F'_{1_S} \xrightarrow{\alpha'} F'_{2_S} \longrightarrow M_S \longrightarrow 0$$

be another presentation. Then there exists a corresponding map $\beta' : F'_1 \longrightarrow F'_2$ and $(\mathrm{Coker}\,\beta)^o_S \overset{\varphi}{\cong} (\mathrm{Coker}\,\beta')^o_S \cong M_S$. So we can find a Λ-monomorphism

$$\psi : (\mathrm{Coker}\,\beta)^o \longrightarrow (\mathrm{Coker}\,\beta')^o$$

such that $1_{R_S} \otimes \psi = \varphi$. Then obviously, $\mathrm{Coker}\,\psi \in {}_{\Lambda}\underline{\underline{M}}^T_{\underline{\underline{S}}_o}$, and

$$[(\mathrm{Coker}\,\beta')^o] - [(\mathrm{Coker}\,\beta)^o] = \rho_{\underline{\underline{S}}_o}[\mathrm{Coker}\,\psi]^{*)};$$

i.e., κ is independent of the presentation. In particular this shows that $[N_1] - [N_2] \in \mathrm{Im}\,\rho_{\underline{\underline{S}}_o}$, if $N_{1_S} \cong N_{2_S}$.

(ii) κ <u>preserves relations</u>: Given an exact sequence of Λ_S-lattices

$$0 \longrightarrow M'_S \xrightarrow{\psi} M_S \longrightarrow M''_S \longrightarrow 0.$$

Let $N, N' \in {}_{\Lambda}\underline{\underline{M}}^o$ be such that $N_S \cong M_S$ and $N'_S \cong M'_S$. Since $\mathrm{Hom}_{\Lambda_S}(M'_S, M_S) \cong R_S \otimes_R \mathrm{Hom}_{\Lambda}(N', N)$ (cf. III, 1.2), we can find $\chi \in \mathrm{Hom}_{\Lambda}(N', N)$ such that $1_{R_S} \otimes_R \chi = \psi$ (this we always can assume; if necessary after replacing ψ by $s\psi$ for some $0 \neq s \in S$). Then $(\mathrm{Coker}\,\chi)^o_S \cong M''_S$, and we have in $\underline{\underline{G}}^f_o(\Lambda)$, the Grothendieck group of all Λ-modules, the relation

$$[N]_f - [N']_f = [(\mathrm{Coker}\,\chi)^o]_f + [(\mathrm{Coker}\,\chi)^T]_f.$$

But $[(\mathrm{Coker}\,\chi)^T]_o \in \underline{\underline{G}}_o{}^T_{\underline{\underline{S}}_o}(\Lambda)$. Thus, in $\underline{\underline{G}}_o(\Lambda)$ we have

$$[N]_o - [N']_o - [(\mathrm{Coker}\,\chi)^o] \in \mathrm{Im}\,\rho_{\underline{\underline{S}}_o}.$$

Since $[N]_o + \mathrm{Im}\,\rho_{\underline{\underline{S}}_o} = \kappa[M_S]$ and $[N']_o + \mathrm{Im}\,\rho_{\underline{\underline{S}}_o} = \kappa[M'_S]$, we have $[(\mathrm{Coker}\,\chi)^o] + \mathrm{Im}\,\rho_{\underline{\underline{S}}_o} = \kappa[M''_S]$; κ preserves relations.

<u>(iii)</u> $\kappa\iota' : [M] + \mathrm{Im}\,\rho_{\underline{\underline{S}}_o} \overset{\iota'}{\longmapsto} [M_S] \overset{\kappa}{\longmapsto} [M] + \mathrm{Im}\,\rho_{\underline{\underline{S}}_o}$ and

$$\iota'\kappa : [M_S] \overset{\kappa}{\longmapsto} [N] + \mathrm{Im}\,\rho_{\underline{\underline{S}}_o} \overset{\iota'}{\longmapsto} [M_S],$$

---

*) If we have two exact sequences $0 \to M' \to M \to T \to 0$ and $0 \to N' \to N \to T \to 0$, where $M', M, N', N \in {}_{\Lambda}\underline{\underline{M}}^o$ and $T \in {}_{\Lambda}\underline{\underline{M}}^f$, then (3.1) shows $[M] - [M'] = [N] - [N']$ in $\underline{\underline{G}}_o(\Lambda)$.

as is easily seen; i.e., ι' is an isomorphism, and the sequence (3) is exact.

To finish the proof of (3.2) it remains to show that

$$\operatorname{Ker} \bigoplus_{\mathfrak{q} \in S_0} \iota_{\mathfrak{q}} \subset \operatorname{Im} \bigoplus_{\mathfrak{p} \notin S_0} \varrho_{\mathfrak{p}}.$$

Let $[M] - [N] \in \operatorname{Ker} \bigoplus_{\mathfrak{q} \in S_0} \iota_{\mathfrak{q}} = \bigcap_{\mathfrak{q} \in S_0} \operatorname{Ker} \iota_{\mathfrak{q}}$.

Now we apply the fact that the sequence (3) is already proven to be exact with $S = R \setminus \{\mathfrak{p}_0\}$ for every prime ideal $\mathfrak{p}_0$. Then $\operatorname{Ker} \iota_{\mathfrak{p}_0} =$

$= \sum_{\mathfrak{p} \neq \mathfrak{p}_0} \operatorname{Im} \varrho_{\mathfrak{p}}$ and

$$[M] - [N] \in \bigcap_{\mathfrak{q} \in S_0} \operatorname{Ker} \iota_{\mathfrak{q}} = \bigcap_{\mathfrak{q} \in S_0} \Big(\sum_{\mathfrak{p} \neq \mathfrak{q}} \operatorname{Im} \varrho_{\mathfrak{p}}\Big) = \sum_{\mathfrak{p} \notin S_0} \operatorname{Im} \varrho_{\mathfrak{p}},$$

and the diagram in (3.2) is commutative with exact rows. #

3.3 <u>Corollary</u>: We have an exact sequence

$$\bigoplus_{\mathfrak{p} \in \operatorname{spec} R} G_0(\Lambda/\mathfrak{p}\Lambda) \xrightarrow{\oplus \varrho_{\mathfrak{p}}} G_0(\Lambda) \xrightarrow{\iota} G_0(A) \longrightarrow 0. \text{ *)}$$

<u>Proof</u>: This follows from (3.2) for $S_0 = \{0\}$. #

3.4 <u>Theorem</u> (Brauer [6], Swan [2], Strooker [1]): Let $\mathfrak{a}$ be any non-zero ideal of R, then we have a group homomorphism

$$\mu_{\mathfrak{a}} : G_0(\Lambda) \longrightarrow G_0({}_{\Lambda/\mathfrak{a}\Lambda}M^f) = G_0(\Lambda/\mathfrak{a}\Lambda),$$

$$[M] \longmapsto [R/\mathfrak{a} \otimes_R M],$$

and there exists a unique group homomorphism

$$\delta_{\mathfrak{a}} : G_0(A) \longrightarrow G_0(\Lambda/\mathfrak{a}\Lambda)$$

making the following diagram commute

$$\begin{array}{ccc} G_0(\Lambda) & \xrightarrow{\iota} & G_0(A) \\ & {\scriptstyle \mu_{\mathfrak{a}}} \searrow \quad \swarrow {\scriptstyle \delta_{\mathfrak{a}}} & \\ & G_0(\Lambda/\mathfrak{a}\Lambda) & \end{array}.$$

*) spec R = set of all maximal ideals; i.e., prime ideals $\neq 0$.

$\delta_{\underline{\underline{a}}}$ is called the <u>decomposition map with respect to $\underline{\underline{a}}$</u>.

<u>Proof</u>: (cf. also Ex. 7) Observe that $\mu_{\underline{\underline{a}}}$ is well-defined, since $R/\underline{\underline{a}} \otimes_R -$ is an exact functor on the category of Λ-lattices (cf. 1.8). According to (I, Ex. 2,3) it suffices to show that $\operatorname{Ker}\iota \subset \operatorname{Ker}\mu_{\underline{\underline{a}}}$. In (3.3) we have shown $\operatorname{Ker}\iota = \bigoplus_{\underline{\underline{p}} \in \operatorname{spec} R} \operatorname{Im} \rho_{\underline{\underline{p}}}$. And it obviously suffices to prove that

$$\mu_{\underline{\underline{a}}}([M] - [N]) = 0,$$

whenever we have an exact sequence

$$E : 0 \longrightarrow N \longrightarrow M \longrightarrow T \longrightarrow 0,$$

where T is an R-torsion Λ-module. We may even assume that T is simple. Namely, a composition series of T corresponds to a composition series between M and N

$$M = M_0 \supsetneqq M_1 \supsetneqq \dots \supsetneqq M_{s+1} = N$$

and we have the exact sequences

$$0 \longrightarrow M_{i+1} \longrightarrow M_i \longrightarrow M_i/M_{i+1} \longrightarrow 0,$$

where each $M_i/M_{i+1}, 0 \le i \le s$, is a simple Λ-module. Then

$$\mu_{\underline{\underline{a}}}([M] - [N]) = \sum_{i=0}^{s} (\mu_{\underline{\underline{a}}}([M_i] - [M_{i+1}])).$$

Applying $- \otimes_R R/\underline{\underline{a}}$ to the sequence E, we obtain (cf. II, 3.10)

$$\operatorname{Tor}_1^R(M,R/\underline{\underline{a}}) \longrightarrow \operatorname{Tor}_1^R(T,R/\underline{\underline{a}}) \longrightarrow N \otimes_R R/\underline{\underline{a}} \longrightarrow M \otimes_R R/\underline{\underline{a}} \longrightarrow T \otimes_R R/\underline{\underline{a}} \longrightarrow 0.$$

However $\operatorname{Tor}_1^R(M,R/\underline{\underline{a}}) = 0$, M being an R-lattice; in fact $M \in {}_{\Lambda}\underline{\underline{M}}^o \cap {}_R\underline{\underline{P}}^f$ and thus $\operatorname{Tor}_1^R(M,X) = 0$ for every $X \in {}_R\underline{\underline{M}}^f$ (cf. II, §§ 2,3). $T \in {}_{\Lambda}\underline{\underline{M}}^f$ and $\operatorname{Tor}_1^R(-,R/\underline{\underline{a}})$ is a covariant functor (cf. II, 3.5); consequently

$$\operatorname{Tor}_1^R(T,R/\underline{\underline{a}}) \in {}_{\Lambda}\underline{\underline{M}}^f \quad \text{(cf. II, 1.12)},$$

and we have an exact sequence of Λ-modules

$$E : 0 \longrightarrow \operatorname{Tor}_1^R(T,R/\underline{\underline{a}}) \longrightarrow N \otimes_R R/\underline{\underline{a}} \longrightarrow M \otimes_R R/\underline{\underline{a}} \longrightarrow T \otimes_R R/\underline{\underline{a}} \longrightarrow 0,$$

which gives rise to the relation (cf. Ex. 3,1)

$$R_1 : [N \otimes_R R/\underline{\underline{a}}] - [M \otimes_R R/\underline{\underline{a}}] = [\operatorname{Tor}_1^R(T,R/\underline{\underline{a}})] - [T \otimes_R R/\underline{\underline{a}}] \text{ in } \underline{\underline{G}}_o^T(\Lambda).$$

Using a similar argument for the exact sequence

$$E_1 : 0 \longrightarrow \underline{\underline{a}} \longrightarrow R \longrightarrow R/\underline{\underline{a}} \longrightarrow 0,$$

we obtain the exact sequence of Λ-modules

$$E_2 : 0 \longrightarrow \mathrm{Tor}_1^R(R/\underline{\underline{a}},T) \longrightarrow \underline{\underline{a}} \otimes_R T \longrightarrow T \longrightarrow R/\underline{\underline{a}} \otimes_R T \longrightarrow 0.$$

However, $\mathrm{Tor}_1^R(R/\underline{\underline{a}},T) \overset{\text{nat}}{\cong} \mathrm{Tor}_1^R(T,R/\underline{\underline{a}})$ (cf. Ex. 3,2) and $\underline{\underline{a}} \otimes_R T \overset{\text{nat}}{\cong}$ $\overset{\text{nat}}{\cong} T \otimes_R \underline{\underline{a}}$, R being commutative. Thus we obtain the relation

$$R_2 : [T \otimes_R \underline{\underline{a}}] - [T] = [\mathrm{Tor}_1^R(T,R/\underline{\underline{a}})] - [T \otimes_R R/\underline{\underline{a}}]$$

in $\underline{\underline{G}}_o^T(\Lambda)$. Combining this with the relation R_1, we have

$$[N \otimes_R R/\underline{\underline{a}}] - [M \otimes_R R/\underline{\underline{a}}] = [T \otimes_R \underline{\underline{a}}] - [T] \text{ in } \underline{\underline{G}}_o^T(\Lambda).$$

However $[T \otimes_R \underline{\underline{a}}] = [T]$ in $\underline{\underline{G}}_o^T(\Lambda)$. In fact, let $\underline{\underline{b}} = \mathrm{ann}_R(T)$. Then

$$R/\underline{\underline{b}} \otimes_R (T \otimes_R \underline{\underline{a}}) \cong (R/\underline{\underline{b}} \otimes_R T) \otimes_R \underline{\underline{a}} \cong T \otimes_R \underline{\underline{a}}.$$

On the other hand,

$$R/\underline{\underline{b}} \otimes_R (T \otimes_R \underline{\underline{a}}) \cong T \otimes_R (R/\underline{\underline{b}} \otimes_R \underline{\underline{a}}) \cong T \otimes_R \underline{\underline{a}}/\underline{\underline{ab}} \cong T \otimes_R R/\underline{\underline{b}} \cong T,$$

and it should be observed that all these isomorphisms are Λ-isomorphisms. Thus $T \cong T \otimes_R \underline{\underline{a}}$. Hence

$$[N \otimes_R R/\underline{\underline{a}}] - [M \otimes_R R/\underline{\underline{a}}] = 0 \quad \text{in } \underline{\underline{G}}_o^T(\Lambda),$$

where $\underline{\underline{G}}_o^T(\Lambda)$ is the Grothendieck group of the category ${}_\Lambda\underline{\underline{M}}^T$ of all finitely generated Λ-modules which are R-torsion modules.

Given any non-zero ideal $\underline{\underline{a}}$ of R, we have the natural map

$$\varphi : \underline{\underline{G}}_o(\Lambda/\underline{\underline{a}}\Lambda) \longrightarrow \underline{\underline{G}}_o^T(\Lambda),$$
$$[T]_{\underline{\underline{a}}} \longmapsto [T]_T,$$

and we claim that this is a monomorphism. $\underline{\underline{G}}_o^T(\Lambda)$ is free on symbols $[U]$ one for each isomorphism class of simple Λ-modules U. But since $\Lambda/\underline{\underline{a}}\Lambda$ is an artinian and noetherian ring, $\underline{\underline{G}}_o(\Lambda/\underline{\underline{a}}\Lambda)$ is free on symbols $[T]$, one for each isomorphism class of simple $\Lambda/\underline{\underline{a}}\Lambda$-modules T. Moreover, every simple $\Lambda/\underline{\underline{a}}\Lambda$-module is also a simple Λ-module, and consequently φ is monic.

Above we have seen $[N \otimes_R R/\underline{\underline{a}}] = [M \otimes_R R/\underline{\underline{a}}]$ in $\underline{\underline{G}}_o^T(\Lambda)$, consequently, φ being monic,

$$[M] - [N] \in \mathrm{Ker}\, \mu_{\underline{\underline{a}}}. \qquad \#$$

3.5 __Theorem__ (Strooker [1], Takahashi): For any non-zero ideal $\underline{\underline{a}}$ of R we have the following commutative diagram

$$\begin{array}{ccccc} \underline{\underline{K}}_0(\Lambda) & \xrightarrow{\kappa} & \underline{\underline{G}}_0(\Lambda) & \xrightarrow{\iota} & \underline{\underline{G}}_0(A) \\ \eta\downarrow & & & \searrow^{\mu_{\underline{\underline{a}}}} \quad \swarrow_{\delta_{\underline{\underline{a}}}} & \\ \underline{\underline{K}}_0(\Lambda/\underline{\underline{a}}\Lambda) & & \xrightarrow{\kappa_{\underline{\underline{a}}}} & \underline{\underline{G}}_0(\Lambda/\underline{\underline{a}}\Lambda) & \end{array},$$

where $\underline{\underline{K}}_0(\Lambda)$ and $\underline{\underline{K}}_0(\Lambda/\underline{\underline{a}}\Lambda)$ stand for the Grothendieck groups of projective Λ-lattices and finitely generated projective $\Lambda/\underline{\underline{a}}\Lambda$-modules reps. κ and $\kappa_{\underline{\underline{a}}}$ are called __Cartan-maps.__ Moreover, if $\underline{\underline{a}} = \underline{\underline{p}}$ is maximal and if Λ is semi-perfect, then $\delta_{\underline{\underline{p}}}$ is monic on $\mathrm{Im}(\iota\kappa)$.

__Proof:__ We define the maps

$$\begin{array}{rlcl} \kappa : & \underline{\underline{K}}_0(\Lambda) & \longrightarrow & \underline{\underline{G}}_0(\Lambda), \\ & [P]_{\underline{\underline{K}}} & \longmapsto & [P]_{\underline{\underline{G}}}, \\ \kappa_{\underline{\underline{a}}} : & \underline{\underline{K}}_0(\Lambda/\underline{\underline{a}}\Lambda) & \longrightarrow & \underline{\underline{G}}_0(\Lambda/\underline{\underline{a}}\Lambda), \\ & [P_{\underline{\underline{a}}}]_{\underline{\underline{K}}} & \longmapsto & [P_{\underline{\underline{a}}}]_{\underline{\underline{G}}}, \\ \eta : & \underline{\underline{K}}_0(\Lambda) & \longrightarrow & \underline{\underline{K}}_0(\Lambda/\underline{\underline{a}}\Lambda), \\ & [P] & \longmapsto & [P/\underline{\underline{a}}P]. \end{array}$$

Here the subscript indicates whether the element lies in $\underline{\underline{G}}_0$ or in $\underline{\underline{K}}_0$. Obviously, all these maps are group homomorphisms and $P/\underline{\underline{a}}P \cong R/\underline{\underline{a}} \otimes_R P$ is a projective $\Lambda/\underline{\underline{a}}\Lambda$-module. In (3.4) we have seen, that the triangle above is commutative. As for the square we have $\mu_{\underline{\underline{a}}}\kappa = \kappa_{\underline{\underline{a}}}\eta$.

To prove the second part of the statement, let $\underline{\underline{a}} = \underline{\underline{p}}$ be a prime ideal, and assume that Λ is semi-perfect. Then the Krull-Schmidt theorem is valid for the projective Λ-lattices, and the idempotents of $\Lambda/\underline{\underline{p}}\Lambda$ can be lifted to idempotents in Λ (cf. III, 7.7).

__Notation:__

$\{E_i\}_{1\le i\le n}$ = the non-isomorphic indecomposable projective Λ-lattices,

$\{E_i/\underline{\underline{p}}E_i\}_{1\le i\le n} = \{U_i\}_{1\le i\le n}$ = the non-isomorphic indecomposable projective $\Lambda/\underline{\underline{p}}\Lambda$-modules,

$\{E_i/\mathrm{rad}\,\Lambda\, E_i\} = \{F_i\}_{1\le i\le n}$ = the non-isomorphic simple $\Lambda/\underline{\underline{p}}\Lambda$-modules.

(Observe that the $\{F_i\}_{1\le i\le n}$ are also the non-isomorphic indecomposable projective $\Lambda/\mathrm{rad}\Lambda$-modules, and that idempotents from $\Lambda/\mathrm{rad}\Lambda$ can be lifted to idempotents in $\Lambda/\underline{\underline{p}}\Lambda$.)

$\{Z_i\}_{1\le i\le m}$ are Λ-lattices such that the $\{KZ_i\}_{1\le i\le m}$ are the non-isomorphic simple A-modules.

Then

$\underline{\underline{K}}_o(\Lambda)$ is free on the basis $\{[E_i]\}_{1\le i\le n}$,

$\underline{\underline{K}}_o(\Lambda/\underline{\underline{p}}\Lambda)$ is free on the basis $\{[U_i]\}_{1\le i\le n}$,

$\underline{\underline{G}}_o(A)$ is free on the basis $\{[KZ_i]\}_{1\le i\le m}$,

$\underline{\underline{G}}_o(\Lambda/\underline{\underline{p}}\Lambda)$ is free on the basis $\{[F_i]\}_{1\le i\le n}$.

Moreover, in terms of these bases, the maps in the above diagram can be expressed by matrices.

$\eta \longrightarrow \underline{\underline{E}}_n$, the $(n\times n)$ identity matrix, since Λ is semi-perfect,

$\iota_K \longrightarrow \underline{\underline{F}} = (f_{ij})$,

$K_{\underline{\underline{p}}} \longrightarrow \underline{\underline{C}} = (c_{ij})$,

$\delta_{\underline{\underline{p}}} \longrightarrow \underline{\underline{D}} = (d_{ij})$.

The matrix $\underline{\underline{C}}$ is called the <u>Cartan-matrix</u> of $\Lambda/\underline{\underline{p}}\Lambda$, and the matrix $\underline{\underline{D}}$ is called the <u>decomposition matrix</u> modulo $\underline{\underline{p}}$.

The commutativity of the above diagram shows $\underline{\underline{C}} = \underline{\underline{DF}}$.

3.5' <u>Claim</u>: There are integers $a_j, r_i > 0$ such that

$$d_{ij}r_i = f_{ji}a_j\,,\quad 1\le i\le m;\ 1\le j\le n.$$

<u>Proof</u>: Let

$$s_{ij} = \dim_K(\mathrm{Hom}_A(KE_i, KZ_j)),\ 1\le i\le n,\ 1\le j\le m$$

$$= \dim_{\bar{R}}(\mathrm{Hom}_{\bar{\Lambda}}(U_i, \bar{Z}_j)),$$

where "$-$" denotes reduction modulo $\underline{\underline{p}}$.

We have

$$[\bar{Z}_j] = \sum_{k=1}^{n} d_{kj}[F_k],\ 1\le j\le m.$$

Since U_i is a projective $\overline{\Lambda}$-module, $\mathrm{Hom}_{\overline{\Lambda}}(U_i,-)$ is an exact functor, and we have

$$\dim_{\overline{R}}(\mathrm{Hom}_{\overline{\Lambda}}(U_i,\overline{Z}_j)) = \sum_{k=1}^{m} d_{kj} \dim_{\overline{R}}(\mathrm{Hom}_{\overline{\Lambda}}(U_i,F_k)), 1\le i\le n, 1\le j\le m.$$

But if

$$\varphi : U_i \longrightarrow F_k$$

is a non-zero map, then φ is an epimorphism; i.e., $U_i/\mathrm{Ker}\varphi \cong F_k$. But $\mathrm{Ker}\varphi \supset \mathrm{rad}\,\overline{\Lambda}\cdot U_i$; thus $i = k$, and $\mathrm{Hom}_{\overline{\Lambda}}(U_i,F_k) = 0$ if $i \ne k$. Whence

$$s_{ij} = \dim_{\overline{R}}(\mathrm{Hom}_{\overline{\Lambda}}(U_i,\overline{Z}_j)) = d_{ij} \dim_{\overline{R}}(\mathrm{Hom}\ (U_i,F_i)), 1\le i\le n, 1\le j\le m.$$

Since $\mathrm{Hom}_{\overline{\Lambda}}(U_i,F_i) \cong \mathrm{Hom}_{\overline{\Lambda}}(F_i,F_i)$, we put $r_i = \dim_{\overline{R}}(\mathrm{End}_{\overline{\Lambda}}(F_i))$, and obtain

$$s_{ij} = \dim_{\overline{R}}(\mathrm{Hom}_{\overline{\Lambda}}(U_i,\overline{Z}_j)) = d_{ij}r_i.$$

On the other hand,

$$[KE_i] = \sum_{k=1}^{n} f_{ki}[KZ_k],$$

and if we put

$$a_j = \dim_K(\mathrm{End}_A(KZ_j)), 1\le j\le m,$$

then

$$s_{ij} = \dim_K(\mathrm{Hom}_A(KE_i,KZ_j)) = \sum_{k=1}^{m} f_{ki} \dim_K(\mathrm{Hom}_A(KZ_k,KZ_j)) = $$
$$= f_{ji} \dim_K(\mathrm{End}_A(KZ_j)) = f_{ji}a_j, 1\le i\le n, 1\le j\le m.$$

Comparison of both computations shows $d_{ij}r_i = f_{ji}a_j$. <u>This proves (3.5').</u>

3.5" <u>Claim</u>: If P_1 and P_2 are projective Λ-lattices such that $\overline{P}_1$ and $\overline{P}_2$ have isomorphic composition factors then $KP_1 \cong KP_2$.

<u>Proof</u>: Since Λ is semi-perfect, η is injective, and we shall show that the rank d of $\underline{\underline{D}}$ is the same as the rank c of $\underline{\underline{C}}$. The formula of (3.5') can be written as

$$\mathrm{diag}(a_1,\dots,a_n)\underline{\underline{F}} = \underline{\underline{D}}^t\mathrm{diag}(r_1,\dots,r_m),$$

where $\underline{\underline{D}}^t$ denotes the transpose of $\underline{\underline{D}}$. Thus

$$\underline{\underline{F}} = \mathrm{diag}(a_1,\dots,a_n)^{-1}\underline{\underline{D}}^t\mathrm{diag}(r_1,\dots,r_m)$$

and so

$$\underline{\underline{DF}} = \underline{\underline{C}} = \underline{\underline{D}}\ \mathrm{diag}(a_1,\dots,a_n)^{-1}\underline{\underline{D}}^t\mathrm{diag}(r_1,\dots,r_m).$$

Hence

$$c = \text{rank}(\underline{C}) = \text{rank}(\underline{D} \text{ diag}(a_1, \ldots, a_n)^{-1}\underline{D}^t).$$

We put

$$\underline{A} = \text{diag}(\sqrt{a_1}^{-1}, \ldots, \sqrt{a_n}^{-1}) \text{ (observe } a_i > 0).$$

Then

$$C = \text{rank}(\underline{DA} \cdot \underline{A}^t\underline{D}^t).$$

However, for a real matrix $\underline{X}$, we have $\text{rank}(\underline{X}) = \text{rank}(\underline{X}^t\underline{X})$ _(cf. Ex. 3.5)_
and consequently, $c = \text{rank}(\underline{AD})$; but $\underline{A}$ is invertible, and thus
$c = \text{rank}(\underline{D}) = d$. But this means that δ_p is injective on $\text{Im} \, \iota\kappa$. This
proves the claim and the last part of (3.5).   #

3.6 Lemma: Let $\underline{a}$ be a non-zero ideal of R, and let $\underline{S}_0 = \{\underline{q} \, \epsilon \, \text{spec } R : \underline{q}|\underline{a}\}$
and $S = R \setminus \{ \underset{\underline{q} \, \epsilon \, \underline{S}_0}{\cup} \underline{q}\}$. Then the map η of (3.5) can be factored as

$$
\begin{array}{ccc}
\underline{K}_0(\Lambda) & \xrightarrow{\eta} & \underline{K}_0(\Lambda/\underline{a}\Lambda) \\
& & \\
\iota'_{\underline{S}_0} \searrow & & \nearrow \eta_S \\
& \underline{K}_0(\Lambda_S) &
\end{array}
$$

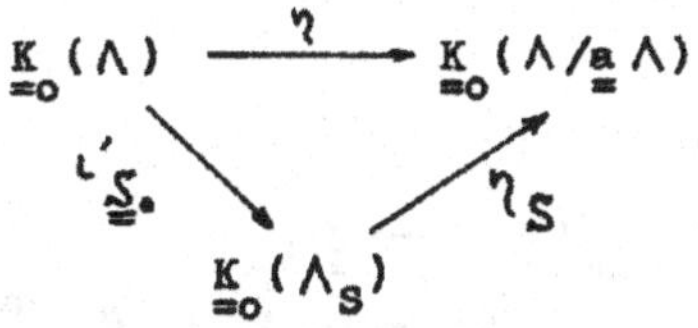

where η_S is a monomorphism. Here Λ is any R-order in A.

Proof: Since $\Lambda/\underline{a}\Lambda \overset{\text{nat}}{\cong} \Lambda_S/\underline{a}\Lambda_S$, η can be factored in the indicated
way, where

$$\iota'_{\underline{S}_0} : \underline{K}_0(\Lambda) \longrightarrow \underline{K}_0(\Lambda_S),$$

$$[P] \longmapsto [P_S],$$

and

$$\eta_S : \underline{K}_0(\Lambda_S) \longrightarrow \underline{K}_0(\Lambda/\underline{a}\Lambda),$$

$$[P_S] \longmapsto [P_S/\underline{a}P_S].$$

Moreover, $\underline{a}\Lambda_S \subset \text{rad } \Lambda_S$, and thus one shows as in (IV, 3.5) $P_S \cong P'_S$
if and only if $P_S/\underline{a}P_S \cong P'_S/\underline{a}P'_S$. Thus

$$\eta : [P_S] - [P'_S] \longmapsto 0$$

implies $P_S/\underline{\underline{a}}P_S \oplus \bar{Q} \cong P'_S/\underline{\underline{a}}P'_S \oplus \bar{Q}$, where $\bar{Q}$ is a projective $\Lambda/\underline{\underline{a}}\Lambda$-module of finite type, which can be assumed to be free; i.e., there exists $Q_S \in {}_{\Lambda_S}\underline{\underline{P}}^f$ such that $Q_S/\underline{\underline{a}}Q_S \cong \bar{Q}$. Thus $P_S \oplus Q_S \cong P'_S \oplus Q_S$ and $[P_S] = [P'_S]$; in $\underline{\underline{K}}_o(\Lambda_S)$; i.e., η_S is monic. #

3.7 __Definition__: A Λ-lattice P is called a __special projective Λ-lattice__, if $P \in {}_{\Lambda}\underline{\underline{P}}^f$ is such that KP is A-free. By ${}_{\Lambda}\underline{\underline{SP}}^f$ we denote the category of special projective Λ-lattices and $\underline{\underline{P}}_o(\Lambda)$ is its Grothendieck group.

3.8 __Lemma__: There is a natural epimorphism

$$\vartheta : \underline{\underline{P}}_o(\Lambda) \longrightarrow \underline{\underline{Z}},$$

which is split, and $\operatorname{Ker}\vartheta = \underline{\underline{C}}_o(\Lambda)$ is called the __reduced projective class group__. Every element $x \in \underline{\underline{C}}_o(\Lambda)$ has the form $x = [\Lambda^{(n)}] - [P]$, where $P \in {}_{\Lambda}\underline{\underline{P}}^f$ with $KP \cong A^{(n)}$.

__Proof__: We define

$$\begin{aligned} \vartheta : \underline{\underline{P}}_o(\Lambda) &\longrightarrow \underline{\underline{Z}}, \\ [P] &\longmapsto n, \text{ if } KP \cong A^{(n)}. \end{aligned}$$

This is a well-defined natural group-homomorphism, and the map

$$\begin{aligned} \eta : \underline{\underline{Z}} &\longrightarrow \underline{\underline{P}}_o(\Lambda) \text{ induced by} \\ 1 &\longmapsto [\Lambda] \end{aligned}$$

is such that $\vartheta\eta = 1_{\underline{\underline{Z}}}$; i.e., ϑ is a split epimorphism.

If $x \in \operatorname{Ker}\vartheta$, say $x = [P] - [Q]$, then $KP \cong KQ \cong A^{(n)}$. However since P is projective, there exists P' such that $P \oplus P' \cong \Lambda^{(m)}$. Then $x = [\Lambda^{(m)}] - [Q \oplus P']$. #

3.9 __Lemma__ (Swan [4]): We have an exact sequence

$$0 \longrightarrow \underline{\underline{C}}_o(\Lambda) \xrightarrow{\alpha} \underline{\underline{K}}_o(\Lambda) \xrightarrow{\iota'} \underline{\underline{G}}_o(A).$$

__Proof__: We define the maps

$$\begin{aligned} \iota' : \underline{\underline{K}}_o(\Lambda) &\longrightarrow \underline{\underline{G}}_o(A), \\ [P] &\longmapsto [KP], \end{aligned}$$

$$\alpha : \underline{\underline{C}}_0(\Lambda) \longrightarrow \underline{\underline{K}}_0(\Lambda),$$
$$[\Lambda^{(n)}]-[P_2] \longmapsto [\Lambda^{(n)}]-[P_2].$$

As to the exactness of the above sequence, we have $\iota'\alpha = 0$. Conversely, if $\iota' : [P_1] - [P_2] \longmapsto 0$, we may assume that P_1 is Λ-free. Then $\iota' : [P_1] - [P_2] \longmapsto 0$ implies $KP_1 \oplus X \cong KP_2 \oplus X$ (cf. 1.7). Because of the validity of the Krull-Schmidt theorem for A-modules, we have $KP_1 \cong KP_2$; i.e., $[P_1] - [P_2] \in \operatorname{Im}\alpha$. To show that α is monic, let

$$\alpha : [F] - [P] \longmapsto 0;$$

i.e., there exists $Q \in {}_{\Lambda}\underline{\underline{P}}^f$ such that

$$F \oplus Q \cong P \oplus Q \quad \text{(cf. 1.7).}$$

But we may as well assume that Q is Λ-free; i.e., $[F] - [P] = 0$ in $\underline{\underline{C}}_0(\Lambda)$. #

3.10 <u>Corollary</u>: We have the following commutative diagram with exact rows

$$\begin{array}{ccccccccc}
0 & \longrightarrow & \widetilde{\underline{\underline{G}}_0(\Lambda)} & \xrightarrow{\beta} & \underline{\underline{G}}_0(\Lambda) & \xrightarrow{\iota} & \underline{\underline{G}}_0(A) & \longrightarrow & 0 \\
 & & \uparrow{\scriptstyle\mu} & & \uparrow{\scriptstyle\kappa} & & \uparrow{\scriptstyle 1_{\underline{\underline{G}}_0(A)}} & & \\
0 & \longrightarrow & \underline{\underline{C}}_0(\Lambda) & \xrightarrow{\alpha} & \underline{\underline{K}}_0(\Lambda) & \xrightarrow{\iota'} & \underline{\underline{G}}_0(A), & &
\end{array}$$

where $\widetilde{\underline{\underline{G}}_0(\Lambda)} = \operatorname{Ker}\iota$, and β is the injection.

<u>Proof</u>: The bottom and top rows are exact and by definition, $\iota' = \iota\kappa$. Also, it is clear that $\beta\mu = \kappa\alpha$, where

$$\mu : \underline{\underline{C}}_0(\Lambda) \longrightarrow \widetilde{\underline{\underline{G}}_0(\Lambda)},$$
$$[\Lambda^{(n)}]-[P] \longmapsto [\Lambda^{(n)}]_0-[P]_0. \qquad \#$$

3.11 <u>Corollary</u> (Swan [4]): Let Λ be a hereditary R-order in A. Then there is an exact sequence

$$0 \longrightarrow \underline{\underline{C}}_0(\Lambda) \longrightarrow \underline{\underline{G}}_0(\Lambda) \longrightarrow \underline{\underline{G}}_0(A) \longrightarrow 0,$$

which is split; i.e.,

$$G_0(\wedge) \cong \underset{=}{G}_0(A) \oplus \underset{=}{C}_0(\wedge).$$

Proof: Since $\wedge$ is hereditary, $\underset{=}{K}_0(\wedge) = \underset{=}{G}_0(\wedge)$, and the result follows from (3.10), since $\underset{=}{G}_0(A)$ is a free $\underset{=}{Z}$-module on a finite basis.   #

3.12 **Theorem** (Heller-Reiner [4,5], Strooker): If $\delta_p : \underset{=}{G}_0(A) \longrightarrow$ $\underset{=}{G}_0(\wedge/p\wedge)$ is an epimorphism for every $\underset{=}{p} \in \text{spec } R$, then there is an exact sequence

$$\underset{=}{C}_0(\wedge) \xrightarrow{\beta\mu} \underset{=}{G}_0(\wedge) \xrightarrow{\iota} \underset{=}{G}_0(A) \longrightarrow 0.$$

(For the definition of the maps (cf. 3.4;3.10).)

Proof: Because of the commutative diagram of (3.10), it suffices to show that μ is an epimorphism, and for this we only have to show that every $x \in \text{Ker } \iota$ lies in $\text{Im } \kappa$. Let $x \in \text{Ker } \iota$; by (3.3),

$$x = \sum_{\underset{=}{p} \in \text{spec } R} x_{\underset{=}{p}} \in \bigoplus_{\underset{=}{p} \in \text{spec } R} \underset{\underset{=}{p}\neq 0}{G}_0(\wedge/\underset{=}{p}\wedge).$$

There are only finitely many $x_{\underset{=}{p}} \neq 0$, and it suffices to show that $\mathcal{G}_p(y) \in \text{Im } \kappa, y \in \underset{=}{G}_0(\wedge/\underset{=}{p}\wedge)$, for every $\underset{=}{p} \in \text{spec } R$. Hence we may assume $x = \mathcal{G}_p(y)$ with $y = \sum_{i=1}^{n} \alpha_i [\overline{Y}_i]$, where $\{\overline{Y}_i\}_{1 \leq i \leq n}$ are representatives of the isomorphism classes of the simple $\wedge/\underset{=}{p}\wedge$-modules. Since δ_p is an epimorphism, so is $\mu_p : \underset{=}{G}_0(\wedge) \longrightarrow \underset{=}{G}_0(\wedge/\underset{=}{p}\wedge)$ (cf. 3.4), and there are $\wedge$-lattices $\{M_i, N_i\}_{1 \leq i \leq n}$ such that $[\overline{Y}_i] = \mu_p([M_i]-[N_i])$; i.e., $[\overline{Y}_i] = [M_i/\underset{=}{p}M_i] - [N_i/\underset{=}{p}N_i], 1 \leq i \leq n$.

Consequently

$$x = \sum_{i=1}^{n} \alpha_i (\mathcal{G}_p[M_i/\underset{=}{p}M_i] - \mathcal{G}_p[N_i/\underset{=}{p}N_i]),$$

and it suffices to show that $\mathcal{G}_p[M_i/\underset{=}{p}M_i] \in \text{Im } \kappa$; we omit the index i. We choose an ideal $\underset{=}{a}$ of R which is coprime to the Higman ideal $\underset{=}{H}(\wedge)$ and which lies in the same ideal class as $\underset{=}{p}$ (cf. VII, 1.6). Then

$$\mathcal{G}_p([M/\underset{=}{p}M]) = [M]_0 - [\underset{=}{p}M]_0 = [M]_0 - [\underset{=}{a}M]_0.$$

Taking a projective presentation for $M/\underset{=}{a}M$,

$$0 \longrightarrow Q \longrightarrow P \longrightarrow M/\underset{=}{a}M \longrightarrow 0$$

with $P \in {}_{\Lambda}\underline{\underline{P}}^f$, we conclude $Q \in {}_{\Lambda}\underline{\underline{P}}^f$. In fact, for $\underline{\underline{q}} \nmid \underline{\underline{H}}(\Lambda)$, $Q_{\underline{\underline{q}}} \in {}_{\Lambda_{\underline{\underline{q}}}}\underline{\underline{P}}^f$ and for $\underline{\underline{q}} \mid \underline{\underline{H}}(\Lambda)$, $P_{\underline{\underline{q}}} \cong Q_{\underline{\underline{q}}}$ since $(M/\underline{\underline{a}}M)_{\underline{\underline{q}}} = 0$, and Q is projective. Since we have an isomorphism between $\underline{\underline{G}}_o(\Lambda)$, the Grothendieck group of all Λ-lattices and $\underline{\underline{G}}_o^f(\Lambda)$, the Grothendieck group of all finitely generated Λ-modules (cf. 3.1), the above sequence together with the sequence

$$0 \longrightarrow \underline{\underline{a}}M \longrightarrow M \longrightarrow M/\underline{\underline{a}}M \longrightarrow 0$$

shows

$$\varrho_{\underline{\underline{p}}}([M/\underline{\underline{p}}M]) = [M]_o - [\underline{\underline{a}}M]_o = [P]_o - [Q]_o,$$

and $\varrho_{\underline{\underline{p}}}([M/\underline{\underline{p}}M]) \in \operatorname{Im}\kappa$; i.e., μ is an epimorphism, and the above sequence is exact. #

3.13 <u>Theorem</u> (Heller-Reiner [5]): There is an exact sequence

$$\underline{\underline{K}}_1(A) \xrightarrow{\vartheta} \underline{\underline{G}}_o^T(\Lambda) \xrightarrow{\varrho} \underline{\underline{G}}_o(\Lambda) \xrightarrow{\iota} \underline{\underline{G}}_o(A) \longrightarrow 0.$$

<u>Proof</u>: From (2.9) it follows that

$$\underline{\underline{K}}_1(A) \cong \underline{\underline{G}}_1({}_A\underline{\underline{P}}^f) \cong \underline{\underline{G}}_1({}_A\underline{\underline{M}}^f),$$

and we shall define $\vartheta : \underline{\underline{G}}_1({}_A\underline{\underline{M}}^f) \longrightarrow \underline{\underline{G}}_o^T(\Lambda)$. Given a pair $[L,\alpha] \in \underline{\underline{G}}_1({}_A\underline{\underline{M}}^f)$; i.e., $L \in {}_A\underline{\underline{M}}^f$ and an automorphism α of L. We pick $M \in {}_{\Lambda}\underline{\underline{M}}^o$, such that $KM = L$ $(M \subset L)$. Then there exists $0 \neq r \in R$ such that $r\alpha|_M : M \longrightarrow M$ is a monomorphism.

We now define

$$\vartheta : [L,\alpha] \longmapsto [\operatorname{Coker}(r\alpha|_M)] - [\operatorname{Coker} r1_M].$$

Observe that both $\operatorname{Coker}(r\alpha|_M)$ and $\operatorname{Coker}(r1_M)$ are R-torsion Λ-modules.

(i) <u>ϑ is independent of the choice of M and r.</u>

We recall (cf. II, Ex. 2,1; 5) that for any ring S, given two monomorphisms of left S-modules

$$\sigma : M'' \longrightarrow M'; \ \tau : M' \longrightarrow M,$$

we have an exact sequence

(1) $$0 \longrightarrow \text{Coker}\,\sigma \longrightarrow \text{Coker}\,\sigma\tau \longrightarrow \text{Coker}\,\tau \longrightarrow 0.$$

If now, in the definition of ϑ we take the same M, but a different $0 \neq r_1 \in R$ such that $r_1\alpha|_M \in \text{End}_\Lambda(M)$, then the exact sequences

$$0 \longrightarrow \text{Coker}(r\alpha|_M) \longrightarrow \text{Coker}(rr_1\alpha|_M) \longrightarrow \text{Coker}(r_1 1_M) \longrightarrow 0,$$

$$0 \longrightarrow \text{Coker}(r1_M) \longrightarrow \text{Coker}(rr_1 1_M) \longrightarrow \text{Coker}(r_1 1_M) \longrightarrow 0,$$

lead to the relation

$$[\text{Coker}(r\alpha|_M)] - [\text{Coker}(r\cdot 1_M)] = [\text{Coker}(rr_1\alpha|_M)] - [\text{Coker}(rr_1 1_M)].$$

Similarly one shows

$$[\text{Coker}(r_1\alpha|_M)] - [\text{Coker}(r_1 1_M)] = [\text{Coker}(rr_1\alpha|_M)] - [\text{Coker}(rr_1 1_M)].$$

Thus ϑ is independent of the choice of $0 \neq r \in R$.

If now for $N \in {}_\Lambda\underline{\underline{M}}^0$ also KN = KM = L, $N \subset L$, we can pick $0 \neq r \in R$ such that

$$r\alpha|_M : M \longrightarrow M \text{ and } r\alpha|_N : N \longrightarrow N,$$

by the above result. It suffices to show (then take $\alpha = 1_L$)

$$[\text{Coker}(r\alpha|_M)] = [\text{Coker}(r\alpha|_N)].$$

Replacing N by r'N if necessary, we may assume $M \supset N$, since $[\text{Coker}(r\alpha|_N)] = [\text{Coker}(r\alpha|_{r'N})]$, where $r\alpha|_{r'N} : r'N \longrightarrow r'N$. The embedding $\iota : N \longrightarrow M$ and the monomorphism $r\alpha|_M : M \longrightarrow M$ give rise to the exact sequence

$$0 \longrightarrow M/N \longrightarrow M/Nr\alpha \longrightarrow M/Mr\alpha \longrightarrow 0;$$

thus to the relation

$$[\text{Coker}(r\alpha|_M)] = [M/Nr\alpha] - [M/N].$$

Similarly, the exact sequence

$$0 \longrightarrow N/Nr\alpha \longrightarrow M/Nr\alpha \longrightarrow M/N \longrightarrow 0,$$

gives

$$[N/Nr\alpha] = [M/Nr\alpha] - [M/N]; \text{ i.e.,}$$

$$[\text{Coker}(r\alpha|_M)] = [\text{Coker}(r\alpha|_N)],$$

and ϑ is independent of M and r.

(ii) ϑ <u>preserves relations</u>: Given an exact sequence

(2) $$0 \longrightarrow (L', \alpha') \xrightarrow{\sigma} (L, \alpha) \xrightarrow{\tau} (L", \alpha") \longrightarrow 0$$

in $\sum_{\underline{\underline{A}}} \underline{\underline{M}}^f$ (cf. 1.10). Let $M \in {}_{\Lambda}\underline{\underline{M}}^0$ be such that $KM = L$. We put $M' = L'\sigma \cap M$; then M' is an R-pure submodule of M such that $KM' = L'\sigma \cong L'$. We then have the exact sequence of Λ-lattices

$$0 \longrightarrow M' \xrightarrow{\sigma'} M \xrightarrow{\tau'} M" \longrightarrow 0,$$

where $M" = M/M'$, and σ', τ' are the canonical homomorphisms. W.l.o.g. we may replace L' by $L'\sigma$ and $L"$ by L/L' such that α' and $\alpha"$ are induced from the automorphism $\alpha: L \longrightarrow L$.

The exact sequence (2) gives the commutative diagram with exact rows

$$\begin{array}{ccccccccc} 0 & \longrightarrow & L' & \xrightarrow{\sigma} & L & \xrightarrow{\tau} & L" & \longrightarrow & 0 \\ & & \downarrow{\scriptstyle \alpha'} & & \downarrow{\scriptstyle \alpha} & & \downarrow{\scriptstyle \alpha"} & & \\ 0 & \longrightarrow & L' & \xrightarrow{\sigma} & L & \xrightarrow{\tau} & L" & \longrightarrow & 0. \end{array}$$

We choose $0 \neq r \in R$ such that $r\alpha|_M : M \longrightarrow M$; then $r\alpha'|_{M'} : M' \longrightarrow M'$ and $r\alpha"|_{M"} : M" \longrightarrow M"$.

We claim that we can complete the following diagram with exact rows and columns commutatively:

$$\begin{array}{ccccccccc} & & 0 & & 0 & & 0 & & \\ & & \downarrow & & \downarrow & & \downarrow & & \\ 0 & \longrightarrow & M' & \xrightarrow{\sigma'} & M & \xrightarrow{\tau'} & M" & \longrightarrow & 0 \\ & & \downarrow{\scriptstyle r\alpha'} & & \downarrow{\scriptstyle r\alpha} & & \downarrow{\scriptstyle r\alpha"} & & \\ 0 & \longrightarrow & M' & \xrightarrow{\sigma'} & M & \xrightarrow{\tau'} & M" & \longrightarrow & 0 \\ & & \downarrow{\scriptstyle \beta'} & & \downarrow{\scriptstyle \beta} & & \downarrow{\scriptstyle \beta"} & & \\ 0 & \dashrightarrow & T' & \overset{\gamma}{\dashrightarrow} & T & \overset{\delta}{\dashrightarrow} & T" & \dashrightarrow & 0 \\ & & \downarrow & & \downarrow & & \downarrow & & \\ & & 0 & & 0 & & 0 & & \end{array} ,$$

where $T = M/\operatorname{Im} r\alpha$, $T' = M'/\operatorname{Im} r\alpha'$ $T" = M"/\operatorname{Im} r\alpha"$, and where β', β and $\beta"$ are the canonical homomorphisms. Let us show first $Mr\alpha \cap M' = M'r\alpha'$.

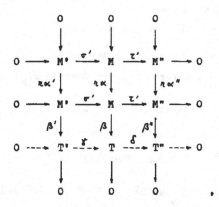

In fact, we have the inclusion $Mr\alpha \cap M' \supset M'r\alpha'$. Let now $m' = mr\alpha$ with $m' \in M'$, $m \in M$, and assume $m \notin M'$. Since M' is an R-pure submodule of M, $mr \notin M'$, and $mr \notin L'$. So $mr + L' \neq 0$ in L''; but $(mr + L')\alpha'' = 0$, a contradiction to the fact, that α'' is an automorphism. Hence

$$T' = M'/M'r\alpha' \subset M/Mr\alpha = T,$$

and we let γ be the injection $T' \hookrightarrow T$. But then

$$T'' = (M/M')/[(Mr\alpha + M')/M'] \cong M/(Mr\alpha + M'),$$

and

$$T/T' = (M/Mr\alpha)/(M'/M'r\alpha') \cong (M/Mr\alpha)/(M'/Mr\alpha' \cap M')$$

$$\cong (M/Mr\alpha)/(M' + Mr\alpha)/Mr\alpha \cong M/(Mr\alpha + M') \cong T'',$$

and we let δ be the canonical epimorphism $T \longrightarrow T''$. Thus the above diagram is commutative with exact rows and columns. Similarly one finds such a commutative diagram with $r1_{M'}$ in place of $r\alpha'|_{M'}$, $r1_M$ in place of $r\alpha|_M$ and $r1_{M''}$ in place of $r\alpha''|_{M''}$. Thus

$$\vartheta : ([L,\alpha] - [L',\alpha'] - [L'',\alpha'']) \longmapsto 0.$$

If now α and β are automorphisms of $L \in {}_{\Lambda}\underline{\underline{M}}^f$, then, choosing $M \in {}_{\Lambda}\underline{\underline{M}}^o$ with $KM = L$ and $0 \neq r \in R$ such that

$$r\alpha|_M : M \longrightarrow M \text{ and } r\beta|_M : M \longrightarrow M$$

are monomorphisms, we obtain the exact sequences (cf. (1)),

$$0 \longrightarrow \mathrm{Coker}(r\alpha|_M) \longrightarrow \mathrm{Coker}(rr\alpha\beta|_M) \longrightarrow \mathrm{Coker}(r\beta|_M) \longrightarrow 0,$$

$$0 \longrightarrow \mathrm{Coker}(r1_M) \longrightarrow \mathrm{Coker}(rr1_M) \longrightarrow \mathrm{Coker}(r1_M) \longrightarrow 0; \text{ i.e.,}$$

$$\vartheta : [L,\alpha\beta] - [L,\alpha] - [L,\beta] \longmapsto 0.$$

Thus we have shown that ϑ <u>is a group homomorphism</u>. *)

(iii) <u>The sequence in (3.13) is exact</u>. In view of (3.2;3.3) it suffices to establish exactness at $\underline{\underline{G}}_o^T(\Lambda)$. Let $x \in \mathrm{Im}\,\vartheta$; i.e., $x = [T_1] - [T_2]$, where

$$0 \longrightarrow M \longrightarrow M \longrightarrow T_1 \longrightarrow 0 \quad \text{and}$$

$$0 \longrightarrow M \longrightarrow M \longrightarrow T_2 \longrightarrow 0$$

are exact sequences. Then obviously $\varrho : x \longmapsto 0$; i.e., $\mathrm{Im}\,\vartheta \subset \mathrm{Ker}\,\varrho$.

---

*) In the proof of the existence of δ, we have not used that A is semisimple, and Heller [2][2] has generalized this sequence in (3.13) to certain classes of categories (cf. Bass [8]).

__Conversely__, let $\varrho : [T_1] - [T_2] \longmapsto 0$, and choose $M', M, N', N \in {}_{\Lambda}\underline{\underline{M}}^{o}$ such that

$$0 \longrightarrow M' \longrightarrow M \longrightarrow T_1 \longrightarrow 0 \text{ and}$$

$$0 \longrightarrow N' \longrightarrow N \longrightarrow T_2 \longrightarrow 0$$

are exact sequences. Moreover, we may assume $KM = KN$, $M' \subset M$, $N' \subset N$ and $T_1 = M/M'$, $T_2 = N/N'$. Since $\varrho[T_1] = \varrho[T_2]$, we have

$$[M' \oplus N] = [M \oplus N'] \text{ in } \underline{\underline{G}}_o(\Lambda).$$

Now we apply (1.5) and conclude that there exist $X, X', X'' \in {}_{\Lambda}\underline{\underline{M}}^{o}$ such that

$$0 \longrightarrow X' \longrightarrow M' \oplus N \oplus X \longrightarrow X'' \longrightarrow 0 \quad \text{and}$$

$$0 \longrightarrow X' \longrightarrow M \oplus N' \oplus X \longrightarrow X'' \longrightarrow 0$$

are exact sequences of Λ-lattices. We tensor these sequences with K over R and observe that any two extension of KX' by KX'' are congruent, A being semi-simple. Hence we obtain the commutative diagram with exact rows (cf. II, 5.1)

$$\begin{array}{ccccccccc} 0 & \longrightarrow & KX' & \longrightarrow & K(M' \oplus N \oplus X) & \longrightarrow & KX'' & \longrightarrow & 0 \\ & & \downarrow 1_{KX'} & & \downarrow \alpha & & \downarrow 1_{KX''} & & \\ 0 & \longrightarrow & KX' & \longrightarrow & K(M \oplus N' \oplus X) & \longrightarrow & KX'' & \longrightarrow & 0 \end{array} .$$

Observing $K(M' \oplus N \oplus X) = K(M \oplus N' \oplus X)$, we can choose $0 \neq r \in R$ such that

$$\beta = r\alpha|_{M' \oplus N \oplus X} : M' \oplus N \oplus X \longrightarrow M \oplus N' \oplus X,$$

is a monomorphism. As above we obtain a commutative diagram with exact rows and columns

$$\begin{array}{ccccccccc} & & 0 & & 0 & & 0 & & \\ & & \downarrow & & \downarrow & & \downarrow & & \\ 0 & \longrightarrow & X' & \xrightarrow{\sigma} & M' \oplus N \oplus X & \xrightarrow{\tau} & X'' & \longrightarrow & 0 \\ & & \downarrow r1_{X'} & & \downarrow \beta & & \downarrow r1_{X''} & & \\ 0 & \longrightarrow & X' & \xrightarrow{\varphi} & M \oplus N' \oplus X & \xrightarrow{\psi} & X'' & \longrightarrow & 0 \\ & & \downarrow & & \downarrow & & \downarrow & & \\ 0 & \longrightarrow & T' & \longrightarrow & T & \longrightarrow & T'' & \longrightarrow & 0 \\ & & \downarrow & & \downarrow & & \downarrow & & \\ & & 0 & & 0 & & 0 & & \end{array} .$$

This gives rise to the relation

$$[\mathrm{Coker}\beta] = [\mathrm{Coker}(r1_{X'})] + [\mathrm{Coker}(r\cdot 1_{X''})].$$

We also have the relation

$$[\mathrm{Coker}(r1_{M'\oplus N\oplus X})] = [\mathrm{Coker}(r1_{X'})] + [\mathrm{Coker}(r1_{X''})].$$

Thus

$$[\mathrm{Coker}\beta] = [\mathrm{Coker}(r1_{M'\oplus N\oplus X})] \in \mathrm{Im}\,\vartheta.$$

Let $0 \neq r_1 \in R$ with

$$r_1(M\oplus N'\oplus X) \hookrightarrow M' \oplus N \oplus X, \text{ componentwise.}$$

Then the two maps

$$\beta: M' \oplus N \oplus X \longrightarrow M \oplus N' \oplus X$$

and

$$\gamma_{r_1} : M \oplus N' \oplus X \longrightarrow M' \oplus N \oplus X,$$

where γ_{r_1} is multiplication with r_1, yield the relation

$$[\mathrm{Coker}\,\gamma_{r_1}\beta] = [\mathrm{Coker}\,\gamma_{r_1}] + [\mathrm{Coker}\,\beta] = [\mathrm{Coker}\,\beta] + [M'/Mr_1] + [N/N'r_1] + [X/r_1X].$$

Thus:

$$\vartheta(\alpha) = [\mathrm{Coker}\,\gamma_{r_1}\beta] - [\mathrm{Coker}(rr_1 1_{M'\oplus N\oplus X})]$$

$$= [\mathrm{Coker}\beta] - [\mathrm{Coker}(rr_1 1_{M'\oplus N\oplus X})] + [M'/Mr_1] + [N/N'r_1] + [X/r_1X].$$

However, the two pairs of maps

$$M \xrightarrow{r_1} M' \; ; \; M' \hookrightarrow M \text{ and } N' \xrightarrow{r_1} N' \; ; \; N' \hookrightarrow N$$

yield

$$[M/r_1M] = [M'/r_1M] + [M/M'] = [M'/r_1M] + [T_1],$$

$$[N/r_1N'] = [N'/r_1N'] + [N/N'] = [N'/r_1N'] + [T_2].$$

Summarizing, we obtain

$$[T_1] - [T_2] = -\vartheta(\alpha) + [\mathrm{Coker}\beta] - [\mathrm{Coker}(rr_1 1_{M'\otimes N\otimes X})]$$
$$+ [M/r_1M] + [N'/r_1N'] + [X/r_1X] \in \mathrm{Im}\,\vartheta,$$

since $[\mathrm{Coker}\beta] \in \mathrm{Im}\,\vartheta$. This finishes the proof of the exactness of our sequence. #

3.14 Corollary: Let U denote the group of units in A and let $[U,U]$ be its commutator subgroup, then we have an exact sequence

$$U/[U,U] \xrightarrow{\vartheta'} \underline{\underline{G}}_o^T(\Lambda) \xrightarrow{\varrho} \underline{\underline{G}}_o(\Lambda) \xrightarrow{\iota} \underline{\underline{G}}_o(A) \longrightarrow 0.$$

Proof: A is semi-perfect, and the statement follows from (2.5) and (3.13), if one observes that $\underline{\underline{G}}_o^T(\Lambda)$ is commutative. #

3.15 Corollary: Let K be an algebraic number field with Dedekind domain R. If $A = \oplus_{i=1}^n A_i$ is the decomposition of A into simple K-algebras A_i with center $(A_i) = F_i$, then we have an exact sequence

$$\prod_{i=1}^n St_{F_i}(A_i) \xrightarrow{\vartheta''} \underline{\underline{G}}_o^T(\Lambda) \xrightarrow{\varrho} \underline{\underline{G}}_o(\Lambda) \xrightarrow{\iota} \underline{\underline{G}}_o(A) \longrightarrow 0,$$

where $St_{F_i}(A_i)$ is the ray of A_i over F_i.

Proof: This follows from (3.13) together with (2.12). #

Exercises §3:

We retain the notation of the previous sections.

1.) Let $\underline{\underline{C}}$ be an admissible subcategory of ${}_S\underline{\underline{M}}^f$, where S is a ring (cf. 1.1). If

$$0 \longrightarrow M_1 \longrightarrow M_2 \longrightarrow M_3 \longrightarrow M_4 \longrightarrow 0$$

is an exact sequence with $M_i \in \underline{\underline{C}}, 1 \le i \le 4$, show that in $\underline{\underline{G}}_o(\underline{\underline{C}})$,

$$[M_2] - [M_3] = [M_1] - [M_4].$$

2.) Let R be a commutative ring and $M,N \in {}_R\underline{\underline{M}}^f$; show that $\mathrm{Tor}_1^R(M,N) \overset{\mathrm{nat}}{\cong} \mathrm{Tor}_1^R(N,M)$.

3.) Heller-Reiner [5]: Define a map

$$\vartheta_1 : \underline{\underline{G}}_1({}_A\underline{\underline{M}}^f) \longrightarrow \underline{\underline{G}}_o^T(\Lambda)$$

as follows: Given $[L, \alpha] \in \underline{\underline{G}}_1({}_{\Lambda}\underline{\underline{M}}^f)$, choose $M \in {}_{\Lambda}\underline{\underline{M}}^o$ such that $KM = L$, and put

$$\vartheta_1 : [L, \alpha] \longmapsto [M/(\mathrm{Im}\,\alpha|_M \cap M)] - [\mathrm{Im}(\alpha|_M)/(\overline{\mathrm{Im}}\alpha|_M \cap M) \quad \text{in } \underline{\underline{G}}_o^T(\Lambda).$$

Show $\vartheta = \vartheta_1$, where ϑ is defined in the proof of (3.13).

4.) Construct explicitely a map

$$\vartheta_o : \underline{\underline{K}}_1(A) \longrightarrow \underline{\underline{G}}_o^T(\Lambda)$$

such that the following diagram commutes

$$\begin{array}{ccc} \underline{\underline{G}}_1({}_{A}\underline{\underline{M}}^f) & \xrightarrow{\vartheta} & \\ \wr\,| & & \underline{\underline{G}}_o^T(\Lambda) \\ \underline{\underline{K}}_1(A) & \xrightarrow{\vartheta_o} & \end{array} .$$

5.) Let $\underline{\underline{A}}$ be a real matrix. Show that rank $\underline{\underline{A}}$ = rank$(\underline{\underline{A}}^t\underline{\underline{A}})$.

6.) Prove the <u>9-lemma</u>. Given a commutative diagram

$$\begin{array}{ccccccccc} & & 0 & & 0 & & 0 & & \\ & & \uparrow & & \uparrow & & \uparrow & & \\ 0 & \longrightarrow & X' & \longrightarrow & X & \longrightarrow & X'' & \longrightarrow & 0 \\ & & \uparrow & & \uparrow & & \uparrow & & \\ 0 & \longrightarrow & Y' & \longrightarrow & Y & \longrightarrow & Y'' & \longrightarrow & 0 \\ & & \uparrow & & \uparrow & & \uparrow & & \\ 0 & \dashrightarrow & Z' & \dashrightarrow & Z & \dashrightarrow & Z'' & \dashrightarrow & 0 \\ & & \uparrow & & \uparrow & & \uparrow & & \\ & & 0 & & 0 & & 0 & & \end{array}$$

with exact rows and columns; show that one can fill in the diagram commutatively such that the bottom tow is exact. Reverse the arrows and prove the dual statement.

7.) Give a <u>different proof of 3.4</u>: If $\underline{\underline{a}}$ is a non-zero ideal in R. Then the functor $- \boxtimes_R \underline{\underline{a}}$ induces an automorphism

$$\alpha : \underline{\underline{G}}_o^f(\Lambda) \longrightarrow \underline{\underline{G}}_o^f(\Lambda),$$
$$[M] \longmapsto [M \boxtimes_R \underline{\underline{a}}].$$

Moreover, α acts as the identity on the subgroup $\underline{\underline{G}}_o^T(\Lambda)$ consisting of the equivalence classes of R-torsion Λ-modules.

<u>Proof</u>: Obviously, $- \otimes_R \underline{\underline{a}}$ induces a group-homomorphism

$$\alpha : \underline{\underline{G}}_o^f(\Lambda) \longrightarrow \underline{\underline{G}}_o^f(\Lambda),$$

$$[M] \longmapsto [M \otimes_R \underline{\underline{a}}],$$

$\underline{\underline{a}}$ being R-projective. If T is an R-torsion Λ-module, we have $\alpha[T] = [T]$ (cf. proof of 3.4). Thus α leaves $\underline{\underline{G}}_o^T(\Lambda)$ elementwise fixed. Now, let $M \in {}_\Lambda\underline{\underline{M}}^o$, then M is R-flat and applying $M \otimes_R -$ to the exact sequence

$$0 \longrightarrow \underline{\underline{a}} \longrightarrow R \longrightarrow R/\underline{\underline{a}} \longrightarrow 0$$

we conclude $[M] - \alpha[M] = [M/\underline{\underline{a}}M]$; since $\alpha[M/\underline{\underline{a}}M] = [M/\underline{\underline{a}}M]$ we find

$$\alpha[M] - \alpha^2[M] = [M] - \alpha[M];$$

i.e., $\alpha(2-\alpha)[M] = [M]$. However, $\underline{\underline{G}}_o^f(\Lambda)$ is generated by Λ-lattices, and so

$$1_{\underline{\underline{G}}_o^f(\Lambda)} = \alpha(2-\alpha) = (2-\alpha)\alpha;$$

i.e., α must be both an epimorphism and a monomorphism; whence it is an automorphism leaving $\underline{\underline{G}}_o^T(\Lambda)$ elementwise fixed.

Now $\mu'_{\underline{\underline{a}}} = 1-\alpha : \underline{\underline{G}}_o^f(\Lambda) \longrightarrow \underline{\underline{G}}_o(\Lambda/\underline{\underline{a}}\Lambda)$, where we consider $\underline{\underline{G}}_o(\Lambda/\underline{\underline{a}}\Lambda) \subset \underline{\underline{G}}_o^f(\Lambda)$. Now, using the isomorphism (3.1), we have $\mu_{\underline{\underline{a}}} = 1-\alpha'$, where α' is induced by α and it is clear that $\operatorname{Ker}\iota \subset \operatorname{Ker}\mu_{\underline{\underline{a}}}$; whence the existence of $\delta_{\underline{\underline{a}}}$.

8.) Simplify the proof of (3.2) by using the isomorphism $\underline{\underline{G}}_o(\Lambda) \cong \underline{\underline{G}}_o^f(\Lambda)$!

§4 Grothendieck groups and genera

For a Λ-lattice M we study the Grothendieck group $\underline{\underline{D}}_M$, and derive that $M \vee N$ if and only if $M^{(s)} \cong N^{(s)}$ for some integer s. Moreover, there are only finitely many non-isomorphic indecomposable direct summands of $M^{(s)}$ for all $s \in \underline{\underline{N}}$. Prime ideals are assumed to be different from zero.

Let K be an $\underline{\underline{A}}$-field with Dedekind domain R; A is a finite dimensional separable K-algebra and Λ is an R-order in A. We remark, that all results in this section hold if R is any Dedekind domain with quotient field K such that the Jordan-Zassenhaus theorem is valid for Λ-lattices.

4.1 Proposition (Jacobinski [4]): For $M \in {}_{\Lambda}\underline{\underline{M}}^{o}$, $\underline{\underline{D}}_M$ is a finitely generated $\underline{\underline{Z}}$-module.

Proof: We recall that $\underline{\underline{D}}_M$ is the Grothendieck group generated by all direct summands of $M^{(n)}$ for some $n \in \underline{\underline{N}}$ (cf. 1.9). Let Γ be a maximal R-order in A containing Λ and let $\underline{\underline{S}}_o$ be a finite non-empty subset of spec R containing all prime ideals $\underline{\underline{p}}$ for which $\Gamma_{\underline{\underline{p}}} \neq \Lambda_{\underline{\underline{p}}}$. Then we have a $\underline{\underline{Z}}$-homomorphism

$$\varphi : \underline{\underline{D}}_M \longrightarrow \oplus_{\underline{\underline{p}} \in \underline{\underline{S}}_o} \underline{\underline{D}}_{\hat{M}_{\underline{\underline{p}}}} ,$$

$$\langle N \rangle \longmapsto (\langle \hat{N}_{\underline{\underline{p}}} \rangle)_{\underline{\underline{p}} \in \underline{\underline{S}}_o} .$$

However, the Krull-Schmidt theorem is valid for ${}_{\hat{\Lambda}_{\underline{\underline{p}}}}\underline{\underline{M}}^{o}$ (cf. VI, 3.1), and so every $\hat{N}_{\underline{\underline{p}}}$ has a unique decomposition into indecomposable modules; i.e., $\underline{\underline{D}}_{\hat{M}_{\underline{\underline{p}}}}$ is a free abelian group on $\{\langle \hat{N}_{i_{\underline{\underline{p}}}} \rangle\}_{1 \le i \le n_{\underline{\underline{p}}}}$, where $\{\hat{N}_{i_{\underline{\underline{p}}}}\}_{1 \le i \le n_{\underline{\underline{p}}}}$ are the non-isomorphic indecomposable direct summands of $\hat{M}_{\underline{\underline{p}}}$, $\underline{\underline{p}} \in \underline{\underline{S}}_o$. Thus, $\oplus_{\underline{\underline{p}} \in \underline{\underline{S}}_o} \underline{\underline{D}}_{\hat{M}_{\underline{\underline{p}}}}$ is a free abelian group with a finite basis.

We have

$$\operatorname{Ker} \varphi = \{\langle N_1 \rangle - \langle N_2 \rangle : N_1 \vee N_2;\ N_1, N_2 \underline{\underline{|}}\ M^{(s)}, s \in \underline{\underline{N}}\}$$

by (1.7), (VII, 1.9) and because of the validity of the Krull-Schmidt theorem. (We recall that $N_1 \vee N_2$ indicates that N_1 and N_2 lie in the same genus and $X_{\cong} | Y$ means that X is isomorphic to a direct summand of Y. Also we would like to remind that all results on genera remain valid if the localizations are replaced by the completions (cf. VII, 1.2).)

Given $\langle N_1\rangle - \langle N_2\rangle \in \operatorname{Ker}\varphi$; i.e., $N_1 \vee N_2$. Then $e_M e_{N_i} = e_{N_i}, i=1,2$ (cf. VII, 2.4), where e_M is the unique minimal central idempotent in A such that $e_M M = M$. Therefore we may apply (VII, 3.4) to conclude that there exists $M_1 \vee M$ such that

$$M \oplus N_1 \cong M_1 \oplus N_2; \text{ i.e.,}$$

$$\langle M\rangle - \langle M_1\rangle = \langle N_1\rangle - \langle N_2\rangle .$$

By the Jordan-Zassenhaus theorem (VI, 4.5;4.7) there are only finitely many non-isomorphic lattices in the same genus as M; i.e.,

* Ker φ is a finite abelian group.

The exact sequence

$$0 \longrightarrow \operatorname{Ker}\varphi \longrightarrow \underline{\underline{D}}_M \xrightarrow{\varphi} \bigoplus_{\underline{\underline{p}} \in \underline{\underline{S}}_o} \underline{\underline{D}}_{\hat{M}_{\underline{\underline{p}}}}$$

shows that $\underline{\underline{D}}_M$ is of finite type (cf. I, 4.3), $\underline{\underline{Z}}$ being noetherian. #

4.2 __Theorem__ (Jacobinski [4], Roiter [4]): For $M,N \in {}_\Lambda\underline{\underline{M}}^o$, we have

$$M \vee N \text{ if and only if } M^{(s)} \cong N^{(s)}$$

for some positive integer s.

__Proof:__ In (VI, 3.6) we have shown that $M^{(s)} \cong N^{(s)}$ implies $M \vee N$. Let us therefore assume $M \vee N$. Then $\langle N\rangle \in \underline{\underline{D}}_M$; in fact, it follows from (VII, 3.4) that there exists $N_1 \vee M$ such that

$$M^{(2)} \cong N \oplus N_1 .$$

Thus $\langle M\rangle - \langle N\rangle \in \operatorname{Ker}\varphi$ (cf. proof of 4.1), and Ker φ being a finite abelian group (cf. *), there exists a positive integer s' such that

$$\langle M^{(s')}\rangle = \langle N^{(s')}\rangle \text{ in } \underline{\underline{D}}_M;$$

by (1.7) we can find a positive integer $t \geq s'$ with

$$M^{(t)} \cong N^{(s')} \oplus M^{(t-s')}.$$

Applying (VII, 3.4) successively, we find for every positive integer n,

$$N^{(n)} \cong M^{(n-1)} \oplus N_n,\ N_n \vee M.$$

However, up to isomorphism, there are only finitely many lattices $N_i \vee M$. Consequently there exists $N_o \vee M$ and an infinite increasing sequence of natural numbers $\{n_i\}_{i=1,2,\ldots}$, where each n_i is a multiple of s', with

$$N^{(n_i)} \cong M^{(n_i-1)} \oplus N_o,\ i=1,2\ldots .$$

Let $n_1 = s'm$ and choose $n_i > tm$. Then

$$N^{(n_i)} \cong M^{(n_i-tm)} \oplus M^{([t-s']m)} \oplus M^{(n_1-1)} \oplus N_o$$

$$\cong M^{(n_i-tm)} \oplus M^{([t-s']m)} \oplus N^{(s'm)}$$

$$\cong M^{(n_i-tm)} \oplus M^{(tm)} = M^{(n_i)}.$$

Thus $s = n_i$ has the required property. #

4.3 <u>Theorem</u> (Jacobinski [4], Jones [1]): Let $M \in {}_{\Lambda}\underline{\underline{M}}^o$. Then there are only finitely many non-isomorphic indecomposable Λ-lattices $N \in {}_{\Lambda}\underline{\underline{M}}^o$ such that $N_{\cong} \mid M^{(s)}$ for some natural number s; i.e., the class

$$\underline{\underline{E}}_M = \{N \in {}_{\Lambda}\underline{\underline{M}}^o : N_{\cong} \mid M^{(s)},\ s \in \underline{\underline{N}}\}$$

has only finitely many non-isomorphic indecomposable objects.

We <u>remark</u> that this result has not actually been proved by Jones; however, once the concept of $\underline{\underline{D}}_M$ is defined, (4.3) follows from a technique introduced by A. Jones.

<u>Proof</u>: For a fixed positive integer t, let $\underline{\underline{F}}_t$ be the set of all t-tuples of natural numbers, not all zero. We order $\underline{\underline{F}}_t$ partially by

$$(n_i)_{1\leq i\leq t} \leq (n_i')_{1\leq i\leq t} \text{ if } n_i \leq n_i',\ 1\leq i\leq t.$$

<u>Claim</u>: Every non-empty subset $\underline{\underline{F}}$ of $\underline{\underline{F}}_t$ has only a finite number of minimal elements with respect to "$\leq$".

Proof: We use induction on t. The result is trivial for t = 1. Let $(n_i')_{1 \le i \le t}$ be a fixed element of $\underline{\underline{F}}$. For each $1 \le k \le t$ and for each $0 \le s \le n_k'$, the set

$$\underline{\underline{F}}_k^s = \{(n_i) \in \underline{\underline{F}} : n_k = s\}$$

has only a finite number of minimal elements by the induction hypothesis. Let $\underline{\underline{F}}_{min}$ be the set consisting of all minimal elements in $\bigcup_{\substack{1 \le k \le t \\ 0 \le s \le n_k'}} \underline{\underline{F}}_k^s$. Then $\underline{\underline{F}}_{min}$ is a finite set. It remains to show that every minimal element of $\underline{\underline{F}}$ lies in $\underline{\underline{F}}_{min}$. Let $(n_i)_{1 \le i \le t}$ be a minimal element of $\underline{\underline{F}}$. Then $(n_i)_{1 \le i \le t} \not> (n_i')_{1 \le i \le t}$, and thus, for some $1 \le k \le t, 0 \le s \le n_k'$ we have $(n_i)_{1 \le i \le t} \in \underline{\underline{F}}_k^s$; hence $\underline{\underline{F}}$ has only finitely many minimal elements. This proves the claim.

We now turn to the proof of (4.3) and, retaining the notation of the proof of (4.1), we consider the map:

$$\varphi : \underline{\underline{D}}_M \longrightarrow \oplus_{\underline{\underline{p}} \in \underline{\underline{S}}_o} \underline{\underline{D}}_{\hat{M}_{\underline{\underline{p}}}} .$$

Since $\underline{\underline{D}}_{\hat{M}_{\underline{\underline{p}}}} \xrightarrow{\sim} \underline{\underline{Z}}^{(n_{\underline{\underline{p}}})}$, $\langle \hat{N}_{\underline{\underline{p}}} \rangle \longmapsto (n_i)_{1 \le i \le n_{\underline{\underline{p}}}}$, if $\hat{N}_{\underline{\underline{p}}} \cong \oplus_{i=1}^{n_{\underline{\underline{p}}}} \hat{N}_{i_{\underline{\underline{p}}}}^{(n_i)}$, we obtain a homomorphism

$$\varphi' : \underline{\underline{D}}_M \longrightarrow \oplus_{\underline{\underline{p}} \in \underline{\underline{S}}_o} \underline{\underline{Z}}^{(n_{\underline{\underline{p}}})} .$$

We now consider $\underline{\underline{F}} = \{ \varphi'\langle N\rangle \;:\; N_{\cong} \,|\, M^{(s)}$, for some natural number $s\}$. According to the claim, $\underline{\underline{F}}$ has only a finite number of minimal elements under the partial ordering "$\le$". To complete the proof, it suffices to show that N decomposes if $\varphi'\langle N\rangle$ is not minimal. But if for a given $N_{\cong} \,|\, M^{(s)}$ there exists $N_1 \,|\, M^{(s')}$ such that $\varphi'\langle N\rangle > \varphi'\langle N_1\rangle$, then N_1 is locally a direct summand of N (cf. Ex. 4,1) and thus N decomposes by (VII, 3.8). Hence there are only finitely many non-isomorphic indecomposable Λ-lattices N, with $N_{\cong} \,|\, M^{(s)}$ for some natural number s.#

4.4 Corollary: There are only finitely many non-isomorphic indecomposable projective Λ-lattices.

<u>Proof</u>: This follows readily from (4.3) if we take ${}_\Lambda\Lambda$ for M. #

4.5 <u>Notation and Remark</u>: For the rest of this section we assume that R is any Dedekind domain with quotient field K. Then $({}_\Lambda\underline{\underline{M}}^o, \oplus)$ is a category satisfying (1.1,ii,iii) and we write

$\underline{\underline{K}}_o({}_\Lambda\underline{\underline{M}}^o)$ for its Grothendieck group relative to $\oplus$.

This should not be confused with $\underline{\underline{K}}_o(\Lambda)$ which is the Grothendieck group of all projective Λ-lattices. Similarly for the localizations and completions.

We have seen in (VII, 3.10) that every Λ-lattice X in the genus of $M_1 \oplus M_2$ decomposes as $X = N_1 \oplus N_2$ with $N_1 \vee M_1$ and $N_2 \vee M_2$. Thus we have the commutative semi-group (with cancellation cf. VI, 3.5)

$$\mathcal{G}({}_\Lambda\underline{\underline{M}}^o) = \{\mathcal{G}(M) : M \in {}_\Lambda\underline{\underline{M}}^o\}$$

with addition

$$\mathcal{G}(M) + \mathcal{G}(N) = \mathcal{G}(M \oplus N).$$

We embed $\mathcal{G}({}_\Lambda\underline{\underline{M}}^o)$ into a universal abelian group denoted by $\underline{\underline{K}}_o({}_\Lambda\mathcal{G})$.

4.6 <u>Lemma</u> (Faddeev [1]): We have for every $\underline{\underline{p}} \in$ spec R the following commutative diagram

$$\begin{array}{ccccccccc} & & & & & & \underline{\underline{K}}_o({}_{\hat\Lambda_{\underline{\underline{p}}}}\underline{\underline{M}}^o) & & \\ & & & & & \overset{\varphi_3}{\nearrow} & & \overset{\varphi_4}{\searrow} & \\ \underline{\underline{K}}_o({}_\Lambda\underline{\underline{M}}^o) & \xrightarrow{\varphi_1} & \underline{\underline{K}}_o({}_\Lambda\mathcal{G}) & \xrightarrow{\varphi_2} & \underline{\underline{K}}_o({}_{\Lambda_{\underline{\underline{p}}}}\underline{\underline{M}}^o) & & & & \underline{\underline{K}}_o(\hat\Lambda_{\underline{\underline{p}}}), \\ & & & & & \underset{\varphi_5}{\searrow} & & \underset{\varphi_6}{\nearrow} & \\ & & & & & & \underline{\underline{K}}_o(A) & & \end{array}$$

where

$$\varphi_1 : \underline{\underline{K}}_o({}_\Lambda\underline{\underline{M}}^o) \longrightarrow \underline{\underline{K}}_o({}_\Lambda\mathcal{G}),$$

$$\langle M \rangle \longmapsto \langle \mathcal{G}(M) \rangle \text{ is epic,}$$

$$\varphi_2 : \underline{\underline{K}}_o({}_\Lambda\mathcal{G}) \longrightarrow \underline{\underline{K}}_o({}_{\Lambda_{\underline{\underline{p}}}}\underline{\underline{M}}^o),$$

$$\langle \mathcal{G}(M) \rangle \longmapsto \langle M_{\underline{\underline{p}}} \rangle \quad \text{is epic.}$$

φ_i, $3 \leq i \leq 6$, are defined by means of tensoring with the appropriate rings. φ_3 and φ_6 are monic, φ_4 and φ_5 are epic. In particular,

$$\varphi_3(\underline{\underline{K}}_o({}_{\Lambda_{\underline{\underline{p}}}}\underline{\underline{M}}^o)) = \varphi_4^{-1}(\varphi_6\underline{\underline{K}}_o(A)).$$

Proof: Since A is semi-simple, $\underline{\underline{K}}_o(A) = \underline{\underline{G}}_o(A)$, and it is clear, that all the above maps are group homomorphisms. The last statement is precisely the reformulation of (IV, 1.9). The other facts are easily proved. To show that φ_2 is epic, let $M(\underline{\underline{p}}) \in {}_{\Lambda_{\underline{\underline{p}}}}\underline{\underline{M}}^o$ be given. By (IV, 1.12) there exists $M \in {}_\Lambda\underline{\underline{M}}^o$ such that $M_{\underline{\underline{p}}} \cong M(\underline{\underline{p}})$, and hence $\varphi_2 : \langle \mathfrak{g}(M)\rangle \longmapsto \langle M(\underline{\underline{p}})\rangle$. #

4.7 Lemma (Faddeev [1]): If Γ is a maximal R-order in A, then

$$\underline{\underline{K}}_o({}_\Gamma\mathfrak{g}) \cong \underline{\underline{K}}_o({}_{\Gamma_{\underline{\underline{p}}}}\underline{\underline{M}}^o) \cong \underline{\underline{K}}_o(A)$$

and

$$\underline{\underline{K}}_o({}_{\hat{\Gamma}_{\underline{\underline{p}}}}\underline{\underline{M}}^o) \cong \underline{\underline{K}}_o(\hat{A}_{\underline{\underline{p}}}).$$

Proof: Obviously, $\underline{\underline{K}}_o({}_{\Gamma_{\underline{\underline{p}}}}\underline{\underline{M}}^o) \cong \underline{\underline{K}}_o(A)$ and $\underline{\underline{K}}_o({}_{\hat{\Gamma}_{\underline{\underline{p}}}}\underline{\underline{M}}^o) \cong \underline{\underline{K}}_o(\hat{A}_{\underline{\underline{p}}})$ (cf. IV, 5.6; IV, 5.7), and in view of (4.6) it remains to show that φ_2 is monic. But $\varphi_2 \langle \mathfrak{g}(M_1)\rangle = \varphi_2 \langle \mathfrak{g}(M_2)\rangle$ implies $M_{1_{\underline{\underline{p}}}} \cong M_{2_{\underline{\underline{p}}}}$, since locally cancellation is allowed (cf. VI, 3.5). But then $KM_1 \cong KM_2$ and $M_1 \vee M_2$; i.e., $\mathfrak{g}(M_1) = \mathfrak{g}(M_2)$. #

Remark: If we consider only the semi-groups $\underline{\underline{K}}_o({}_\Lambda\underline{\underline{M}}^o)^+$ generated by the isomorphism classes (M) of $M \in {}_\Lambda\underline{\underline{M}}^o$ with addition $(M) + (N) = (M \oplus N)$ and, similarly $\underline{\underline{K}}_o({}_\Lambda\mathfrak{g})^+$ etc. then (4.6) and (4.7) remain valid for these semi-groups. In addition, in $\underline{\underline{K}}_o({}_{\Lambda_{\underline{\underline{p}}}}\underline{\underline{M}}^o)^+$ and in $\underline{\underline{K}}^o({}_\Lambda\mathfrak{g})^+$ cancellation is allowed (cf. VI, 3.5).

4.8 Theorem (Faddeev [1]): Let $\{L_i\}_{1 \leq i \leq s}$ be a complete set of non-isomorphic simple left A-modules. For every $1 \le i \le s$, let $M_i \in {}_\Lambda\underline{\underline{M}}^o$ be such that $L_i \cong KM_i$. Denote by $\underline{\underline{K}}_o(\{M_i\}_{1 \le i \le s})$ the subgroup of $\underline{\underline{K}}_o({}_\Lambda\underline{\underline{M}}^o)$ generated by the $\langle M_i \rangle$, $1 \le i \le s$. We define

$$\underline{\underline{K}}_o^I({}_{\Lambda_{\underline{\underline{p}}}}\underline{\underline{M}}^o) = \varphi_2\varphi_1\underline{\underline{K}}_o(\{M_i\}_{1 \le i \le s})$$

and

$$\underline{\underline{K}}_o^I({}_\Lambda\mathfrak{g}) = \varphi_1\underline{\underline{K}}_o(\{M_i\}_{1 \le i \le s}).$$

Then we have an isomorphism of abelian groups

$$\psi : \underline{\underline{K}}_o({}_\Lambda\mathfrak{g})/\underline{\underline{K}}_o^I({}_\Lambda\mathfrak{g}) \longrightarrow \oplus_{\underline{\underline{p}} \in \underline{\underline{S}}_o} \underline{\underline{K}}_o({}_{\Lambda_{\underline{\underline{p}}}}\underline{\underline{M}}_o)/\underline{\underline{K}}_o^I({}_{\Lambda_{\underline{\underline{p}}}}\underline{\underline{M}}^o),$$

where $\underline{\underline{S}}_o$ is the set of all prime ideals in R, such that $\Lambda_{\underline{\underline{p}}}$ is not maximal.

Proof: If Λ is maximal, then $\underline{\underline{K}}_o({}_\Lambda\mathfrak{g}) = \underline{\underline{K}}_o^I({}_\Lambda\mathfrak{g})$ by (4.7), and $\underline{\underline{K}}_o({}_{\Lambda_{\underline{\underline{p}}}}\underline{\underline{M}}^o) = \underline{\underline{K}}_o^I({}_{\Lambda_{\underline{\underline{p}}}}\underline{\underline{M}}^o)$, and there is nothing to prove. Thus we may assume that Λ is not maximal and $\underline{\underline{S}}_o$ is not empty. To define ψ, let $\langle\mathfrak{g}(M)\rangle \in \underline{\underline{K}}_o({}_\Lambda\mathfrak{g})$ be given. Then there exists exactly one $\langle\mathfrak{g}(N)\rangle \in \underline{\underline{K}}_o^I({}_\Lambda\mathfrak{g})$ such that $KM \cong KN$. Moreover, for almost all $\underline{\underline{p}} \in$ spec R we have $M_{\underline{\underline{p}}} \cong N_{\underline{\underline{p}}}$ (cf. IV, 1.8); in particular, this isomorphism exists for all $\underline{\underline{p}} \notin \underline{\underline{S}}_o$ (cf. IV, 5.7). We now define

$$\psi' : \underline{\underline{K}}_o({}_\Lambda\mathfrak{g}) \longrightarrow \oplus_{\underline{\underline{p}} \in \underline{\underline{S}}_o} \underline{\underline{K}}_o({}_{\Lambda_{\underline{\underline{p}}}}\underline{\underline{M}}^o)/\underline{\underline{K}}_o^I({}_{\Lambda_{\underline{\underline{p}}}}\underline{\underline{M}}^o),$$

$$\langle\mathfrak{g}(M)\rangle \longmapsto \langle M_{\underline{\underline{p}}}\rangle_{M_{\underline{\underline{p}}} \not\cong N_{\underline{\underline{p}}}} + (\oplus_{\underline{\underline{p}} \in \underline{\underline{S}}_o} \underline{\underline{K}}_o^I({}_{\Lambda_{\underline{\underline{p}}}}\underline{\underline{M}}^o)).$$

Then ψ' is a well-defined group homomorphism. Moreover, it is an epimorphism. To prove this, it suffices to show that for $\langle M(\underline{\underline{p}})\rangle \in \underline{\underline{K}}_o({}_{\Lambda_{\underline{\underline{p}}}}\underline{\underline{M}}^o)$ there exists $M \in {}_\Lambda\underline{\underline{M}}^o$ such that

$$\psi' : \langle\mathfrak{g}(M)\rangle \longmapsto \langle M(\underline{\underline{p}})\rangle + (\oplus_{\underline{\underline{p}} \in \underline{\underline{S}}_o} \underline{\underline{K}}_o^I({}_{\Lambda_{\underline{\underline{p}}}}\underline{\underline{M}}^o)).$$

We choose the unique $\langle\mathfrak{g}(N)\rangle \in \underline{\underline{K}}_o^I({}_\Lambda\mathfrak{g})$ such that $KN \cong KM(\underline{\underline{p}})$. By

(IV, 1.8) there exists $M \in {}_{\Lambda}\underline{\underline{M}}^{o}$ such that $M_{\underline{\underline{p}}} \cong M(\underline{\underline{p}})$ and $M_{\underline{\underline{q}}} \cong N_{\underline{\underline{q}}}$ for all $\underline{\underline{q}} \neq \underline{\underline{p}}$. Then

$$\psi' : \langle \mathfrak{g}(M)\rangle \longmapsto \langle M(\underline{\underline{p}})\rangle + (\oplus_{\underline{\underline{p}} \in \underline{\underline{S}}_o} \underline{\underline{K}}_o^{I}({}_{\Lambda_{\underline{\underline{p}}}}\underline{\underline{M}}^{o})),$$

and ψ' is epic. Let now

$$\langle \mathfrak{g}(M')\rangle - \langle \mathfrak{g}(M)\rangle \in \operatorname{Ker}\psi; \text{ i.e., } \langle M'_{\underline{\underline{p}}}\rangle - \langle M_{\underline{\underline{p}}}\rangle \in \underline{\underline{K}}_o^{I}({}_{\Lambda_{\underline{\underline{p}}}}\underline{\underline{M}}^{o})$$

for every $\underline{\underline{p}} \in \underline{\underline{S}}_o$. Since the elements in $\underline{\underline{K}}_o(\{M_i\}_{1\leq i\leq s})$ are uniquely determined by the central idempotents and their rank, there exists $\langle N\rangle - \langle N'\rangle \in \underline{\underline{K}}_o(\{M_i\}_{1\leq i\leq s})$ such that

$$\langle M'_{\underline{\underline{p}}}\rangle - \langle M_{\underline{\underline{p}}}\rangle = \langle N'_{\underline{\underline{p}}}\rangle - \langle N_{\underline{\underline{p}}}\rangle \quad \text{for every } \underline{\underline{p}} \in \underline{\underline{S}}_o.$$

But the map $\varphi_2 = \oplus_{\underline{\underline{p}} \in \underline{\underline{S}}_o} \varphi_2(\underline{\underline{p}})$ (cf. 4.6) is a monomorphism, and so

$$\langle \mathfrak{g}(M')\rangle - \langle \mathfrak{g}(M)\rangle = \langle \mathfrak{g}(N')\rangle - \langle \mathfrak{g}(N)\rangle,$$

and $\langle \mathfrak{g}(M')\rangle - \langle \mathfrak{g}(M)\rangle \in \underline{\underline{K}}_o^{I}({}_{\Lambda}\mathfrak{g})$. Since obviously all elements in $\underline{\underline{K}}_o^{I}({}_{\Lambda}\mathfrak{g})$ lie in the kernel of ψ', we obtain the induced isomorphism

$$\psi : \underline{\underline{K}}_o({}_{\Lambda}\mathfrak{g})/\underline{\underline{K}}_o^{I}({}_{\Lambda}\mathfrak{g}) \xrightarrow{\sim} \oplus_{\underline{\underline{p}} \in \underline{\underline{S}}_o} \underline{\underline{K}}_o({}_{\Lambda_{\underline{\underline{p}}}}\underline{\underline{M}}^{o})/\underline{\underline{K}}_o^{I}({}_{\Lambda_{\underline{\underline{p}}}}\underline{\underline{M}}_o). \qquad \#$$

Remark: We point out, that in $\underline{\underline{K}}_o({}_{\Lambda}\mathfrak{g})^{+}$ we in general do not have unique decomposition, since a Krull-Schmidt theorem does not hold locally. The question,under which conditions $\underline{\underline{K}}_o({}_{\Lambda}\mathfrak{g})^{+}$ has unique decomposition is still open (cf. VI, 3.2). If we turn to the question of the uniqueness of global decomposition, we see that given $M \in {}_{\Lambda}\underline{\underline{M}}^{o}$, we can decompose the genus of M into indecomposable genera

$$\mathfrak{g}(M) = \oplus_{i=1}^{n} \mathfrak{g}(M_i),$$

and with every such decomposition we can associate decompositions of M:

$$M = \oplus_{i=1}^{n} N_i, \; N_i \vee M_i, 1\leq i\leq n.$$

However, in (VII, 3.4), we have shown that not even this latter decomposition is unique. For more details on this we refer to Jacobinski [4].

<u>Exercise §4</u>:

1.) With the notation of the proof of (4.3), show that $N \underset{\cong}{|} M^{(s)}$, $N_{1} \underset{\cong}{|} M^{(s)}$ and $\varphi'\langle N\rangle \geq \varphi'\langle N_1\rangle$ implies $N_1 \underset{loc}{|} N$; i.e., N_1 is a local direct summand of N.

§5 Jacobinski's cancellation theorem

We give a necessary condition, for a cancellation law to hold globally. Prime ideals means maximal ideals.

In this section we shall assume that K is an $\underline{\underline{A}}$-field with Dedekind domain R. The aim is to prove the following cancellation rule of Jacobinski.

5.1 **Theorem** (Jacobinski [4]): Let $M \in {}_{\Lambda}\underline{\underline{M}}^{o}$ satisfy Eichler's condition. If $M \oplus X \cong N \oplus X$ for some $X_{\cong}|M^{(s)}$, for some $s \in \underline{\underline{N}}$, then $M \cong N$. *)

Remarks: (i) This result is "best possible" in this generality. In fact, if one drops the hypothesis that M satisfy Eichler's condition, Swan [4] has given a counterexample: Let G be the generalized quaternion group of order 32. Then there exists a projective non-free $\underline{\underline{Z}}G$-lattice M such that

$$\underline{\underline{Z}}G \oplus \underline{\underline{Z}}G \cong \underline{\underline{Z}}G \oplus M.$$

However, also the condition $X_{\cong}|M^{(s)}$ is necessary as can be seen from the concept of restricted genera (cf. VII, 4.9, 4.10).

(ii) Serre (Bass [5]) has shown that $M \oplus P \cong N \oplus P$ implies $M \cong N$ in case $\Lambda^{(2)}{}_{loc}|M$ and $P \in {}_{\Lambda}\underline{\underline{P}}^{f}$. (We point out here that Serre's result is valid for a much wider class of rings than orders and that it recently has been generalized by Dress [1][2].) It should be observed that here $P_{\cong}|M^{(n)}$ for some n.

(iii) Instead of presenting here Jacobinski's original proof we develop Swan's version, which uses Grothendieck groups. It is based on an exact sequence in K-theory, similar to the sequence of (3.13).

5.2 **Notation**:

$\underline{\underline{S}}_o$ = finite non-empty set of prime ideals of R containing all primes $\underline{\underline{p}}$ dividing the Higman ideal $\underline{\underline{H}}(\Lambda)$ (cf. V, 3.1),

$M \in {}_{\Lambda}\underline{\underline{M}}^{o}$ a faithful Λ-lattice,

${}_{\Lambda}\underline{\underline{M}}^{T}_{\underline{\underline{S}}_o}$ = the class of R-torsion Λ-modules of finite type with $(\mathrm{ann}_R(T),\underline{\underline{p}}) = 1$ for every $\underline{\underline{p}} \in \underline{\underline{S}}_o$; in short $(\mathrm{ann}_R T, \underline{\underline{S}}_o) = 1$.

*) Recently Ju.A. Drozd (Izv. Akad. Nauk SSSR, 33 (1969), 1080-1088) has given a different proof of (5.1) using the ring of adèles and some results of Bass on congruence subgroups.

Then ${}_{\Lambda}\underline{\underline{M}}^{T}_{\underline{\underline{S}}_o}$ is an admissible subcategory (cf. 3.2) of ${}_{\Lambda}\underline{\underline{M}}^{f}$ and $\underline{\underline{G}}_{o\underline{\underline{S}}_o}^{T}(\Lambda)$ is its Grothendieck group.

5.3 Proposition: We have an exact sequence of abelian groups

$$\underline{\underline{G}}_{o\underline{\underline{S}}_o}^{T}(\Lambda) \xrightarrow{\sigma} \underline{\underline{D}}_M \xrightarrow{\varphi} \oplus_{\underline{\underline{p}} \in \underline{\underline{S}}_o} \underline{\underline{D}}_{\hat{M}_{\underline{\underline{p}}}} ,$$

where φ is defined as in (4.1) and

$$\sigma : \underline{\underline{G}}_{o\underline{\underline{S}}_o}^{T}(\Lambda) \longrightarrow \underline{\underline{D}}_M ,$$

$$[T] \longmapsto \langle N\rangle - \langle N'\rangle ,$$

if

$$0 \longrightarrow N' \longrightarrow N \longrightarrow T \longrightarrow 0$$

is an exact sequence of Λ-modules with $N', N \in \underline{\underline{E}}_M = \{ N \in {}_{\Lambda}\underline{\underline{M}}^{o} : N \underset{\cong}{|} M^{(s)}$ for some $s \in \underline{\underline{N}}\}$.

Proof: We show first how to choose N', N for a given T. Since $\Lambda/\mathrm{ann}_R(T)\cdot\Lambda$ is an artinian and noetherian ring, T has a composition series,

$$T = T_o \underset{\neq}{\supset} T_1 \underset{\neq}{\supset} \dots \underset{\neq}{\supset} T_s \underset{\neq}{\supset} T_{s+1} = 0$$

and in $\underline{\underline{G}}_{o\underline{\underline{S}}_o}^{T}(\Lambda)$ we have $[T] = \sum_{i=o}^{s}[T_i/T_{i+1}]$. Thus, to define σ it suffices to define σ on $[T]$, where T is a simple Λ-module in ${}_{\Lambda}\underline{\underline{M}}^{T}_{\underline{\underline{S}}_o}$, and then we extend σ $\underline{\underline{Z}}$-linearly. Since $(\mathrm{ann}_R T, \underline{\underline{H}}(\Lambda)) = 1$, and since M is faithful, we may apply (VII, 3.3): There exists an exact sequence of Λ-modules

$$0 \longrightarrow N \longrightarrow M \xrightarrow{\alpha} T \longrightarrow 0.$$

Moreover, tensoring this sequence with $R_{\underline{\underline{p}}}$, $\underline{\underline{p}} \in \underline{\underline{S}}_o$ shows $N \vee M$ (cf. VII, 1.9), and by (4.2), $\langle N\rangle \in \underline{\underline{D}}_M$. We now put

$$\sigma : \underline{\underline{G}}_{o\underline{\underline{S}}_o}^{T}(\Lambda) \longrightarrow \underline{\underline{D}}_M ,$$

$$\sigma : [T] \longmapsto \langle M\rangle - \langle N\rangle .$$

The proof that σ is well-defined can be given as the one of (3.1), (3.2). However for the sake of completeness we shall give here a short direct proof.

σ <u>is a well-defined group homomorphism</u>.

(i) If we have another presentation

$$0 \longrightarrow N_1 \longrightarrow N_2 \xrightarrow{\beta} T \longrightarrow 0, \ N_1, N_2 \in \underline{\underline{E}}_M,$$

in ${}_\Lambda\underline{\underline{M}}^f$, then α and β are projective homomorphisms by (VII, 3.5) (cf. V, § 2) and we conclude from Schanuel's lemma (V, 2.6) that

$$M \oplus N_1 \cong N \oplus N_2;$$

i.e., $\langle M \rangle - \langle N \rangle = \langle N_2 \rangle - \langle N_1 \rangle$ in $\underline{\underline{D}}_M$, and σ is independent of the presentation.

(ii) σ preserves relations. Given an exact sequence

$$0 \longrightarrow T' \xrightarrow{\alpha} T \xrightarrow{\beta} T'' \longrightarrow 0 \text{ in } {}_\Lambda\underline{\underline{M}}^T_{\underline{\underline{S}}_0},$$

we take presentations of T' and T''

$$0 \longrightarrow N_1 \xrightarrow{\kappa} N_1' \xrightarrow{\gamma} T' \longrightarrow 0$$

and

$$0 \longrightarrow N_2 \xrightarrow{\lambda} N_2' \xrightarrow{\delta} T'' \longrightarrow 0$$

respectively, $N_i, N_i' \in \underline{\underline{E}}_M$, $i=1,2$. Since δ and γ are projective homomorphism, we can complete the following diagram

$$\begin{array}{cccccccccc}
 & & & 0 & & 0 & & 0 & & \\
 & & & \uparrow & & \uparrow & & \uparrow & & \\
 & & 0 \longrightarrow & T' & \longrightarrow & T & \longrightarrow & T'' & \longrightarrow 0 & \\
 & & & \uparrow & & \uparrow & & \uparrow & & \\
 & & 0 \longrightarrow & N_1' & \longrightarrow & N_1' \oplus N_2' & \longrightarrow & N_2' & \longrightarrow 0 & \\
 & & & \uparrow & & \vdots & & \uparrow & & \\
E : & & 0 \longrightarrow & N_1 & \dashrightarrow & X & \dashrightarrow & N_2 & \longrightarrow 0 & \\
 & & & \uparrow & & \uparrow & & \uparrow & & \\
 & & & 0 & & 0 & & 0 & . &
\end{array}$$

The construction is done as the one in the proof of (3.1) (cf. Ex. 3.6).

Since $T',T,T'' \in {}_{\Lambda}\underline{\underline{M}}^{T}_{\underline{\underline{S}}_o}$ we have for every $\underline{\underline{p}} \mid \underline{\underline{H}}(\Lambda)$, $E_{\underline{\underline{p}}} = R_{\underline{\underline{p}}} \otimes_R E \equiv 0$, where E is the bottom sequence in the diagram, and we conclude

$$[E] \longmapsto 0$$

under the natural isomorphism

$$\mathrm{Ext}^1_{\Lambda}(N_2,N_1) \xrightarrow{\sim} \oplus_{\underline{\underline{p}} \mid \underline{\underline{H}}(\Lambda)} \mathrm{Ext}^1_{\Lambda_{\underline{\underline{p}}}}(N_{2_{\underline{\underline{p}}}},N_{1_{\underline{\underline{p}}}})$$

(cf. V, 3.7). Thus, E is split exact and consequently $X \cong N_1 \oplus N_2$; i.e.,

$$\sigma([T]) = \sigma([T']) + \sigma([T'']),$$

and σ is a well-defined group homomorphism.

(iii) To show the exactness of the sequence

$$\underline{\underline{G}}_{o_{\underline{\underline{S}}_o}}^{T}(\Lambda) \xrightarrow{\sigma} \underline{\underline{D}}_M \xrightarrow{\varphi} \oplus_{\underline{\underline{p}} \in \underline{\underline{S}}_o} \underline{\underline{D}}_{\hat{M}_{\underline{\underline{p}}}},$$

we recall that $\mathrm{Ker}\,\varphi = \{\langle N_1\rangle - \langle N_2\rangle ; N_1 \vee N_2,\ N_1,N_2 \in \underline{\underline{E}}_M\}$ (cf. proof of 4.1). Thus,

$$[T] \overset{\sigma}{\longmapsto} \langle N'\rangle - \langle N\rangle \overset{\varphi}{\longmapsto} 0,$$

since $T_{\underline{\underline{p}}} = 0$ for every $\underline{\underline{p}} \in \underline{\underline{S}}_o$. Conversely, if $\langle N_1\rangle - \langle N_2\rangle \overset{\varphi}{\longmapsto} 0$, then we can embed N_1 into N_2 such that $N_2/N_1 \in {}_{\Lambda}\underline{\underline{M}}^{T}_{\underline{\underline{S}}_o}$ (cf. VII, 3.1); i.e., $\langle N_1\rangle - \langle N_2\rangle \in \mathrm{Im}\,\sigma$, and the above sequence is exact. #

5.4 __Lemma__: Let $S = R \setminus \{\bigcup_{\underline{\underline{p}} \in \underline{\underline{S}}_o} \underline{\underline{p}}\}$, and let "$-_S$" denote the localization at S. If for every natural number n, we denote by $\mathrm{Aut}_{\Lambda_S}(M_S^{(n)})$ the group of Λ_S-automorphisms of $M_S^{(n)}$ (written multiplicatively), then we have a group-homomorphism

$$\vartheta_n : \mathrm{Aut}_{\Lambda_S}(M_S^{(n)}) \longrightarrow \underline{\underline{G}}_{o_{\underline{\underline{S}}_o}}^{T}(\Lambda),$$

$$\alpha \longmapsto [\mathrm{Coker}\; s\alpha] - [\mathrm{Coker}\; s1_{M^{(n)}}],$$

where $0 \neq s \in S$ is such that

$s\alpha : M^{(n)} \longrightarrow M^{(n)}$ is a monomorphism.

Proof: Given $\alpha \in \mathrm{Aut}_{\Lambda_S}(M_S^{(n)})$ and $0 \neq s \in S$ such that

$$s\alpha : M^{(n)} \longrightarrow M^{(n)},$$

then we have the two exact sequences

$$E_1 : 0 \longrightarrow M^{(n)} \xrightarrow{s\alpha} M^{(n)} \longrightarrow T_1 \longrightarrow 0$$

and

$$E_2 : 0 \longrightarrow M^{(n)} \xrightarrow{s1_{M^{(n)}}} M^{(n)} \longrightarrow T_2 \longrightarrow 0.$$

Since s is a unit in $R_{\underline{p}}$ for every $\underline{p} \in \underline{S}_o$, it is readily seen that $T_1, T_2 \in {}_\Lambda M^T_{\underline{S}_o}$; i.e.,

$$\vartheta_n : \mathrm{Aut}_{\Lambda_S}(M_S^{(n)}) \longrightarrow \underline{G}_{o_{\underline{S}_o}}^{T}(\Lambda).$$

The proof that ϑ_n is a well-defined group homomorphism is similar to the demonstration for ϑ in (3.13) and is left as an exercise to the reader. #

5.5 Theorem: We have an exact sequence

$$\mathrm{Aut}_{\Lambda_S}(M_S) \xrightarrow{\vartheta} \underline{G}_{o_{\underline{S}_o}}^{T}(\Lambda) \xrightarrow{\sigma} \underline{D}_M \xrightarrow{\varphi} \oplus_{\underline{p} \in \underline{S}_o} \underline{D}_{\hat{M}_{\underline{p}}},$$

where σ and φ are defined as in (5.3) and

$$\vartheta : \alpha \longmapsto [\mathrm{Coker}\ s\alpha] - [\mathrm{Coker}\ s1_M],$$

whith $0 \neq s \in S$ such that $s\alpha : M \longrightarrow M$.

Proof: To simplify the notation, we put $\Omega_S = \mathrm{End}_{\Lambda_S}(M_S)$. We shall show first that we have an exact sequence

$$* \quad GL(1, \Omega_S) \xrightarrow{\vartheta} \underline{G}_{o_{\underline{S}_o}}^{T}(\Lambda) \xrightarrow{\sigma} \underline{D}_M \xrightarrow{\varphi} \oplus_{\underline{p} \in \underline{S}_o} \underline{D}_{\hat{M}_{\underline{p}}}.$$

We observe that $\mathrm{Aut}_{\Lambda_S}(M_S^{(n)}) \overset{nat}{\cong} GL(n, \Omega_S)$, and hence, by (5.4) we have a group homomorphism

$$\vartheta_n : GL(n, \Omega_S) \longrightarrow \underline{G}_{o_{\underline{S}_o}}^{T}(\Lambda) \text{ for every } n \in \underline{N}.$$

Let

$$\iota_{n,n+1} : GL(n, \Omega_S) \longrightarrow GL(n+1, \Omega_S)$$

be the natural embedding (cf. 2.1). We claim that the diagram

$$\begin{array}{ccc} GL(n,\Omega_S) & \xrightarrow{\vartheta_n} & \underset{=}{G}^{T}_{o_{\underset{=}{S}_o}}(\Lambda) \\ & \searrow^{\iota_{n,n+1}} \quad \nearrow_{\vartheta_{n+1}} & \\ & GL(n+1,\Omega_S) & \end{array}$$

is commutative. In fact, let $\alpha \in \mathrm{Aut}_{\Lambda_S}(M_S^{(n)})$; then

$$\alpha \xrightarrow{\iota_{n,n+1}} \alpha \oplus 1_{M_S} \xrightarrow{\vartheta_{n+1}} [\mathrm{Coker}\ s(\alpha \oplus 1_{M_S})] - [\mathrm{Coker}\ s1_{M^{(n+1)}}]$$

$$= [\mathrm{Coker}\ s\alpha] - [\mathrm{Coker}\ s1_{M^{(n)}}] = \vartheta_n(\alpha).$$

Thus the above diagram is commutative. By the universal property of the injective limit (cf. I, Ex. 9,3) there exists a unique map

$$\tilde{\vartheta} : GL(\Omega_S) = \underrightarrow{\lim}\, GL(n,\Omega_S) \longrightarrow \underset{=}{G}^{T}_{o_{\underset{=}{S}_o}}(\Lambda).$$

Because of (5.4), exactness need only be shown at $\underset{=}{G}^{T}_{o_{\underset{=}{S}_o}}(\Lambda)$. Given $\alpha \in GL(\Omega_S)$, then $\alpha : M_S^{(n)} \longrightarrow M_S^{(n)}$ for some n, and

$$\tilde{\vartheta} : \alpha \longmapsto [\mathrm{Coker}\ s\alpha] - [\mathrm{Coker}\ s1_{M^{(n)}}] \text{, where}$$

$$0 \longrightarrow M^{(n)} \xrightarrow{s\alpha} M^{(n)} \longrightarrow T_1 \longrightarrow 0$$

$$0 \longrightarrow M^{(n)} \xrightarrow{s1_{M^{(n)}}} M^{(n)} \longrightarrow T_2 \longrightarrow 0$$

are exact sequences of Λ-modules. But

$$\sigma : [T_i] \longmapsto \langle M^{(n)}\rangle - \langle M^{(n)}\rangle = 0,\ i=1,2,$$

and $\mathrm{Im}\,\tilde{\vartheta} \subset \mathrm{Ker}\,\sigma$. Then $[T_1] = [T']$ where T' is the direct sum of the composition factors of T_1. Hence we may assume that T_1 and T_2 are direct sums of simple modules in ${}_{T}\underset{=}{M}^{T}_{\underset{=}{S}_o}$. According to (VII, 3.3), we have two exact sequences for some $n \in \underset{=}{N}$

$$0 \longrightarrow N_1 \xrightarrow{\alpha} M^{(n)} \longrightarrow T_1 \longrightarrow 0$$

$$0 \longrightarrow N_2 \xrightarrow{\beta} M^{(n)} \longrightarrow T_2 \longrightarrow 0.$$

Since $\sigma : [T_1]-[T_2] \longmapsto 0$, we have in $\underset{=}{D}_M$ the relation $\langle N_1\rangle - \langle N_2\rangle = 0$; i.e., $N_1 \oplus X = N_2 \oplus X$ (cf. 1.7) where $X_{\cong}|M^{(t')}$ for some t'. And we may as well assume that

$$N_1 \oplus X \cong N_2 \oplus X \cong M^{(t)} \text{ for some } t \geq n.$$

Thus we obtain the exact sequences

$$0 \longrightarrow N_1 \oplus X \xrightarrow{\alpha \oplus 1_X} M \oplus X \longrightarrow T_1 \longrightarrow 0,$$

$$0 \longrightarrow N_2 \oplus X \xrightarrow{\beta \oplus 1_X} M \oplus X \longrightarrow T_2 \longrightarrow 0; \text{ i.e.,}$$

exact sequences

$$0 \longrightarrow M^{(t)} \xrightarrow{\alpha'} M \oplus X \longrightarrow T_1 \longrightarrow 0,$$

$$0 \longrightarrow M^{(t)} \xrightarrow{\beta'} M \oplus X \longrightarrow T_2 \longrightarrow 0.$$

Since $X_S \cong M_S^{(t-n)}$ (cf. Ex. 5.2), we can find an element $0 \neq s \in S$ and a Λ_S-isomorphism $\delta : X_S \xrightarrow{\sim} M_S^{(t-n)}$, such that $(1_M \oplus s\delta) : M^{(n)} \oplus X \longrightarrow M^{(t)}$. We write $\gamma = (1_{M^{(n)}} \oplus s\delta)$. The sequences of monomorphisms

$$M^{(t)} \xrightarrow{\alpha'} M^{(n)} \oplus X \longrightarrow M^{(t)},$$

$$M^{(t)} \xrightarrow{\beta'} M^{(n)} \oplus X \longrightarrow M^{(t)},$$

show that (cf. (1), 3.13)

$$[\text{Coker } \alpha'\gamma] = [\text{Coker } \alpha'] + [\text{Coker } \gamma] \text{ and}$$
$$[\text{Coker } \beta'\gamma] = [\text{Coker } \beta'] + [\text{Coker } \gamma].$$

Thus

$$[T_1] - [T_2] = [\text{Coker } \alpha'\gamma] - [\text{Coker } \beta'\gamma].$$

However, $\alpha_1 = 1_{R_S} \otimes \alpha'\gamma : M_S^{(t)} \longrightarrow M_S^{(t)}$ and $\alpha_2 = 1_{R_S} \otimes \beta'\gamma : M_S^{(t)} \longrightarrow M_S^{(t)}$ are automorphisms of $M_S^{(t)}$. Thus

$$\alpha_1 \cdot \alpha_2^{-1} \xmapsto{\tilde{\vartheta}} [T_1] - [T_2].$$

This shows that the sequence (*) is exact. We observe that $\underline{\underline{G}}_{\underline{\underline{o}}_S}^{T}(\Lambda)$ is commutative and so $[GL(\Omega_S), GL(\Omega_S)] \subset \text{Ker}\,\tilde{\vartheta}$, and $\tilde{\vartheta}$ induces a map $\vartheta' : \underline{\underline{K}}_1(\Omega_S) \longrightarrow \underline{\underline{G}}_{\underline{\underline{o}}_S}^{T}(\Lambda)$ such that the sequence

$$\underline{\underline{K}}_1(\Omega_S) \xrightarrow{\vartheta'} \underline{\underline{G}}_{\underline{\underline{o}}_S}^{T}(\Lambda) \xrightarrow{\sigma} \underline{\underline{D}}_M$$

is an exact sequence of abelian groups. However, $\Omega_S = \text{End}_{\Lambda_S}(M_S) = R_S \otimes_R \text{End}_\Lambda(M)$ is semi-primary. We recall that we have to show

$\overline{\Omega}_S = \Omega_S/\text{rad}\ \Omega_S$ is artinian and noetherian. It obviously is noetherian. Observe that $\text{rad}\ R_S = \bigcap_{\underline{\underline{p}} \in \underline{\underline{S}}_0} \underline{\underline{p}} R_S$ (cf. I, 8.2) and

$$R_S/\text{rad}\ R_S = \bigoplus_{\underline{\underline{p}} \in \underline{\underline{S}}_1} R/\underline{\underline{p}} \quad (\text{cf. I, 8.9}).$$

Since $R/\underline{\underline{p}}$ is a field, $R_S/\text{rad}\ R_S$ is artinian; thus so is

$$(R_S \otimes_R \text{End}_\Lambda(M))/(\text{rad}\ R_S)(R_S \otimes_R \text{End}_\Lambda(M)).$$

Now one shows as in the proof of (IV, 2.6) that $\text{rad}(R_S \otimes_R \text{End}_\Lambda(M)) \supset (\text{rad}\ R_S)(R_S \otimes_R \text{End}_\Lambda(M))$; i.e., Ω_S is semi-primary.
Hence, we may apply (2.5) to conclude, that we have an epimorphism

$$GL(1,\Omega_S) = \text{Aut}_{\Lambda_S}(M_S) \longrightarrow \underline{\underline{K}}_1(\Omega_S), \text{ and we set}$$

$$\vartheta : \alpha \longmapsto [\text{Coker}\ s\alpha] - [\text{Coker}\ s1_M],$$

where $0 \neq s \in S$ is such that $s\alpha : M \longrightarrow M$. Combining all this, we obtain the desired exact sequence. #

Remark: We point out that for the proof of (5.5) we have not used that K is an $\underline{\underline{A}}$-field. It suffices to assume that R is a Dedekind domain with quotient field K and Λ an R-order in the separable finite dimensional K-algebra A. We also do not have to assume the validity of the Jordan-Zassenhaus theorem.
We now turn to the proof of (5.1):

5.6 Lemma: Let $M \in {}_\Lambda\underline{\underline{M}}^0$. Then $M \oplus X \cong N \oplus X$ with $X_{\cong}|M^{(s)}$ is equivalent to the existence of two exact sequences of Λ-modules

$$0 \longrightarrow M \longrightarrow M \longrightarrow T_1 \longrightarrow 0,$$

$$0 \longrightarrow N \longrightarrow M \longrightarrow T_2 \longrightarrow 0,$$

where $T_1, T_2 \in {}_\Lambda\underline{\underline{M}}^T_{\underline{\underline{S}}_0}$, and $[T_1] = [T_2]$ in $\underline{\underline{K}}_0{}^T_{\underline{\underline{S}}_0}(\Lambda)$.

Proof: We first assume that M is a faithful Λ-lattice. $M \oplus X \cong N \oplus X$ with $X_{\cong}|M^{(n)}$ implies $N_{\cong}|M^{(n')}$. Thus the above relation is equivalent to $\langle M\rangle = \langle N\rangle$ in $\underline{\underline{D}}_M$. Choose an exact sequence

$$0 \longrightarrow N \xrightarrow{\beta} M \longrightarrow T' \longrightarrow 0$$

with $T' \in {}_{\Lambda}\underline{\underline{M}}^{T}_{\underline{\underline{S}}_o}$ (cf. VII, 3.1). Then $[T'] \in \operatorname{Ker}\sigma = \operatorname{Im}\vartheta$ (cf. 5.5). Thus, there exists $\alpha \in \operatorname{Aut}_{\Lambda_S}(M_S)$, $0 \neq s \in S$ and two exact sequences

$$0 \longrightarrow M \xrightarrow{s\alpha} M \longrightarrow T_1 \longrightarrow 0,$$

$$0 \longrightarrow M \xrightarrow{s\,1_M} M \longrightarrow T'' \longrightarrow 0,$$

such that $[T_1] - [T''] = [T']$ in $\underline{\underline{G}}_o{}^{T}_{\underline{\underline{S}}_o}(\Lambda)$. By ((1), 3.13), $[\operatorname{Coker} s\beta] = [T'] + [T''] = [T_1]$ and we have two exact sequences

$$0 \longrightarrow N \xrightarrow{s\beta} M \longrightarrow T_2 \longrightarrow 0\ ,$$

$$0 \longrightarrow M \xrightarrow{s\alpha} M \longrightarrow T_1 \longrightarrow 0\ ,$$

with $[T_2] = [T_1]$ in $\underline{\underline{G}}_o{}^{T}_{\underline{\underline{S}}_o}(\Lambda)$. <u>Conversely</u>, if we have two exact sequences

$$0 \longrightarrow M \longrightarrow M \longrightarrow T_1 \longrightarrow 0\ ,$$

$$0 \longrightarrow N \longrightarrow M \longrightarrow T_2 \longrightarrow 0\ ,$$

with $[T_1] = [T_2]$ in $\underline{\underline{G}}_o{}^{T}_{\underline{\underline{S}}_o}(\Lambda)$, then $\langle M\rangle = \langle N\rangle$ in $\underline{\underline{D}}_M$; i.e., $M \oplus X \cong N \oplus X$ for $X_{\cong} \,|\, M^{(s)}$. If now M is not a faithful Λ-lattice, then M is a faithful Λe_M-lattice (cf. VII, 3.2), where e_M is the central idempotent corresponding to M. Then $N, X \in {}_{\Lambda e_M}\underline{\underline{M}}^{o}$, and we obtain the above relations for Λe_M-modules. However, T_1, T_2 are necessarily Λe_M-modules, and (5.6) is true for Λ-modules if and only if it is true for Λe_M-modules. #

We now assume that $M \in {}_{\Lambda}\underline{\underline{M}}^{o}$ satisfies Eichler's condition, and that K is an $\underline{\underline{A}}$-field. We have to show (cf. 5.6) that the existence of two exact sequences of Λ-modules

$$0 \longrightarrow M \longrightarrow M \xrightarrow{\varphi} T_1 \longrightarrow 0\ ,$$

$$0 \longrightarrow N \longrightarrow M \xrightarrow{\psi} T_2 \longrightarrow 0\ ,$$

with $[T_1] = [T_2]$ in $\underline{\underline{G}}_{o_{\underline{\underline{S}}_o}}^{T}(\Lambda)$ implies $N \cong M$. We choose $\underline{\underline{S}}_o$ such that it contains all the maximal ideals dividing $\underline{\underline{H}}(\Lambda)$ and all the primes of $\underline{\underline{S}}(M)$ of the Swan-Eichler theorem (VI, 7.2). We <u>claim</u> that for T_1 we can find a composition series where the factors occur in any prescribed order. Since T_1 is an R-torsion module, it is - as Λ-module - the direct sum of its $\underline{\underline{p}}$-primary components (cf. I, 8.9)

$$T_1 = \bigoplus_{\underline{\underline{p}}} X_{\underline{\underline{p}}},$$

and since $(\operatorname{ann}_R T_1, \underline{\underline{S}}_o) = 1$, $\underline{\underline{p}} \notin \underline{\underline{S}}_o$; in particular, $\underline{\underline{p}} \nmid \underline{\underline{H}}(\Lambda)$; and $X_{\underline{\underline{p}}}$ is a module over the separable $\hat{R}_{\underline{\underline{p}}}$-order $\hat{\Lambda}_{\underline{\underline{p}}}$ (cf. V, 3.7). But $\hat{\Lambda}_{\underline{\underline{p}}}$ is maximal (cf. VI, 2.5) and if $\hat{\Lambda}_{\underline{\underline{p}}} = \oplus_{i=1}^{n} \hat{\Lambda}_i$ is the decomposition into separable $\hat{R}_{\underline{\underline{p}}}$-orders in simple algebras, then $X_{\underline{\underline{p}}}$ decomposes accordingly: $X_{\underline{\underline{p}}} = \oplus_{i=1}^{n} X_i$ and each X_i is a module over the separable order $\hat{\Lambda}_i$ in a simple algebra. But $\hat{\Lambda}_i$ has - up to isomorphism - only one simple module (cf. VI, Ex. 2,4). Hence we can find a composition series for T where the factors occur in any prescribed order. We can choose composition series

$$T_1 = X_o \underset{\neq}{\supset} X_1 \underset{\neq}{\supset} \cdots \underset{\neq}{\supset} X_n \underset{\neq}{\supset} X_{n+1} = 0,$$

$$T_2 = Y_o \underset{\neq}{\supset} Y_1 \underset{\neq}{\supset} \cdots \underset{\neq}{\supset} Y_n \underset{\neq}{\supset} Y_{n+1} = 0,$$

such that $X_i/X_{i+1} \cong Y_i/Y_{i+1}$, $0 \le i \le n$, since $[T_1] = [T_2]$ in $\underline{\underline{G}}_{o_{\underline{\underline{S}}_o}}^{T}(\Lambda)$.

Putting

$$M_i = X_i \varphi^{-1} \text{ and } N_i = Y_i \psi^{-1}, \quad i=0,1,\ldots,n+1,$$

we shall use induction on i to show that $M_i \cong N_i$. For i=0, $M_o = M = N_o$. Assume now that $M_i \cong N_i$. We then have the two exact sequences

$$0 \longrightarrow M_{i+1} \longrightarrow M_i \xrightarrow{\varphi_i} X_i/X_{i+1} \longrightarrow 0,$$

$$0 \longrightarrow N_{i+1} \longrightarrow N_i \xrightarrow{\psi_i} Y_i/Y_{i+1} \longrightarrow 0.$$

Since $\sigma_i : M_i \xrightarrow{\cong} N_i$ and $\varrho_i : X_i/X_{i+1} \xrightarrow{\cong} Y_i/Y_{i+1}$, and since with M also M_i satisfies Eichler's condition, we can apply (VI, 7.2), the theorem of Eichler-Swan: We have two epimorphisms

$$M_i \xrightarrow{\varphi_i \varrho_i} Y_i/Y_{i+1} ,$$

$$M_i \xrightarrow{\sigma_i \psi_i} Y_i/Y_{i+1},$$

and $\operatorname{ann}_R(Y_i/Y_{i+1}) \not\subset \underline{\underline{S}}(M)$. Thus

$$\operatorname{Ker} \varphi_i \varrho_i \cong \operatorname{Ker} \sigma_i \psi_i ; \text{ i.e.,}$$

$$M_{i+1} \cong N_{i+1}.$$

Now for $i = s + 1$ we obtain

$$\operatorname{Ker} \varphi \cong \operatorname{Ker} \psi ; \text{ i.e., } M \cong N. \quad \#$$

This concludes the proof of (5.1). #

5.7 <u>Corollary</u>: Let $M \in {}_\Lambda\underline{\underline{M}}^o$. If

$$M \oplus X \cong N \oplus X,\ X_{\cong} | M^{(s)},$$

then

$$M^{(2)} \cong M \oplus N.$$

<u>Proof</u>: Since $X_{\cong} | M^{(s)}$, we obtain

$$M \oplus M^{(s)} \cong N \oplus M^{(s)}; \text{ i.e.,}$$

$$(M \oplus M) \oplus M^{(n-1)} \cong (N \oplus M) \oplus M^{(n-1)}.$$

But $M \oplus M$ satisfies Eichler's condition (cf. VI, 7.1) and thus by (5.1)

$$M \oplus M \cong N \oplus M. \quad \#$$

5.8 <u>Corollary</u>: Assume that no simple component of A is a totally definite quaternion algebra, and assume that $P \in {}_\Lambda\underline{\underline{P}}^f$ is locally free. Then $P \oplus X \cong Q \oplus X$, $X \in {}_\Lambda\underline{\underline{P}}^f$ implies $P \cong Q$.

<u>Proof</u>: Since P is locally free, P satisfies Eichler's condition, and $P \vee \Lambda^{(m)}$ i.e., $P^{(t)} \cong \Lambda^{(tm)}$ for some t (cf. 4.2). Thus $X_{\cong} | \Lambda^{(s)}$ implies

$X_{\approx} | P^{(n)}$ for some n; now the result follows from (5.1). #

5.9 <u>Lemma</u>: Let $M \in {}_{\Lambda}\underline{M}^{o}$ satisfy Eichler's condition, and let there be given an epimorphism

$$\varphi : M \longrightarrow U,$$

where U is an R-torsion Λ-module with $(\mathrm{ann}_R U, \underline{H}(\Lambda)) = 1$. Then for every $N \vee M$ there exists an epimorphism

$$\varphi_N : N \longrightarrow U.$$

If $\mathrm{Ker}\,\varphi \not\cong M$, then $\mathrm{Ker}\,\varphi_N \not\cong N$. Moreover, then the correspondence $N \longmapsto \mathrm{Ker}\,\varphi_N$ is a fixpoint free permutation on the non-isomorphic lattices is the genus of M.

<u>Proof</u>: We decompose U into its $\underline{p}$-primary components $U = \oplus_{i=1}^{n} U_i$ with $\mathrm{ann}_R U_i = \underline{p}_i^{s_i}$. If we have epimorphisms $\varphi_{N_i} : N \longrightarrow U_i$, then we construct an epimorphism $\varphi_N : N \longrightarrow U$ as in the proof of (VII, 3.3). Thus we may for the moment assume $\mathrm{ann}_R U = \underline{p}^s$. Let $N \vee M$. Then $M_{\underline{p}} \cong N_{\underline{p}}$ and thus $M/\underline{p}^s M \cong M_{\underline{p}}/\underline{p}^s M_{\underline{p}} \cong N_{\underline{p}}/\underline{p}^s M_{\underline{p}} \cong N/\underline{p}^s N$, and since $\varphi : M \longrightarrow U$ induces an epimorphism $\overline{\varphi} : M/\underline{p}^s M \longrightarrow U$, we get an epimorphism $\overline{\varphi}_N : N/\underline{p}^s N \longrightarrow U$, and consequently an epimorphism $\varphi_N : N \longrightarrow U$. Let us now return to an arbitrary U with $(\mathrm{ann}_R U, \underline{H}(\Lambda)) = 1$.

Since $(\mathrm{ann}_R U, \underline{H}(\Lambda)) = 1$, φ and φ_N are projective homomorphisms (cf. VII, 3.5), and hence by Schanuel's lemma (V, 2.6), we conclude

$$M \oplus \mathrm{Ker}\,\varphi_N \cong N \oplus \mathrm{Ker}\,\varphi_M.$$

If $N \cong \mathrm{Ker}\,\varphi_N$ then by (5.1)

$$M \cong \mathrm{Ker}\,\varphi_M,$$

since $\mathrm{Ker}\,\varphi_N \vee M$ and $\mathrm{Ker}\,\varphi_M \vee M$. But this was excluded.

Hence the map

$$\varrho : N \longmapsto \mathrm{Ker}\,\varphi_N$$

on the set of non-isomorphic lattices in the genus of M has no fixpoints. It should be observed that the map ϱ is well-defined. In fact, if we have two epimorphisms

$$N \xrightarrow{\varphi_1} U,$$

$$N \xrightarrow{\varphi_2} U,$$

then $\operatorname{Ker}\varphi_1 \cong \operatorname{Ker}\varphi_2$ by (5.1). It remains to show, that it is a bijection. By the Jordan-Zassenhaus theorem there are only finitely many non-isomorphic lattices N in $\mathfrak{g}(M)$, and so it suffices to show that ρ is injective. But if $\operatorname{Ker}\varphi_{N_1} \cong \operatorname{Ker}\varphi_{N_2}$, then the same reasoning as above shows

$$N_1 \oplus \operatorname{Ker}\varphi_{N_2} \cong N_2 \oplus \operatorname{Ker}\varphi_{N_2};$$

i.e., $N_1 \cong N_2$. #

5.10 **Definition**: Given a genus $\mathfrak{g}(M)$ and a Λ-module U with $(\operatorname{ann}_R U, \underline{\underline{H}}(\Lambda)) = 1$. If M satisfies Eichler's condition, and if we have an epimorphism

$$\varphi : M \longrightarrow U,$$

then we call M <u>U-positive</u> if $\operatorname{Ker}\varphi \not\cong M$, <u>U-negative</u> if $\operatorname{Ker}\varphi \cong M$.

5.11 **Lemma**: Let $X, M \in {}_{\Lambda}\underline{\underline{M}}^{o}$, such that M,X satisfy Eichler's condition. If there exists a Λ-module U, $(\operatorname{ann}_R(U), \underline{\underline{H}}(\Lambda)) = 1$, such that X is U-negative and M is U-positive, then there exists $N \vee M$, $N \not\cong M$ such that $N \oplus X \cong M \oplus X$.

Moreover for every $Y \vee X$ and $N_1 \vee M$ there exists $N_2 \vee M$, $N_1 \not\cong N_2$ such that

$$N_1 \oplus Y \cong N_2 \oplus Y.$$

Proof: By hypothesis, we have two exact sequences

$$0 \longrightarrow N \longrightarrow M \longrightarrow U \longrightarrow 0, \quad N \not\cong M$$

$$0 \longrightarrow X \longrightarrow X \longrightarrow U \longrightarrow 0,$$

and Schanuel's lemma implies

$$N \oplus X \cong M \oplus X, \ N \not\cong M.$$

If now $N_1 \vee M$, then by (5.9) there exists $N_2 \vee M$, $N_1 \not\cong N_2$ and an embedding

$$0 \longrightarrow N_2 \longrightarrow N_1 \longrightarrow U \longrightarrow 0$$

which implies $N_1 \oplus X \cong N_2 \oplus X$. And if $Y \vee X$, then we have an exact sequence

$$0 \longrightarrow Y \longrightarrow Y \longrightarrow U \longrightarrow 0,$$ since X is U-negative,

which implies $N_1 \oplus Y \cong N_2 \oplus Y$, $N_1 \not\cong N_2$. #

5.12 <u>Theorem</u> (Jacobinski [4]): There exists a positive integer s depending only on Λ such that for $M,N \in {}_{\Lambda}\underline{\underline{M}}^{o}$,

$$M \vee N \text{ if and only if } M^{(s)} \cong N^{(s)}.$$

<u>Proof</u>: This actually is the same proof as that of (4.2) but using (VII, 2.13) and (5.11). Let t be such that $g(M) < t$ for every $M \in {}_{\Lambda}\underline{\underline{M}}^{o}$ (cf. VII, 2.13), where $g(M)$ denotes the number of non-isomorphic lattices in the genus of M. In the notation of the proof of (4.1) $|\operatorname{Ker} \varphi_M| < t$. Thus given $M \vee N$, we have $\langle M^{(r)} \rangle = \langle N^{(r)} \rangle$ in $\underline{\underline{D}}_M$, where $r = |\operatorname{Ker} \varphi_M|$. By (1.7), we get

$$M^{(r)} \oplus X \cong N^{(r)} \oplus X$$

for some $X_{\cong}|M^{(k)}$. Now we apply (5.7) to conclude $M^{(2r)} \cong N^{(r)} \oplus M^{(r)}$. Since $g(M) < t$ there exist positive integers $m < t$, and $2m < n < 2t$ such that

$$N^{(nr)} \cong M^{(nr-1)} \oplus N_o;\ N^{(mr)} \cong M^{(mr-1)} \oplus N_o.$$

Then

$$N^{(nr)} \cong M^{(nr-2mr)} \oplus M^{(mr)} \oplus N^{(mr)} \cong M^{(nr)}.$$

Now $s_M = nt < 2t^2$;

Thus, if we put $s = \prod_{i=t}^{2t^2} i$, then s has the desired properties. #

<u>Exercises §5</u>:

1.) Let $\underline{\underline{S}}_o$ be a finite non-empty set of maximal ideals in an R. Show that for an R-torsion Λ-module T, $(\operatorname{ann}_R T, \underline{\underline{S}}_o) = 1$ if and only if $R_{S_o} \otimes_R T = 0$ where

$$S_o = R \setminus \{ \bigcup_{\underline{\underline{p}} \in \underline{\underline{S}}_o} \underline{\underline{p}} \} .$$

2.) Let R_S be a semi-local Dedekind domain (cf. I, 7.8). Show that two Λ_S-lattices lie in the same genus if and only if they are isomorphic. Prove (VI, 3.5, 3.6) for Λ_S-lattices!

CHAPTER IX

SPECIAL TYPES OF ORDERS

§1 Clean orders

An order is clean if every special projective Λ-lattice is locally free. Group rings are clean; hereditary clean orders are maximal. Projective lattices over group rings are locally free. If Λ is clean and if $\Lambda' \subset \Lambda$ then the map $\varphi : \underline{\underline{P}}_o(\Lambda') \longrightarrow \underline{\underline{P}}_o(\Lambda)$ is an epimorphism.

Let R be a Dedekind domain with quotient field K and Λ an R-order in the separable finite dimensional K-algebra A.

1.1 Definition: Λ is called a clean R-order if every special projective Λ-lattice P (i.e., KP is A-free) is a progenerator.

1.2 Theorem (Strooker [1], Lam [1]): The following conditions for Λ are equivalent:

(i) Λ is clean.

(ii) $\Lambda_{\underline{\underline{p}}}$ is clean for every $\underline{\underline{p}} \in$ spec R.

(iii) $\hat{\Lambda}_{\underline{\underline{p}}}$ is clean for every $\underline{\underline{p}} \in$ spec R.

(iv) Every special projective Λ-lattice is locally free.

(v) The Cartan-map

$$\kappa_{\underline{\underline{p}}} : \underline{\underline{K}}_o(\Lambda/\underline{\underline{p}}\Lambda) \longrightarrow \underline{\underline{G}}_o(\Lambda/\underline{\underline{p}}\Lambda)$$

is injective for every $\underline{\underline{p}} \in$ spec R.

(vi) For $P, P' \in {}_{\Lambda}\underline{\underline{P}}^f$, $KP \cong KP'$ if and only if $P \vee P'$. (We recall that $P \vee P'$ means that P and P' lie in the same genus; i.e., they are locally isomorphic.)

If the genus of Λ, $\mathfrak{g}(\Lambda)$ contains only finitely many non-isomorphic lattices, then the following condition is equivalent to the previous ones:

(vii) The reduced projective class group of Λ, $\underline{\underline{C}}_o(\Lambda)$ is a finite group.

Proof: (i)$\Longleftrightarrow$(ii) Let $P(\underline{\underline{p}}) \in {}_{\Lambda_{\underline{\underline{p}}}}\underline{\underline{P}}^f$ be special projective, say $KP(\underline{\underline{p}}) \cong A^{(\alpha)}$. Then there exists a Λ-lattice P such that $P_{\underline{\underline{q}}} \cong \Lambda_{\underline{\underline{q}}}^{(\alpha)}$ for all $\underline{\underline{q}} \neq \underline{\underline{p}}$ and $P_{\underline{\underline{p}}} \cong P(\underline{\underline{p}})$ (cf. IV, 1.8). Moreover, P is special projective (cf. IV, 3.1). Since Λ is clean, P is a progenerator and consequently $P(\underline{\underline{p}})$ is a progenerator; i.e., $\Lambda_{\underline{\underline{p}}}$ is clean. Obviously, if $\Lambda_{\underline{\underline{p}}}$ is clean for every $\underline{\underline{p}}$, then so is Λ (cf. IV, 3.1).

(ii)$\Longleftrightarrow$(iii) If $\hat{\Lambda}_{\underline{\underline{p}}}$ is clean so is $\Lambda_{\underline{\underline{p}}}$ (cf. IV, 3.2). Conversely, if $\Lambda_{\underline{\underline{p}}}$ is clean and if $\hat{P} \in {}_{\hat{\Lambda}_{\underline{\underline{p}}}}\underline{\underline{P}}^f$ is such that $\hat{K}_{\underline{\underline{p}}}\hat{P} \cong \hat{A}_{\underline{\underline{p}}}^{(\alpha)}$, then $P^{*} = \hat{P} \cap A^{(\alpha)} \in {}_{\Lambda_{\underline{\underline{p}}}}\underline{\underline{P}}^f$ is special projective, whence a progenerator. Then $\hat{P}$ is a progenerator by (IV, 3.2).

(iii) $\Longrightarrow$ (iv) Let P be special projective, say $KP \cong A^{(\alpha)}$. Since Λ is clean, P is a progenerator and we shall show $P_{\underline{\underline{p}}} \cong \Lambda_{\underline{\underline{p}}}^{(\alpha)}$ for every $\underline{\underline{p}} \in$ spec R. According to (VI, 1.2) this is equivalent to showing that $\hat{P}_{\underline{\underline{p}}} \cong \hat{\Lambda}_{\underline{\underline{p}}}^{(\alpha)}$.

Let $\{\hat{E}_i\}_{1 \le i \le n}$ be the set of non-isomorphic indecomposable projective $\hat{\Lambda}_{\underline{\underline{p}}}$-lattices. Then

$$\hat{P}_{\underline{\underline{p}}} \cong \oplus_{i=1}^{n} \hat{E}_i^{(n_i)},$$

and $\hat{P}_{\underline{\underline{p}}}$ is a progenerator if and only if $n_i \ge 1, 1 \le i \le n$. Let

$$\hat{\Lambda}_{\underline{\underline{p}}} \cong \oplus_{i=1}^{n} \hat{E}_i^{(m_i)}, \quad m_i \ge 1, 1 \le i \le n.$$

We now consider $\underline{\underline{K}}_o(\hat{\Lambda}_{\underline{\underline{p}}})$, the Grothendieck group of projective $\hat{\Lambda}_{\underline{\underline{p}}}$-lattices.

There are two non-negative integers s,t not both zero, such that in $\underline{\underline{K}}_o(\hat{\Lambda}_{\underline{\underline{p}}})$,

$$s(\textstyle\sum_{i=1}^{n} m_i[\hat{E}_i]) - t(\sum_{i=1}^{n} n_i[\hat{E}_i]) = \sum_{i=1}^{n} z_i[E_i]$$

with $z_i \ge 0$ and $z_i = 0$ for at least one i.

In fact, since $m_i > 0, 1 \leq i \leq n$, we choose j such that

$$n_j m_j^{-1} = \max_{1 \leq i \leq n} \{n_i m_i^{-1}\}; \; s = n_j, \; t = m_j.$$

Then in $\mathbf{K}_0(\hat{\Lambda}_{\mathfrak{p}})$,

$$\sum_{i=1}^{n} (sm_i - tn_i)[\hat{E}_i] = \sum_{i=1}^{n} z_i[\hat{E}_i]$$

has the desired properties.

Let

$$\hat{X} = \oplus_{i=1}^{n} \hat{E}_i^{(z_i)} \in {}_{\hat{\Lambda}_{\mathfrak{p}}}\mathfrak{P}^f.$$

Then

$$\hat{\Lambda}_{\mathfrak{p}}^{(s)} \cong \hat{P}_{\mathfrak{p}}^{(t)} \oplus \hat{X},$$

and we conclude from $\hat{\Lambda}_{\mathfrak{p}}^{(s)} \cong \hat{\Lambda}_{\mathfrak{p}}^{(t)} \oplus \hat{K}_{\mathfrak{p}}\hat{X}$, that $\hat{K}_{\mathfrak{p}}\hat{X}$ is $\hat{\Lambda}_{\mathfrak{p}}$-free; i.e., $\hat{X}$ is a progenerator. But we have constructed $\hat{X}$ such that $z_i = 0$ for at least one i. Consequently $\hat{X} = 0$ and $\hat{P}_{\mathfrak{p}}$ is $\hat{\Lambda}_{\mathfrak{p}}$-free. It follows from (IV, 3.6) that P is locally free.

<u>(iv) $\Longrightarrow$ (v)</u> Assume that the Cartan-map

$$\kappa_{\mathfrak{p}} : \mathbf{K}_0(\Lambda/\mathfrak{p}\Lambda) \longrightarrow \mathbf{G}_0(\Lambda/\mathfrak{p}\Lambda),$$

$$[P]_{K_0} \longmapsto [P]_{G_0}$$

is not injective for some $\mathfrak{p} \in \operatorname{spec} R$; i.e., there exist $\bar{P}, \bar{P}' \in {}_{\Lambda/\mathfrak{p}\Lambda}\mathfrak{P}^f$ such that

$$\kappa_{\mathfrak{p}} : [\bar{P}] - [\bar{P}'] \longmapsto 0 \quad \text{for } \bar{P} \not\cong \bar{P}'.$$

Consequently $\bar{P}$ and $\bar{P}'$ have isomorphic composition factors as $\Lambda/\mathfrak{p}\Lambda$-modules. We recall that $\Lambda/\mathfrak{p}\Lambda \cong \hat{\Lambda}_{\mathfrak{p}}/\mathfrak{p}\hat{\Lambda}_{\mathfrak{p}}$. In (VIII, 3.5) we have established the commutativity of the following diagram

$$\begin{array}{ccc}
\mathbf{K}_0(\hat{\Lambda}_{\mathfrak{p}}) & \xrightarrow{\ \iota\kappa\ } & \mathbf{K}_0(\hat{\Lambda}_{\mathfrak{p}}) \\
\eta_{\mathfrak{p}} \downarrow & & \downarrow \delta_{\mathfrak{p}} \\
\mathbf{K}_0(\Lambda/\mathfrak{p}\Lambda) & \xrightarrow{\ \kappa_{\mathfrak{p}}\ } & \mathbf{G}_0(\Lambda/\mathfrak{p}\Lambda),
\end{array}$$

where $\eta_{\underline{\underline{p}}}$ is an isomorphism, $\hat{\Lambda}_{\underline{\underline{p}}}$ being semi-perfect (cf. IV, 2.1, 3.5). Thus, there exist two $\hat{\Lambda}_{\underline{\underline{p}}}$-lattices $\hat{P}, \hat{P}' \in {}_{\hat{\Lambda}_{\underline{\underline{p}}}}\underline{\underline{P}}^f$ such that

$$\delta_{\underline{\underline{p}}}\iota\kappa : [\hat{P}] - [\hat{P}'] \longmapsto 0.$$

However, we have seen that $\delta_{\underline{\underline{p}}}$ is injective on Im $\iota\kappa$ (cf. VIII, 3.5), and so $\hat{K}_{\underline{\underline{p}}}\hat{P} \cong \hat{K}_{\underline{\underline{p}}}\hat{P}'$. If now $\hat{X} \in {}_{\hat{\Lambda}_{\underline{\underline{p}}}}\underline{\underline{P}}^f$ is such that $\hat{P} \oplus \hat{X} \cong \hat{\Lambda}^{(n)}_{\underline{\underline{p}}}$, then it follows from (iv) that $\hat{P} \oplus \hat{X} \cong \hat{P}' \oplus \hat{X}$. But here we can cancel (cf. VI, 3.5). Thus $\hat{P} \cong \hat{P}'$. But we had chosen $\hat{P}$ and $\hat{P}'$ such that $\hat{P}/\underline{\underline{\hat{p}}}\hat{P} \cong \overline{P}$ and $\hat{P}'/\underline{\underline{\hat{p}}}\hat{P}' = \overline{P'}$ a contradiction, and so $\kappa_{\underline{\underline{p}}}$ is injective.

<u>(v) $\Longrightarrow$ (vi)</u> Given two projective Λ-lattices P, P' with $KP \cong KP'$. We must show $P \vee P'$; i.e., $P_{\underline{\underline{p}}} \cong P'_{\underline{\underline{p}}}$ for every $\underline{\underline{p}} \in$ spec R. Because of (VI, 1.2) it suffices to show $\hat{P}_{\underline{\underline{p}}} \cong \hat{P}'_{\underline{\underline{p}}}$. However, thanks to (v), $\iota\kappa$ is injective in the above diagram D and so $\hat{P}_{\underline{\underline{p}}} \cong \hat{P}'_{\underline{\underline{p}}}$.

<u>(vi) $\Longrightarrow$ (iii)</u> This is trivial.

We assume now that $\mathcal{G}(\Lambda)$ contains only finitely many non-isomorphic lattices.

<u>(i) $\Longleftrightarrow$ (vii)</u> Assume that Λ is clean and let $x \in \underline{\underline{C}}_o(\Lambda)$ (cf. VIII, 3.8) then $x = [F] - [P]$ where F is a free Λ-lattice and $P \in {}_{\Lambda}\underline{\underline{P}}^f$ with $KP \cong KF$. Since Λ is clean F and P lie in the same genus, and by (VII, 3.4) we can find $Q \vee \Lambda$ such that $[F]-[P] = [\Lambda] - [Q]$ in $\underline{\underline{C}}_o(\Lambda)$. However, there are only finitely many possibilities for Q and so $\underline{\underline{C}}_o(\Lambda)$ is a finite group.

<u>Conversely</u>, let us assume that $\underline{\underline{C}}_o(\Lambda)$ is a finite group. (It suffices to assume that every element in $\underline{\underline{C}}_o(\Lambda)$ has finite order.) If $P \in {}_{\Lambda}\underline{\underline{P}}^f$ is special projective, say $KP \cong A^{(n)}$, then

$$[\Lambda^{(n)}] - [P] \in \underline{\underline{C}}_o(\Lambda) \text{ has finite order;}$$

i.e., $m([\Lambda^{(n)}] - [P]) = 0$ for some $m \in \underline{\underline{N}}$. Then there exists a projective Λ-lattice Q such that $\Lambda^{(nm)} \oplus Q = P^{(m)} \oplus Q$ (cf. VIII, 1.7). By (VI, 3.5, 3.6) $\Lambda^{(n)}$ and P lie in the same genus, and Λ is clean. #

1.3 <u>Corollary</u>: A hereditary order is clean if and only if it is maximal.

<u>Proof</u>: If Λ is maximal, then it is hereditary and clean by (1.2, vi) (cf. IV, 5.7). Conversely, if Λ is a non-maximal hereditary R-order, then we may as well assume that A is simple. It will be shown in (2.15) that Λ is contained in two maximal R-orders $\Lambda_1 \neq \Lambda_2$. Let M_1 and M_2 be irreducible Λ_1- and Λ_2-lattices resp. Then $KM_1 \cong KM_2$ but M_1 and M_2 do not lie in the same genus (cf. VI, 4.8). Hence Λ is not clean.

We shall show next that <u>group rings of finite groups are clean</u>. However, we shall prove a more general theorem, which does not use as many properties of the group ring as the older proofs do (cf. e.g. Curtis-Reiner [1], Swan [2]).
The following statement gives a handy test to decide whether an order is clean or not.

1.4 <u>Theorem</u> (Hattori [1]): Let $\hat{R}$ be a complete Dedekind domain with quotient field $\hat{K}$ of characteristic zero. Assume furthermore that $\hat{R}/\pi\hat{R}$ is a finite field, where $\pi\hat{R} = \mathrm{rad}\,\hat{R}$. For the $\hat{R}$-order $\hat{\Lambda}$ in the semi-simple $\hat{K}$-algebra $\hat{A}$ we put

$$[\hat{\Lambda},\hat{\Lambda}] = \{\sum_{\text{finite}} a_i b_i - b_i a_i : a_i, b_i \in \hat{\Lambda}\}.$$

If the images of a full set of non-equivalent primitive idempotents of $\hat{\Lambda}$ lie in the torsion-free part of $\hat{\Lambda}/[\hat{\Lambda},\hat{\Lambda}]$, then $\hat{\Lambda}$ is a clean $\hat{R}$-order.

<u>Proof</u>: Let $\hat{P} \in {}_{\hat{\Lambda}}\underline{P}^f$. Then we have the isomorphism

$$\mu : \hat{P}^* \otimes_{\hat{\Lambda}} \hat{P} \longrightarrow \mathrm{End}_{\hat{\Lambda}}(\hat{P}),$$
$$\alpha \otimes p \longmapsto \eta, \text{ where } p'\eta = (p'\alpha)p.$$

Here $\hat{P}^* = \mathrm{Hom}_{\hat{\Lambda}}(\hat{P}, \hat{\Lambda})$ (cf. III, 1.5).
We define

$$\tilde{\tau} : \hat{P}^* \otimes_{\hat{\Lambda}} \hat{P} \longrightarrow \hat{\Lambda}/[\hat{\Lambda}, \hat{\Lambda}],$$
$$\alpha \otimes p \longmapsto (p)\alpha + [\hat{\Lambda}, \hat{\Lambda}].$$

(We point out that $\tilde{\tau}$ should not be confused with the trace map, which is defined by

$$\tau : \hat{P} \otimes_{\mathrm{End}_{\hat{\Lambda}}(\hat{P})} \hat{P}^* \longrightarrow \hat{\Lambda} \text{ (cf. III, 1.4)}$$

followed by the canonical R-homomorphism $\hat{\Lambda} \longrightarrow \hat{\Lambda}/[\hat{\Lambda}, \hat{\Lambda}]$.)

For $\varphi \in \mathrm{End}_{\hat{\Lambda}}(\hat{P})$, we define the <u>trace of</u> φ

$$\mathrm{tr}\,\varphi = (\varphi)\mu^{-1}\tilde{\tau} \in \hat{\Lambda}/[\hat{\Lambda}, \hat{\Lambda}] = \tilde{\Lambda}.$$

Since all maps are $\hat{R}$-homomorphism, we have

$$\mathrm{tr}(\varphi + \psi) = \mathrm{tr}(\varphi) + \mathrm{tr}(\psi).$$

Moreover, μ is a ring homomorphism (cf. I, Ex. 3,5) and it is easily verified that

$$\mathrm{tr}(\varphi\psi) = \mathrm{tr}(\psi\varphi).$$

The trace map is indeed well-defined, since

$$\tilde{\tau} : \alpha\lambda \otimes p - \alpha \otimes \lambda p \longmapsto 0 \text{ in } \tilde{\Lambda}.$$

The <u>rank element of P</u> is defined as

$$r_{\hat{\Lambda}}(\hat{P}) = \mathrm{tr}\, 1_{\hat{P}} \in \tilde{\Lambda}.$$

The following properties of the rank element are easily checked:

(i) $r_{\hat{\Lambda}}(\hat{\Lambda}) = \tilde{1}$,

(ii) $r_{\hat{\Lambda}}(-)$ is independent of the representative in an isomorphism class, since replacing $\hat{P}$ by an isomorphic copy amounts to conjugation: we have the following commutative diagram for $\sigma : \hat{P} \xrightarrow{\sim} \hat{Q}$:

$$\begin{array}{ccccc}
\mathrm{End}_{\hat{\Lambda}}(\hat{P}) & \longrightarrow & \hat{P}^* \otimes_{\hat{\Lambda}} \hat{P} & \longrightarrow & \tilde{\Lambda} \\
\varphi \downarrow \quad \downarrow & & \downarrow (\sigma^{-1})^* \otimes \sigma & & \downarrow 1_{\tilde{\Lambda}} \\
\sigma^{-1}\varphi\sigma \quad \mathrm{End}_{\hat{\Lambda}}(\hat{Q}) & \longrightarrow & \hat{Q}^* \otimes_{\hat{\Lambda}} \hat{Q} & \longrightarrow & \tilde{\Lambda}.
\end{array}$$

(iii) $r_{\hat{\Lambda}}(\hat{P} \oplus \hat{Q}) = r_{\hat{\Lambda}}(\hat{P}) + r_{\hat{\Lambda}}(\hat{Q})$.

(iv) If $\hat{P} = \hat{\Lambda}\hat{e}$ for an idempotent $\hat{e}$ of $\hat{\Lambda}$, then $r_{\hat{\Lambda}}(\hat{P}) = \tilde{e}$, where $\tilde{e}$ is the image of $\hat{e}$ in $\tilde{\Lambda}$; in fact,

$$\mu^{-1} : 1_{\hat{P}} \longmapsto \alpha_{\hat{e}} \otimes \hat{e},$$

where $\alpha_{\hat{e}} : \hat{\Lambda}\hat{e} \longrightarrow \hat{\Lambda}$ is the injection.

After having set up the machinery we come to the proof of our theorem: Assume that projective $\hat{\Lambda}$-lattices $\hat{P},\hat{Q}$ with $\hat{K}\hat{P} \cong \hat{K}\hat{Q}$ are given. Let $\{\hat{e}_i\}_{1\le i\le n}$ be a set of non-equivalent primitive idempotents in $\hat{\Lambda}$, the images of which lie in the $\hat{R}$-torsion-free part of $\widetilde{\Lambda}$. If

$$\hat{P} = \oplus_{i=1}^{n} \hat{\Lambda}\hat{e}_i^{(n_i)}, \quad \hat{Q} = \oplus_{i=1}^{n} \hat{\Lambda}\hat{e}_i^{(m_i)},$$

then

$$r_{\hat{\Lambda}}(\hat{P}) = \sum_{i=1}^{n} n_i\widetilde{e}_i, \quad r_{\hat{\Lambda}}(\hat{Q}) = \sum_{i=1}^{n} m_i\widetilde{e}_i.$$

We have $r_{\hat{\Lambda}}(\hat{P}) = r_{\hat{\Lambda}}(\hat{Q})$. In fact, we may assume $\hat{K}\hat{P} = \hat{K}\hat{Q}$. Then

$$r_{\hat{A}}(\hat{K}\hat{P}) = r_{\hat{A}}(\hat{K}\hat{Q}),$$

and the map $\hat{\Lambda} \hookrightarrow \hat{K} \otimes_{\hat{R}} \hat{\Lambda}$ induces a map

$$\psi : \hat{\Lambda}/[\hat{\Lambda},\hat{\Lambda}] \longrightarrow (\hat{K} \otimes_{\hat{R}} \hat{\Lambda})/[\hat{K} \otimes_{\hat{R}} \hat{\Lambda}, \hat{K} \otimes_{\hat{R}} \hat{\Lambda}],$$

since $\hat{K} \otimes_{\hat{R}} [\hat{\Lambda},\hat{\Lambda}] = [\hat{K} \otimes_{\hat{R}} \hat{\Lambda}, \hat{K} \otimes_{\hat{R}} \hat{\Lambda}]$. Then, for $\varphi \in \mathrm{End}_{\hat{\Lambda}}(\hat{P})$, we have

$$\psi(\mathrm{tr}\,\varphi) = \mathrm{tr}(1_{\hat{K}} \otimes \varphi).$$

Applying this here, we conclude

$$\psi(r_{\hat{\Lambda}}(\hat{P})) = \psi(r_{\hat{\Lambda}}(\hat{Q})).$$

However $r_{\hat{\Lambda}}(\hat{P})$ and $r_{\hat{\Lambda}}(\hat{Q})$ lie in the $\hat{R}$-torsion-free part of $\widetilde{\Lambda}$, and ψ maps the torsion-free part of $\widetilde{\Lambda}$ monically into $\hat{A}/[\hat{A},\hat{A}]$. Thus $r_{\hat{\Lambda}}(\hat{P}) = r_{\hat{\Lambda}}(\hat{Q})$.

We assume now that $\hat{P} \not\cong \hat{Q}$, say $m_1 \ne n_1$. In the $\hat{R}$-torsion-free part $\widetilde{\Lambda}_0$ of $\widetilde{\Lambda}$ we therefore have the relation

$$\sum_{i=1}^{n} (m_i - n_i)\widetilde{e}_i = 0; \text{ i.e., a relation}$$

$$\sum_{i=1}^{n} k_i\widetilde{e}_i = 0, \quad k_i \in \hat{R}.$$

Since $\hat{R}$ is local and char $\hat{K} = 0$, we may assume that $k_1 = 1$ if $\hat{P} \not\cong \hat{Q}$. The canonical homomorphism

$$\hat{\Lambda} \longrightarrow \hat{\Lambda}/\mathrm{rad}\,\hat{\Lambda} = \overline{\Lambda}$$

induces an $\hat{R}$-epimorphism

$$\varrho : \hat{\Lambda}/[\hat{\Lambda},\hat{\Lambda}] \longrightarrow \bar{\Lambda}/[\bar{\Lambda},\bar{\Lambda}].$$

$\bar{\Lambda} = \hat{\Lambda}/\mathrm{rad}\hat{\Lambda}$ is a semi-simple $\bar{R} = \hat{R}/\hat{\pi}\hat{R}$-algebra and a decomposition of $\bar{\Lambda}$ into $\bar{R}$-algebras $\bar{\Lambda} = \bar{\Lambda}_1 \oplus \bar{\Lambda}_2$ induces an $\bar{R}$-decomposition

$$\bar{\Lambda}/[\bar{\Lambda},\bar{\Lambda}] = \bar{\Lambda}_1/[\bar{\Lambda}_1,\bar{\Lambda}_1] \oplus \bar{\Lambda}_2/[\bar{\Lambda}_2,\bar{\Lambda}_2].$$

Since the idempotents $\{\hat{e}_i\}_{1\le i\le n}$ are primitive and non-equivalent, they all lie in different $\bar{R}$-summands of $\bar{\Lambda}/[\bar{\Lambda},\bar{\Lambda}]$. Hence the relation $\sum_{i=1}^{n} k_i\tilde{e}_i = 0$ implies $\varrho(\tilde{e}_1) = 0$ in the summand $\bar{\Lambda}_1/[\bar{\Lambda}_1,\bar{\Lambda}_1]$, where $\bar{\Lambda}_1$ is the simple component of $\bar{\Lambda}$ corresponding to $\bar{e}_1$. Since $\bar{R}$ is a finite field, $\bar{\Lambda}_1 = (\underline{\underline{k}})_n$, where $\underline{\underline{k}}$ is a finite extension of $\bar{R}$, since there are no finite skewfields (cf. III, 6.7). Hence $\bar{e}_1 \in [\bar{\Lambda}_1,\bar{\Lambda}_1]$. On the one hand, it follows from (III, 6.11) that

$$\mathrm{Trd}_{\bar{\Lambda}_1/\underline{\underline{k}}}(e_1) = 1;$$

on the other hand $\mathrm{Trd}_{\bar{\Lambda}_1/\underline{\underline{k}}}(x) = 0$ for every $x \in [\bar{\Lambda}_1,\bar{\Lambda}_1]$. Thus we have obtained a contradiction to the assumption $\hat{P} \ncong \hat{Q}$. Consequently $\hat{K}\hat{P} \cong \hat{K}\hat{Q}$ implies $\hat{P} \cong \hat{Q}$ and $\hat{\Lambda}$ is clean by (1.2,vi). #

1.5 __Corollary:__ If K has characteristic zero and if R has finite residue class fields, then commutative R-orders are clean.

__Proof:__ This follows from (1.4) and (1.2). #

1.6 __Corollary:__ Let R be a Dedekind domain with quotient field of characteristic zero. If R has finite residue class fields for all $\underline{\underline{p}} \in$ spec R that divide $|G|R$, then RG is a clean R-order.

__Proof:__ In view of (1.2), and (1.3) it suffices to show that $\hat{\Lambda}_{\underline{\underline{p}}} = \hat{R}_{\underline{\underline{p}}}G$ is clean for all $\underline{\underline{p}} \in$ spec R that divide $|G|R$; since, for $\underline{\underline{p}} \nmid |G|R$, $\hat{\Lambda}_{\underline{\underline{p}}}$ is separable (cf. III, 4.8) and thus maximal by (VI, 2.5). Now, for the group ring RG we have $RG/[RG,RG] \cong$ center(RG) (cf. Ex. 1,1) and consequently $\hat{R}_{\underline{\underline{p}}}G/[\hat{R}_{\underline{\underline{p}}}G,\hat{R}_{\underline{\underline{p}}}G]$ is $\hat{R}_{\underline{\underline{p}}}$-torsion-free, and the statement follows from (1.4) and (1.2). #

1.7 <u>Theorem</u> (Swan [2]): Let Λ be an R-order in the separable K-algebra A, and assume that Λ has only one genus of indecomposable projective lattices. Given $P \in {}_{\Lambda}\underline{\underline{P}}^f$ and any indecomposable projective Λ-lattice Q and a non-zero ideal $\underline{\underline{a}}$ of R. Then there exists $Q_o \in {}_{\Lambda}\underline{\underline{P}}^f$, $Q_o \subset Q$ such that for some $n \in \underline{\underline{N}}$,

$$P \cong Q^{(n)} \oplus Q_o, \quad (\mathrm{ann}_R(Q/Q_o), \underline{\underline{a}}) = 1.$$

<u>Proof</u>: Decomposing P into indecomposable summands

$$P = \oplus_{i=1}^{n+1} P_i,$$

the hypotheses imply $P_i \vee Q$, $1 \leq i \leq n+1$. Now the statement follows from (VII, 4.4). #

1.8 <u>Lemma</u> (Roggenkamp [7]): The hypotheses of (1.7) are satisfied if Λ is a clean R-order in the simple K-algebra A. In particular, the Krull-Schmidt theorem is valid locally for ${}_{\Lambda}\underline{\underline{P}}^f$.

<u>Proof</u>: Let $\underline{\underline{p}} \in$ spec R and assume $A = (D)_n$, D a skewfield. We shall show first that there exists exactly one indecomposable projective $\Lambda_{\underline{\underline{p}}}$-lattice. Assume that e is a primitive idempotent in $\Lambda_{\underline{\underline{p}}}$ and let $P_{\underline{\underline{p}}} \in {}_{\Lambda_{\underline{\underline{p}}}}\underline{\underline{P}}^f$ be indecomposable. If L is the - up to isomorphism unique - simple A-module, then

$$K\Lambda_{\underline{\underline{p}}}e \cong L^{(t)} \text{ and } KP_{\underline{\underline{p}}} \cong L^{(s)}.$$

If $s = t$, then $\Lambda_{\underline{\underline{p}}}e \cong P_{\underline{\underline{p}}}$, Λ being clean. Thus we may assume $s < t$. We pick two positive integers s_1 and t_1 such that $ss_1 = tt_1$. It follows from the cleanness of $\Lambda_{\underline{\underline{p}}}$ that

$$P_{\underline{\underline{p}}}^{(s_1)} \not\cong \Lambda_{\underline{\underline{p}}}e^{(t_1)},$$

moreover, $s_1 > t_1$. Passing to the completion, we decompose $\hat{\Lambda}_{\underline{p}}e$ and $\hat{P}_{\underline{p}}$ into indecomposable modules

$$\hat{P}_{\underline{p}} = \oplus_{i=1}^{\sigma} \hat{P}_i, \quad \hat{\Lambda}_{\underline{p}}e = \oplus_{i=1}^{\tau} \hat{Q}_i.$$

Then

$$\oplus_{i=1}^{\sigma} \hat{P}_i^{(s_1)} \cong \oplus_{i=1}^{\tau} \hat{Q}_i^{(t_1)}.$$

The Krull-Schmidt theorem for ${}_{\hat{\Lambda}_{\underline{p}}}\underline{M}^o$ and $s_1 > t_1$ imply $\sigma < \tau$. We assume

$$\hat{P}_i \cong \hat{Q}_i, \quad 1 \le i \le \sigma.$$

Then

$$\hat{K}_{\underline{p}}(\oplus_{i=\sigma}^{\tau} \hat{Q}_i) \cong \hat{K}_{\underline{p}}L^{(t-s)},$$

as is easily seen. By the familiar argument, there exists $0 \ne X_{\underline{p}} \in {}_{\Lambda_{\underline{p}}}\underline{P}^f$ such that

$$\hat{X}_{\underline{p}} \cong \oplus_{i=\sigma}^{\tau} \hat{Q}_i,$$

i.e., $\Lambda_{\underline{p}}e$ decomposes (cf. VI, 1.2), since $\hat{X}_{\underline{p}} \oplus \hat{P}_{\underline{p}} \cong \hat{\Lambda}_{\underline{p}}e$ implies $X_{\underline{p}} \oplus P_{\underline{p}} \cong \Lambda_{\underline{p}}e$. Similarly one derives a contradiction if one assumes $s > t$. Hence $P_{\underline{p}} \cong \Lambda_{\underline{p}}e$.

Thus there exists only one indecomposable projective $\Lambda_{\underline{p}}$-lattice. This obviously implies the validity of the Krull-Schmidt theorem for projective $\Lambda_{\underline{p}}$-lattices.

We now turn to the global situation: Let $P, Q \in {}_{\Lambda}\underline{P}^f$ be indecomposable, say

$$KP = L^{(s)}, \quad KQ = L^{(t)}.$$

If $s = t$ then $P \vee Q$ by (1.2). Let us therefore assume $s > t$. The first part of the proof implies

$$P_{\underline{p}} \cong \Lambda_{\underline{p}}e^{(s_{\underline{p}})}, \quad Q_{\underline{p}} \cong \Lambda_{\underline{p}}e^{(t_{\underline{p}})}, \text{ for every } \underline{p} \in \operatorname{spec} R,$$

where $\Lambda_{\underline{\underline{p}}}e$ is the indecomposable projective $\Lambda_{\underline{\underline{p}}}$-lattice. Since $s > t$ we must have $s_{\underline{\underline{p}}} > t_{\underline{\underline{p}}}$. This must hold for every $\underline{\underline{p}} \in$ spec R and so P is a local direct summand of Q. By (VII, 3.8) Q decomposes. Thus $P \vee Q$ and the hypotheses of (1.7) are satisfied. #

Notation: Given a Dedekind domain R with quotient field K and a finite group G. Without stating it explicitely, we shall assume that char K $\nmid |G|$. We say that R is nice for G if char K = 0, R has finite residue class fields for all $\underline{\underline{p}} \in$ spec R that divide $|G|R$ and no rational prime dividing $|G|$ is a unit in R.

1.9 Theorem (Swan [2]): If R is nice for G, then every projective RG-lattice is locally free.

The proof is done in several steps:

1.10 Lemma: Let p be a rational prime number and G a p-group. If for $\underline{\underline{p}} \in$ spec R, $\underline{\underline{p}} \supset pR$, then every projective $R_{\underline{\underline{p}}}G$-lattice is free; i.e., every projective RG-lattice is locally free.

Proof: We put $\bar{R} = R/\underline{\underline{p}}R$ and we shall show that $\bar{R}G$ is indecomposable. Taking this for granted, let $P_{\underline{\underline{p}}} \in {}_{R_{\underline{\underline{p}}}G}\underline{\underline{P}}^f$. Then $\bar{P} = P_{\underline{\underline{p}}}/\underline{\underline{p}}P_{\underline{\underline{p}}}$ is $\bar{R}G$-free, as follows from the validity of the Krull-Schmidt theorem, say $\bar{P} \cong \bar{R}G^{(n)}$. Then $P_{\underline{\underline{p}}}$ and $(RG)_{\underline{\underline{p}}}^{(n)}$ are both projective covers for $\bar{P}$ (cf. III, 7.1) and thus they are isomorphic by (III, 7.3). Hence $P_{\underline{\underline{p}}}$ is free. Now if $P \in {}_{RG}\underline{\underline{P}}^f$, then for every $\underline{\underline{q}} \neq \underline{\underline{p}}$, $P_{\underline{\underline{q}}}$ is projective, $R_{\underline{\underline{q}}}G$ being separable; and $P_{\underline{\underline{q}}} \cong R_{\underline{\underline{q}}}G^{(n)}$; i.e., P is locally free.

Now, to show that $\bar{R}G$ is indecomposable, it suffices to show that $\underline{\underline{k}}G$ is indecomposable, where $\underline{\underline{k}}$ is an algebraically closed field containing $\bar{R}$. Since $\underline{\underline{k}}$ has characteristic p, we have $(1-g)^{p^s} = 0$, for every $g \in G$, where $|G| = p^s$. If N is the $\underline{\underline{k}}$-module generated by $\{g-1\}_{g \in G}$, then N is a two-sided $\underline{\underline{k}}G$-ideal, since

$$g_1(g-1) = (g_1g-1) - (g_1-1) \text{ and}$$

$$(g-1)g_1 = (gg_1-1) - (g_1-1).$$

Then it follows from Wedderburn's theorem (cf. Ex. 1,5) that $N \subset \text{rad } \underline{\underline{k}}G$, since N has a basis of nilpotent elements. However, $\dim_{\underline{\underline{k}}}(N) = p^s - 1$ implies $\underline{\underline{k}}G/N \cong \underline{\underline{k}}$ and so $N = \text{rad } \underline{\underline{k}}G$. Thus $\underline{\underline{k}}G/\text{rad } \underline{\underline{k}}G$ is indecomposable, and the method of lifting idempotents shows that $\underline{\underline{k}}G$ is indecomposable. #

1.11 __Lemma:__ If no rational prime dividing $|G|$ is a unit in R, then the rank of every $P \in {}_{RG}\underline{\underline{P}}^f$ is a multiple of the order of G.

__Proof:__ Let p^s be the highest power of the rational prime p dividing $|G|$ and let G_p be a p-Sylow subgroup of G. Then RG_p is a subring of RG and every RG-lattice is also an RG_p-lattice. Moreover, RG is RG_p-free and thus every projective RG-lattice is also RG_p-projective. Let $P \in {}_{RG}\underline{\underline{P}}^f$ and denote by P_{G_p} this lattice considered as RG_p-lattice. Then (1.10) implies that $\text{rank}(P_{G_p}) = n \cdot p^s$, for some n.

Using this argument for all rational prime divisors of $|G|$, we conclude

$$\text{rank}(P) = m \prod_{i=1}^{t} p_i^{s_i}, \text{ for some m.}$$

But $\prod_{i=1}^{t} p_i^{s_i} = |G|$, and so the statement is proved. #

1.12 __Lemma:__ Let G be commutative and assume that no rational prime divisor of $|G|$ is a unit in R. Then every $P \in {}_{RG}\underline{\underline{P}}^f$ is locally free.

__Proof:__ $A = KG$ is commutative, G being commutative and according to (VI, 3.6) the Krull-Schmidt theorem is locally valid for the projective Λ-lattices, $\Lambda = RG$. Let $\{\underline{\underline{p}}_i\}_{1 \le i \le n}$ be the prime ideals in R dividing $|G|R$, and decompose

$$R_{\underline{\underline{p}}_i} G = \oplus_{j=1}^{t_i} P_{ij}$$

into indecomposable modules. Since A is commutative, every simple A-module occurs with multiplicity one in A. Consequently $P_{ij} \not\cong P_{ik}$ for $j \neq k$. Let P be an indecomposable Λ-lattice, and write

$$P_{\underset{=}{p}_i} \cong \oplus_{j=1}^{t_i} P_{ij}^{(\alpha_{ij})},$$

where $\{\alpha_{ij}\}$ are non-negative integers. If $\alpha_{ij} > 0$ for all i,j then P has Λ as local direct summand and P decomposes by (VII, 3.8); a contradiction unless P is locally isomorphic to Λ. Therefore we may assume $\alpha_{i1},\ldots,\alpha_{ik} = 0$ for some $1 \leq k < t_i$. Then no simple component of $K(\oplus_{j=1}^{k} P_{ij})$ can occur in $KP_{\underset{=}{p}_l}$, $l \geq 2$, A being commutative. Now we choose $M \in {}_{\Lambda}\underline{\underline{M}}^{o}$ such that $KM \cong K(\oplus_{j=1}^{k} P_{ij})$. The family $M_{\underset{=}{q}}$ for $\underset{=}{q} \nmid |G|R$, $N_{\underset{=}{p}_1} = \oplus_{j:\alpha_{1j}=0} P_{1j}$ satisfies the hypotheses of (IV, 1.8); i.e., there exists $P' \in {}_{\Lambda}\underline{\underline{M}}^{o}$ such that $P'_{\underset{=}{q}} = M_{\underset{=}{q}}$ for $\underset{=}{q} \nmid |G|R$ and $P'_{\underset{=}{p}_1} \cong N_{\underset{=}{p}_1}$. Moreover, P' is projective. Since not all α_{1i} are zero, rank $P' < |G|$, a contradiction to (1.11). Thus $\alpha_{ij} \neq 0$ and P is locally isomorphic to Λ. #

We now come to the proof of (1.9): Up to now we have not used the fact that R is nice for G. But this is needed in order to apply (1.4). Let $P \in {}_{\Lambda}\underline{\underline{P}}^{f}$, and let χ be the character of KP (cf. Ex. 1,2). Then

$$\chi(1) = \dim_K(KP) = n \cdot |G| \text{ by } (1.11).$$

For a fixed $1 \neq g \in G$, let $H = \langle g \rangle$ be the cyclic subgroup of G generated by g. Then P_H is RH-projective and by (1.12) it is locally free. Thus $KP_H \cong KH^{(m)}$ and the characters of KP_H and of KP coincide on g. Thus $\chi(g) = 0$ since $1 \neq g \in G$. This must hold for all $1 \neq g \in G$. Thus χ must be a multiple of the regular character afforded by KG. Hence $KP \cong KG^{(n)}$ (cf. Ex. 1.2). Since RG is clean by (1.6) the statement of (1.9) follows from (1.2). #

1.13 Corollary (Swan [2]): If R is nice for G, then (1.7) is applicable with Q = RG.

The proof follows from (1.9). #

1.14 <u>Corollary</u>: If R is nice for G, then RG is indecomposable; i.e., it does not contain non-trivial idempotents.

<u>Proof</u>: This is a consequence of (1.9). However, there is a short proof due to Jacobinski, if K is an algebraic number field. We put A = KG. For $g \in G$, we have

$$\mathrm{Tr}_{A/K}(g) = \begin{cases} |G| & \text{if } g = 1 \\ 0 & \text{otherwise,} \end{cases}$$

since $gg' = g'$ if and only if $g = 1$. Thus for every $x \in RG$, $\mathrm{Tr}_{A/K}(x) \in |G|R$. We decompose A into simple components $A = \oplus_{i=1}^{s} A_i$. Let K_i be the center of A_i, and $r_i^2 = [A_i : K_i]$, $s_i = [K_i : K]$. If e is a non-trivial idempotent in A, then

$$\mathrm{Tr}_{A/K}(e) = \sum_{i=1}^{s} r_i \, \mathrm{Tr}_{K_i/K}(\mathrm{Trd}_{A_i/K_i}(e))$$

$$= \sum_{i=1}^{s} r_i \, \mathrm{Tr}_{K_i/K}(n_i) = \sum_{i=1}^{s} r_i s_i n_i ,$$

where $0 \leq n_i \leq r_i$ and for at least one i_o we have $0 \leq n_{i_o} < r_{i_o}$, and for at least one j_o we have $0 < n_{j_o} \leq r_{j_o}$. Thus $0 < \mathrm{Tr}_{A/K}(e) < n = |G|$. If $\mathrm{Tr}_{A/K}(e) = z$ then $z \in |G|R$ and $z/|G| \in \mathbb{Q} \cap R = \mathbb{Z}$, a contradiction.#

We now return to the general set-up where R is a Dedekind domain with quotient field K.

1.15 <u>Theorem</u> (Swan [5], Roggenkamp [7]): Let Λ be a clean R-order in A. By $\underline{\underline{P}}_o(\Lambda)$ we denote the Grothendieck group of the special projective Λ-lattices. If Λ' is an R-order in A contained in Λ, then the homomorphism

$$\varphi : \underline{\underline{P}}_o(\Lambda') \longrightarrow \underline{\underline{P}}_o(\Lambda),$$

$$[P] \longmapsto [\Lambda \otimes_{\Lambda'} P]$$

is an epimorphism.

<u>Proof</u>: For $P' \in {}_{\Lambda'}\underline{\underline{P}}^f$, $\Lambda \otimes_{\Lambda'} P'$ is projective and if $KP' \cong A^{(n)}$ then $K(\Lambda \otimes_{\Lambda'} P') \cong A^{(n)}$ and $\Lambda \otimes_{\Lambda'} P'$ is special projective. Since the relations in $\underline{\underline{P}}_o(-)$ are induced from direct sums, φ is a well-defined group homomorphism. To show that φ is an epimorphism, let a special projective Λ-lattice P be given. Since Λ is clean, $P \vee \Lambda \otimes_{\Lambda'} F'$, where F' is a free Λ'-lattice such that $KP \cong KF'$. We embed P into $\Lambda \otimes_{\Lambda'} F'$ such that the factor module $(\Lambda \otimes_{\Lambda'} F')/P = U$ satisfies the following conditions (cf. VII, 3.1):

(i) $U = \oplus_{i=1}^{n} U_i$, U_i is a simple left Λ-module,

(ii) $(\mathrm{ann}_R U_i, \underline{\underline{H}}(\Lambda')\underline{\underline{a}}) = 1$, $1 \le i \le n$, where $\underline{\underline{H}}(\Lambda')$ is the Higman ideal of Λ' and $\underline{\underline{a}}$ is the product of all $\underline{\underline{p}} \in \mathrm{spec}\, R$ for which $\Lambda_{\underline{\underline{p}}} \ne \Lambda'_{\underline{\underline{p}}}$,

(iii) $(\mathrm{ann}_R U_i, \mathrm{ann}_R U_j) = 1$ for $1 \le i \ne j \le n$.

Since $\Lambda' \subset \Lambda$, and since $(\mathrm{ann}_R U_i, \underline{\underline{a}}) = 1$, U_i is also a simple Λ'-module, and by (VII, 3.3) we can find an epimorphism

$$F' \xrightarrow{\sigma} U \longrightarrow 0,$$

F' being faithful. Moreover, $P' = \mathrm{Ker}\, \sigma \in {}_{\Lambda'}\underline{\underline{P}}^f$, since for all $\underline{\underline{p}} | \underline{\underline{H}}(\Lambda')$, $P'_{\underline{\underline{p}}} \cong F'_{\underline{\underline{p}}}$ and for the other primes $\Lambda'_{\underline{\underline{q}}}$ is maximal (cf. VI, 2.5). P' obviously is special projective. Tensoring the exact sequence

$$0 \longrightarrow P' \longrightarrow F' \longrightarrow U \longrightarrow 0$$

with Λ over Λ' yields

$$\mathrm{Tor}_1^{\Lambda'}(\Lambda, U) \xrightarrow{\delta} \Lambda \otimes_{\Lambda'} P' \longrightarrow \Lambda \otimes_{\Lambda'} F' \longrightarrow \Lambda \otimes_{\Lambda'} U \longrightarrow 0.$$

However, $\mathrm{Tor}_1^{\Lambda'}(\Lambda, U)$ is an R-torsion module since U is (cf. the proof of (VIII, 3.4). On the other hand, $\Lambda \otimes_{\Lambda'} P'$ is torsion-free, since it is projective; thus $\mathrm{Im}\, \delta = 0$ and we have an exact sequence of Λ-modules

$$0 \longrightarrow \Lambda \otimes_{\Lambda'} P' \longrightarrow \Lambda \otimes_{\Lambda'} F' \longrightarrow \Lambda \otimes_{\Lambda'} U \longrightarrow 0.$$

We now consider the exact sequence of Λ'-modules

$$0 \longrightarrow \Lambda' \longrightarrow \Lambda \longrightarrow \Lambda/\Lambda' \longrightarrow 0$$

and tensor it with U over Λ':

$$\mathrm{Tor}_1^{\Lambda'}(\Lambda/\Lambda', U) \longrightarrow U \longrightarrow \Lambda \otimes_{\Lambda'} U \longrightarrow \Lambda/\Lambda' \otimes_{\Lambda'} U \longrightarrow 0.$$

Since $(\mathrm{ann}_R(\Lambda/\Lambda'), \mathrm{ann}_R U) = 1$ (cf. ii) we conclude as in the proof of (VIII, 3.4) $U \cong \Lambda \otimes_{\Lambda'} U$. Thus we have the two exact sequences

$$0 \longrightarrow \Lambda \otimes_{\Lambda'} P' \longrightarrow \Lambda \otimes_{\Lambda'} F' \xrightarrow{\alpha} U \longrightarrow 0$$

and

$$0 \longrightarrow P \longrightarrow \Lambda \otimes_{\Lambda'} F' \xrightarrow{\beta} U \longrightarrow 0.$$

Since $(\mathrm{ann}_R U, \underline{\underline{H}}(\Lambda)) = 1$, α and β are projective homomorphisms (cf. V, §§2,3) and an application of Schanuel's lemma (V, 2.6, VII, 3.5) implies

$$(\Lambda \otimes_{\Lambda'} P') \oplus (\Lambda \otimes_{\Lambda'} F') \cong P \oplus (\Lambda \otimes_{\Lambda'} F').$$

Hence

$$\varphi : [P'] \longmapsto [P] ,$$

and φ is epic. #

<u>Remark</u>: In Ch. II, §3 we have only defined the functors $\mathrm{Tor}_1^S(-,N)$ for a ring S; but it is an easy exercise to show that $\mathrm{Tor}_1^S(M,N)$ is naturally isomorphic to $\mathrm{Tor}_1^{S^{op}}(N^{op}, M^{op})$. Thus we define $\mathrm{Tor}_1^S(M,-) = \mathrm{Tor}_1^{S^{op}}(-^{op}, M^{op})$. This is then also a covariant right exact functor, and we have connecting homomorphisms etc.

1.16 <u>Corollary</u>: Let Λ be a clean R-order in A, where no simple component of A is a totally definite quaternion algebra. If Λ' is any R-order in A contained in Λ, then every special projective Λ-lattice P can be written as

$$P \cong \Lambda \otimes_{\Lambda'} P'$$

where P' is a special projective Λ'-lattice.

<u>Proof</u>: The proof of (1.15) shows

$$(\Lambda \otimes_{\Lambda'} P') \oplus (\Lambda \otimes_{\Lambda'} F') \cong P \oplus (\Lambda \otimes_{\Lambda'} F'),$$

where $\Lambda \otimes_{\Lambda'} P \vee \Lambda \otimes_{\Lambda'} F'$. By (VIII, 4.2, 5.1) we can cancel. #

Exercises §1:

1.) Let G be a finite group and R a commutative ring. Show that as R-modules

$$RG/[RG,RG] \cong C = \text{center of } RG.$$

(Hint: The center C of RG is generated over R by the class sums $[g]$ for $g \in G$, where $[g] = \sum_{x \in S} x^{-1}gx$, x ranging over a set S of right coset representatives of $G/C(g)$, where $C(g)$ is the centralizer of g. $[RG,RG]$ is generated over R by all elements of the form $gh - hg$, $g,h \in G$. We now define an R-homomorphism $\overline{\varphi}: RG \longrightarrow C$; $\sum r_g g \longmapsto \sum r_g[g]$. To show that $[RG,RG] \subset \operatorname{Ker}\overline{\varphi}$ it suffices to show $[gh] = [hg]$. Thus we get an R-homomorphism $\varphi : RG/[RG,RG] \longrightarrow C$, $\bar{g} = g + [RG,RG] \longmapsto [g]$. φ is an epimorphism, and to show that it is monic, we construct an inverse map $\psi : C \longrightarrow RG/[RG,RG]$; $[g] \longmapsto g + [RG,RG] = \bar{g}$. Now show that ψ is well-defined and $\varphi\psi = 1_C$, $\psi\varphi = 1_{RG/[RG,RG]}$.)

2.) Let K be a field and G a finite group. With each KG-module L - this always means finitely generated - we associate a matrix representation

$$\varphi_L : G \longrightarrow GL(n,K).$$

For $g \in G$ we put $\chi_L(g) = \operatorname{Tr}(\varphi(g))$, and the function $\chi_L : G \longrightarrow K$ is called the <u>character of L</u>. The following facts are easily checked:

(i) χ_L is independent of the chosen basis of L,

(ii) $\chi_L(1) = \dim_K(L)$,

(iii) χ_L is K-linear and symmetric,

(iv) χ_L is a class function on G; i.e., $\chi_L(g) = \chi_L(g'gg'^{-1})$,

(v) χ_L is the sum of the characters of the composition factors of L,

(vi) $\chi_{KG}(g) = \begin{cases} |G| & \text{if } g=1 \\ 0 & \text{otherwise.} \end{cases}$

Let $\{L_i\}_{1 \le i \le s}$ be a full set of non-isomorphic simple KG-modules and denote their characters by $\{\chi_i\}_{1 \le i \le s}$. For a KG-module L we have $\chi_L = \sum_{i=1}^{s} \alpha_i \chi_i$, if L_i occurs with multiplicity α_i as composition factor in L. Show that if char K = 0 the $\{\chi_i\}_{1\leq i\leq s}$ are linearly independent. Use this to show that the character χ_L determines L up to isomorphism.

3.) Let $\hat{\mathbb{Z}}_2$ be the ring of 2-adic integers and let $\hat{\Lambda}$ be the quaternion order with $\hat{\mathbb{Z}}_2$-basis 1,i,j,k. Then $[\hat{\Lambda},\hat{\Lambda}] = 2(\hat{\mathbb{Z}}_2 i + \hat{\mathbb{Z}}_2 j + \hat{\mathbb{Z}}_2 k)$ and $\hat{\Lambda}/[\hat{\Lambda},\hat{\Lambda}]$ has $\hat{\mathbb{Z}}_2$-torsion; but the image of the idempotent 1 lies in the torsion-free part of $\hat{\Lambda}/[\hat{\Lambda},\hat{\Lambda}]$.

4.) In the notation of (1.4) show that $tr(\varphi\psi) = tr(\psi\varphi)$.

5.) Wedderburn: Let K be an algebraically closed field, A a finite dimensional K-algebra and N a two-sided A-ideal which has a K-basis consisting of nilpotent elements. Show $N \subset$ rad A. (<u>Hint</u>: The problem is easily reduced to the case where A is semi-simple, and since N is a two-sided ideal we may assume that A itself has a basis consisting of nilpotent elements. Moreover, we may assume that A is simple. Since K is algebraically closed, we must have $A \cong (K)_n$. If $a \in A$ is nilpotent, then $Tr_{A/K}(a) = 0$. Hence the hypotheses on A imply that $Tr_{A/K}(a) = 0$ for every $a \in A$. But this can not be since A is separable.)

§2 Hereditary orders

Hereditary orders are classified. If $\hat{R}$ is complete and if $\hat{A} = (\hat{D})_n$, $\hat{D}$ a skewfield with maximal order $\hat{\Omega}$, then every hereditary order in A is Morita equivalent to

$$\hat{\Lambda} = \begin{pmatrix} \hat{\Omega} & \hat{\Omega} & \cdots & & \hat{\Omega} \\ \text{rad}\,\hat{\Omega} & \hat{\Omega} & & & \vdots \\ \vdots & & \ddots & & \\ & & & \hat{\Omega} & \hat{\Omega} \\ \text{rad}\,\hat{\Omega} & \cdots & & \text{rad}\,\hat{\Omega} & \hat{\Omega} \end{pmatrix} \; t \times t, \quad 1 \leq t \leq n.$$

There are exactly t maximal $\hat{R}$-orders containing $\hat{\Lambda}$, and $\hat{\Lambda}$ has exactly t non-isomorphic irreducible lattices.

Let R be a Dedekind domain with quotient field K. By $\hat{X}$ and $X^{\#}$ we denote the completion and localization resp. of X at a fixed $\underline{\underline{p}} \in \operatorname{spec} R$. We recall that an R-order Λ in the separable finite dimensional K-algebra A is called <u>hereditary</u> if every left Λ-lattice is projective. For the moment we shall call such an order left hereditary.

2.1 <u>Theorem</u> (Auslander [1]): An R-order Λ in A is left hereditary if and only if it is right hereditary.
Before we come to the <u>proof</u>, we have to introduce some machinery.

2.2 <u>Lemma</u>: The functor

$$* = \hom_R(-,R) : {}_{\Lambda}\underline{\underline{M}}^{o} \longrightarrow \underline{\underline{M}}^{o}_{\Lambda}$$

is contravariant and exact. Moreover, for $M \in {}_{\Lambda}\underline{\underline{M}}^{o}$,

$$M^{**} = \operatorname{Hom}_R(\operatorname{Hom}_R(M,R),R) \overset{\text{nat}}{\cong} M$$

as left Λ-modules.

<u>Proof</u>: $\hom_R(-,R) : {}_{R}\underline{\underline{M}}^{o} \longrightarrow \underline{\underline{M}}^{o}_{R}$ is contravariant and exact, since every $M \in {}_{R}\underline{\underline{M}}^{o}$ is R-projective (cf. I, 2.11). Moreover, if $M \in {}_{\Lambda}\underline{\underline{M}}^{o} \subset {}_{R}\underline{\underline{M}}^{o}$, then $M^* = \operatorname{Hom}_R(M,R) \in \underline{\underline{M}}^{o}_{\Lambda}$ by (II, 1.12). Thus $* : {}_{\Lambda}\underline{\underline{M}}^{o} \longrightarrow \underline{\underline{M}}^{o}_{\Lambda}$ is a contravariant exact functor. For $M \in {}_{R}\underline{\underline{M}}^{o}$ we have a natural isomorphism

$M^{**} \cong M$. The naturality of this isomorphism implies that for $M \in {}_{\Lambda}\underline{\underline{M}}^{o}$, $M^{**} \overset{nat.}{\cong} M$ as left Λ-modules. #

2.3 Definition: Let Λ be an R-order in A. For $M \in {}_{\Lambda}\underline{\underline{M}}^{o}$, $M^{*} = \mathrm{Hom}_R(M,R) \in \underline{\underline{M}}^{o}_{\Lambda}$ is called the dual of M with respect to R.

2.4 Lemma: The $R^{\#}$-order $\Lambda^{\#}$ is left hereditary if and only if $\mathrm{rad}\,\Lambda^{\#} \in {}_{\Lambda^{\#}}\underline{\underline{P}}^{f}$.

The proof is done exactly as the one of (IV, 4.19). #

Now we come to the proof of (2.1). Since Λ is hereditary if and only if $\Lambda_{\underline{\underline{p}}}$ is hereditary for every $\underline{\underline{p}} \in \mathrm{spec}\, R$, it suffices to prove the theorem locally. Let $\Lambda^{\#}$ be left hereditary. We choose a short exact sequence of right $\Lambda^{\#}$-modules

$$E : 0 \longrightarrow M^{\#} \longrightarrow (\Lambda^{\#})^{(n)} \longrightarrow \mathrm{rad}\,\Lambda^{\#} \longrightarrow 0.$$

Taking duals we obtain the short exact sequence of left $\Lambda^{\#}$-lattices (cf. 2.2)

$$E^{*} : 0 \longrightarrow (\mathrm{rad}\,\Lambda^{\#})^{*} \longrightarrow (\Lambda^{\#})^{*(n)} \longrightarrow (M^{\#})^{*} \longrightarrow 0,$$

which is split since $(M^{\#})^{*} \in {}_{\Lambda^{\#}}\underline{\underline{P}}^{f}$, $\Lambda^{\#}$ being left hereditary. But then E^{**} is also split and $(\mathrm{rad}\,\Lambda^{\#})^{**} \in {}_{\Lambda^{\#}}\underline{\underline{P}}^{f}$. However, $(\mathrm{rad}\,\Lambda^{\#})^{**} \cong \mathrm{rad}\,\Lambda^{\#}$ as right $\Lambda^{\#}$-lattice and so $\Lambda^{\#}$ is right hereditary by (2.4). #

Remark: In view of (2.1) we may simply talk about hereditary R-orders.

2.5 Lemma (Harada [1]): If Λ is a hereditary R-order in A, then every R-order in A containing Λ is hereditary.

Proof: It obviously suffices to prove the statement locally: let $\Lambda_1^{\#}$ contain the hereditary order $\Lambda^{\#}$. Then $\mathrm{rad}\,\Lambda_1^{\#} \in {}_{\Lambda^{\#}}\underline{\underline{P}}^{f}$, and there exist $\varphi_i \in \mathrm{Hom}_{\Lambda^{\#}}(\mathrm{rad}\,\Lambda_1^{\#}, \Lambda^{\#})$ and $m_i \in \mathrm{rad}\,\Lambda_1^{\#}$, $1 \le i \le n$ such that

$$x\, 1_{\mathrm{rad}\,\Lambda_1^{\#}} = \sum_{i=1}^{n} (x\varphi_i) m_i, \; x \in \mathrm{rad}\,\Lambda_1^{\#} \text{ (cf. III, 1.5).}$$

But $\varphi_i \in \mathrm{Hom}_{\Lambda^{\#}}(\mathrm{rad}\,\Lambda_1^{\#}, \Lambda^{\#}) \subset \mathrm{Hom}_{\Lambda_1^{\#}}(\mathrm{rad}\,\Lambda_1^{\#}, \Lambda_1^{\#})$ and so $\mathrm{rad}\,\Lambda_1^{\#} \in {}_{\Lambda_1^{\#}}\underline{\underline{P}}^{f}$ (cf. III, 1.5). Thus $\Lambda_1^{\#}$ is hereditary (cf. 2.4). #

2.6 Lemma: Let $\Lambda_1 \supset \Lambda$ be R-orders in A. If $M \in {}_{\Lambda_1}\underline{\underline{M}}^o$ is Λ-projective, then it is Λ_1-projective.

Proof: The proof is exactly as that of (2.5). #

Remark: (i) The class of hereditary orders is invariant under Morita equivalence (cf. IV, 3.7).

(ii) In (IV, 4.3) we have shown that every hereditary R-order Λ in A contains all central idempotents of A. Since a direct sum of orders is hereditary if and only if each summand is hereditary, we may restrict our attention to hereditary orders in simple algebras, and we assume from now on that $A = (D)_n$.

2.8 Lemma (Harada [1]): Let $\hat{\Lambda}$ be a hereditary $\hat{R}$-order in the simple separable $\hat{K}$-algebra $\hat{A} = (\hat{D})_n$, $\hat{D}$ a skewfield. Let $\hat{M}$ be a two-sided $\hat{\Lambda}$-ideal in $\hat{\Lambda}$ *) (i.e., $\hat{K}\hat{M} = \hat{A}$), properly containing rad $\hat{\Lambda}$. Then

(i) $\hat{M}^2 + \text{rad}\,\hat{\Lambda} = \hat{M}$.

(ii) Some power of M is idempotent: $(\hat{M}^m)^2 = \hat{M}^m$ and $\hat{M} = \hat{M}^m + \text{rad}\,\hat{\Lambda}$.

In addition, if $\{\hat{N}_i\}_{1\le i\le s}$ are the non-isomorphic irreducible $\hat{\Lambda}$-lattices, and if $\hat{N}_i$ occurs with multiplicity t_i in ${}_{\hat{\Lambda}}\hat{\Lambda}$, then $\hat{\Lambda}/\text{rad}\,\hat{\Lambda} \cong \oplus_{i=1}^{s} (\bar{D}_i)_{t_i}$, where $\bar{D}_i$ is a finite dimensional skewfield over $\hat{R}/\text{rad}\,\hat{R}, 1\le i\le s$.

Proof: Because of the validity of the Krull-Schmidt theorem for ${}_{\hat{\Lambda}}\underline{\underline{M}}^o$ and since $\hat{\Lambda}$ is hereditary, in the decomposition of ${}_{\hat{\Lambda}}\hat{\Lambda}$ into indecomposable $\hat{\Lambda}$-lattices,

$${}_{\hat{\Lambda}}\hat{\Lambda} = \oplus_{i=1}^{s} \hat{N}_i^{(t_i)}, \quad \hat{N}_i \not\cong \hat{N}_j \text{ for } i \ne j,$$

the $\{\hat{N}_i\}_{1\le i\le s} = \underline{\underline{Ir}}(\hat{\Lambda})$ are the irreducible $\hat{\Lambda}$-lattices, and $n = \sum_{i=1}^{s} t_i$. However, $\hat{\Lambda}$ is semi-perfect, (cf. IV, 2.1) and $\hat{X} \cong \hat{Y}$, $\hat{X},\hat{Y} \in {}_{\hat{\Lambda}}\underline{\underline{M}}^o$ if and only if $\hat{X}/(\text{rad}\,\hat{\Lambda})\hat{X} \cong \hat{Y}/(\text{rad}\,\hat{\Lambda})\hat{Y}$ (this is proved as IV, 3.5). We conclude thus

$$\hat{\Lambda}/\text{rad}\,\hat{\Lambda} \cong \oplus_{i=1}^{s} (\bar{D}_i)_{t_i}, \quad \bar{D}_i = \text{End}_{\hat{\Lambda}/\text{rad}\,\hat{\Lambda}} (\hat{N}_i/(\text{rad}\,\hat{\Lambda})\hat{N}_i).$$

Now let $\hat{M}$ be a two-sided $\hat{\Lambda}$-ideal properly containing rad $\hat{\Lambda}$. Then

*) We remind that a two-sided ideal is always assumed to span $\hat{A}$.

$0 \neq \bar{M} = \hat{M}/\mathrm{rad}\hat{\Lambda}$ is a two-sided ideal in $\bar{\Lambda} = \hat{\Lambda}/\mathrm{rad}\hat{\Lambda}$, and thus it is of the form $\bar{\Lambda}\bar{e}$, where $\bar{e}$ is a central idempotent in $\bar{\Lambda}$. Consequently $\hat{M}^2 + \mathrm{rad}\,\hat{\Lambda} = \hat{M}$, since $\bar{M}$ is idempotent. However, $\hat{\Lambda}$ is semi-perfect and there exists an idempotent $\hat{e} \in \hat{M} \subset \hat{\Lambda}$ which maps onto $\bar{e}$ (cf. IV, 2.2). Hence $\hat{M}$ contains the idempotent two-sided $\hat{\Lambda}$-ideal $\hat{\Lambda}\hat{e}\hat{\Lambda}$. (We point out that, though $\bar{e}$ is central, $\hat{e}$ need not be central; in fact $\hat{\Lambda}$ does not contain non-trivial central idempotents, $\hat{A}$ being simple.) Since $\hat{\Lambda}\hat{e}\hat{\Lambda}$ is a $\hat{\Lambda}$-ideal in $\hat{A}$, $\hat{\Lambda}/\hat{\Lambda}\hat{e}\hat{\Lambda}$ is artinian (cf. proof of IV, 2.2). But $\hat{\Lambda}\hat{e}\hat{\Lambda} \subset \hat{M}^m$ for all $m \in \underline{\underline{N}}$ and so $\hat{M}^m$ is idempotent for some $m \in \underline{\underline{N}}$, and $M^m + \mathrm{rad}\,\hat{\Lambda} \supset \hat{\Lambda}\hat{e}\hat{\Lambda} + \mathrm{rad}\hat{\Lambda} = \hat{M}$. #

2.9 <u>Theorem</u> (Harada [1]): Let $\hat{\Lambda}$ be a hereditary $\hat{R}$-order in $\hat{A}$. Then there is a one-to-one, inclusions reversing correspondence between the two-sided idempotent $\hat{\Lambda}$-ideals $\hat{I}$ and the $\hat{R}$-orders $\hat{\Sigma}$ containing $\hat{\Lambda}$. $\hat{I}$ corresponds to $\hat{\Sigma}$, where

$$\hat{I} = (\hat{\Lambda} : \hat{\Sigma})_r = \mathrm{Hom}_{\hat{\Lambda}}(\hat{\Sigma}, {}_{\hat{\Lambda}}\hat{\Lambda}) \text{ and } \hat{\Sigma} = \Lambda_r(\hat{I}) = \mathrm{End}_{\hat{\Lambda}}({}_{\hat{\Lambda}}\hat{I}).$$

<u>Proof</u>: For $\hat{M} \in {}_{\hat{\Lambda}}\underline{\underline{M}}^o$ we have the <u>trace map</u>

$$\tau_{\hat{M}} : \hat{M} \otimes_{\mathrm{End}_{\hat{\Lambda}}(\hat{M})} \mathrm{Hom}_{\hat{\Lambda}}({}_{\hat{\Lambda}}\hat{M}, \hat{\Lambda}) \longrightarrow \hat{\Lambda},$$
$$m \otimes \varphi \longmapsto m\varphi .$$

Then $\mathrm{Im}\tau_{\hat{M}}$ is a two-sided $\hat{\Lambda}$-ideal in $\hat{\Lambda}$ (cf. III, 1.4). We shall first verify the following statements.

2.10 <u>Lemma</u>: Let $\hat{\Lambda}$ be a hereditary $\hat{R}$-order in $\hat{A}$. A two-sided $\hat{\Lambda}$-ideal $\hat{M}$ in $\hat{\Lambda}$ is idempotent if and only if $\mathrm{Im}\tau_{\hat{M}} = \hat{M}$.

<u>Proof</u>: Since $\hat{M} \subset \hat{\Lambda}$, we always have $\mathrm{Im}\,\tau_{\hat{M}} \supset \hat{M}$. If now $\hat{M}$ is idempotent, then for every $\varphi \in \mathrm{Hom}_{\hat{\Lambda}}(M, {}_{\hat{\Lambda}}\hat{\Lambda})$ we have

$$\hat{M}\varphi = (\hat{M}^2)\varphi = \hat{M}(\hat{M}\varphi) \subset \hat{M},$$

since $\hat{M}$ is two-sided. Hence $\mathrm{Im}\,\tau_{\hat{M}} \subset \hat{M}$ and so $\hat{M} = \mathrm{Im}\,\tau_{\hat{M}}$ for every two-sided idempotent ideal in $\hat{\Lambda}$.

<u>Conversely</u>, let us assume $\mathrm{Im}\,\tau_{\hat{M}} = \hat{M}$. Since $\hat{M} \in {}_{\hat{\Lambda}}\underline{\underline{P}}^f$, $\mathrm{Im}\,\tau_{\hat{M}} \cdot \hat{M} = \hat{M}$ (cf. III, 1.7) and so $\hat{M}^2 = \hat{M}$ and $\hat{M}$ is idempotent. #

2.11 <u>Lemma</u>: Let $\hat{\Lambda}$ be a hereditary $\hat{R}$-order in $\hat{A}$. A two-sided $\hat{\Lambda}$-ideal $\hat{M}$ in $\hat{\Lambda}$ is idempotent if and only if $\Lambda_r(\hat{M}) = \mathrm{End}_{\hat{\Lambda}}({}_{\hat{\Lambda}}\hat{M}) = \mathrm{Hom}_{\hat{\Lambda}}(\hat{M}, {}_{\hat{\Lambda}}\hat{\Lambda})$.

<u>Proof</u>: We always have $\mathrm{Hom}_{\hat{\Lambda}}(\hat{M}, {}_{\hat{\Lambda}}\hat{\Lambda}) \supset \mathrm{End}_{\hat{\Lambda}}({}_{\hat{\Lambda}}\hat{M})$, since $\hat{M} \subset \hat{\Lambda}$. If now $\hat{M}$ is idempotent, then $\mathrm{Im}\,\tau_{\hat{M}} = \hat{M}$; i.e., $\hat{M}\varphi \subset \hat{M}$ for every $\varphi \in \mathrm{Hom}_{\hat{\Lambda}}(\hat{M}, {}_{\hat{\Lambda}}\hat{\Lambda})$. Hence $\mathrm{Hom}_{\hat{\Lambda}}(\hat{M}, {}_{\hat{\Lambda}}\hat{\Lambda}) = \mathrm{End}_{\hat{\Lambda}}({}_{\hat{\Lambda}}M)$.

<u>Conversely</u>, if $\mathrm{Hom}_{\hat{\Lambda}}(\hat{M}, {}_{\hat{\Lambda}}\hat{\Lambda}) = \mathrm{End}_{\hat{\Lambda}}({}_{\hat{\Lambda}}\hat{M})$, then $\mathrm{Im}\,\tau_{\hat{M}} \subset \hat{M}$; i.e., $\mathrm{Im}\,\tau_{\hat{M}} = \hat{M}$ and $\hat{M}$ is idempotent by (2.10). #

2.12 <u>Lemma</u>: Let $\hat{\Lambda}_1 \supset \hat{\Lambda}$ be hereditary orders. Then $\mathrm{Hom}_{\hat{\Lambda}}(\hat{\Lambda}_1, {}_{\hat{\Lambda}}\hat{\Lambda})$ and $\mathrm{Hom}_{\hat{\Lambda}}(\hat{\Lambda}_1, \hat{\Lambda}_{\hat{\Lambda}})$ are two-sided idempotent $\hat{\Lambda}$-ideals in $\hat{\Lambda}$.

<u>Proof</u>: It suffices to show that $\mathrm{Hom}_{\hat{\Lambda}}(\hat{\Lambda}_1, {}_{\hat{\Lambda}}\hat{\Lambda}) = \hat{I}$ is idempotent. Since $\hat{\Lambda}$ is hereditary, we have $\hat{I}\,\hat{\Lambda}_1 = \hat{\Lambda}_1$ (cf. V, 4.9): in fact, since $\hat{\Lambda}_1 \in {}_{\hat{\Lambda}}\underline{\underline{P}}^f$, we have $\mathrm{Hom}_{\hat{\Lambda}}(\hat{\Lambda}_1, {}_{\hat{\Lambda}}\hat{\Lambda})\,\hat{\Lambda}_1 = \hat{\Lambda}_1$ via the map $\mu_{\hat{\Lambda}_1}$. Hence $\hat{I}^2 = \hat{I}(\hat{\Lambda}_1\hat{I}) = (\hat{I}\,\hat{\Lambda}_1)\hat{I} = \hat{\Lambda}_1\hat{I} = \hat{I}$ and $\hat{I}$ is idempotent. #

Now we turn to the <u>proof of (2.9)</u>.

Given a two-sided idempotent $\hat{\Lambda}$-ideal $\hat{I}$ in $\hat{\Lambda}$. By (2.11) $\hat{\Sigma} = \mathrm{Hom}_{\hat{\Lambda}}(\hat{I}, {}_{\hat{\Lambda}}\hat{\Lambda})$ is an $\hat{R}$-order in $\hat{\Lambda}$ containing $\hat{\Lambda}$ and $\mathrm{Hom}_{\hat{\Lambda}}(\hat{\Sigma}, \hat{\Lambda}_{\hat{\Lambda}}) = \mathrm{Hom}_{\hat{\Lambda}}(\mathrm{Hom}_{\hat{\Lambda}}(\hat{I}, {}_{\hat{\Lambda}}\hat{\Lambda}), \hat{\Lambda}_{\hat{\Lambda}}) \cong \hat{I}$, $\hat{\Lambda}$ being hereditary (I, 2.12). All these isomorphisms are natural, and identifying naturally isomorphic structures, we get actual equalities.

<u>Conversely</u>, if $\hat{\Sigma}$ is an $\hat{R}$-order containing $\hat{\Lambda}$, then $\hat{I} = \mathrm{Hom}_{\hat{\Lambda}}(\hat{\Sigma}, \hat{\Lambda}_{\hat{\Lambda}})$ is idempotent by (2.12) and $\hat{\Sigma} = \mathrm{End}_{\hat{\Lambda}}({}_{\hat{\Lambda}}\hat{I}) = \mathrm{Hom}_{\hat{\Lambda}}(\hat{I}, {}_{\hat{\Lambda}}\hat{\Lambda})$ by (2.11). Thus we have a one-to-one correspondence between the $\hat{R}$-orders containing $\hat{\Lambda}$ and the idempotent two-sided $\hat{\Lambda}$-ideals in $\hat{\Lambda}$. Obviously, this correspondence is inclusion reversing. #

We <u>remark</u> that one gets a similar one-to-one correspondence (not the same!) if one takes $\Lambda_l(\hat{I})$ and $(\hat{\Lambda} : \hat{\Sigma})_l$ for $\Lambda_r(\hat{I})$ and $(\hat{\Lambda} : \hat{\Sigma})_r$.

2.13 <u>Corollary</u>: Let $\hat{\Lambda}$ be a hereditary $\hat{R}$-order in $\hat{A}$. If $\hat{I}_1$ and $\hat{I}_2$ are idempotent two-sided ideals in $\hat{\Lambda}$, then

$$\Lambda_r(\hat{I}_1 + \hat{I}_2) = \Lambda_r(\hat{I}_1) \cap \Lambda_r(\hat{I}_2).$$

<u>Proof</u>: Trivially, $\Lambda_r(\hat{I}_1) \cap \Lambda_r(\hat{I}_2) \subset \Lambda_r(\hat{I}_1 + \hat{I}_2)$. But (2.9) implies that $\Lambda_r(\hat{I}_1 + \hat{I}_2) \subset \Lambda_r(\hat{I}_i)$, $i=1,2$ and so we have equality. #

2.14 <u>Definition</u>: We say that a hereditary $\hat{R}$-order <u>$\hat{\Lambda}$ is of type s</u>, if in the decomposition

$${}_{\hat{\Lambda}}\hat{\Lambda} = \oplus_{i=1}^{s} \hat{N}_i^{(t_i)}, \quad \hat{N}_i \not\cong \hat{N}_j \text{ for } i \neq j, \ \hat{N}_i \text{ indecomposable},$$

there occur s non-isomorphic modules.

We <u>remark</u>, that this actually is the type from the left, however, the method of lifting idempotents and the structure of $\hat{\Lambda}/\text{rad}\,\hat{\Lambda}$ show that $\hat{\Lambda}$ has type s for left modules if and only if it has type s for right modules.

2.15 <u>Theorem</u> (Harada [1]): Let $\hat{\Lambda}$ be a hereditary $\hat{R}$-order in $\hat{A}$ of type s. Then there are exactly s maximal $\hat{R}$-orders in $\hat{A}$ containing $\hat{\Lambda}$, say $\{\hat{\Gamma}_i\}_{1\le i\le s}$. Moreover, $\hat{\Lambda} = \bigcap_{i=1}^{s} \hat{\Gamma}_i$ and every irreducible $\hat{\Lambda}$-lattice is a $\hat{\Gamma}_i$-lattice for some $1\le i\le s$. There are exactly 2^s-1 hereditary $\hat{R}$-orders in $\hat{A}$ containing $\hat{\Lambda}$ and every maximal, strictly ascending chain of hereditary orders starting with $\hat{\Lambda}$, has length s.

<u>Proof</u>: If $\hat{\Lambda}$ has type s, then $\bar{\Lambda} = \hat{\Lambda}/\text{rad}\,\hat{\Lambda} = \oplus_{i=1}^{s} (\bar{D}_i)_{t_i}$ (cf. 2.8) and in $\bar{\Lambda}$ there are exactly s minimal non-zero two-sided ideals. Let $\{\bar{e}_i\}_{1\le i\le s}$ be the primitive orthogonal central idempotents of $\bar{\Lambda}$ and lift these idempotents to orthogonal idempotents $\{\hat{e}_i\}_{1\le i\le s}$ of $\hat{\Lambda}$ (cf. IV, 2.3). $\{\hat{\Lambda}\hat{e}_i\hat{\Lambda}\}_{1\le i\le s}$ are non-zero idempotent two-sided $\hat{\Lambda}$-ideals in $\hat{\Lambda}$. Moreover, these ideals are all different, since $(\hat{\Lambda}\hat{e}_i\hat{\Lambda} + \text{rad}\,\hat{\Lambda})/\text{rad}\,\hat{\Lambda} = \bar{\Lambda}\bar{e}_i$, $1\le i\le s$. In addition, these ideals are minimal two-sided idempotent ideals. In fact, let $\hat{I}$ be a two-sided idempotent ideal in $\hat{\Lambda}$ with $\hat{I} \subset \hat{\Lambda}\hat{e}_i\hat{\Lambda}$. Observe

that $\hat{I} \not\subset \operatorname{rad}\hat{\Lambda}$ by Nakayama's lemma. Then $\hat{e}_1 \in \hat{I} + \operatorname{rad}\hat{\Lambda}$; i.e., $\hat{e}_1 = \alpha + \beta$, $\alpha \in \hat{I}$, $\beta \in \operatorname{rad}\hat{\Lambda}$. Then $(\alpha+\beta)^2 = (\alpha+\beta)$; i.e., $\alpha^2 + \alpha\beta + \beta\alpha - \alpha = \beta - \beta^2 \in \hat{I}$, $\hat{I}$ being two-sided. Hence $\beta(1-\beta) \in \hat{I}$; but $1-\beta$ is a unit in $\hat{\Lambda}$ since $\beta \in \operatorname{rad}\hat{\Lambda}$ and so $\beta \in \hat{I}$, and $\hat{e}_1 \in \hat{I}$. Hence $\hat{\Lambda}\hat{e}_1\hat{\Lambda} = \hat{I}$, and $\hat{\Lambda}\hat{e}_1\hat{\Lambda}$ is minimal. We therefore have found s different minimal two-sided idempotent ideals in $\hat{\Lambda}$, and by (2.9), there are at least s different maximal $\hat{R}$-orders in $\hat{A}$ containing $\hat{\Lambda}$, say $\{\hat{\Gamma}_i\}_{1\le i\le s}$. If $\hat{M}_i$ is an irreducible $\hat{\Gamma}_i$-lattice, then $\hat{M}_i \not\cong \hat{M}_j$ for $i \ne j$ (cf. VI, 5.8) and there can not be more than s maximal $\hat{R}$-orders containing $\hat{\Lambda}$, since $\hat{\Lambda}$ has precisely s non-isomorphic irreducible lattices. Moreover, we see that every irreducible $\hat{\Lambda}$-lattice is a lattice for some $\hat{\Gamma}_i$, $1\le i\le s$. Since $\sum_{i=1}^{n} \hat{\Lambda}\hat{e}_i\hat{\Lambda} = \hat{\Lambda}$ we conclude with (2.13) that $\hat{\Lambda} = \bigcap_{i=1}^{n} \hat{\Gamma}_i$. The same argument as above shows that every $\hat{R}$-order $\hat{\Sigma}$ containing $\hat{\Lambda}$ is the intersection of some of the $\hat{\Gamma}_i$, since the irreducible $\hat{\Sigma}$-lattices are also irreducible $\hat{\Lambda}$-lattices. Thus there are $\sum_{i=o}^{s} \binom{s}{i} - 1 = 2^s - 1$ hereditary $\hat{R}$-orders in $\hat{A}$ containing $\hat{\Lambda}$. It should be noted that the intersections of different maximal orders yield different hereditary orders, since the irreducible lattices of a hereditary order determine the maximal orders lying above it. The remainder of the statements is now clear. #

2.16 <u>Corollary</u>: There are exactly $\binom{s}{i}$ hereditary $\hat{R}$-orders of type $i \le s$ in $\hat{A}$ containing a given hereditary $\hat{R}$-order of type s, $s > i$.

<u>Proof</u>: If $\hat{\Lambda} = \bigcap_{j=1}^{s} \hat{\Gamma}_j$, $\hat{\Gamma}_j$ maximal, $1\le j\le s$, then the orders of type i containing $\hat{\Lambda}$ are exactly those which are intersections of i orders of the $\{\hat{\Gamma}_j\}_{1\le j\le s}$. #

2.17 <u>Corollary</u>: In a hereditary $\hat{R}$-order in $\hat{A}$ of type s, there are exactly 2^s-1 different idempotent two-sided ideals. There are s maximal ones and s minimal ones.

<u>Proof</u>: This is an immediate consequence of (2.16) and (2.9). #

2.18 Corollary: Let $\hat{\Lambda}$ be a hereditary $\hat{R}$-order in $\hat{A}$, and denote the center of $\hat{A}$ by $\hat{F}$. Then $\hat{\Lambda}$ contains the integral closure of $\hat{R}$ in $\hat{F}$.

Proof: This is true for maximal $\hat{R}$-orders in $\hat{A}$ and so the statement follows from (2.15). #

2.19 Theorem (Harada [11]): If $\hat{\Lambda}$ is a hereditary $\hat{R}$-order in $\hat{A}$, then $\mathrm{rad}\,\hat{\Lambda}$ is a progenerator for ${}_{\hat{\Lambda}}\underline{M}^o$ and for $\underline{M}^o_{\hat{\Lambda}}$.

Proof: It suffices to show that $\Lambda_l(\mathrm{rad}\,\hat{\Lambda}) = \Lambda_r(\mathrm{rad}\,\hat{\Lambda}) = \hat{\Lambda}$. In fact, if these conditions are satisfied, then

$$\mathrm{Hom}_{\hat{\Lambda}}(\mathrm{rad}\,\hat{\Lambda}, {}_{\hat{\Lambda}}\hat{\Lambda})\mathrm{rad}\,\hat{\Lambda} = \hat{\Lambda}$$

and

$$\mathrm{rad}\,\hat{\Lambda}\,\mathrm{Hom}_{\hat{\Lambda}}(\mathrm{rad}\,\hat{\Lambda}, \hat{\Lambda}_{\hat{\Lambda}}) = \hat{\Lambda},$$

since $\mathrm{rad}\,\hat{\Lambda}$ is a projective left $\hat{\Lambda}$-module as well as a projective right $\hat{\Lambda}$-module. With (IV, 4.13) we conclude $\mathrm{Hom}_{\hat{\Lambda}}(\mathrm{rad}\,\hat{\Lambda}, {}_{\hat{\Lambda}}\hat{\Lambda}) =$ $= \mathrm{Hom}_{\hat{\Lambda}}(\mathrm{rad}\,\hat{\Lambda}, \hat{\Lambda}_{\hat{\Lambda}}) = (\mathrm{rad}\,\hat{\Lambda})^{-1}$ and $\mathrm{rad}\,\hat{\Lambda}$ is an invertible two-sided $\hat{\Lambda}$-ideal; hence it is a progenerator by (IV, 4.18). (We note that it is not sufficient for an ideal to be a progenerator that it be left and right projective; e.g., the maximal orders containing a hereditary order are two-sided projective; but they are not progenerators!)

Assume that $(\mathrm{rad}\,\hat{\Lambda})\hat{\Lambda}_1 \subset \mathrm{rad}\,\hat{\Lambda}$, where $\hat{\Lambda}_1$ is a minimal proper over-order of $\hat{\Lambda}$. Then $\mathrm{Hom}_{\hat{\Lambda}}(\hat{\Lambda}_1, \hat{\Lambda}_{\hat{\Lambda}}) = \hat{I}_1$ is a maximal idempotent two-sided $\hat{\Lambda}$-ideal, which is a right $\hat{\Lambda}_1$-module. Moreover, every right $\hat{\Lambda}_1$-ideal in $\hat{\Lambda}$ is contained in $\hat{I}_1$. Let $\hat{I}_2, \dots, \hat{I}_s$ be the other maximal two-sided idempotent ideals in $\hat{\Lambda}$ - $\hat{\Lambda}$ is of type s. Now we set $\hat{J} = \bigcap_{i=2}^{s} \hat{I}_i$; then $\hat{I}_1\hat{J} \subset \mathrm{rad}\,\hat{\Lambda} \subset \hat{J}$; i.e., $\hat{I}_1\hat{J} = \hat{I}_1\mathrm{rad}\,\hat{\Lambda}$. Since $\hat{\Lambda}_1\hat{I}_1 = \hat{\Lambda}_1$ (cf. proof of 2.12), we obtain $\hat{\Lambda}_1\hat{J}\hat{\Lambda}_1 = \hat{\Lambda}_1(\mathrm{rad}\,\hat{\Lambda})\hat{\Lambda}_1 = \hat{\Lambda}_1\,\mathrm{rad}\,\hat{\Lambda}$. But $\hat{\Lambda} = \hat{I}_1 + \hat{J}$ as is easily seen; hence $\hat{\Lambda}_1 = \hat{I}_1 + \hat{J}\hat{\Lambda}_1$; thus $\hat{\Lambda}_1\hat{J}\hat{\Lambda}_1 = \hat{I}_1\mathrm{rad}\,\hat{\Lambda} + \hat{J}\hat{\Lambda}_1\mathrm{rad}\,\hat{\Lambda}$. However, $\hat{J}\hat{\Lambda}_1 \subset \hat{\Lambda}_1\hat{J}\hat{\Lambda}_1$ and so $\hat{\Lambda}_1\hat{J}\hat{\Lambda}_1 = \hat{I}_1\mathrm{rad}\,\hat{\Lambda} = \hat{I}_1(\hat{I}_1\hat{J}) = \mathrm{rad}\,\hat{\Lambda}$ by Nakayama's lemma. But $\hat{J} \supsetneq \mathrm{rad}\,\hat{\Lambda}$, and so we have obtained a contradiction. Consequently $\Lambda_r(\mathrm{rad}\,\hat{\Lambda}) = \hat{\Lambda}$; similarly one shows $\Lambda_l(\mathrm{rad}\,\hat{\Lambda}) = \hat{\Lambda}$.#

2.20 <u>Definition</u>: A hereditary $\hat{R}$-order in $\hat{A} = (\hat{D})_n$ is said to be <u>minimal</u> if it is of type n. For, then there are no hereditary $\hat{R}$-orders in $\hat{A}$ properly contained in $\hat{\Lambda}$ (cf. 2.16).

2.21 <u>Theorem</u> (Harada [2]): Let $\hat{\Lambda}_1$ and $\hat{\Lambda}_2$ be hereditary $\hat{R}$-orders in $\hat{A}$. If $\hat{\Lambda}_1$ and $\hat{\Lambda}_2$ are of the same type, they are Morita equivalent.

<u>Proof</u>: 1.) <u>Reduction to minimal hereditary orders</u>.
To prove the theorem we shall use induction on the type. If $\hat{\Lambda}_1$ and $\hat{\Lambda}_2$ are of type 1, they are maximal and thus conjugate (cf. IV, 5.8). Let us assume that all hereditary R-orders of type s-1 are Morita equivalent. We recall that $\hat{\Lambda}$ is of type s-1 if and only if $\hat{\Lambda}$ is the intersection of s-1 maximal orders, if and only if $\hat{\Lambda}$ has exactly s-1 non-isomorphic irreducible lattices. Assume that $\hat{\Lambda}_1$ and $\hat{\Lambda}_2$ are of type s, and let

$$\hat{\Lambda}_1 = \oplus_{i=1}^{s} \hat{M}_i^{(t_i)}, \quad \hat{M}_i \not\cong \hat{M}_j \text{ for } i \neq j.$$
$$\hat{\Lambda}_2 = \oplus_{i=1}^{s} \hat{N}_i^{(t_i)}, \quad \hat{N}_i \not\cong \hat{N}_j \text{ for } i \neq j.$$

Then $\hat{E}_1 = \oplus_{i=1}^{s} \hat{M}_i$ and $\hat{E}_2 = \oplus_{i=1}^{s} \hat{N}_i$ are progenerators for ${}_{\hat{\Lambda}_1}\underline{\underline{M}}^o$ and ${}_{\hat{\Lambda}_2}\underline{\underline{M}}^o$ resp., and we have a Morita equivalence between ${}_{\hat{\Lambda}_1}\underline{\underline{M}}^o$ and ${}_{\hat{\Omega}_1}\underline{\underline{M}}^o$, ${}_{\hat{\Lambda}_2}\underline{\underline{M}}^o$ and ${}_{\hat{\Omega}_2}\underline{\underline{M}}^o$, where $\hat{\Omega}_i = \mathrm{End}_{\hat{\Lambda}_i}(\hat{E}_i)$, i=1,2 are hereditary $\hat{R}$-orders in $(\hat{D})_s$ of type s. Thus $\hat{\Omega}_1$ and $\hat{\Omega}_2$ are minimal hereditary $\hat{R}$-orders in $(\hat{D})_s$. If we can show that $\hat{\Omega}_1$ and $\hat{\Omega}_2$ are Morita equivalent, then so are $\hat{\Lambda}_1$ and $\hat{\Lambda}_2$, since being Morita equivalent is a transitive relation.
We thus return to our old notation and assume that $\hat{\Lambda}_1$ and $\hat{\Lambda}_2$ are minimal hereditary $\hat{R}$-orders in $\hat{A} = (\hat{D})_n$, and that all hereditary $\hat{R}$-orders of type $s < n$ are Morita equivalent.
2.) <u>All minimal hereditary $\hat{R}$-orders in $\hat{A}$ are Morita equivalent</u>.
Let
$$\hat{\Lambda}_1 = \oplus_{i=1}^{n} \hat{M}_i, \ \hat{M}_i \not\cong \hat{M}_j, i \neq j; \ \hat{\Lambda}_2 = \oplus_{i=1}^{n} \hat{N}_i, \ \hat{N}_i \not\cong \hat{N}_j, i \neq j.$$

$$\hat{\Lambda}_1 = \bigcap_{i=1}^{n} \hat{\Gamma}_i, \; \hat{\Lambda}_2 = \bigcap_{i=1}^{n} \hat{\Gamma}_i',$$

where $\{\hat{\Gamma}_i\}_{1 \le i \le n}$ and $\{\hat{\Gamma}_i'\}_{1 \le i \le n}$ are maximal $\hat{R}$-orders in $\hat{A}$ and $\hat{M}_i \in {}_{\hat{\Gamma}_i}\underline{M}^o$, $\hat{N}_i \in {}_{\hat{\Gamma}_i'}\underline{M}^o$. The hereditary $\hat{R}$-orders

$$\hat{\Omega}_1 = \bigcap_{i=1}^{n-1} \hat{\Gamma}_i \text{ and } \hat{\Omega}_2 = \bigcap_{i=1}^{n-1} \hat{\Gamma}_i'$$

are Morita equivalent according to the induction hypothesis. Thus there exists a progenerator $\hat{E} \in {}_{\hat{\Omega}_1}\underline{M}^o$ such that $\hat{\Omega}_2 = \mathrm{End}_{\hat{\Omega}_1}(\hat{E})$, say

$$\hat{E} = \hat{X}_1 \oplus \ldots \oplus \hat{X}_{n-1} \oplus \hat{X}_{n-1},$$

where - if necessary after renumbering the $\{\hat{\Gamma}_i\}_{1 \le i \le n}$ - $\hat{X}_i \cong \hat{M}_i, 1 \le i \le n-1$.
We consider now the $\hat{\Lambda}_1$-lattice

$$\hat{E}_1 = \hat{X}_1 \oplus \ldots \oplus \hat{X}_{n-1} \oplus \hat{M}_n,$$

which is obviously a progenerator for ${}_{\hat{\Lambda}_1}\underline{M}^o$. Moreover, the minimal hereditary $\hat{R}$-orders $\hat{\Lambda}_1$ and $\mathrm{End}_{\hat{\Lambda}_1}(\hat{E}_1)$ are Morita equivalent. We <u>claim</u>

$$\hat{\Lambda}_1' = \mathrm{End}_{\hat{\Lambda}_1}(\hat{E}_1) = \left(\bigcap_{i=1}^{n-1} \hat{\Gamma}_i'\right) \cap \hat{\Gamma},$$

where $\hat{\Gamma}$ is some maximal $\hat{R}$-order in $\hat{A}$. In fact, the irreducible $\hat{\Lambda}_1'$-lattices are

$$\hat{Y}_i = \mathrm{Hom}_{\hat{\Lambda}_1}(\hat{X}_1 \oplus \ldots \oplus \hat{X}_{n-1} \oplus \hat{M}_n, \hat{M}_i), 1 \le i \le n.$$

But $\hat{M}_i \in {}_{\hat{\Gamma}_i}\underline{M}^o$ and we get

$$\hat{Y}_i = \mathrm{Hom}_{\hat{\Gamma}_i}(\hat{\Gamma}_i\hat{X}_1 \oplus \ldots \oplus \hat{\Gamma}_i\hat{X}_{n-1} \oplus \hat{\Gamma}_i\hat{M}_n, \hat{M}_i).$$

Indeed, if $\Lambda \subset \Lambda_1$ are two orders and if $M \in {}_{\Lambda}\underline{M}^o$ and if $N \in {}_{\Lambda_1}\underline{M}^o$, then $\mathrm{Hom}_{\Lambda}(M,N) = \mathrm{Hom}_{\Lambda_1}(\Lambda_1 M, N)$. Obviously we have a map

$$\mathrm{Hom}_{\Lambda_1}(\Lambda_1 M, N) \longrightarrow \mathrm{Hom}_{\Lambda}(M,N),$$

$$\varphi \longmapsto \varphi|_M,$$

which is monic, since there exists $0 \neq r \in R$ such that $r\Lambda_1 \subset \Lambda$. Thus $\mathrm{Hom}_{\Lambda_1}(\Lambda_1 M, N) \subset \mathrm{Hom}_{\Lambda}(M,N)$. On the other hand, every $\varphi \in \mathrm{Hom}_{\Lambda}(M,N)$ can be extended to $\varphi_{\Lambda_1} \in \mathrm{Hom}_{\Lambda_1}(\Lambda_1 M, N)$. Hence

$$\mathrm{Hom}_{\Lambda_1}(\Lambda_1 M, N) = \mathrm{Hom}_{\Lambda}(M, N).$$

Now, back to the above situation: Since $\hat{\Gamma}_1$ is maximal, all the modules $\hat{\Gamma}_1\hat{X}_j, 1 \le j \le n-1$ and $\hat{\Gamma}_1\hat{M}_n$ are isomorphic left $\hat{\Gamma}_1$-lattices. Thus, for $1 \le i \le n-1$, $\hat{Y}_i$ is an irreducible $\hat{\Omega}_2$-lattice, and we conclude $\hat{\Omega}_2 \supset \hat{\Lambda}_1'$; i.e., $\hat{\Lambda}_1'$ has the desired form.

It therefore remains to show that two minimal hereditary $\hat{R}$-orders of the form

$$\hat{\Lambda}_1 = \bigcap_{i=1}^{n} \hat{\Gamma}_i \text{ and } \hat{\Lambda}_2 = (\bigcap_{i=1}^{n-1} \hat{\Gamma}_i) \cap \hat{\Gamma}_n'$$

are Morita equivalent. We put

$$\hat{\Omega} = \bigcap_{i=1}^{n-1} \hat{\Gamma}_i.$$

Then $\hat{\Omega}$ is a minimal over-order of $\hat{\Lambda}_1$ and $\hat{\Lambda}_2$. According to (2.9, 2.15 and 2.17)

$$\hat{I}_1 = \mathrm{Hom}_{\hat{\Lambda}_1}(\hat{\Omega}, \hat{\Lambda}_{1\hat{\Lambda}_1}) \text{ and } \hat{I}_2 = \mathrm{Hom}_{\hat{\Lambda}_2}(\hat{\Omega}, \hat{\Lambda}_{2\hat{\Lambda}_2})$$

are maximal two-sided idempotent ideals in $\hat{\Lambda}_1$ and $\hat{\Lambda}_2$ resp. From (V, 4.9) we conclude

$$\hat{\Omega}\hat{I}_1 = \hat{\Omega} \text{ and } \hat{\Omega}\hat{I}_2 = \hat{\Omega}.$$

Let $\hat{J}_i$ be a maximal two-sided $\hat{\Lambda}_i$-ideal in $\hat{\Lambda}_i$ containing $\hat{I}_i, i=1,2$; i.e., $\hat{J}_i = \hat{I}_i + \mathrm{rad}\,\hat{\Lambda}_i$. Then $\hat{\Lambda}_r(\hat{J}_i) \neq \hat{\Lambda}_i, i=1,2$. In fact, if $\Lambda_r(\hat{J}_1) = \hat{\Lambda}_1$, then $\mathrm{Hom}_{\hat{\Lambda}_1}(\hat{J}_{1\,\hat{\Lambda}_1}, \hat{\Lambda}_1)\hat{J}_1 = \hat{\Lambda}_1$. Since $\hat{J}_1^m$ is idempotent for some $m \in \underline{\underline{N}}$ (cf. 2.8), this means

$$\hat{\Lambda}_1 = \mathrm{Hom}_{\hat{\Lambda}_1}(\hat{J}_{1\,\hat{\Lambda}_1}, \hat{\Lambda}_1)^m \hat{J}_1^m = \mathrm{Hom}_{\hat{\Lambda}_1}(\hat{J}_{1\,\hat{\Lambda}_1}, \hat{\Lambda}_1)^m \hat{J}_1^{2m} = \hat{J}_1^m,$$

a contradiction since $\hat{J}^m \subset \hat{J} \subsetneq \hat{\Lambda}_1$.

We next observe that $\hat{J}_1^m = \hat{I}_1$; in fact, $\hat{J}_1$ can only contain one maximal idempotent ideal and so $\hat{J}_1^m = \hat{I}_1$, $\hat{I}_1$ being maximal. Thus $\Lambda_r(\hat{J}_1) \supsetneq \hat{\Lambda}_1$ implies $\Lambda_r(\hat{J}_1) = \hat{\Omega}$. However, $\hat{I}_1$ is the maximal right $\hat{\Omega}$-ideal in $\hat{\Lambda}_1$ and

$\hat{I}_i = \hat{\mathfrak{J}}_i$, i=1,2.

We now __claim__ $\hat{I}_i \supset \operatorname{rad} \hat{\Omega}$, i=1,2. Assume $\hat{I}_1 \not\supset \operatorname{rad} \hat{\Omega}$. Then $\hat{\Omega} \supset \hat{\Lambda}_1 + \operatorname{rad} \hat{\Omega} \underset{\neq}{\supset} \hat{\Lambda}_1$ and so $\hat{\Omega} = \hat{\Lambda}_1 + \operatorname{rad} \hat{\Omega}$ by the minimality of $\hat{\Omega}$. Hence

$$\begin{aligned} \hat{\Omega} = \hat{\Omega}\hat{I}_1 &= \hat{\Lambda}_1\hat{I}_1 + (\operatorname{rad} \hat{\Omega})\hat{I}_1 \\ &= \hat{\Lambda}_1\hat{I}_1 + \operatorname{rad} \hat{\Omega} \cdot \hat{\Omega}\hat{I}_1 \\ &= \hat{I}_1 + \operatorname{rad} \hat{\Omega}, \end{aligned}$$

and Nakayama's lemma implies $\hat{I}_1 = \hat{\Omega}$, a contradiction. Thus $\hat{I}_i \supset \operatorname{rad} \hat{\Omega}$, i=1,2; and we shall show next

$$\hat{I}_1/\operatorname{rad} \hat{\Omega} \cong \hat{I}_2/\operatorname{rad} \hat{\Omega}$$

as right $\hat{\Omega}$-modules. From (2.8) we conclude $\overline{\Omega} = \hat{\Omega}/\operatorname{rad} \hat{\Omega} = \overline{D}_1 \oplus \ldots \oplus \overline{D}_{n-2} \oplus (\overline{D}_{n-1})_2$, where $\overline{D}_i$, $1 \le i \le n-1$ are skewfields over $\hat{R}/\operatorname{rad} \hat{R}$. However, $\hat{I}_1/\operatorname{rad} \hat{\Omega}$ and $\hat{I}_2/\operatorname{rad} \hat{\Omega}$ are right $\hat{\Omega}$-modules which are not left $\hat{\Omega}$-modules, and

$$\overline{\Omega}(\hat{I}_i/\operatorname{rad} \hat{\Omega}) = \hat{\Omega}/\operatorname{rad} \hat{\Omega}, i=1,2.$$

Thus

$$\begin{aligned} \hat{I}_1/\operatorname{rad} \hat{\Omega} &\cong \overline{D}_1 \oplus \ldots \oplus \overline{D}_{n-2} \oplus \begin{pmatrix} 1 & 0 \\ 0 & 0 \end{pmatrix}(\overline{D}_{n-1})_2 \\ &\cong \hat{I}_2/\operatorname{rad} \hat{\Omega}, \end{aligned}$$

as right $\overline{\Omega}$-modules. Let

$$\overline{\varphi} : \hat{I}_1/\operatorname{rad} \hat{\Omega} \xrightarrow{\sim} \hat{I}_2/\operatorname{rad} \hat{\Omega}$$

be the established isomorphism. If $\psi_i : \hat{I}_i \longmapsto \hat{I}_i/\operatorname{rad} \hat{\Omega}$, i=1,2, are the canonical epimorphisms, then we can complete the following diagram:

$$\begin{array}{ccc} \hat{I}_1 & \overset{\varphi}{\dashrightarrow} & \hat{I}_2 \\ \psi_1 \downarrow & & \downarrow \psi_2 \\ \hat{I}_1/\operatorname{rad} \hat{\Omega} & \xrightarrow{\overline{\varphi}} & \hat{I}_2/\operatorname{rad} \hat{\Omega} \end{array},$$

$\hat{I}_1$ being a projective right $\hat{\Omega}$-module. φ is then given by left multiplication with some $0 \neq a \in \hat{A}$. However, the commutativity of the above diagram shows $a\cdot\text{rad}\,\hat{\Omega} \subset \text{rad}\,\hat{\Omega}$. In the proof of (2.19) we have established $\Lambda_1(\text{rad}\,\hat{\Omega}) = \hat{\Omega}$ and so $a \in \hat{\Omega}$ and we can extend φ to a homomorphism of right $\hat{\Omega}$-modules

$$\varphi_1 : \hat{\Omega} \longrightarrow \hat{\Omega},$$

which is also given by left multiplication with a. φ_1 induces $\bar{\varphi}_1 : \hat{\Omega}/\text{rad}\,\hat{\Omega} \longrightarrow \hat{\Omega}/\text{rad}\,\hat{\Omega}$ which restricts to the isomorphism $\bar{\varphi}$. If $\bar{\varphi}_1$ is not an isomorphism then $\text{Ker}\,\bar{\varphi}_1 = \hat{\Omega}/\text{rad}\,\hat{\Omega}/\hat{I}_1/\text{rad}\,\hat{\Omega}$; i.e., $a\hat{\Omega} \subset \hat{I}_1$ and $a\hat{\Omega} \subset \hat{I}_2$ since $\bar{\varphi}_1$ maps onto $\hat{I}_2/\text{rad}\,\hat{\Omega}$. Hence $\hat{I}_2 = a\hat{\Omega} + \text{rad}\,\hat{\Omega} \subset \hat{I}_1 \cap \hat{I}_2$. Consequently $\hat{I}_1 = \hat{I}_2$ as follows from the maximality of $\hat{I}_1$ and $\hat{I}_2$. But then we can choose as φ and $\bar{\varphi}$ the identity map. Consequently, we may assume that φ_1 induces an isomorphism $\bar{\varphi}_1 : \bar{\Omega} \longrightarrow \bar{\Omega}$; i.e., $a\hat{\Omega} + \text{rad}\,\hat{\Omega} = \hat{\Omega}$ and thus $a\hat{\Omega} = \hat{\Omega}$ and a is a unit in $\hat{\Omega}$. This implies in particular, that $a\cdot\text{rad}\,\hat{\Omega} = \text{rad}\,\hat{\Omega}$ and so $a\cdot\hat{I}_1 = \hat{I}_2$; i.e., φ is an isomorphism.

We shall show finally that

$$a\hat{\Lambda}_1 a^{-1} = \hat{\Lambda}_2.$$

Since $a\hat{\Omega}a^{-1} = \hat{\Omega}$, we conclude

$$\bigcap_{i=1}^{n-1} \hat{\Gamma}_i = a\Big(\bigcap_{i=1}^{n-1} \hat{\Gamma}_i\Big)a^{-1} = \bigcap_{i=1}^{n-1} a\hat{\Gamma}_i a^{-1}.$$

Consequently,

$$a\hat{\Lambda}_1 a^{-1} = \hat{\Omega} \cap a\hat{\Gamma}_n a^{-1}.$$

On the other hand,

$$\Lambda_2 \subset \Lambda_1(\hat{I}_2) = \Lambda_1(a\hat{I}_1) = a\,\Lambda_1(\hat{I}_1)a^{-1} \subset a\hat{\Gamma}_n a^{-1}.$$

Thus $a\hat{\Lambda}_1 a^{-1} \supset \hat{\Lambda}_2$. But both $\hat{\Lambda}_2$ and $a\hat{\Lambda}_1 a^{-1}$ are minimal hereditary orders and so $a\hat{\Lambda}_1 a^{-1} = \hat{\Lambda}_2$ and $\hat{\Lambda}_1$ is Morita equivalent to $\hat{\Lambda}_2$. #

2.22 <u>Corollary</u>: All minimal hereditary $\hat{R}$-orders in $\hat{A}$ are conjugate.

<u>Proof</u>: According to (2.21) two minimal hereditary $\hat{R}$-orders $\hat{\Lambda}_1$ and $\hat{\Lambda}_2$

are Morita equivalent via a progenerator $\hat{E}_1 \in {}_{\hat{\Lambda}_1}\underline{\underline{M}}^o$. Since $\hat{\Lambda}_1$ is the direct sum of non-isomorphic irreducible lattices it follows from the Krull-Schmidt theorem that ${}_{\hat{\Lambda}_1}\hat{\Lambda}_1 \cong \hat{E}_1$; i.e., $\hat{E}_1 = \hat{\Lambda}_1 a$ for some regular element $a \in \hat{A}$. Hence $\hat{\Lambda}_2 = \mathrm{End}_{\hat{\Lambda}_1}(\hat{E}_1) = a^{-1}\hat{\Lambda}_1 a$. #

2.23 __Corollary__: Let $\hat{\Lambda}_1$ and $\hat{\Lambda}_2$ be minimal hereditary $\hat{R}$-orders in $\hat{A}$. If $\hat{\Lambda}$ is an $\hat{R}$-order containing $\hat{\Lambda}_1$, then $\hat{\Lambda}$ is conjugate to an $\hat{R}$-order containing $\hat{\Lambda}_2$.

__Proof__: This is an immediate consequence of (2.22).

2.24 __Theorem__ (Harada [2]): There exist minimal hereditary $\hat{R}$-orders in $\hat{A}$, and every hereditary $\hat{R}$-order in $\hat{A}$ contains a minimal one.

__Proof__: Let $\hat{A} = (\hat{D})_n$, $\hat{D}$ a skewfield with maximal $\hat{R}$-order $\hat{\underline{\Omega}}$ (cf. IV, 5.2) rad $\hat{\underline{\Omega}} = \omega_o\hat{\underline{\Omega}}$.

$$\hat{\Lambda} = \begin{pmatrix} \hat{\underline{\Omega}} & \hat{\underline{\Omega}} & \cdots & \hat{\underline{\Omega}} \\ \omega_o\hat{\underline{\Omega}} & \hat{\underline{\Omega}} & \ddots & \vdots \\ \vdots & \ddots & \hat{\underline{\Omega}} & \hat{\underline{\Omega}} \\ \omega_o\hat{\underline{\Omega}} & \cdots & \omega_o\hat{\underline{\Omega}} & \hat{\underline{\Omega}} \end{pmatrix}_{n\times n}$$

is a minimal hereditary $\hat{R}$-order in $\hat{A}$. It is easily seen that

$$\mathrm{rad}\,\hat{\Lambda} = \begin{pmatrix} \omega_o\hat{\underline{\Omega}} & \hat{\underline{\Omega}} & \cdots & \hat{\underline{\Omega}} & \hat{\underline{\Omega}} \\ \omega_o\hat{\underline{\Omega}} & \omega_o\hat{\underline{\Omega}} & \ddots & & \vdots \\ \vdots & & \ddots & \ddots & \vdots \\ \vdots & & & \omega_o\hat{\underline{\Omega}} & \hat{\underline{\Omega}} \\ \omega_o\hat{\underline{\Omega}} & \cdots & \cdots & \omega_o\hat{\underline{\Omega}} & \omega_o\hat{\underline{\Omega}} \end{pmatrix}_{n\times n}.$$

But rad $\hat{\Lambda} \in {}_{\hat{\Lambda}}\underline{\underline{P}}^f$, since

$$\mathrm{rad}\,\hat{\Lambda}\begin{pmatrix} 0 & \cdots & \cdots & 0 & \omega_o^{-1} \\ 1 & \ddots & & \vdots & 0 \\ 0 & \ddots & \ddots & \vdots & \vdots \\ \vdots & \ddots & \ddots & 0 & \vdots \\ 0 & \cdots & 0 & 1 & 0 \end{pmatrix}_{n\times n} = \hat{\Lambda}.$$

Consequently $\hat{\Lambda}$ is hereditary by (2.4). Moreover, $\hat{\Lambda}/\mathrm{rad}\,\hat{\Lambda} = \bigoplus_1^n \hat{\underline{\Omega}}/\omega_o\hat{\underline{\Omega}}$

shows that $\hat{\Lambda}$ is minimal (cf. 2.20). If now $\hat{\Lambda}_1$ is any hereditary $\hat{R}$-order in $\hat{A}$, then there exists a Morita equivalence between ${}_{\hat{\Lambda}_1}\underline{M}^o$ and ${}_{\hat{\Lambda}_2}\underline{M}^o$, where $\hat{\Lambda}_2$ is an $\hat{R}$-order containing $\hat{\Lambda}$ (cf. 2.21). As in the proof of (2.21) we get a Morita equivalence between $\hat{\Lambda}$ and a hereditary $\hat{R}$-order $\hat{\Lambda}_o$ contained in $\hat{\Lambda}_1$. Then $\hat{\Lambda}_o$ is necessarily minimal. #

__Remark__: Let $\hat{\Gamma} = (\hat{\Omega})_n$ and let $\hat{\Lambda}$ be as in the proof of (2.24). Then $\operatorname{rad}\hat{\Gamma} \subsetneq \operatorname{rad}\hat{\Lambda}$, though $\hat{\Gamma} \supsetneq \hat{\Lambda}$.

2.25 __Theorem__ (Harada [2]): Let $\hat{A} = (\hat{D})_n$. Then every hereditary $\hat{R}$-order in $\hat{A}$ is conjugate to an hereditary $\hat{R}$-order of the following type:

$$\hat{\Lambda}(m_1,\dots m_r) = \begin{pmatrix} (\hat{\Omega})_{m_1} & (\hat{\Omega})_{m_1\times m_2} & \cdots & \cdots & (\hat{\Omega})_{m_1\times m_r} \\ \omega_o(\hat{\Omega})_{m_2\times m_1} & (\hat{\Omega})_{m_2} & \ddots & & \vdots \\ \vdots & & \ddots & & \vdots \\ \vdots & & & (\hat{\Omega})_{m_{r-1}} & \vdots \\ \omega_o(\hat{\Omega})_{m_r\times m_1} & \cdots & \cdots & \omega_o(\hat{\Omega})_{m_r\times m_{r-1}} & (\hat{\Omega})_{m_r} \end{pmatrix}$$

with $\sum_{i=1}^{r} m_i = n$. Here $\hat{\Omega}$ with radical $\omega_o\hat{\Omega}$ is the maximal $\hat{R}$-order in $\hat{D}$ and $(\hat{\Omega})_{m_i \times m_j}$ denotes the set of $(m_i \times m_j)$-matrices with entries in $\hat{\Omega}$.

__Proof__: It is easily seen that there are 2^n-1 different $\hat{R}$-orders of the above type, all of which contain $\hat{\Lambda}$ from the proof of (2.24). Thus, these are all the hereditary $\hat{R}$-orders in $\hat{A}$ containing $\hat{\Lambda}$ (cf. 2.15). Now the statement follows from (2.23) and (2.24). #

__Remark__: Most of the structure theorems for hereditary $\hat{R}$-order have been obtained independently also by Brumer [1,2] and Drozd-Kirichenko-Roiter [1].

2.26 __Lemma__: Let $\hat{\Lambda}$ be a hereditary $\hat{R}$-order of type s in $\hat{A} = (\hat{D})_n$. If $\hat{\Omega}$ with radical $\omega_o\hat{\Omega}$ is the maximal $\hat{R}$-order in $\hat{D}$, then $(\operatorname{rad}\hat{\Lambda})^s = \omega_o\hat{\Lambda}$.

__Proof__: Because of (2.25) we may assume $\hat{\Lambda}$ to be of the form $\hat{\Lambda}(m_1,\dots,m_s)$.

Then the statement follows from a simple matrix calculation. #

2.27 <u>Theorem</u> (Harada [2]): Let $\hat{\Lambda}$ be a hereditary $\hat{R}$-order of type s in $\hat{A}$. For every maximal (minimal) two-sided $\hat{\Lambda}$-ideal $\hat{J}$ in $\hat{A}$, properly containing rad $\hat{\Lambda}$, the ideals $(\text{rad}\,\hat{\Lambda})^{-1}\hat{J}(\text{rad}\,\hat{\Lambda})^{i}$, $1 \leq i \leq s-1$, are all different and $(\text{rad}\,\hat{\Lambda})^{-s}\hat{J}(\text{rad}\,\hat{\Lambda})^{s} = \hat{J}$.

<u>Proof</u>: Putting $\hat{N} = \text{rad}\,\hat{\Lambda}$, we have (cf. proof of 2.24) $\hat{N}=\hat{\Lambda}a=b\hat{\Lambda}$ for some regular elements a and b in $\hat{A}$ and $\hat{N}$ is invertible, $N^{-1} = a^{-1}\hat{\Lambda} = \hat{\Lambda}b^{-1}$. Since $\hat{J} \supset \text{rad}\,\hat{\Lambda}$, $\hat{\Lambda}/\hat{N} \cong \hat{\Lambda}/\hat{J} \oplus \hat{J}/\hat{N}$ as two-sided $\hat{\Lambda}/\hat{N}$-module. We now assume $\hat{N}^{-i}\hat{J}\hat{N}^{i} = \hat{J}$ for some i. Then $\hat{J}_o = \sum_{j=0}^{i-1} \hat{N}^{-j}\hat{J}\hat{N}^{j}$ is a two-sided $\hat{\Lambda}$-ideal in $\hat{\Lambda}$ properly containing $\hat{N}$; moreover, $\hat{N}^{-1}\hat{J}_o\hat{N} = \hat{J}_o$. Thus

$$(*) \qquad \hat{\Lambda}/\hat{N} \cong \hat{\Lambda}/\hat{J}_o \oplus \hat{J}_o/\hat{N}.$$

Since $\hat{N}\hat{J}_o = \hat{J}_o\hat{N}$ we have an isomorphism of right $\hat{\Lambda}$-modules

$$\hat{N}/\hat{J}_o\hat{N} = \hat{N}/\hat{N}\hat{J}_o \cong \hat{\Lambda}/\hat{J}_o.$$

On the other hand, we have an isomorphism of right $\hat{\Lambda}$-modules

$$\hat{J}_o/\hat{N} \cong \hat{J}_o/\hat{J}_o\hat{N}/\hat{N}/\hat{N}\hat{J}_o.$$

Because of (*) this shows $\hat{\Lambda}/\hat{N} \cong \hat{J}_o/\hat{J}_o\hat{N}$, whence $\hat{J}_o \cong \hat{\Lambda}$ as right $\hat{\Lambda}$-module (cf. 2.8). Thus $\Lambda_r(\hat{J}_o) = \hat{\Lambda}$, and there exists a left $\hat{\Lambda}$-lattice $\hat{J}_o^{-1}$ such that $\hat{J}_o^{-1}\hat{J}_o = \hat{\Lambda}$, $\hat{J}_o$ being projective. Since a power of $\hat{J}_o$ is idempotent, this implies $\hat{J}_o = \hat{\Lambda}$ (cf. proof of 2.21).
We now assume that $\hat{J}$ is a minimal two-sided ideal properly containing $\hat{N}$. Then $\hat{N}^{-1}\hat{J}\hat{N}$ is also minimal, and the relation $\sum_{j=0}^{i-1} \hat{N}^{-j}\hat{J}\hat{N}^{j} = \hat{\Lambda}$ implies $i \geq s$. On the other hand, there are only s minimal two-sided $\hat{\Lambda}$-ideals in $\hat{\Lambda}$, properly containing $\hat{N}$; thus $\hat{N}^{-s}\hat{J}\hat{N}^{s} = \hat{J}$. If now $\hat{J}$ is a maximal two-sided ideal in $\hat{\Lambda}$, then we take $\hat{J}_o = \bigcap_{j=0}^{i-1} \hat{N}^{-j}\hat{J}\hat{N}^{-j}$ and the same argument as above shows $\hat{J}_o = \hat{N}$ and hence $i = s$. #

2.28 <u>Corollary</u>: If $\hat{\Lambda}$ is a hereditary $\hat{R}$-order of type s say $\hat{\Lambda} = \bigcap_{i=1}^{s} \hat{\Gamma}_i$, where $\hat{\Gamma}_i$ are maximal $\hat{R}$-orders in $\hat{A}$, then conjugation with rad $\hat{\Lambda}$ induces a fixpoint free permutation of order s on $\{\hat{\Gamma}_i\}_{1 \leq i \leq s}$.

The <u>proof</u> is left as an exercise. #

2.29 <u>Lemma</u> (Roggenkamp [5]): Let $R^{\#}$ with completion $\hat{R}$ be the localization of a Dedekind domain. If $A = (D)_n$ is a central simple K-algebra, we write $\hat{A} = (\hat{D}_1)_{rn}$, where D and $\hat{D}_1$ are finite dimensional central skewfields. If $\Lambda^{\#}$ is a non-maximal hereditary $R^{\#}$-order in A, then the Krull-Schmidt theorem is valid for ${}_{\Lambda^{\#}}\underline{\underline{M}}^{o}$ if and only if r = 1.

<u>Proof</u>: If r = 1, then the Krull-Schmidt theorem is valid for ${}_{\Lambda^{\#}}\underline{\underline{M}}^{o}$ by (VI, 3.3). Now let r > 1. If $M \in {}_{\Lambda^{\#}}\underline{\underline{M}}^{o}$ is irreducible, then $\hat{M} = \oplus_{i=1}^{r} \hat{M}_i$, where the $\{\hat{M}_i\}_{1 \le i \le r}$ are irreducible $\hat{\Lambda}$-lattices. Since $\hat{\Lambda}$ is not maximal, there exist $\hat{\Gamma}_1, \hat{\Gamma}_2$, different maximal $\hat{R}$-order in $\hat{A}$ containing $\hat{\Lambda}$ (cf. 2.15). Let $\hat{M}_1$ and $\hat{M}_2$ be the irreducible $\hat{\Gamma}_1$- and $\hat{\Gamma}_2$-lattice resp. We consider the $\hat{\Lambda}$-lattices

$$\hat{N}_1 = \hat{M}_1^{(r-1)} \oplus \hat{M}_2, \quad \hat{N}_2 = \hat{M}_1^{(r)}, \quad \hat{N}_3 = \hat{M}_1^{(r-2)} \oplus \hat{M}_2^{(2)}.$$

Then $\hat{K}\hat{N}_1 \cong \hat{K}\hat{N}_2 \cong \hat{K}\hat{N}_3 \cong \hat{L}_1^{(r)}$, where $\hat{L}_1$ is the simple $\hat{A}$-module. If L is the simple A-module, then $\hat{L} \cong \hat{L}_1^{(r)}$ and thus by (IV, 1.9) there are $N_i \in {}_{\Lambda^{\#}}\underline{\underline{M}}^{o}$ such that $\hat{R}N_i = \hat{N}_i, 1 \le i \le 3$. Then

$$N_1 \oplus N_1 \cong N_3 \oplus N_2$$

since $X^{\#} \cong Y^{\#}$ if and only if $\hat{X} \cong \hat{Y}$ (cf. VI, 1.2) But obviously $N_1 \not\cong N_2$ and $N_1 \not\cong N_3$, since $\hat{N}_1 \not\cong \hat{N}_2$ and $\hat{N}_1 \not\cong \hat{N}_3$ (cf. VI, 5.8). Thus the Krull-Schmidt theorem can not be valid for ${}_{\Lambda^{\#}}\underline{\underline{M}}^{o}$. #

<u>Exercises §2</u>:

1.) Let

$$\Gamma_i = \left\{ \begin{pmatrix} \alpha & p^{-i}\beta \\ p^{i}\gamma & \delta \end{pmatrix}, \alpha, \beta, \gamma, \delta \in \underline{\underline{Z}} \right\}, \text{ p a rational prime number.}$$

Then $\{\Gamma_i\}_{i=0,1,\dots}$ are maximal $\underline{\underline{Z}}$-orders in $(\underline{\underline{Q}})_2$. Show that $\Gamma_o \cap \Gamma_1$ is a hereditary $\underline{\underline{Z}}$-order, but $\Gamma_o \cap \Gamma_2$ is not a hereditary $\underline{\underline{Z}}$-order in $(\underline{\underline{Q}})_2$.

2.) Let $R^{\#}$ be the localization of the Dedekind domain R at some prime, and let A be a simple separable K-algebra. State and prove theorems similar to (2.10) and (2.11) for $\Lambda^{\#}$. (Caution: A need not be central.)

3.) Let $\Lambda^{\#}$ be a hereditary $R^{\#}$-order in the simple separable K-algebra A. Give a necessary and sufficient condition for the Krull-Schmidt theorem to hold for ${}_{\Lambda^{\#}}\underline{\underline{M}}^{o}$!

4.) Prove (2.28)! Let $\hat{\Lambda}$ be a hereditary $\hat{R}$-order oftype s in the simple separable $\hat{K}$-algebra $\hat{A}$. If $\hat{I}$ is a minimal (maximal) two-sided idempotent $\hat{\Lambda}$-ideal in $\hat{\Lambda}$, show that $\{\hat{N}^{-i}\hat{I}\hat{N}^{i}\}_{0\le i\le s-1}$ where $\hat{N} = \mathrm{rad}\,\hat{\Lambda}$ are all minimal (maximal) idempotent ideals in $\hat{\Lambda}$.

5.) Let Λ be a hereditary R-order in the central simple K-algebra A. Compute $\underline{\underline{Ir}}(\Lambda)$ where $\underline{\underline{Ir}}(\Lambda)$ is the set of irreducible non-isomorphic Λ-lattices. (__Hint:__ Use (VI, 5.7)). Then Compute $\underline{\underline{Ir}}(\Lambda)$ dropping the hypothesis that A is central simple. (__Hint:__ Use (2.18)).

6.) (Michler [1]) Let $\hat{\Lambda}$ be a hereditary $\hat{R}$-order in the central simple $\hat{K}$-algebra $\hat{A}$. If $\hat{\Lambda}$ is of type s, show that there are exactly s simple components in $\hat{\Lambda}/\mathrm{rad}\,\hat{\Lambda}$. (__Hint:__ Show that there is a one-to-one correspondence between the idempotent two-sided ideals in $\hat{\Lambda}$ and the idempotent two-sided ideals of $\hat{\Lambda}/\mathrm{rad}\,\hat{\Lambda}$. If B is a semi-simple artinian ring, then there are 2^s idempotent two-sided ideals in B (including 0 and B), if B has s simple components.)

§3 Grothendieck rings of finite groups

For an integral group ring RG of a finite group we have $\underline{\underline{G}}_o(R_SG) \cong \underline{\underline{K}}_o(KG)$, where R_S is semi-local. This isomorphism is applied to clarify the additive structure of $\underline{\underline{G}}_o(RG)$. We take the Berman-Witt induction theorem for granted and follow the systematic approach of Lam, using Frobenius functors and Frobenius modules. In this section we can only cover little of the theory of integral representations, and for a survey on most of the results we refer the interested reader to Reiner's exposition [18][2]. *)

Remark: Let K be an algebraic number field and R a Dedekind domain with quotient field K; G is a finite group of order $|G|$ such that char $K \nmid |G|$. We put $A = KG$ and $\Lambda = RG$. $\underline{\underline{G}}_o(A) = \underline{\underline{K}}_o(A)$ is isomorphic to the classical character ring (cf. Ex. 1.2, Ex. 3.1, Curtis-Reiner [1, Ch. IV, §30]), and Grothendieck groups seem to be the proper generralization of the character ring.

3.1 Definition: Let $M, N \in {}_\Lambda\underline{\underline{M}}^o$, then $M \otimes_R N$ becomes a Λ-lattice, called the outer tensor product of M and N, denoted by $M \#_R N$, if we define

$$g(m \otimes n) = gm \otimes gn,\ g \in G,\ m \otimes n \in M \otimes_R N,$$

and extend this action R-linearly. The elements in $M \#_R N$ are denoted by $\sum_i m_i \# n_i$.

3.2 Theorem (Frobenius reciprocity law, Swan [2], Lam [1]): Let H be a subgroup of G and let $j : H \longrightarrow G$ be the canonical embedding. Then we have two functors

$$j^* : {}_{RH}\underline{\underline{M}}^o \longrightarrow {}_{RG}\underline{\underline{M}}^o,$$

$$M \longmapsto RG \otimes_{RH} M;$$

$$j_* : {}_{RG}\underline{\underline{M}}^o \longrightarrow {}_{RH}\underline{\underline{M}}^o,$$

$$M \longmapsto M_H,$$

where M_H is obtained from M by restriction of the operators to RH.

*) Cf. also J.P. Serre, Introduction a la théorie de Brauer, Sém. I.H.E.S. 1965/66.

j^* is called the <u>induction functor</u> and j_* is called the <u>restriction functor</u>. Both functors are covariant and exact. They are transitive in the following sense: If $H' \leq H$ and if $j' : H' \longrightarrow H$ is the natural embedding, then $(jj')^* = j^*j'^*$ and $(jj')_* = j_*j'_*$ (we write these functors on the left). Moreover, we have the reciprocity law: There is a natural isomorphism of RG-lattices,

$$\varphi : j^*(N \#_R j_*(M)) \longrightarrow j^*(N) \#_R M,\ M \in {}_{RG}\underline{M}^o, N \in {}_{RH}\underline{M}^o.$$

<u>Proof</u>: Since RG is a free right RH-module on (G:H) elements, j^* is an exact covariant functor. The restriction functor is obviously covariant, exact and transitive. j^* is transitive because of the transitivity of the tensor product. Now, to prove the reciprocity law we shall show that the map

$$\varphi : RG \otimes_{RH} (N \#_R M_H) \longrightarrow (RG \otimes_{RH} N) \#_R M,$$

$$x \otimes (n \# m) \longmapsto (x \otimes n) \# xm\ ,\ x \in G$$

induces an RG-isomorphism. For a fixed $g \in G$, we define the map

$$\varphi_g : N \#_R M_H \longrightarrow (RG \otimes_{RH} N) \#_R M,$$

$$n \# m \longmapsto (g \otimes n) \# gm,$$

which is a well-defined R-homomorphism, M being an RG-lattice. The map

$$\tilde{\varphi} : RG \times (N \#_R M_H) \longrightarrow (RG \otimes_{RH} N) \#_R M,$$

$$(g\ ,\ n \# m) \longmapsto (g \otimes n) \# gm$$

is biadditive and RH-balanced; in fact, if $h \in H$, then

$$\tilde{\varphi} : (gh, n \# m) \longmapsto (gh \otimes n) \# ghm = (g \otimes hn) \otimes ghm, \text{ and}$$

$$\tilde{\varphi} : (g, hn \# hm) \longmapsto (g \otimes hn) \# ghm.$$

Thus we obtain an R-homomorphism

$$\varphi : RG \otimes_{RH} (N \#_R M_H) \longrightarrow (RG \otimes_{RH} N) \#_R M.$$

However, this is even RG-linear: Let $g_o \in G$. Then

$$g_0\{(g \otimes n) \# gm\} = (g_0 g \otimes n) \# g_0 gm = (g_0 g \otimes (n \# m))\varphi .$$

To show that φ is an isomorphism, we set up its inverse:

$$\psi : (RG \otimes_{RH} N) \#_R M \longrightarrow RG \otimes_{RH} (N \#_R M_H),$$

$$(g \otimes n) \# m \longmapsto g \otimes (n \# g^{-1}m).$$

As above one shows that ψ is a well-defined RG-homomorphism. It is now easily seen that φ and ψ are inverse to each other.
As for the naturality, let $\sigma : M \longrightarrow M'$ be an RG-homomorphism and $\tau : N \longrightarrow N'$ an RH-homomorphism; we have to show that the following diagram commutes

$$\begin{array}{ccc} RG \otimes_{RH} (N \#_R M_H) & \xrightarrow{1 \otimes (\tau \otimes \sigma)} & RG \otimes_{RH} (N' \#_R M'_H) \\ \varphi_{N,M} \downarrow & & \downarrow \varphi_{N',M'} \\ (RG \otimes_{RH} N) \# M & \xrightarrow{(1 \otimes \tau) \otimes \sigma} & (RG \otimes_{RH} N') \#_R M' ; \end{array}$$

but

$$\begin{array}{ccc} g \otimes (n \# m) & \longmapsto & g \otimes (n\tau \# m\sigma) \\ \downarrow & & \downarrow \\ (g \otimes n) \# gm & \longmapsto & (g \otimes n\tau) \# gm\sigma \end{array}$$

is commutative, σ being RG-linear. #

Remark: (i) If $R = K$ is a field, then the above reciprocity law is equivalent to the reciprocity law of characters.

(ii) One obtains also a reciprocity law by considering as $j : H \rightarrow G$ any group homomorphism from a group H into a group G.

(iii) The reciprocity law is valid for any commutative ring, if one restricts oneself to RG-modules which are finitely generated and R-projective.

3.3 Lemma: $\underline{\underline{G}}_0(\Lambda)$ is a commutative ring under the outer tensor product with $[R_G]$ acting as identity; moreover, it is even a $\underline{\underline{K}}_0(R)$-algebra, and $\underline{\underline{K}}_0(\Lambda)$ is a $\underline{\underline{G}}_0(\Lambda)$-module.

Proof: We recall that $\Lambda = RG$ is the group ring of a finite group and char $K \nmid |G|$. $\underline{\underline{G}}_0(\Lambda)$ is the Grothendieck group of all Λ-lattices with respect to short exact sequences, and $\underline{\underline{K}}_0(\Lambda)$ is the Grothendieck group of the projective Λ-lattices. To show that $\underline{\underline{G}}_0(\Lambda)$ is a ring, we need only observe that for $M \in {}_\Lambda\underline{\underline{M}}^0$, $N \in {}_\Lambda\underline{\underline{M}}^0$ both $M \#_R -$ and $- \#_R N$ are exact functors; but this is clear, since lattices are R-projective. Obviously R_G, the trivial Λ-module, serves as identity in $\underline{\underline{G}}_0(\Lambda)$. ($R$ becomes an RG-lattice if we define for $g \in G$, $gr = r$, $r \in R$; this is the trivial R-module R_G.) It should be observed that the outer tensor product is commutative.*) Given now $N \in \underline{\underline{K}}_0(R)$, then N can be considered as trivial RG-module, and $N \otimes_R M \in {}_{RG}\underline{\underline{M}}^0$ for $M \in {}_{RG}\underline{\underline{M}}^0$. Thus $\underline{\underline{G}}_0(\Lambda)$ is a $\underline{\underline{K}}_0(R)$-algebra. To show that $\underline{\underline{K}}_0(\Lambda)$ is a $\underline{\underline{G}}_0(\Lambda)$-module, we have to show that for $M \in {}_\Lambda\underline{\underline{M}}^0$, $P \in {}_\Lambda\underline{\underline{P}}^f$, $M \#_R P \in {}_\Lambda\underline{\underline{P}}^f$. Since the tensor product is additive it suffices to show that $M \#_R \Lambda \in {}_\Lambda\underline{\underline{P}}^f$. But $M \#_R \Lambda \cong \Lambda \#_R M = j^*(R) \#_R M \cong j^*(R \#_{R1} j_*(M))$ as follows from (3.2) for $H = \{1\}$. Thus the outer tensor product $\Lambda \#_R M$ is the same as the ordinary tensor product $\Lambda \otimes_R M$ considering M as R-module and G acting on Λ; but this module is obviously Λ-projective, M being R-projective. Thus $\underline{\underline{K}}_0(\Lambda)$, the Grothendieck group of projective Λ-lattices is a $\underline{\underline{G}}_0(\Lambda)$-module, as is easily checked. #

3.4 Definitions: (1) We define the universal Frobenius category, $\underline{\underline{\text{Frob}}}$, the objects of which are commutative rings S. A morphism in $\underline{\underline{\text{Frob}}}$, $S \longrightarrow T$ is a pair of maps (I^*, I_*), where $I^* : S \longrightarrow T$ is a $\underline{\underline{Z}}$-homomorphism and $I_* : T \longrightarrow S$ is a ring homomorphism. In addition, we require the pair (I^*, I_*) to satisfy the following reciprocity law

$$I^*(s \cdot I_* t) = (I^* s)t, \quad s \in S, \ t \in T;$$

i.e., we have the following commutative diagram

$$\begin{array}{ccccc} S \times S & \xleftarrow{1_S \times I_*} & S \times T & \xrightarrow{I^* \times 1_T} & T \times T \\ \downarrow{\scriptstyle \mu_S} & & & & \downarrow{\scriptstyle \mu_T} \\ S & & \xrightarrow{\quad I^* \quad} & & T \end{array}$$

*) In the sense of tensor products; i.e., up to isomorphisms.

where the maps μ_S and μ_T are the respective multiplications in S and T. The composition of two morphisms

$$(I^*, I_*) : S \longrightarrow T \text{ and } (J^*, J_*) : T \longrightarrow U$$

is defined as

$$(J^*, J_*)(I^*, I_*) = (J^*I^*, I_*J_*) : S \longrightarrow U.$$

(We remark, that here we write the morphisms on the left, because we have written homomorphisms of Grothendieck groups on the left.) It is easily checked that this composite satisfies also the reciprocity law and that $\mathbf{Frob}$ becomes a category.

(ii) A *Frobenius functor* is a functor $\mathbf{F}$ from a category $\mathbf{C}$ to $\mathbf{Frob}$, and a morphism between two Frobenius functors $\mathbf{F}$ and $\mathbf{F}'$ is a natural transformation $\mathbf{F} \longrightarrow \mathbf{F}'$.

(iii) Given a Frobenius functor $\mathbf{F} : \mathbf{C} \longrightarrow \mathbf{Frob}$, A *Frobenius $\mathbf{F}$-module* B is a function such that

(1) B assigns to each $C \in ob(\mathbf{C})$ an $\mathbf{F}(C)$-module $B(C)$.

(2) B assigns to each morphism $\alpha : C \longrightarrow C'$ in $\mathbf{C}$ a pair of additive maps $B(\alpha) = (B(\alpha)^*, B(\alpha)_*)$,

$$B(\alpha)^* : B(C) \longrightarrow B(C'),$$

$$B(\alpha)_* : B(C') \longrightarrow B(C),$$

such that the following diagrams are commutative

$$\text{(I)} \qquad \begin{array}{ccc} \mathbf{F}(C') \times B(C') & \xrightarrow{\mathbf{F}(\alpha)_* \times B(\alpha)_*} & \mathbf{F}(C) \times B(C) \\ \mu' \downarrow & & \downarrow \mu \\ B(C') & \xrightarrow{B(\alpha)_*} & B(C), \end{array}$$

where $\mu : \mathbf{F}(C) \times B(C) \longrightarrow B(C)$ is induced from the $\mathbf{F}(C)$-module structure of $B(C)$;

$$\text{(II)}\qquad \begin{array}{ccccc} & & \underline{F}(C) \times B(C') & & \\ & {\scriptstyle 1_{F(C)} \times B(\alpha)_*} \swarrow & & \searrow {\scriptstyle \underline{F}(\alpha)^* \times 1_{B(C')}} & \\ \underline{F}(C) \times B(C) & & & & \underline{F}(C') \times B(C') \\ \downarrow {\scriptstyle \mu} & & & & \downarrow {\scriptstyle \mu'} \\ B(C) & & \xrightarrow{\quad B(\alpha)^* \quad} & & B(C'), \end{array}$$

$$\text{(III)}\qquad \begin{array}{ccccc} & & \underline{F}(C') \times B(C) & & \\ & {\scriptstyle \underline{F}(\alpha)_* \times 1_{B(C)}} \swarrow & & \searrow {\scriptstyle 1_{\underline{F}(C')} \times B(\alpha)^*} & \\ \underline{F}(C) \times B(C) & & & & \underline{F}(C') \times B(C') \\ \downarrow {\scriptstyle \mu} & & & & \downarrow {\scriptstyle \mu'} \\ B(C) & & \xrightarrow{\quad B(\alpha)^* \quad} & & B(C'). \end{array}$$

In addition, it is required that the correspondence

$$\alpha \longmapsto B(\alpha)^*, (\alpha \longmapsto B(\alpha)_*), \quad \alpha \in \operatorname{morph}_{\underline{C}}(C,C')$$

makes B into a covariant (contravariant) functor from $\underline{C}$ to $\underline{\underline{Ab}}$.

(iv) A __morphism__ $\varphi : B \longrightarrow B'$ __of Frobenius__ $\underline{F}$__-modules__ is a family $\{\varphi(C)\}$ of $\underline{F}(C)$-homomorphisms

$$\varphi(C) : B(C) \longrightarrow B'(C)$$

such that for $\alpha \in \operatorname{morph}_{\underline{C}}(C,C')$ the following diagrams are commutative.

$$\begin{array}{ccc} B(C) & \xrightarrow{\varphi(C)} & B'(C) \\ {\scriptstyle B(\alpha)^*} \downarrow & & \downarrow {\scriptstyle B'(\alpha)^*} \\ B(C') & \xrightarrow{\varphi(C')} & B'(C') \end{array} \quad , \qquad \begin{array}{ccc} B(C') & \xrightarrow{\varphi(C')} & B'(C') \\ {\scriptstyle B(\alpha)_*} \downarrow & & \downarrow {\scriptstyle B'(\alpha)_*} \\ B(C) & \xrightarrow{\varphi(C)} & B'(C). \end{array}$$

(v) For two Frobenius $\underline{F}$-modules B, B', we define the direct sum $B \oplus B'$ by

$$(B \oplus B')(C) = B(C) \oplus B'(C)$$

and

$$(B \oplus B')(\alpha) = (B(\alpha)^* \oplus B'(\alpha)^*, B(\alpha)_* \oplus B'(\alpha)_*).$$

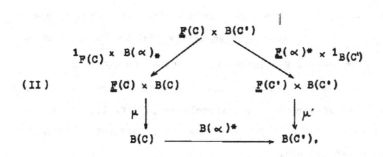

This makes the category of Frobenius $\underline{\underline{F}}$-modules into an additive category ${}_{\underline{\underline{F}}}\underline{\underline{M}}$. But ${}_{\underline{\underline{F}}}\underline{\underline{M}}$ is even an <u>abelian category</u>: we define for a morphism $\varphi: B \longrightarrow B'$ of Frobenius $\underline{\underline{F}}$-modules $(\mathrm{Ker}\,\varphi)(C) = \mathrm{Ker}\,\varphi(C)$ and $(\mathrm{Coker}\,\varphi)(C) = \mathrm{Coker}\,\varphi(C)$.[*] It is easily checked that $\mathrm{Ker}\,\varphi$ and $\mathrm{Coker}\,\varphi$ have the properties of kernels and cokernels resp. (cf. II, Ex. 1,3). We <u>remark</u> that every Frobenius functor $\underline{\underline{F}}$ can be considered as Frobenius $\underline{\underline{F}}$-module in a natural way.

3.5 <u>Lemma</u>: Let $\underline{\underline{C}}$ be the category of finite groups and group monomorphisms, and let R be a Dedekind domain (it suffices to take any commutative ring and consider only R-projective finitely generated RG-modules). Then $\underline{\underline{G}}_R : \underline{\underline{C}} \longrightarrow \underline{\underline{\mathrm{Frob}}}$,

$$\underline{\underline{G}} \longmapsto \underline{\underline{G}}_o(RG), \qquad \begin{array}{l} G_R^*(\alpha) : [M] \longrightarrow [RG \otimes_{RH} M], \\ G_{R*}(\alpha) : [M] \longrightarrow [M_H] \end{array}$$

for a monomorphism $\alpha : H \longrightarrow G$, is a Frobenius functor. And K:

(i) $K(G) = \underline{\underline{K}}_o(RG)$;

(ii) if $j : H \longrightarrow G$ is a monomorphism of groups then $K(j) = (j^*, j_*)$,

is a Frobenius $\underline{\underline{G}}_R$-module, in the notation of (3.2).

<u>Remark</u>: This example led Lam [1] to the definition of Frobenius modules and Frobenius functors; however, there are other examples of Frobenius functors and Frobenius modules (cf. Lam [1]).

<u>Proof</u>: This follows immediately from (3.2) and (3.3). However as an illustration we shall go into some of the details. Let $j : H \longrightarrow G$ be a monomorphism of groups, then we get two induced maps

$$j^* : \underline{\underline{G}}_o(RH) \longrightarrow \underline{\underline{G}}_o(RG),$$
$$[M] \longmapsto [RG \otimes_{RH} M] = [j^*(M)];$$

$$j_* : \underline{\underline{G}}_o(RG) \longrightarrow \underline{\underline{G}}_o(RH),$$
$$[M] \longmapsto [M_H] = [j_*(M)],$$

[*] It is clear how $(\mathrm{Ker}\,\varphi)^*$ and $(\mathrm{Ker}\,\varphi)_*$ are induced, and we leave it as an exercise to verify that they satisfy (I,II,III). Similarly for $\mathrm{Coker}\,\varphi$.

induced by induction and restriction (cf. 3.2). Thus $\underline{\underline{G}}_R$ is a Frobenius functor. From (3.2) it also follows that j induces two maps

$$\alpha^* : \underline{\underline{K}}_o(RH) \longrightarrow \underline{\underline{K}}_o(RG),$$
$$[P] \longmapsto [RG \otimes_{RH} P] = [j^*P];$$
$$\alpha_* : \underline{\underline{K}}_o(RG) \longrightarrow \underline{\underline{K}}_o(RH),$$
$$[P] \longmapsto [P_H] = [j_*P].$$

To show that K is a Frobenius $\underline{\underline{G}}_R$-module, we observe that $\underline{\underline{K}}_o(RG)$ is a $\underline{\underline{G}}_o(RG)$-module (cf. 3.3) and we have to verify the formulae (3.4,iii,2):

$$\alpha_*([M][P]) = [(M \otimes_R P)_H] = [M_H \otimes_R P_H] = j_*([M]) \cdot \alpha_*([P]),$$

$$j^*([M]) \cdot [P] = \alpha^*([M][P_H]) = \alpha^*([M][\alpha_*(P)]),$$

$$[M][\alpha^*(P)] = \alpha^*(j_*([M])[P]).$$

From the statements in (3.2) it also follows that for $j : H \longrightarrow G$, $j \longmapsto \alpha^*$ and $j \longmapsto \alpha_*$ are functors. #

3.6 __Examples:__ The groups $\underline{\underline{K}}_1(RG)$ (cf. VIII, 2.4) and $\underline{\underline{G}}_1(RG) = \underline{\underline{G}}_1({}_{RG}\underline{\underline{M}}^o)$ (cf. VIII, 1.10) are $\underline{\underline{G}}_o(RG)$-modules and hence also Frobenius $\underline{\underline{G}}_R$-modules. (e.g., $[M,\alpha] \in \underline{\underline{G}}_1({}_{RG}\underline{\underline{M}}^o)$, then $[N][M,\alpha] = [N \otimes_R M, 1 \otimes \alpha]$.)

We have the following morphisms of Frobenius modules:

(i) The Cartan map (cf. VIII, 3.5)

$\kappa : \underline{\underline{K}}_o(RG) \longrightarrow \underline{\underline{G}}_o(RG)$,

$\kappa_{\underline{p}} : \underline{\underline{K}}_o((R/\underline{p})G) \longrightarrow \underline{\underline{G}}_o((R/\underline{p})G)$.

(ii) The map $\iota_{\underline{\underline{S}}_o} : \underline{\underline{G}}_o(RG) \longrightarrow \underline{\underline{G}}_o(R_SG)$ (cf. VIII, 3.2).

The verification of these facts is left as an exercise (cf. Ex. 3,2). Moreover, since the category of Frobenius $\underline{\underline{G}}_R$-modules is abelian, we conclude that the kernels and cokernels of all Frobenius maps are Frobenius modules..

3.7 __Definition:__ Let $\underline{F} : \underline{C} \longrightarrow \underline{\underline{Frob}}$ be a Frobenius functor, B a Frobenius module, and let $\underline{\underline{C}}_o$ be a subclass of objects in $\underline{C}$. For each

$C \in ob(\underline{\underline{C}})$ we define

$$B_{\underline{\underline{C}}_0}(C) = \bigcap_{\substack{\alpha \in morph_{\underline{\underline{C}}}(C',C) \\ C' \in \underline{\underline{C}}_0}} Ker(B(\alpha)_*)$$

and

$$B^{\underline{\underline{C}}_0}(C) = \sum_{\substack{\alpha \in morph_{\underline{\underline{C}}}(C',C) \\ C' \in \underline{\underline{C}}_0}} Im(B(\alpha)^*).$$

3.8 <u>Theorem</u> (Swan [2], Lam [1]):

(i) $\underline{\underline{F}}(C)B^{\underline{\underline{C}}_0}(C) + \underline{\underline{F}}^{\underline{\underline{C}}_0}(C)B(C) \subset B^{\underline{\underline{C}}_0}(C)$,

$\underline{\underline{F}}(C)B_{\underline{\underline{C}}_0}(C) + \underline{\underline{F}}_{\underline{\underline{C}}_0}(C)B(C) \subset B_{\underline{\underline{C}}_0}(C)$,

(ii) $B^{\underline{\underline{C}}_0}(C)$ and $B_{\underline{\underline{C}}_0}(C)$ are $\underline{\underline{F}}(C)$ submodules of $B(C)$,

(iii) $\underline{\underline{F}}_{\underline{\underline{C}}_0}(C)B^{\underline{\underline{C}}_0}(C) = 0$; $\underline{\underline{F}}^{\underline{\underline{C}}_0}(C)B_{\underline{\underline{C}}_0}(C) = 0$.

(iv) If $\varphi : B \longrightarrow B'$ is a morphism of Frobenius $\underline{\underline{F}}$-modules, then

$$\varphi(C) : B_{\underline{\underline{C}}_0}(C) \longrightarrow B'_{\underline{\underline{C}}_0}(C),$$

$$\varphi(C) : B^{\underline{\underline{C}}_0}(C) \longrightarrow B'^{\underline{\underline{C}}_0}(C).$$

(v) If for each morphism $\alpha : C \longrightarrow C_1$ in $\underline{\underline{C}}$, $B(\alpha)_*(B_{\underline{\underline{C}}_0}(C_1)) \subset B_{\underline{\underline{C}}_0}(C)$ resp. $B(\alpha)^*(B^{\underline{\underline{C}}_0}(C)) \subset B^{\underline{\underline{C}}_0}(C_1)$, then $B_{\underline{\underline{C}}_0}$ resp. $B^{\underline{\underline{C}}_0}$ is a Frobenius $\underline{\underline{F}}$-module.

<u>Proof</u>: (i) This follows readily from the properties of a Frobenius $\underline{\underline{F}}$-module (3.4); it should be observed that $\underline{\underline{F}}$ is itself a Frobenius $\underline{\underline{F}}$-module.

(ii) obviously follows from (i). As for (iii), let $x \in \underline{\underline{F}}_{\underline{\underline{C}}_0}(C)B^{\underline{\underline{C}}_0}(C)$, then $x = \sum_{i=1}^{n} y_i z_i$, $y_i \in \underline{\underline{F}}_{\underline{\underline{C}}_0}(C)$, $z_i \in B^{\underline{\underline{C}}_0}(C)$; i.e., $z_i = \sum_{j=1}^{n_i} B(\alpha_j)^* z_{ij}$.

Now we apply the reciprocity formulae (3.4,iii)

$$x = (\sum_{i=1}^{n} y_i)(\sum_{j=1}^{n_i} B(\alpha_j)^* z_{ji}) = \sum_{i=1}^{n} \sum_{j=1}^{n_i} y_i B(\alpha_j)^* z_{ij} =$$

$$= \sum_{i=1}^{n} \sum_{j=1}^{n_i} B(\alpha_j)^* (\underline{\underline{F}}(\alpha_j)_* y_i \cdot z_{ij}) = 0.$$

Similarly one shows $\underline{\underline{F}}^{\underline{\underline{C}}_o}(C) B_{\underline{\underline{C}}_o}(C) = 0.$

(iv) Follows from the commutative diagrams in (3.4, iv).

(v) Follows from the definition of a Frobenius module, since the reciprocity laws for $B_{\underline{\underline{C}}_o}$ and $B^{\underline{\underline{C}}_o}$ are automatically satisfied. #

3.9 <u>Definition</u>: Let $N \subset M$ be $\underline{\underline{Z}}$-modules, we say that N has <u>order ε in M</u> if $\varepsilon M \subset N$. M has order ε if $\varepsilon M = 0$. The set of orders of N in M - if not empty - contains a unique minimal element $\exp(M/N) = e$.

3.10 <u>Theorem</u> (<u>Induction-Restriction Principle</u>; Swan [2], Lam [1]): Let $\underline{\underline{F}} : \underline{\underline{C}} \longrightarrow \underline{\underline{Prob}}$ be a Frobenius functor and B a Frobenius $\underline{\underline{F}}$-module. Assume that for a fixed $C \in$ ob $\underline{\underline{C}}$, and for some subclass $\underline{\underline{C}}_o$ of C, $\underline{\underline{F}}^{\underline{\underline{C}}_o}(C)$ has exponent ε in $\underline{\underline{F}}(C)$. Then we have:

(i) <u>Principle of restriction</u>: $B_{\underline{\underline{C}}_o}(C)$ has exponent ε.

(ii) <u>Principle of induction</u>: If for every $C' \in \underline{\underline{C}}_o$, $B(C')$ has exponent e, then $B(C)$ has exponent $e\varepsilon$.

(iii) $B^{\underline{\underline{C}}_o}(C)$ has exponent ε in $B(C)$.

<u>Proof</u>: (i) $\varepsilon \cdot B_{\underline{\underline{C}}_o}(C) = \varepsilon \cdot 1B_{\underline{\underline{C}}_o}(C)$, where 1 is the identity in $\underline{\underline{F}}(C)$. But then $\varepsilon \cdot 1 \in \underline{\underline{F}}^{\underline{\underline{C}}_o}(C)$ and it follows from (3.8,iii) that $\varepsilon \cdot 1B_{\underline{\underline{C}}_o}(C) = 0.$

(ii) If the hypotheses of (ii) are satisfied, then

$$e\varepsilon B(C) = e(\varepsilon 1 P(C)) \subset e(\underline{\underline{F}}^{\underline{\underline{C}}_o}(C)B(C)) \subset eB^{\underline{\underline{C}}_o}(C) = e \sum_{\substack{\alpha \in \mathrm{morph}_{\underline{\underline{C}}}(C';C) \\ C' \in \underline{\underline{C}}_o}} \mathrm{Im}(B(\alpha)^*) = 0$$

since $eB(C') = 0$ by (3.8,i).

(iii) $\varepsilon B(C) = \varepsilon(1B(C)) \subset \underline{\underline{F}}^{\underline{\underline{C}}_o}(C)B(C) \subset B^{\underline{\underline{C}}_o}(C)$ by (3.8,i). #

3.11 <u>Corollary</u>: (i) Under the hypotheses of (3.10), assume that for a fixed $C \in ob(\underline{\underline{C}})$, $B(C)$ is a free abelian group. Given a collection $\underline{\underline{C}}_o \subset ob(\underline{\underline{C}})$ such that $\underline{\underline{F}}_o^{\underline{\underline{C}}_o}(C)$ has finite index in $\underline{\underline{F}}(C)$, then $B_{\underline{\underline{C}}_o}(C) = 0$.

(ii) In addition to (i) assume that φ is a morphism of Frobenius $\underline{\underline{F}}$-modules, $\varphi : B \longrightarrow B'$. To show that $\varphi(C) : B(C) \longrightarrow B'(C)$ is injective it suffices to show that $\varphi(C') : B(C') \longrightarrow B'(C')$ is injective for every $C' \in \underline{\underline{C}}_o$ for which there exists a morphism $\alpha : C' \longrightarrow C$.

<u>Proof</u>: (i) This follows from (3.10,i), since $B_{\underline{\underline{C}}_o}(C)$ is a torsion subgroup of $B(C)$.

(ii) We have for every $\alpha : C' \longrightarrow C$, the commutative diagram (cf. 3.4,iv)

$$\begin{array}{ccc} B(C) & \xrightarrow{\varphi(C)} & B'(C) \\ {\scriptstyle B(\alpha)_*}\downarrow & & \downarrow{\scriptstyle B'(\alpha)_*} \\ B(C') & \xrightarrow{\varphi(C')} & B'(C'). \end{array}$$

If $\varphi(C')$ is injective for every $C' \in \underline{\underline{C}}_o$ for which there exists a morphism $\alpha : C' \longrightarrow C$, and if we take $x \in \mathrm{Ker}\,\varphi(C)$, then $B(\alpha)_* : x \longmapsto 0$ for every α; i.e., $x \in B_{\underline{\underline{C}}_o}(C)$. By (i) this is zero; i.e., $x = 0$, and $\varphi(C)$ is injective. #

3.12 <u>Corollary</u>: Suppose that for a fixed $C \in ob(\underline{\underline{C}})$, $\underline{\underline{F}}(C)$ is free abelian and $\underline{\underline{F}}^{\underline{\underline{C}}_o}(C)$ has finite exponent ε in $\underline{\underline{F}}(C)$. If for every $C' \in \underline{\underline{C}}_o$, $\underline{\underline{F}}(C')$ has no nilpotent elements, then $\underline{\underline{F}}(C)$ has no nilpotent elements.

<u>Proof</u>: Let $x \in \underline{\underline{F}}(C)$ be nilpotent, say $x^n = 0$. Then for every $\alpha : C' \longrightarrow C$ with $C' \in \underline{\underline{C}}_o$, we have $F(\alpha)_* : x \longmapsto 0$, since $\underline{\underline{F}}(C')$ has no nilpotent elements and since $F(\alpha)_*$ is a ring homomorphism. Thus $x \in \underline{\underline{F}}_{\underline{\underline{C}}_o}(C)$; the latter group is zero by (3.11,i). Thus $x = 0$. #

Remark: We return now to the examples (3.5, 3.6); i.e., $\underline{\underline{C}}$ is the category of finite groups and monomorphisms, and

$$\begin{aligned} \underline{\underline{G}}_R : \underline{\underline{C}} &\longrightarrow \underline{\underline{\mathrm{Frob}}}\,, \\ G &\longmapsto \underline{\underline{G}}_o(RG) \end{aligned}$$

is the underlying Frobenius functor. Let $\underline{\underline{C}}_o$ be a fixed class of objects in $\underline{\underline{C}}$ for $G \in \underline{\underline{C}}$, we shall write $e^{\underline{\underline{C}}_o}(R,G)$ for the exponent of $\underline{\underline{G}}_o^{\underline{\underline{C}}_o}(RG)$ in $\underline{\underline{G}}_o(RG)$ (cf. 3.9). However, $\underline{\underline{G}}_o(RG)$ is a ring and $\underline{\underline{G}}_o^{\underline{\underline{C}}_o}(RG)$ is an ideal (cf. 3.8). Hence $e^{\underline{\underline{C}}_o}(R,G)$ is the least positive integer e (if it exists) such that $e[R_G] \in \underline{\underline{G}}_o^{\underline{\underline{C}}_o}(RG)$, where R_G denotes the trivial RG-module.

3.13 **Theorem** (Swan [2]): Let $\underline{p} \in$ spec R and let G be a finite group. Then

(i) $e^{\underline{\underline{C}}_o}(R/\underline{p},G)$ divides $e^{\underline{\underline{C}}_o}(K,G)$,

(ii) $e^{\underline{\underline{C}}_o}(R,G)$ divides $e^{\underline{\underline{C}}_o}(K,G)^2$.

Proof: (i) We recall that K is an algebraic number field and that R is a Dedekind domain with quotient field K. By (VIII, 3.4) we have the commutative triangle

$$\begin{array}{ccc} \underline{\underline{G}}_o(RG) & \xrightarrow{\ \iota\ } & \underline{\underline{G}}_o(KG) \\ & {\scriptstyle \mu_{\underline{p}}} \searrow \quad \swarrow {\scriptstyle \delta_{\underline{p}}} & \\ & \underline{\underline{G}}_o(R/\underline{p}\cdot G) & \end{array} \quad .$$

Here ι and $\mu_{\underline{p}}$ are induced by $K \otimes_R -$ and $R/\underline{p} \otimes_R -$ resp. Both functors commute with induction and restriction, and thus $K \otimes_R -$ and $R/\underline{p} \otimes_R -$ are morphisms of Frobenius functors. But then also $\delta_{\underline{p}}$ is a morphism of Frobenius functors, and we also have the commutative diagram

$$\begin{array}{ccc} \underline{\underline{G}}_o^{\underline{\underline{C}}_o}(RG) & \xrightarrow{\ \iota^{\underline{\underline{C}}_o}\ } & \underline{\underline{G}}_o^{\underline{\underline{C}}_o}(KG) \\ & {\scriptstyle \mu_{\underline{p}}^{\underline{\underline{C}}_o}} \searrow \quad \swarrow {\scriptstyle \delta_{\underline{p}}^{\underline{\underline{C}}_o}} & \\ & \underline{\underline{G}}_o^{\underline{\underline{C}}_o}(R/\underline{p}\cdot G) & \end{array} \quad , \text{ (cf. 3.8,iv).}$$

Hence $e^{\underset{=}{C}O}(K,G) \cdot [(R/\underset{=}{p})_G] \varepsilon \underset{=}{G}^{\underset{=}{C}O}((R/\underset{=}{p})G)$.

(ii) In (VIII, 3.3) we had established the following exact sequence

$$\underset{\underset{=}{p} \varepsilon \text{ spec } R}{\oplus} \underset{=0}{G}(R/\underset{=}{p} \cdot G) \overset{\oplus \, \varrho_{\underset{=}{p}}}{\longrightarrow} \underset{=0}{G}(RG) \overset{\iota}{\longrightarrow} \underset{=0}{G}(KG) \longrightarrow 0.$$

Since induction and restriction are exact on the category $_{RG}\underset{=}{M}^f$ of all finitely generated RG-modules, it is easily seen that $\varrho_{\underset{=}{p}} : \underset{=0}{G}(R/\underset{=}{p}\cdot G)$ $\longrightarrow \underset{=0}{G}(RG)$ commutes with induction and restriction. Consequently $\underset{\underset{=}{p} \in \text{ spec } R}{\oplus} \varrho_{\underset{=}{p}}$

and ι are morphisms of Frobenius functors. The map $\iota^{\underset{=}{C}O} : \underset{=0}{G}^{\underset{=}{C}O}(RG) \longrightarrow$ $\underset{=0}{G}^{\underset{=}{C}O}(KG)$ is surjective; in fact for every group monomorphism $j : G' \longrightarrow G$, we have the commutative diagram

$$
\begin{array}{ccc}
\underset{=0}{G}(KG') & \overset{j^*}{\longrightarrow} & \underset{=0}{G}(KG) \\
\iota' \big\uparrow & & \big\uparrow \iota \\
\underset{=0}{G}(RG') & \overset{j^*}{\longrightarrow} & \underset{=0}{G}(RG)
\end{array}
$$

where the vertical maps are epimorphisms. Now, we pick $x \varepsilon \underset{=0}{G}^{\underset{=}{C}O}(RG)$ such that $\iota^{\underset{=}{C}O} : x \longmapsto e \cdot [K_G]$, where $e = e^{\underset{=}{C}O}(K,G)$. Then $(x-e[R_G]) \varepsilon \text{ Ker } \iota$, and we can write $(x-e[R_G]) = \underset{\underset{=}{p}}{\oplus} \varrho_{\underset{=}{p}}(x_{\underset{=}{p}})$, where only finitely many $x_{\underset{=}{p}} \neq 0$. According to (i) we have $ex_{\underset{=}{p}} \varepsilon \underset{=0}{G}^{\underset{=}{C}O}(R/\underset{=}{p} \cdot G)$. Thus

$$e^2[R_G] = ex - \underset{\underset{=}{p}}{\oplus} \varrho_{\underset{=}{p}}(ex_{\underset{=}{p}}) \varepsilon \underset{=0}{G}^{\underset{=}{C}O}(RG). \qquad \#$$

3.14 **Corollary:** Let K be any field and G a finite group. Then $e^{\underset{=}{C}O}(K,G)$ divides $e^{\underset{=}{C}O}(\underset{=}{Q},G)$. For any commutative ring R, $e^{\underset{=}{C}O}(R,G)$ divides $e^{\underset{=}{C}O}(\underset{=}{Q},G)^2$, where $\underset{=}{Q}$ is the field of rational numbers.

Proof: If K is of characteristic zero, then we have a monomorphism $\underset{=}{Q} \longrightarrow K$ which induces a morphism of Frobenius functors $\varphi : \underset{=}{G}_Q \longrightarrow \underset{=}{G}_K$, such that $\varphi^{\underset{=}{C}O} : \underset{=Q}{G}^{\underset{=}{C}O} \longrightarrow \underset{=K}{G}^{\underset{=}{C}O}$, and thus $e^{\underset{=}{C}O}(Q,G) \cdot [G_K] \varepsilon \underset{=K}{G}^{\underset{=}{C}O}$. Similarly if char $K = p > 0$, then we have a map $\underset{=}{Z}/p\underset{=}{Z} \longrightarrow K$, and the statement follows from (3.13). The second part follows from the first one and from (3.13). #

3.15 **Definition**: Let $\mathfrak{C}'$ be the class of cyclic groups, $\mathfrak{E}'$ the class of elementary groups; i.e., $G \in \mathfrak{E}'$ if G is the direct product of a cyclic group and a p-group for some prime p in $\mathbb{Z}$. $\mathfrak{H}'$ is the class of hyper-elementary groups; i.e., $G \in \mathfrak{H}'$ if $G = \langle a \rangle \cdot H$ is the semi-direct product of a cyclic group $\langle a \rangle$, whose order is prime to p, by a p-group H, and $\langle a \rangle$ is automatically normal in G. For a fixed group G we denote by $\mathfrak{C}$, $\mathfrak{E}$, $\mathfrak{H}$ the classes of cyclic subgroups, elementary subgroups and hyper-elementary subgroups of G.

We quote without proof the classical induction theorems of Artin, Berman, Brauer and Witt. The proofs may be found in Swan [2], or more explicitely in Curtis-Reiner [1, Ch. VI].

3.16 **Theorem** (Swan [2]): Let G be a finite group of order n, and let ζ be a primitive n-th root of unity. Then

(i) $\underline{\underline{G}}_o^{\mathfrak{C}}(\underline{\underline{Q}}G)$ has exponent n in $\underline{\underline{G}}_o(\underline{\underline{Q}}G)$,

(ii) $\underline{\underline{G}}_o^{\mathfrak{E}}(\underline{\underline{Q}}(\zeta)G) = \underline{\underline{G}}_o(\underline{\underline{Q}}(\zeta)G)$,

(iii) $\underline{\underline{G}}_o^{\mathfrak{H}}(\underline{\underline{Q}}G) = \underline{\underline{G}}_o(\underline{\underline{Q}}G)$.

Remark: Naturally, this is not the original formulation of the induction theorem.

(i) E. Artin has proved - using generalized L-series - that the n-fold of any rational character of G is a linear combination of rational characters induced from cyclic subgroups. Since $\underline{\underline{G}}_o(\underline{\underline{Q}}G)$ is the rational character ring, this means $n \cdot \underline{\underline{G}}_o(\underline{\underline{Q}}G) \subset \underline{\underline{G}}_o^{\mathfrak{C}}(\underline{\underline{Q}}G)$ in our terminology (cf. Curtis-Reiner [1], 39.1).

(ii) R. Brauer proved that every complex character is a linear combination of characters induced from elementary subgroups. Since every complex character can be realized in $\underline{\underline{Q}}(\zeta)$ our statement follows (cf. Curtis-Reiner [1], 40.1).

(iii) S.D. Berman and E. Witt generalized Brauer's formula: Every rational character of G is a linear combination of characters induced from hyper-elementary subgroups (cf. Curtis-Reiner [1], 42.3).

3.17 <u>Corollary</u>: Let d be the greatest common divisor of n and $\Phi(n)$, where Φ is Euler's Φ-function. Then $\mathbf{G}_0^{\mathfrak{E}}(\mathbb{Q}G)$ has exponent d in $\mathbf{G}_0(\mathbb{Q}G)$.

<u>Proof</u>: $\Phi(n)$ can be characterized as the dimension of $\mathbb{Q}(\zeta)$ over $\mathbb{Q}$. Let

$$\varphi : \mathbb{Q} \longrightarrow \mathbb{Q}(\zeta)$$

be the embedding, then this induces a morphism of Frobenius functors

$$\varphi_+ : \mathbf{G}_{\mathbb{Q}(\zeta)} \longrightarrow \mathbf{G}_{\mathbb{Q}}$$

by considering $\mathbb{Q}(\zeta)$-modules as $\mathbb{Q}$-modules. Then $[\mathbb{Q}(\zeta)_G] \longmapsto \Phi(n)[\mathbb{Q}_G]$. By (3.16,ii), $\Phi(n)[\mathbb{Q}_G] \in \mathbf{G}_{\mathbb{Q}}^{\mathfrak{E}}(\mathbb{Q}G)$. However, $\mathfrak{C} \subset \mathfrak{E}$ and thus $n[\mathbb{Q}_G] \in \mathbf{G}_0^{\mathfrak{E}}(\mathbb{Q}G)$ by (3.16,i); i.e., $d[\mathbb{Q}_G] \in \mathbf{G}_0^{\mathfrak{E}}(\mathbb{Q}G)$. #

3.18 <u>Corollary</u> (Swan [2]): Let S be a commutative ring. Then

(i) $\mathbf{G}_0^{\mathfrak{C}}(SG)$ has exponent n^2 in $\mathbf{G}_0(SG)$;

(ii) $\mathbf{G}_0^{\mathfrak{E}}(SG)$ has exponent d^2 in $\mathbf{G}_0(SG)$;

(iii) $\mathbf{G}_0^{\mathfrak{H}}(SG) = \mathbf{G}_0(SG)$.

If S is a field one can replace n^2 and d^2 by n and d resp; we recall that for an arbitrary commutative ring S we only consider SG-modules that are S-projective and finitely generated.

<u>Proof</u>: (i) This follows for $S = \mathbb{Z}$ from (3.13) and (3.16) and for arbitrary S from (3.14).

(ii) This follows from (3.13),(3.14) and (3.17).

(iii) This follows from (3.13),(3.14) and (3.16). #

3.19 <u>Remark</u>: The induction-restriction principle shows that (3.18) remains valid for every Frobenius $\mathbf{G}_R$-module; i.e., in particular $\mathbf{K}_0(RG)$, $\mathbf{G}_1(RG)$, $\mathbf{K}_1(RG)$ etc.

We obtain some classical results from (3.18).

3.20 <u>Lemma</u>: Let K be an algebraic number field and R a Dedekind domain with quotient field K; then RG is a clean R-order for every finite group G.

<u>Proof</u>: In view of (1.2) it suffices to show that the Cartan map

$$\kappa_{\underline{p}} : \underline{\underline{K}}_0(R/\underline{\underline{p}} \cdot G) \longrightarrow \underline{\underline{G}}_0(R/\underline{\underline{p}} \cdot G)$$

is monic. Because of (3.11) and (3.18) it suffices to show this in case G is cyclic. However, then $R/\underline{\underline{p}} \cdot G$ is commutative; and for a commutative algebra it is easily seen, that the Cartan map is monic (cf. Ex. 3,3).#

<u>Remark</u>: As other consequence of (3.18) one can prove Brauer's theorem on the minors of the decomposition matrix (cf. Curtis-Reiner [1, Ch. XII]), and Brauer's theorem on the cokernel of the Cartan map.

3.21 <u>Theorem</u> (Swan [5]): Let G be a finite group, K an algebraic number field and R a semi-local Dedekind domain with quotient field K. Then we have an isomorphism:

$$\underline{\underline{G}}_0(RG) \cong \underline{\underline{K}}_0(A).$$

<u>Proof</u>: In view of (3.16) and (3.18) it suffices to prove Swan's theorem in case G is a hyperelementary group, say $G = \langle a \rangle B$, where $\langle a \rangle$ is the cyclic group of order m, $\langle a \rangle \lhd G$ and B is a p-group with $p \nmid m$. In view of the exact sequence

$$\bigoplus_{\underline{\underline{p}} \in \operatorname{spec} R} \underline{\underline{G}}_0(R/\underline{\underline{p}} \cdot G) \xrightarrow{\oplus \varrho_{\underline{p}}} \underline{\underline{G}}_0(RG) \xrightarrow{\iota} \underline{\underline{K}}_0(A) \longrightarrow 0 \quad \text{(cf. VIII, 3.2)},$$

it suffices to show that for all $\underline{\underline{p}} \in \operatorname{spec} R$ - this is a finite set, R being semi-local - $\varrho_{\underline{\underline{p}}}(\underline{\underline{G}}_0(R/\underline{\underline{p}}G)) = 0$. We now fix $\underline{\underline{p}} \in \operatorname{spec} R$ and write $\bar{R} = R/\underline{\underline{p}}$.

<u>Case 1</u>: $\underline{p} \nmid |G| R$. Let $\bar{M} \in {}_{\bar{R}G}\underline{\underline{M}}^f$, and choose a free RG-module F which maps onto $\bar{M}$:

$$0 \longrightarrow P \longrightarrow F \longrightarrow \bar{M} \longrightarrow 0.$$

Then $(\operatorname{ann}_R \bar{M}, \underline{\underline{H}}(RG)) = 1$, where $\underline{\underline{H}}(RG) = |G| R$ is the Higman ideal of RG.

Thus $P \vee F$ (cf. VII, 1.9) and since R is semi-local, $P \cong F$ (cf. VIII, Ex. 5,2). Hence $\underline{\underline{\rho}}_p(\bar{M}) = 0$.

<u>Reduction of the remaining cases</u>: Since $\underline{\underline{G}}_0(\bar{R}G)$ is free on the simple $\bar{R}G$-modules it suffices to show that all simple $\bar{R}G$-modules lie in Ker $\underline{\underline{\rho}}_p$.

In addition we may assume that M is a faithful representation module. (This should not be confused with a faithful $\bar{R}G$-module; i.e., $\mathrm{ann}_{\bar{R}G}(\bar{M}) = 0$. The <u>kernel of a representation module</u> $\bar{M}$ is defined as the normal subgroup

$$N = \{g \in G : gm = m \text{ for all } m \in M\},$$

and $\bar{M}$ is called a <u>faithful representation module</u> if $N = \{1\}$.) Assume that $\bar{M}$ is not faithful and let $\{1\} \neq N = \mathrm{Ker}(\bar{M})$. Then $\bar{M}$ is a faithful $\bar{R}(G/N)$-module, and if $\underline{\underline{\rho}}_p([M]) = 0$ in $\underline{\underline{G}}_0(\bar{R}(G/N))$, then $\underline{\underline{\rho}}_p([\bar{M}]) = 0$ in $\underline{\underline{G}}_0(\bar{R}G)$, since the homomorphism $G \to G/N$ induces a map $\underline{\underline{G}}_0(\bar{R}(G/N)) \to \underline{\underline{G}}_0(\bar{R}G)$, $M \mapsto M_G$, where $gm = (g + N)m$.

By induction on $|G|$ we may assume that M is a simple faithful RG-representation module, since for $|G| = 1$, the statement of our theorem is

<u>Case 2</u>: $\underline{\underline{p}}|mR$. We recall that $G = \langle a \rangle B$, $|\langle a \rangle| = m$, B is a p-group. Let q be a rational prime dividing m, $\underline{\underline{p}}|qR$ and let H be q-Sylow subgroup of $\langle a \rangle$. Then H is a non-trivial normal subgroup of G since $(p,m) = 1$ and since $\langle a \rangle$ is cyclic.

We <u>claim that H acts trivially on $\bar{M}$</u>. Let $G' = G/H$ and consider the canonical homomorphism

$$\varphi : \bar{R}G \longrightarrow \bar{R}G'.$$

Then

$$\mathrm{Ker}\ \varphi = \sum \bar{R}y(u - 1) \quad \text{where}$$

$u \in H$, and y ranges over a set of coset representatives of H in G. However, $y(u - 1) = (u' - 1)y$ and since H is a cyclic q-group and char $\bar{R} = q$, rad $\bar{R}H = \sum_{u \in H} \bar{R}(u - 1)$ (cf. proof of 1.10). Since rad $\bar{R}H$ is nilpotent, Ker $\varphi \subset$ rad $\bar{R}G$ and so Ker$\varphi\,\bar{M} = 0$, $\bar{M}$ being simple. In particular $(u - 1)\bar{M} = 0$ for all $u \in H$ and $\bar{M}$ can not be a faithful representation module, since $H \neq \{1\}$.

Case 3: $\underline{p} | pR$ Again $\bar{M}$ is a faithful simple $\bar{R}G$-representation module. (This automatically implies $\underline{p} \nmid mR$, since $(\underline{p}, m) = 1$.)

Case (3i): Assume that $\bar{R}$ contains the m-th roots of unity. Then $\bar{R}\langle a\rangle$ is commutative and separable, and every simple representation is 1-dimensional over $\bar{R}$, since $\bar{R}\langle a\rangle$ is split by $\bar{R}$, $\bar{R}$ containing the m-th roots of unity (cf. Curtis-Reiner [1], 41.1). Then $\bar{M}|_{\langle a\rangle}$ contains a 1-dimensional $\bar{R}\langle a\rangle$-module U and we shall show that $\bar{M} \cong \bar{R}G \otimes_{\bar{R}\langle a\rangle} U = U^G$. This will settle case (3i) by induction, since $m \neq 0$.

We consider the $\bar{R}G$-homomorphism

$$\varphi : \bar{R}G \otimes_{\bar{R}\langle a\rangle} U \longrightarrow \bar{M}$$

$$x \otimes u \longmapsto xu.$$

Since $\bar{M}$ is simple, φ is an epimorphism and hence $\dim_{\bar{R}}(\bar{M}) \leq \dim_{\bar{R}} U^G = m = |B|$. If we can show $\dim_{\bar{R}}(\bar{M}) \geq m$ then $\bar{M} \cong U^G$.

We consider the conjugates $U^{(x)} = x \otimes U$ for $x \in G$. This is an $\bar{R}\langle a\rangle$-submodule of U^G under the action $a(x \otimes u) = ax \otimes u = x \otimes (x^{-1}ax)u$, $\langle a\rangle$ being normal in G. Thus $M|_{\langle a\rangle} \supset U^{(x)}$ for all $x \in G$. But for $x, y \in B$, $x \neq y$, $U^{(x)} \not\cong_{\bar{R}\langle a\rangle} U^{(y)}$. Assume to the contrary that for $x \neq y$ in B, $U^{(x)} \cong U^{(y)}$; i.e., $U^{(z)} \cong U$ for $1 \neq z \in B$. This would mean $au = z^{-1}azu$, for the basis element $u \in U$. However, U is a faithful $\bar{R}\langle a\rangle$-module. In fact, if $\mathrm{Ker}\, U = \langle a^r\rangle \neq 1$, then $\langle a^r\rangle \triangleleft G$, $\langle a^r\rangle$ being a characteristic subgroup of $\langle a\rangle$. But then a^r acts trivially on U^G, $\langle a^r\rangle$ being normal in G and the epimorphism

$$\varphi : U^G \longrightarrow \bar{M} \text{ shows } \langle a^r\rangle \subset \mathrm{Ker}\, \bar{M} = \{1\},$$

a contradiction. Consequently, U is a faithful $\bar{R}\langle a\rangle$-representation module, and z centralizes a; recall that we had assumed $U^{(z)} \cong U$. Now $B_1 = \{z \in B : z^{-1}az = a\}$ is a normal p-subgroup of G. In the proof of Case 2 we have seen that G can not have a faithful simple representation module if $B_1 \neq \{1\}$. Hence $B_1 = \{1\}$ and consequently $z = 1$. But then $\dim_{\bar{R}}\bar{M} = m$ and $\bar{M} \cong U^G$. (We remark that this argument is a consequence of Clifford's theorem (cf. Curtis-Reiner 49.2).)

Case (3ii): $\underline{p}|pR$, $\underline{p}\nmid mR$ and $\bar{M}$ is a faithfull simple $\bar{R}G$-representation module. We shall show that $\bar{M}$ is $\bar{R}G$-projective. Let $\underline{k}$ be a finite extension field of $\bar{R}$ containing the m-th roots of unity. If $\underline{k} \otimes_{\bar{R}} \bar{M} \in {}_{kG}\underline{P}^f$, then $\bar{M} \in {}_{\bar{R}G}\underline{P}^f$, since $\underline{k} \otimes_{\bar{R}} \bar{M}$ is the direct sum of $(\underline{k} : \bar{R})$ copies of $\bar{M}$ as $\bar{R}G$-module. Since $\bar{M}$ is faithful, the same argument shows that the composition factors of $\underline{k} \otimes_{\bar{R}} \bar{M}$ are faithful simple $\underline{k}G$-modules. From case (3,i) we conclude that $\underline{k} \otimes_{\bar{R}} \bar{M}$ is induced from a projective $\underline{k}\langle a\rangle$-module (observe char $\underline{k} \nmid |a|$). Hence $\underline{k} \otimes_{\bar{R}} \bar{M}$ has as composition factors projective modules. But then $\underline{k} \otimes_{\bar{R}} \bar{M}$ is the direct sum of its composition factors and consequently projective.

Hence $\bar{M}$ is $\bar{R}G$-projective. However, this shows $\rho_{\underline{p}}([\bar{M}]) = 0$ as shows the next lemma. #

3.22 Lemma: Let R be a semi-local Dedekind domain with quotient field K and Λ a clean R-order in the separable finite dimensional K-algebra A. If for some $\underline{p} \in \operatorname{spec} R$, $\bar{M} \in {}_{\Lambda/\underline{p}\Lambda}\underline{M}^f$ has finite homological dimension as $\Lambda/\underline{p}\Lambda$-module, then $\rho_{\underline{p}}([\bar{M}]) = 0$.

Proof: We have the exact sequence

$$0 \longrightarrow \underline{p}\Lambda \longrightarrow \Lambda \longrightarrow \Lambda/\underline{p}\Lambda \longrightarrow 0$$

with $\underline{p}\Lambda \cong \Lambda$, R being semi-local, and so from (II, 4.5) $\operatorname{hd}_\Lambda(\Lambda/\underline{p}\Lambda) = 1$ and the change of ring theorem (II, 4.5) shows $\operatorname{hd}_\Lambda(\bar{M}) \leq \operatorname{hd}_{\Lambda/\underline{p}\Lambda}(\bar{M})+1<\infty$.

Hence $\bar{M}$ has a finite projective resolution as Λ-module

$$0 \longrightarrow P_n \longrightarrow P_{n-1} \longrightarrow \dots \longrightarrow P_1 \longrightarrow P_o \longrightarrow \bar{M} \longrightarrow 0,$$

$P_i \in {}_\Lambda\underline{P}^f$. Then it is easily seen that

$$\rho_{\underline{p}}([\bar{M}]) = \sum_{i=0}^{n} (-1)^i [P_i] \text{ in } \underline{G}_o(\Lambda) \text{ (cf. VIII, Ex. 3,1).}$$

However $\sum_{i=0}^{n} (-1)^i [KP_i] = 0$ in $G_0(A)$, Λ is clean and R semi-local; so $\sum_{i=0}^{n} (-1)^i [P_i] = 0$, since

$$\iota' : K_0(\Lambda) \longrightarrow K_0(A)$$

is monic and since we have the following commutative triangle

$$\begin{array}{ccc} G_0(\Lambda) & \xrightarrow{\ \iota\ } & K_0(A) \\ & \nwarrow^{\kappa} \quad \nearrow_{\iota'} & \\ & K_0(\Lambda) & \end{array}$$

(cf. VIII, 3.10). #

Remark: We shall next derive some consequences from (3.21), which we formulate not only for group rings, since presumably there are other orders for which (3.21) holds. Therefore we assume now that R is a Dedekind domain with quotient field K and Λ is an R-order in the finite dimensional separable K-algebra A.

3.23 **Theorem** (Swan [5]): Let $\{\mathfrak{p}_i\}_{1 \le i \le t}$ be a finite subset of spec R and put $S = R \setminus (\bigcup_{i=1}^{t} \mathfrak{p}_i)$. Assume that $G_0(\Lambda_S) \cong K_0(A)$. Then we have the exact sequence

$$\bigoplus_{\substack{\mathfrak{p} \in \operatorname{spec} R \\ \mathfrak{p} \ne \mathfrak{p}_i, 1 \le i \le t}} G_0(\Lambda/\mathfrak{p}\Lambda) \xrightarrow{\oplus \varrho_{\mathfrak{p}}} G_0(\Lambda) \xrightarrow{\ \iota\ } K_0(A) \longrightarrow 0.$$

Proof: In (VIII, 3.2) we have shown the exactness of the sequence

$$\bigoplus_{\substack{\mathfrak{p} \in \operatorname{spec} R \\ \mathfrak{p} \ne \mathfrak{p}_i, 1 \le i \le t}} G_0(\Lambda/\mathfrak{p}\Lambda) \longrightarrow G_0(\Lambda) \rightarrow G_0(\Lambda_S) \longrightarrow 0.$$

Now the statement follows with the isomorphism $G_0(\Lambda_S) \cong K_0(A)$. #

3.24 **Theorem** (Swan [5]): Assume that for every finite set $\{\mathfrak{p}_i\}_{1 \le i \le t}$ we have an isomorphism $G_0(\Lambda_S) \cong K_0(A)$ with $S = R \setminus \{\bigcup_{i=1}^{t} \mathfrak{p}_i\}$. Then the sequence

$$\underline{\underline{C}}_0(\Lambda) \longrightarrow \underline{\underline{G}}_0(\Lambda) \longrightarrow \underline{\underline{K}}_0(A) \longrightarrow 0$$

is exact.

<u>Proof</u>: We recall that $\underline{\underline{C}}_0(\Lambda)$ is the reduced projective class group and $\underline{\underline{K}}_0(\Lambda)$ is the Grothendieck group of the projective Λ-lattices. From (VIII, 3.10) we get the commutative diagram with exact rows

$$\begin{array}{ccccccc} 0 \longrightarrow & \mathrm{Ker}\,\iota & \xrightarrow{\beta} & \underline{\underline{G}}_0(\Lambda) & \xrightarrow{\iota} & \underline{\underline{K}}_0(A) & \longrightarrow 0 \\ & \mu \uparrow & & \uparrow \kappa & & \| & \\ & \underline{\underline{C}}_0(\Lambda) & \xrightarrow{\alpha} & \underline{\underline{K}}_0(\Lambda) & \xrightarrow{\iota'} & \underline{\underline{K}}_0(A). & \end{array}$$

Putting $\nu = \kappa\alpha$ we have to show $\mathrm{Ker}\,\iota \subset \mathrm{Im}\,\nu$, since $\nu\iota = 0$. Because of the commutativity of our diagram it suffices to show that μ is an epimorphism. We apply (3.23) to the finite set of all maximal ideals dividing the Higman ideal $\underline{\underline{H}}(\Lambda)$. Then $\mathrm{Ker}\,\iota = \bigoplus_{\substack{\underline{\underline{p}} \in \mathrm{spec}\, R \\ \underline{\underline{p}} \nmid \underline{\underline{H}}(\Lambda)}} \varrho_{\underline{\underline{p}}}(\underline{\underline{G}}_0(\Lambda/\underline{\underline{p}}\Lambda))$, and it remains to show that for every simple $\Lambda/\underline{\underline{p}}\Lambda$-module $\bar{M}$, $\varrho_{\underline{\underline{p}}}([\bar{M}]) \in \mathrm{Im}\,\mu$. We take a presentation

$$0 \longrightarrow P \longrightarrow F \longrightarrow \bar{M} \longrightarrow 0,$$

where F is a free Λ-module of finite type. Since $(\mathrm{ann}_R \bar{M}, \underline{\underline{H}}(\Lambda)) = 1$, the usual argument shows that $P \in {}_{\Lambda}\underline{\underline{P}}^f$. But then $\varrho_{\underline{\underline{p}}}([\bar{M}]) = \mu([F]-[P])$ since $KP \cong KF$ is A-free. #

3.25 <u>Theorem</u>: Assume that the sequence

$$\underline{\underline{C}}_0(\Lambda) \xrightarrow{\nu} \underline{\underline{G}}_0(\Lambda) \xrightarrow{\iota} \underline{\underline{K}}_0(A) \longrightarrow 0$$

is exact and let Λ_1 be a clean R-order in A containing Λ. Then the sequence

$$\underline{\underline{C}}_0(\Lambda_1) \xrightarrow{\Psi} \underline{\underline{G}}_0(\Lambda) \longrightarrow \underline{\underline{K}}_0(A) \longrightarrow 0$$

is exact.

<u>Proof</u>: If $\underline{\underline{P}}_0(\Lambda)$ denotes the Grothendieck group of the special pro-

jective Λ-lattices, then it follows from (1.15) that the map

$$\varphi : \underline{\underline{P}}_0(\Lambda) \longrightarrow \underline{\underline{P}}_0(\Lambda_1),$$

$$[P] \longmapsto [\Lambda_1 \otimes_\Lambda P]$$

is an epimorphism. But this map then induces an epimorphism

$$\varphi' : \underline{\underline{C}}_0(\Lambda) \longrightarrow \underline{\underline{C}}_0(\Lambda_1),$$

since $\underline{\underline{P}}_0(\Lambda) \cong \underline{\underline{Z}} \oplus \underline{\underline{C}}_0(\Lambda)$ (cf. VIII, 3.8). To prove (3.25) we have to show that $\nu : \underline{\underline{C}}_0(\Lambda) \longrightarrow \underline{\underline{G}}_0(\Lambda)$ factors through φ'; i.e., we have to complete the following diagram

$$\begin{array}{ccc} & & \underline{\underline{C}}_0(\Lambda_1) \\ & \overset{\varphi'}{\nearrow} & \\ \underline{\underline{C}}_0(\Lambda) & & \Big\downarrow \psi \\ & \underset{\nu}{\searrow} & \\ & & \underline{\underline{G}}_0(\Lambda) . \end{array}$$

According to (I, Ex. 2,3) this is the case if $\operatorname{Ker} \varphi' \subset \operatorname{Ker} \nu$. However, given $x = [F] - [P] \in \underline{\underline{C}}_0(\Lambda)$ one shows as in the proof of (1.15) that we have two exact sequences of Λ-modules

$$0 \longrightarrow P \longrightarrow F \xrightarrow{\alpha} U \longrightarrow 0$$

$$0 \longrightarrow \Lambda_1 \otimes_\Lambda P \longrightarrow \Lambda_1 \otimes_\Lambda F \xrightarrow{\beta} U \longrightarrow 0$$

where α and β are projective homomorphisms. Thus

$$x = [F] - [P] = [(\Lambda_1 \otimes_\Lambda F)_\Lambda] - [(\Lambda_1 \otimes_\Lambda P)_\Lambda],$$

where the subscript indicates that these modules should be considered as Λ-lattices. Now it is clear that $\operatorname{Ker} \varphi' \subset \operatorname{Ker} \nu$. #

3.26 <u>Corollary</u>: Let Γ be a maximal R-order in A containing Λ and assume that the sequence

$$\underline{\underline{C}}_0(\Gamma) \longrightarrow \underline{\underline{G}}_0(\Lambda) \longrightarrow \underline{\underline{K}}_0(A) \longrightarrow 0$$

is exact. Then the map

$$\sigma : \underline{\underline{K}}_0(\Gamma) \longrightarrow \underline{\underline{G}}_0(\Lambda),$$

$$[M] \longmapsto [M_\Lambda]$$

is an epimorphism.

Proof: Since Γ is hereditary, the sequence

$$0 \longrightarrow \underline{\underline{C}}_0(\Gamma) \longrightarrow \underline{\underline{K}}_0(\Gamma) \longrightarrow \underline{\underline{K}}_0(A) \longrightarrow 0$$

is exact. On the other hand, Γ is clean and so we can apply (3.25). We have an exact sequence

$$\underline{\underline{C}}_0(\Gamma) \xrightarrow{\psi} \underline{\underline{G}}_0(\Lambda) \longrightarrow \underline{\underline{K}}_0(A) \longrightarrow 0,$$

where ψ is induced from the restriction of the operators. Hence we get the commutative diagram with exact rows

$$\begin{array}{ccccccc}
\underline{\underline{C}}_0(\Gamma) & \longrightarrow & \underline{\underline{K}}_0(\Gamma) & \longrightarrow & \underline{\underline{K}}_0(A) & \longrightarrow & 0 \\
\| & & \downarrow{\scriptstyle \sigma} & & \| & & \\
\underline{\underline{C}}_0(\Gamma) & \xrightarrow{\psi} & \underline{\underline{G}}_0(\Lambda) & \longrightarrow & \underline{\underline{K}}_0(A) & \longrightarrow & 0,
\end{array}$$

and diagram chasing shows that σ is an epimorphism. #

Notation and Remark: We now assume that K is an $\underline{\underline{A}}$-field with Dedekind domain R. If Γ is a maximal R-order in the simple separable K-algebra A, we can describe the structure of $\underline{\underline{G}}_0(\Gamma)$ and $\underline{\underline{C}}_0(\Gamma)$ explicitely. Moreover, if (3.26) holds for an order Λ (e.g., integral group rings), then this can be used to clarify the structure of $\underline{\underline{K}}_0(\Lambda)$. We first assume that A is central simple.

3.27 Theorem (Heller-Reiner [4,5], Swan [4]): Let I(R) denote the group of all fractional R-ideals in K. Then

$$\underline{\underline{G}}_0^T(\Gamma) \cong I(R),$$

where $\underline{\underline{G}}_0^T(\Gamma)$ denotes the Grothendieck group of all finitely generated R-torsion Γ-modules. If $St_K(A)_0 = \{(\alpha) : 0 \neq \alpha \in K$ is positive at all infinite primes of K at which A is ramified$\}$, then

$$\underline{\underline{C}}_0(\Gamma) \cong I(R)/St_K(A)_0.$$

Proof: We define a map

$$\chi : \underline{\underline{G}}_0^T(\Gamma) \longrightarrow I(R)$$

as follows: T is a module over the artinian and noetherian ring $\Gamma/(\mathrm{ann}_R T)\Gamma$, and hence it has a composition series

$$T = T_0 \underset{\neq}{\supset} T_1 \underset{\neq}{\supset} \dots \underset{\neq}{\supset} T_s \underset{\neq}{\supset} T_{s+1} = 0.$$

The composition factors $X_i = T_i/T_{i+1}$, $0 \le i \le s$, are simple Γ-modules and thus of the form $X_i = \Gamma/I_i$, where I_i is a maximal left Γ-ideal. Applying (VI, 8.9) we conclude that the reduced norm $\nu(I_i)$ (cf. VI, 8.7) is a prime ideal $\underline{\underline{p}}_i \in \operatorname{spec} R$, $0 \le i \le s$. we now define

$$\chi : [T] \longmapsto \prod_{i=0}^{s} \nu(I_i).$$

This gives a well-defined group homomorphism since $\nu(I_i) = \mathrm{ann}_R(X_i) = \mathrm{ann}_R(\Gamma/I_i)$ (cf. proof of VI, 8.9) and because of the Jordan-Hölder theorem. In (VI, 8.10) we have shown that every non-zero integral ideal $\underline{\underline{a}}$ of R occurs as norm of some left Γ-ideal I. Hence χ is an epimorphism. (Observe that the addition in $\underline{\underline{G}}_0^T(\Gamma)$ corresponds to the multiplication in I(R).) But χ is also monic, since for every non-zero prime ideal $\underline{\underline{p}} \in \operatorname{spec} R$, there exists - up to isomorphism - exactly one simple Γ-module T with $\mathrm{ann}_R T = \underline{\underline{p}}$. To be more precise,

$$\underline{\underline{G}}_0^T(\Gamma) \cong \bigoplus_{\underline{\underline{p}} \in \operatorname{spec} R} \underline{\underline{G}}_0(\Gamma/\underline{\underline{p}}\Gamma) \text{ (cf. VIII, 3.2)},$$

and to show that χ is monic it suffices to observe that there exists - up to isomorphism - only one simple $\Gamma/\underline{\underline{p}}\Gamma$-module (cf. VI, Ex. 2,4). Hence we have proved

$$\underline{\underline{G}}_0^T(\Gamma) \cong I(R).$$

We can also set up the inverse map

$$\chi^{-1}: I(R) \longrightarrow \underline{\underline{G}}_0^T(\Gamma).$$

Let M be an irreducible Γ-lattice. If $A = (D)_s$, with $(D : K) = n^2$, then we define

$$\chi^{-1} : I(R) \longrightarrow \underline{\underline{G}}_0^T(\Gamma),$$

$$\underline{\underline{p}} \longmapsto n^{-1}[M/\underline{\underline{p}}M],$$

and then we extend the definition linearly. We remark, that $M/\underline{p}M$ has as composition factors n copies of the simple $\Gamma_{\underline{p}}$-module. In fact, if $\hat{D}_{\underline{p}} = (\hat{D}_1)_{s_1}$ with $(\hat{D}_1 : \hat{K}_{\underline{p}}) = n_1^2$, $s_1 \cdot n_1 = n$, then $\hat{\Gamma}_{\underline{p}}/\mathrm{rad}\,\hat{\Gamma}_{\underline{p}} \cong (\hat{\Omega}/\mathrm{rad}\,\hat{\Omega})_{s_1 s}$, where $\hat{\Omega}$ is the maximal $\hat{R}_{\underline{p}}$-order in $\hat{D}_1$. The dimension of a simple $\Gamma_{\underline{p}}$-module over $R/\underline{p}$ is thus $s \cdot s_1 \cdot n_1 = s \cdot n$, since $(\hat{\Omega}/\mathrm{rad}\,\hat{\Omega} : R/\underline{p}) = n_1$ (cf. IV, 6.7), whereas $\dim_{R/\underline{p}}(M/\underline{p}M) = s \cdot n^2$. Whence

the statement. Now it is easily seen that χ^{-1} is the inverse of χ.

In (VIII, 3.13, 3.14) we constructed an exact sequence

$$GL(1,A) \xrightarrow{\vartheta} \underline{G}_0^T(\Gamma) \xrightarrow{\delta} \underline{K}_0(\Gamma) \xrightarrow{\iota} \underline{K}_0(A) \longrightarrow 0;$$

the map ϑ was defined as follows

$$\vartheta : a \longmapsto [\mathrm{Coker}\; ar \cdot 1_\Gamma] - [\mathrm{Coker}\; r1_\Gamma],$$

where $0 \neq r \in R$ is such that $ra \in \Gamma$. Thus

$$\vartheta : a \longmapsto [\Gamma/\Gamma ra] - [\Gamma/\Gamma r],$$

and

$$\chi\vartheta : a \longmapsto R \cdot \mathrm{Nrd}_{A/K}(a) \quad \text{(cf. VI, 8.8)}.$$

On the other hand, we have shown in (VI, 6.9)

$$\mathrm{St}_K(A)_0 = \{R\;\mathrm{Nrd}_{A/K}(a) : a \in A,\ a \text{ regular}\}.$$

Thus $\mathrm{Im}\,\chi\vartheta = \mathrm{St}_K(A)_0$, and the exactness of the above sequence shows

$$\mathrm{Ker}\,\iota \cong \underline{G}_0^T(\Gamma)/\mathrm{Im}\,\vartheta \cong I(R)/\mathrm{St}_K(A)_0.$$

But we also have the exact sequence

$$0 \longrightarrow \underline{C}_0(\Gamma) \longrightarrow \underline{K}_0(\Gamma) \longrightarrow \underline{K}_0(A) \longrightarrow 0 \quad \text{(cf. VIII, 3.11)}$$

and consequently

$$\underline{C}_0(\Gamma) \cong I(R)/\mathrm{St}_K(A)_0.$$

This last statement can also be proved directly (cf. Ex. 1,4). #

3.28 <u>Lemma</u> (Heller-Reiner [4,5]): Let Λ be an R-order in the separable K-algebra A and Γ a maximal R-order in A containing Λ. Assume that the map

$$\sigma : \underline{\underline{K}}_0(\Gamma) \longrightarrow \underline{\underline{G}}_0(\Lambda),$$
$$[M] \longmapsto [M_\Lambda]$$

is an epimorphism. Then we have a commutative diagram with exact rows

$$\begin{array}{ccccccccc}
\underline{\underline{K}}_1(A) & \xrightarrow{\vartheta'} & \underline{\underline{G}}_0^T(\Gamma) & \xrightarrow{\varrho'} & \underline{\underline{K}}_0(\Gamma) & \xrightarrow{\iota'} & \underline{\underline{K}}_0(A) & \longrightarrow & 0 \\
\| & & \downarrow \tau & & \downarrow \sigma & & \| & & \\
\underline{\underline{K}}_1(A) & \xrightarrow{\vartheta} & \underline{\underline{G}}_0^T(\Lambda) & \xrightarrow{\varrho} & \underline{\underline{G}}_0(\Lambda) & \xrightarrow{\iota} & \underline{\underline{K}}_0(A) & \longrightarrow & 0,
\end{array}$$

where the vertical maps are induced by the restriction of the operator domain. In addition,

$$\underline{\underline{G}}_0(\Lambda) \cong \underline{\underline{K}}_0(A) \oplus \underline{\underline{G}}_0^T(\Gamma)/(\mathrm{Ker}\,\tau + \mathrm{Im}\,\vartheta').$$

__Proof:__ The commutativity of this diagram is easily checked and the exactness of the rows follows from (VIII, 3.13).
Since σ is an epimorphism, diagram chasing shows that τ is also an epimorphism. Thus

$$\mathrm{Ker}\,\iota = \mathrm{Im}\,\varrho = \mathrm{Im}\,\varrho\tau.$$

However, $\underline{\underline{K}}_0(A)$ is a free abelian group with a finite basis and so

$$\underline{\underline{G}}_0(\Lambda) \cong \underline{\underline{K}}_0(A) \oplus \mathrm{Ker}\,\iota\,;\ \text{i.e.},$$

$$\underline{\underline{G}}_0(\Lambda) = \underline{\underline{K}}_0(A) \oplus \mathrm{Im}\,\varrho\tau.$$

But $\mathrm{Im}\,\varrho\tau = \underline{\underline{G}}_0^T(\Gamma)/\mathrm{Ker}\,\varrho\tau$ and $\mathrm{Ker}\,\varrho\tau = \mathrm{Ker}\,\tau + \mathrm{Im}\,\vartheta'$, whence the statement follows. #

__Notation:__ Let $A = \oplus_{i=1}^n A_i$ be the decomposition of A into simple K-algebras $A_i = (D_i)_{s_i}$ with center $(A_i) = K_i$ and let R_i be the integral closure of R in K_i, $(D_i : K_i) = n_i^2$. For $\underline{p} \in$ spec R let $\{\underline{\underline{P}}_{ij}(\underline{p})\}_{1 \le j \le t_{i\underline{p}}}$ be the set of different prime ideals in R_i dividing $\underline{p}R_i$. Let $I_{\underline{p}}(R_i)$ be the subgroup of $I(R_i)$ generated by the elements $\{\underline{\underline{P}}_{ij}(\underline{p})\}_{1 \le j \le t_{i\underline{p}}}$.

In the proof of (3.27) we have set up the isomorphism

$$\chi_{\mathfrak{p}}^{-1} : \prod_{i=1}^{n} I_{\mathfrak{p}}(R_i) \longrightarrow G_0^T(\Gamma_{\mathfrak{p}}),$$

$$P_{ij}(\mathfrak{p}) \longmapsto n_i^{-1}[M_i/\mathfrak{P}_{ij}(\mathfrak{p})M_i],$$

where M_i is an irreducible Γ_i-module. $\chi_{\mathfrak{p}}^{-1}$ induces a map

$$\beta_{\mathfrak{p}} : \prod_{i=1}^{n} I_{\mathfrak{p}}(R_i) \longrightarrow G_0^T(\Lambda_{\mathfrak{p}}) \cong G_0(\Lambda/\mathfrak{p}\Lambda),$$

by restriction of the operators.

3.29 <u>Theorem</u> (Heller-Reiner [4,5]): Let Λ be an R-order in the separable K-algebra A and let Γ be a maximal R-order in A containing Λ. Assume that $\sigma : K_0(\Gamma) \longrightarrow G_0(\Lambda)$ is an epimorphism. Then

$$G_0(\Lambda) \cong K_0(A) \oplus \frac{\prod_{i=1}^{n} I(R_i)}{\prod_{i=1}^{n} St_{K_i}(A_i)_0 \prod_{\mathfrak{p}|H(\Lambda)} \operatorname{Ker} \beta_{\mathfrak{p}}}.\ ^{*)}$$

<u>Proof</u>: In view of (3.28) we have to describe $G_0^T(\Gamma)/(\operatorname{Ker}\tau + \operatorname{Im}\vartheta')$. From (3.26) it follows that

$$G_0^T(\Gamma) = \oplus_{i=1}^{n} G_0^T(\Gamma_i) \cong \prod_{i=1}^{n} I(R_i),$$

$$\operatorname{Im}\vartheta' \cong \oplus_{i=1}^{n} \operatorname{Im}\vartheta_i' \cong \prod_{i=1}^{n} St_{K_i}(A_i)_0.$$

As for $\tau : G_0^T(\Gamma) \longrightarrow G_0^T(\Lambda)$, we have the commutative diagram

$$\begin{array}{ccc} G_0^T(\Gamma) & \xrightarrow{\ \tau\ } & G_0^T(\Lambda) \\ \wr\| & & \wr\| \\ \bigoplus_{\mathfrak{p}\in \operatorname{spec} R} G_0(\Gamma/\mathfrak{p}\Gamma) & \xrightarrow{\ \oplus\tau_{\mathfrak{p}}\ } & \bigoplus_{\mathfrak{p}\in \operatorname{spec} R} G_0(\Lambda/\mathfrak{p}\Lambda) \end{array}$$

(cf. VIII, 3.2),

and $\tau_{\mathfrak{p}}$ is also induced from the restriction of the operators. Thus $\operatorname{Ker}\tau = \bigoplus_{\mathfrak{p}\in \operatorname{spec} R} \operatorname{Ker}\tau_{\mathfrak{p}}$. However, if $\mathfrak{q}\nmid H(\Lambda)$, then $\Lambda/\mathfrak{p}\Lambda \cong \Lambda_{\mathfrak{p}}/\mathfrak{p}\Lambda_{\mathfrak{p}} \cong$ $= \Gamma_{\mathfrak{p}}/\mathfrak{p}\Gamma_{\mathfrak{p}} = \Gamma/\mathfrak{p}\Gamma$ and so $\operatorname{Ker}\tau_{\mathfrak{q}} = 0$. Hence

---

*) This last expression should be taken cum grano salis, since it is written as the direct sum of an additive group and a multiplicative group.

$$\operatorname{Ker} \tau = \bigoplus_{\underline{p} | \underline{H}(\Lambda)} \operatorname{Ker} \tau_{\underline{p}}.$$

However, we have identified $\underline{G}_o^T(\Gamma)$ with $\prod_{i=1}^n I(R_i)$ and so $\underline{G}_o(\Gamma/\underline{p}\Gamma)$ must be identified with $\prod_{i=1}^n I_{\underline{p}}(R_i)$. Then $\tau_{\underline{p}}$ must be replaced by $\beta_{\underline{p}}$ and hence

$$\operatorname{Ker} \tau = \prod_{\underline{p} | \underline{H}(\Lambda)} \operatorname{Ker} \beta_{\underline{p}}. \quad \#$$

<u>Exercises §3</u>:

1.) Let K be an algebraic number field and G a finite group. For a KG-module L let χ_L be the character afforded by L. Show that

$$\chi_{L_1 \otimes_K L_2} = \chi_{L_1} \chi_{L_2} \text{ and } \chi_{L_1 \oplus L_2} = \chi_{L_1} + \chi_{L_2}.$$

Use this to establish a ring isomorphism $\underline{K}_o(KG) \cong \mathcal{X}(KG)$, where $\mathcal{X}(KG)$ is the character ring of KG.

2.) Prove 3.6.

3.) Let B be a finite dimensional commutative algebra over a field K. Show directly that the Cartan map $\kappa : \underline{K}_o(B) \longrightarrow \underline{G}_o(B)$ is monic.

4.) Let Γ be a maximal R-order in the central simple K-algebra A. Show directly

$$\underline{C}_o(\Gamma) \cong I(R)/St_K(A)_o.$$

(<u>Hint</u>: Every element in $\underline{C}_o(\Gamma)$ has the form $[\Gamma^{(2)}] - [P']$ where $P' \vee \Gamma^{(2)}$. Moreover, two such elements are equal if and only if $P_1' \cong P_2'$. Now use (VII, 2.2).)

§4 Divisibility of lattices

The module of characters of an R-torsion Λ-module of finite type is introduced and some properties are derived. The concept "M covers N" (M divides N) is developed. All results of this section are needed for the exploration of Bass-orders in §§5,6.

Let R be a Dedekind domain with quotient field K and Λ an R-order in the separable finite dimensional K-algebra A. We recall that ${}_{\Lambda}\underline{\underline{M}}^T$ denotes the category of R-torsion Λ-modules of finite type.

4.1 Lemma: The functor

$$\mathrm{Hom}_R(-,K/R) = -^* : {}_{\Lambda}\underline{\underline{M}}^T \xrightarrow[\text{nat}]{} \underline{\underline{M}}^T_{\Lambda},$$

is contravariant and exact; moreover, $T^{**} \cong T$. T^* is called the <u>module of characters of T</u>.

Proof: The functor "$-^*$" is contravariant and exact since K/R is an injective R-module. However, we have not introduced the concept of an injective module, and so we shall show directly that $-^*$ is an exact functor. Given an exact sequence of R-torsion modules of finite type

$$0 \longrightarrow T' \longrightarrow T \longrightarrow T'' \longrightarrow 0.$$

Then we get the exact sequence

$$0 \longrightarrow T''^* \longrightarrow T^* \longrightarrow T'^* \longrightarrow \mathrm{Ext}^1_R(T'',K/R).$$

And "$-^*$" is exact if we can show $\mathrm{Ext}^1_R(T'',K/R) = 0$.
An R-torsion module of finite type is the direct sum of its $\underline{\underline{p}}$-primary components and so we may assume $\mathrm{ann}_R(T'') = \underline{\underline{p}}^s$ for some $\underline{\underline{p}} \in \mathrm{spec}\, R$. We even may assume that T is indecomposable as R-module; i.e., $T'' = R/\underline{\underline{p}}^s = R_{\underline{\underline{p}}}/\underline{\underline{p}}^s R_{\underline{\underline{p}}}$, as follows easily from the invariant factor theorem for modules over principal ideal rings. Hence we have to show $\mathrm{Ext}^1_R(R/\underline{\underline{p}}^s,K/R) = 0$; i.e., given any R-homomorphism

$$\varphi : \underline{\underline{a}} \longrightarrow K/R,\ \underline{\underline{a}} = \underline{\underline{p}}^s$$

we must find an R-homomorphism $\psi : R \longrightarrow K/R$ such that $\psi|_{\underline{\underline{a}}} = \varphi$. We observe that $\underline{\underline{a}}$ is invertible, and so there exist families $\{k_i\}_{1\leq i\leq n}$, $\{a_i\}_{1\leq i\leq n}$, $k_i \in K, 0 \neq a_i \in \underline{\underline{a}}$ such that

$$1 = \sum_{i=1}^{n} k_i a_i \text{ and } k_i \underline{\underline{a}} \subset R, 1\leq i\leq n.$$

Moreover, there are elements $x_i \in K/R$ such that

$$\varphi(a_i) = a_i x_i, 1\leq i\leq n;$$

take $x_i = \varphi(a_i)/a_i + R$. (This means that K/R is divisible (cf. II, p. 11).) Setting $y = \sum_{i=1}^{n} (k_i a_i)x_i \in K/R$ we conclude

$\varphi(a) = \sum_{i=1}^{n} \varphi(k_i a_i a) = \sum_{i=1}^{n} k_i a\varphi(a_i) = a\sum_{i=1}^{n} k_i a_i x_i = ay$ for every $a \in \underline{\underline{a}}$. Then

$$\psi : R \longrightarrow K/R,$$
$$r \longmapsto r \cdot y$$

is the desired map. We have now shown that "-*" is an exact contravariant functor on R-torsion modules. But then it is also an exact contravariant functor from ${}_{\Lambda}\underline{\underline{M}}^T$ to $\underline{\underline{M}}^T_{\Lambda}$, (cf. II, 1.12). If we can show that for R-torsion modules $T \overset{nat}{\cong} T^{**}$, then the naturality of this isomorphism shows that for $T \in {}_{\Lambda}\underline{\underline{M}}^T$ we have $T \cong T^{**}$ as Λ-module. Hence we may assume $T = R/\underline{\underline{a}}$, $\underline{\underline{a}}$ a non-zero ideal in R. The exact sequence with canonical homomorphisms

$$0 \longrightarrow \underline{\underline{a}} \overset{\alpha}{\longrightarrow} R \longrightarrow R/\underline{\underline{a}} \longrightarrow 0$$

induces the exact sequence

$$0 \longrightarrow \mathrm{Hom}_R(R/\underline{\underline{a}}, K/R) \longrightarrow \mathrm{Hom}_R(R, K/R) \overset{\alpha^*}{\longrightarrow} \mathrm{Hom}_R(\underline{\underline{a}}, K/R) \longrightarrow 0,$$

and $\mathrm{Ker}\,\alpha^* = \{\varphi : R \longrightarrow K/R \;:\; \varphi|_{\underline{\underline{a}}} = 0\}$.

Under the natural epimorphism

$$\mathrm{Hom}_R(R,K) \longrightarrow \mathrm{Hom}_R(R,K/R),$$

$$\mathrm{Ker}\,\alpha^* \cong \{\varphi \in \mathrm{Hom}_R(R,K) \;:\; \varphi : \underline{\underline{a}} \longrightarrow R\}/R = \underline{\underline{a}}^{-1}/R.$$

Thus, $\mathrm{Ker}\,\alpha^* = (R/\underline{\underline{a}})^* \cong \underline{\underline{a}}^{-1}/R$, and consequently $(R/\underline{\underline{a}})^{**} \cong R/\underline{\underline{a}}$. It is easily checked that this isomorphism is natural. The isomorphism $T \xrightarrow{\simeq} T^{**}$ can be given explicitely:

$$T \longrightarrow \mathrm{Hom}_R(\mathrm{Hom}_R(T, K/R), K/R),$$

$$t \longmapsto t^{**}, \text{ where } t^{**}(t^*) = (t)t^* \text{ for } t^* \in T^*. \qquad \#$$

4.2 <u>Corollary</u>: Let $M, N \in {}_{\Lambda}\underline{\underline{M}}^{o}$ be such that $KM = KN$ and $M \supset N$. Then we have a "natural" isomorphism

$$\Phi_{M,N} : N^*/M^* \xrightarrow{\simeq} (M/N)^*,$$

where $M^* = \mathrm{Hom}_R(M,R) \in \underline{\underline{M}}^{o}_{\Lambda}$ is the dual of M.

<u>Proof</u>: We recall that for lattices, the functor "-*" is contravariant and exact with $M^{**} \cong M$. It follows from the theorem on elementary divisors for Dedekind domains (cf. Ex. 4,1) that

$$M \cong \oplus_{i=1}^{n} \underline{\underline{a}}_i \quad , \text{ as } R\text{-module},$$

$$N \cong \oplus_{i=1}^{n} \underline{\underline{b}}_i \underline{\underline{a}}_i \quad , \text{ as } R\text{-module},$$

and

$$M/N \cong \oplus_{i=1}^{n} \underline{\underline{a}}_i / \underline{\underline{b}}_i \underline{\underline{a}}_i \cong \oplus_{i=1}^{n} R/\underline{\underline{b}}_i , \text{ as } R\text{-module},$$

where $\underline{\underline{a}}_i$ and $\underline{\underline{b}}_i$ are integral ideals in R, $1 \le i \le n$. From the proof of (4.1) it follows that

$$(M/N)^* \cong \oplus_{i=1}^{n} \underline{\underline{b}}_i^{-1}/R \text{ as } R\text{-module.}$$

However,

$$M^* = \mathrm{Hom}_R(M,R) \cong \oplus_{i=1}^{n} \underline{\underline{a}}_i^{-1} \text{ as } R\text{-module}$$

and

$$N^* \cong \oplus_{i=1}^{n} \underline{\underline{b}}_i^{-1} \underline{\underline{a}}_i^{-1} \text{ as } R\text{-module.}$$

Then we have an R-isomorphism $(M/N)^* \cong N^*/M^*$. We can describe this isomorphism explicitely:

$$\Phi_{M,N} : N^*/M^* \longrightarrow (M/N)^* = \mathrm{Hom}_R(M/N, K/R),$$

$$n^* + M^* \longmapsto y, \quad \text{where}$$

$$(m + N)y = mn^* + R.$$

Here one should observe that we have an injection $\mathrm{Hom}_R(N,R) \to \mathrm{Hom}_K(KN,K)$

and so n^* can be extended to M.

This isomorphism is natural in the following sense: Given $M_i, N_i \in {}_{R}\underline{\underline{M}}^o$ such that $KM_i = KN_i$ and $M_i \supset N_i$, i=1,2. If now

$$\varphi : M_1 \longrightarrow M_2 \text{ with } \varphi|_{N_1} : N_1 \longrightarrow N_2$$

is an R-homomorphism, then φ induces R-homomorphisms

$$\overline{\varphi} : M_1/N_1 \longrightarrow M_2/N_2, \ \overline{\varphi}^* : (M_2/N_2)^* \longrightarrow (M_1/N_1)^*.$$

On the other hand, we have the R-homomorphism

$$\varphi^* : N_2^* \longrightarrow N_1^* \text{ with } \varphi^*|_{M_2^*} : M_2^* \longrightarrow M_1^*,$$

which induces an R-homomorphism

$$\overline{(\varphi^*)} : N_2^*/M_2^* \longrightarrow N_1^*/M_1^*.$$

It is then easily verified that the following diagram is commutative

$$\begin{array}{ccc} N_2^*/M_2^* & \xrightarrow{\Phi_{M_2,N_2}} & (M_2/N_2)^* \\ \overline{(\varphi^*)} \downarrow & & \downarrow \overline{\varphi}^* \\ N_1^*/M_1^* & \xrightarrow{\Phi_{M_1,N_1}} & (M_1/N_1)^* \end{array} .$$

This shows in particular, that $\Phi_{M,N}$ is a Λ-isomorphism if $M,N \in {}_{\Lambda}\underline{\underline{M}}^o$. #

4.3 <u>Definition</u>: If M and N are R-lattices in A, then the products MN and NM are defined in a natural way. However, if V is a faithful A-module and if M is an R-lattice in V and N an R-lattice in $\mathrm{Hom}_A(V,A)$, we also can define products MN and NM. We define the <u>left order of M</u>

$$\Lambda_l(M) = \{a \in A : aM \subset M\}$$

and the <u>right order of M</u>

$$\Lambda_r(M) = \{\varphi \in \mathrm{End}_A(V) : M\varphi \subset M\}.$$

It is easily seen that these are R-orders in the respective algebras.

As to the definition of the products, we observe that V is a progenerator, V being faithful. Thus we have two natural isomorphisms

$$\mu : \mathrm{Hom}_A(V,A) \otimes_A V \longrightarrow \mathrm{End}_A(V),$$
$$\varphi \otimes v \longmapsto \eta, \text{ where } v'\eta = (v'\varphi)v,$$

and

$$\tau : V \otimes_{\mathrm{End}_A(V)} \mathrm{Hom}_A(V,A) \longrightarrow A,$$
$$v \otimes \varphi \longmapsto v\varphi .$$

We now put

$$MN = \{(\sum_i m_i \otimes n_i)^{\tau} : m_i \in M, n_i \in N\},$$
$$NM = \{(\sum_i n_i \otimes m_i)^{\mu} : m_i \in M, n_i \in N\}.$$

It is easily verified that MN is an R-lattice in A and that NM is an R-lattice in $\mathrm{End}_A(V)$.

4.4 __Lemma:__ Let M and N be R-lattices in A with orders $\Lambda_l(M), \Lambda_l(N)$, $\Lambda_r(M), \Lambda_r(N)$ resp. Then we have a natural isomorphism

$$(MN)^* \cong \mathrm{Hom}_{\Lambda}(M,N^*)$$

as $[\Lambda_r(N), \Lambda_l(M)]$-bimodules; where $\Lambda = \Lambda_r(M) \cap \Lambda_l(N)$.

__Proof:__ We may assume $M \subset \Lambda$. From (I, Ex. 3,6) we obtain the natural isomorphism of $[\Lambda_r(N), \Lambda_l(M)]$-bimodules

$$\mathrm{Hom}_R(M \otimes_{\Lambda} N, R) \cong \mathrm{Hom}_{\Lambda}(M_{\Lambda}, N^*).$$

As R-modules we have

$$M \otimes_{\Lambda} N = (M \otimes_{\Lambda} N)^{o} \oplus (M \otimes_{\Lambda} N)^{t},$$

where $(M \otimes_{\Lambda} N)^t$ is the torsion part of $M \otimes_{\Lambda} N$ and $(M \otimes_{\Lambda} N)^o =$ $(M \otimes_{\Lambda} N)/(M \otimes_{\Lambda} N)^t$ is a Λ-lattice (cf. I, Ex. 8,3). Then (cf. I, Ex. 8,1)

$$\mathrm{Hom}_R(M \otimes_{\Lambda} N, R) = \mathrm{Hom}_R((M \otimes_{\Lambda} N)^o, R).$$

However, the exact sequence

$$\mathrm{Tor}_1^{\Lambda}(\Lambda/M,N) \xrightarrow{\alpha} M \otimes_{\Lambda} N \xrightarrow{\beta} \Lambda \otimes_{\Lambda} N$$

shows that $\mathrm{Im}\,\alpha = (M \otimes_{\Lambda} N)^t$, $\mathrm{Tor}_1^{\Lambda}(\Lambda/M,N)$ being an R-torsion module (cf. proof of VIII, 3.4), and $\mathrm{Im}\,\beta \cong MN \cong (M \otimes_{\Lambda} N)^o$. Hence

$$(MN)^* = \mathrm{Hom}_R(MN,R) \cong \mathrm{Hom}_R((M \otimes_{\Lambda} N)^o, R) \cong \mathrm{Hom}_{\Lambda}(M,N^*). \qquad \#$$

4.5 __Lemma__: Let V be a faithful A-module, M an R-lattice in V and N an R-lattice in $\mathrm{Hom}_A(V,A)$. Then

$$(MN)^* \overset{nat}{\cong} \mathrm{Hom}_{\Lambda_r(M)\cap\Lambda_l(N)}(M,N^*),$$

$$(NM)^* \overset{nat}{\cong} \mathrm{Hom}_{\Lambda_r(N)\cap\Lambda_l(M)}(N,M^*).$$

The __proof__ is done similarly as the one of (4.4), and it uses strongly the fact that V is a progenerator. #

4.6 __Definition__: Let Λ be an R-order in A and $M \in {}_{\Lambda}\underline{\underline{M}}^o$. A Λ-submodule $M' \subset M$ is called a __hypercharacteristic submodule__ if

$$\mathrm{Hom}_{\Lambda}(M',M) \hookrightarrow \mathrm{Hom}\,(M',M');$$

i.e., if for every $\varphi \in \mathrm{Hom}_{\Lambda}(M',M)$, $\mathrm{Im}\,\varphi \subset M'$.

4.7 __Lemma__ (Drozd-Kirichenko-Roiter [1]): The functor

$$\mathrm{Hom}_R(-,R) : {}_{\Lambda}\underline{\underline{M}}^o \longrightarrow \underline{\underline{M}}^o_{\Lambda}$$

establishes a one-to-one, inclusions reversing correspondence between the R-orders in A containing Λ and the hypercharacteristic Λ-submodules M of $\Lambda^* = \mathrm{Hom}_R(\Lambda,R)$ with $\Lambda^*/M \in \underline{\underline{M}}^T_{\Lambda}$.

__Proof__: Let Λ_1 be an R-order in A containing Λ; then ${}_{\Lambda}\Lambda_1 \in {}_{\Lambda}\underline{\underline{M}}^o$ and we shall show that $\Lambda_1^* = \mathrm{Hom}_R({}_{\Lambda}\Lambda_1,R)$ is a hypercharacteristic right Λ-submodule of Λ^* such that $\Lambda^*/\Lambda_1^* \in \underline{\underline{M}}^T_{\Lambda}$. Since $\Lambda_1/\Lambda \in {}_{\Lambda}\underline{\underline{M}}^T$, the same is true for $(\Lambda_1/\Lambda)^*$. But $(\Lambda_1/\Lambda)^* \cong \Lambda^*/\Lambda_1^*$ by (4.2) and thus Λ^*/Λ_1^* is an R-torsion module. Let now $\varphi \in \mathrm{Hom}_{\Lambda}(\Lambda_1^*, \Lambda_{\Lambda}^*)$ be given. Then $\varphi^* \in \mathrm{Hom}_{\Lambda}({}_{\Lambda}\Lambda, \Lambda_1)$. Since φ^* is uniquely determined by $(1)\varphi^*$, it can be factored as $\varphi^* = \iota\psi$ with $\iota : \Lambda \longrightarrow \Lambda_1$ the canonical injection and $\psi \in \mathrm{Hom}_{\Lambda}(\Lambda_1, \Lambda_1)$. Consequently $\varphi^{**} = \varphi = \psi^*\iota^*$ with $\psi^* \in \mathrm{Hom}_{\Lambda}(\Lambda^*_{1\Lambda}, \Lambda_1^*)$. However, $\iota^* : \Lambda_1^* \longrightarrow \Lambda^*$ is the injection, as shows the exact sequence

$$0 = \mathrm{Hom}_R(\Lambda_1/\Lambda,R) \longrightarrow \Lambda_1^* \xrightarrow{\iota^*} \Lambda^*,$$

induced by ι. Thus

$$\operatorname{Im}\varphi = \operatorname{Im}\psi^*\iota^* = \operatorname{Im}\psi^* \subset \Lambda_1^*,$$

and Λ_1^* is a hypercharacteristic right submodule of Λ^*.

Conversely, let M be a hypercharacteristic right Λ-submodule of Λ^* with $\Lambda^*/M \in \underline{\underline{M}}_\Lambda^T$. Then $M^* \in {}_\Lambda\underline{\underline{M}}^o$, $M^* \supset \Lambda$, $M^*/\Lambda \in {}_\Lambda\underline{\underline{M}}^T$ and it remains to show that M^* is a subring of A. Let $\iota : M \longrightarrow \Lambda^*$ be the injection. Since $-^*$ is an exact functor, and since M is hypercharacteristic, we have the following chain of natural isomorphisms

$$M^* \cong \operatorname{Hom}_\Lambda(\Lambda, M^*) \cong \operatorname{Hom}_\Lambda(M, \Lambda^*) = \operatorname{Hom}_\Lambda(M, M).$$

Thus M^* is a ring, and the ring structure of M^* coincides on Λ with the multiplication in Λ. The remainder of the statements is obvious. #

We shall next introduce the concept of divisibility of modules (Roiter [2,3,5,6]). For this we shall depart from the study of modules over orders and consider for a moment unitary left modules over a noetherian ring S with 1.

4.8 Definition: Let $M, N \in {}_S\underline{\underline{M}}^f$. We have the trace map

$$\tau_{N,M} : M \otimes_{\operatorname{End}_S(M)} \operatorname{Hom}_S(M,N) \longrightarrow N,$$

$$m \otimes \varphi \longmapsto m\varphi.$$

If T is a subset of $\operatorname{Hom}_S(M,N)$ we write

$$M \circ T = \{(\sum_i m_i \otimes \varphi_i)^{\tau_{N,M}} : m_i \in M, \varphi_i \in T\}.$$

T is called epimorphic if $M \circ T = N$ and we say that M covers N, notation $M \succ N$ if $\operatorname{Im} \tau_{M,N} = N$.

We remark, that Roiter uses the term "M divides N" and the notation $M \setminus N$ for M covers N. However, since this conflicts with $M|N$ in the sense of direct summands, Reiner [18][2] has introduced the term "M covers N".

4.9 <u>Lemma</u>: For $M,N \in {}_{\underline{S}}\underline{M}^f$, $M \succ N$ if and only if there is an exact sequence

$$M^{(r)} \longrightarrow N \longrightarrow 0$$

for some natural number r.

The <u>proof</u> is the same as (III, 1.10 (i) $\Longleftrightarrow$ (iv)). #

4.10 <u>Lemma</u>: The relation "to cover" is reflexive and transitive. $M,N \in {}_{\underline{S}}\underline{M}^f$ are said to be <u>associated</u> if $M \succ N$ and $N \succ M$. Being associated is an equivalence relation.

<u>Proof</u>: Since $\tau_{M,M}$ is an epimorphism, "$\succ$" is reflexive. As for the transitivity, let $M \succ X$ and $X \succ Y$. Then we have epimorphisms $M^{(r_1)} \longrightarrow X^{(r)}$ and $X^{(r)} \longrightarrow Y$, consequently, $M \succ Y$. Trivially, being associated is an equivalence relation. #

4.11 <u>Definition</u>: A decomposition of $M \in {}_{\underline{S}}\underline{M}^f$, $M = \oplus_{i=1}^n M_i$ is called a <u>normal decomposition</u> if $M_i \succ M_j$ for every $i < j$. If M does not have a normal decomposition, then M is called <u>normally indecomposable</u>.

We now return to the study of lattices over orders.

4.12 <u>Theorem</u> (Roiter [2]): Let R be a Dedekind domain with quotient field K and Λ an R-order in the separable finite dimensional K-algebra A. By "$\hat{\ }$" we denote the completion at some fixed $\underline{p} \in$ spec R. Let $\hat{M},\hat{N} \in {}_{\hat{\Lambda}}\underline{M}^o$ be such that $\hat{N} \succ \hat{M}$ and assume that $\hat{N}$ is normally indecomposable, or that $\mathrm{End}_{\hat{\Lambda}}(\hat{N})$ has only central idempotents. Then every exact sequence

$$0 \longrightarrow \hat{M}' \longrightarrow \hat{M} \xrightarrow{\varphi} \hat{N} \longrightarrow 0$$

splits.

<u>Proof</u>: Let $\hat{\Omega} = \mathrm{End}_{\hat{\Lambda}}(\hat{N})$ and put

$$\hat{T} = \mathrm{Hom}_{\hat{\Lambda}}(\hat{N},\hat{M})\,\varphi = \{\sigma\varphi : \sigma \in \mathrm{Hom}_{\hat{\Lambda}}(\hat{N},\hat{M})\}.$$

Then $\hat{T}$ is a left $\hat{\Omega}$-ideal in $\hat{\Omega}$, $\mathrm{Hom}_{\hat{\Lambda}}(\hat{N},\hat{M})$ being a left $\hat{\Omega}$-module. Moreover, $\hat{N} \succ \hat{M}$ and φ is an epimorphism. Thus $\hat{T}$ is epimorphic; i.e.,

$$\hat{N} \circ \hat{T} = \hat{N}.$$

Let $\overline{\Omega} = \hat{\Omega}/\text{rad }\hat{\Omega}$ and $\overline{T} = (\hat{T} + \text{rad }\hat{\Omega})/\text{rad }\hat{\Omega}$. Then $\overline{T} \neq 0$ or else $\hat{N} = \hat{N} \circ \hat{T} \subset \hat{N}\,\text{rad }\hat{\Omega}$, a contradiction to Nakayama's lemma. $\overline{\Omega}$ is semi-simple and so $\overline{T} = \overline{\Omega}\overline{e}$ for some non-zero idempotent $\overline{e}$ of $\overline{\Omega}$, which can be lifted to an idempotent $\hat{e}$ of $\hat{\Omega}$, $\hat{\Omega}$ being semi-perfect (cf. IV, 2.1). Then

$$\hat{T} \subset \hat{\Omega}\hat{e} + \text{rad }\hat{\Omega},$$

and thus

$$\hat{N} = \hat{N} \circ \hat{T} \subset \hat{N}\hat{e}\hat{\Omega} + \hat{N}\text{rad }\hat{\Omega} \subset \hat{N}; \text{ i.e.,}$$

$\hat{N} = \hat{N}\hat{e}\hat{\Omega}$ by Nakayama's lemma. But $\hat{e}$ is an idempotent in $\hat{\Omega}$ and so $\hat{N} = \hat{N}\hat{e} \oplus \text{Ker }\hat{e}$. Moreover,

$$\hat{N}\hat{e} \circ \text{Hom}_{\hat{\Lambda}}(\hat{N}\hat{e},\hat{N}) = \hat{N}\hat{e}\hat{e}\hat{\Omega} = \hat{N}\hat{e}\hat{\Omega} = \hat{N}$$

and so $\hat{N}\hat{e} \succ \hat{N}$. But then also $\hat{N}\hat{e} \succ \text{Ker }\hat{e}$. If $\hat{N}$ is normally indecomposable, this implies $\text{Ker }\hat{e} = 0$ and $\hat{e} = 1_{\hat{N}}$. On the other hand if $\hat{\Omega}$ has only central idempotents, then the relation $\hat{N}\hat{e}\hat{\Omega} = \hat{N}$ implies $\text{Ker }\hat{e} = 0$ and $\hat{e} = 1_{\hat{N}}$. Then $1_{\hat{N}} = \tau + \varrho$ with $\tau \in \hat{T}$ and $\varrho \in \text{rad }\hat{\Omega}$. But $1_{\hat{N}} - \varrho = \tau$ is a unit in $\hat{\Omega}$ (cf. I, Ex. 4,5) and hence $1_{\hat{N}} = \psi\varphi$ for some $\psi \in \text{Hom}_{\hat{\Lambda}}(\hat{N},\hat{M})$; i.e., the sequence

$$0 \longrightarrow \hat{M}' \longrightarrow \hat{M} \xrightarrow{\varphi} \hat{N} \longrightarrow 0 \text{ splits.} \qquad \#$$

4.13 <u>Corollary</u>: An exact sequence

$$0 \longrightarrow \hat{M}' \longrightarrow \hat{M} \longrightarrow \hat{N} \longrightarrow 0,$$

where $\hat{N}$ is normally indecomposable or $\text{End}_{\hat{\Lambda}}(\hat{N})$ has only central idempotents, is split if and only if $\hat{M}_1\varphi = \hat{N}$ with $\hat{M}_1 = \hat{N} \circ \text{Hom}_{\hat{\Lambda}}(\hat{N},\hat{M})$.

<u>Proof</u>: If this sequence is split, then there exists $\psi \in \text{Hom}_{\hat{\Lambda}}(\hat{N},\hat{M})$ such that $\psi\varphi = 1_{\hat{N}}$ and thus $\hat{M}_1\varphi = \hat{N}$.

<u>Conversely</u>, if $\hat{M}_1\varphi = \hat{N}$, then it follows directly from the proof of (4.12) that the sequence splits. #

4.14 <u>Lemma</u>: For $M,N \in {}_{\Lambda}\underline{\underline{M}}^{o}$, $M \succ N$ if and only if $\hat{M}_{\underline{\underline{p}}} \succ \hat{N}_{\underline{\underline{p}}}$ for every

$\underline{p} \in \operatorname{spec} R$.

<u>Proof</u>: Assume that $M > N$; then we have an epimorphism

$$M^{(r)} \longrightarrow N \longrightarrow 0 \quad (\text{cf. } 4.9).$$

Tensoring with $\hat{R}_{\underline{p}} \otimes_R -$ implies $\hat{M}_{\underline{p}} > \hat{N}_{\underline{p}}$. Conversely, if $\hat{M}_{\underline{p}} > \hat{N}_{\underline{p}}$ for every $\underline{p} \in \operatorname{spec} R$, then $M_{\underline{p}} > N_{\underline{p}}$ for every $\underline{p} \in \operatorname{spec} R$; in fact, we have the commutative diagram

$$\begin{array}{ccc}
M_{\underline{p}} \otimes_{\operatorname{End}_{\Lambda_{\underline{p}}}(M_{\underline{p}})} \operatorname{Hom}_{\Lambda_{\underline{p}}}(M_{\underline{p}}, N_{\underline{p}}) & \longrightarrow & N_{\underline{p}} \\
\downarrow & & \downarrow \\
\hat{M}_{\underline{p}} \otimes_{\operatorname{End}_{\hat{\Lambda}_{\underline{p}}}(\hat{M}_{\underline{p}})} \operatorname{Hom}_{\hat{\Lambda}_{\underline{p}}}(\hat{M}_{\underline{p}}, \hat{N}_{\underline{p}}) & \longrightarrow & \hat{N}_{\underline{p}} ,
\end{array}$$

and $\hat{R}_{\underline{p}} \otimes_{R_{\underline{p}}} -$ is a faithful functor (cf. I, 9.12). But then

$$\operatorname{Im} \tau_{M_{\underline{p}}, N_{\underline{p}}} = (\operatorname{Im} \tau_{M,N})_{\underline{p}} \quad \text{for every } \underline{p} \in \operatorname{spec} R ,$$

and now the statement follows since for an R-lattice X,

$$X = \bigcap_{\underline{p} \in \operatorname{spec} R} X_{\underline{p}} . \qquad \#$$

4.15 <u>Lemma</u>: Let $\hat{\Lambda}$ be an $\hat{R}$-order, which is indecomposable as left $\hat{\Lambda}$-lattice. Given an exact sequence of $\hat{\Lambda}$-lattices

$$0 \longrightarrow \hat{X}' \longrightarrow \hat{\Lambda} \longrightarrow \hat{X}'' \longrightarrow 0.$$

Then $\hat{X}''$ is indecomposable.

<u>Proof</u>: $\hat{X}''$ has a projective cover $\hat{P}$ (cf. III, 7.6), $\hat{\Lambda}$ being semi-perfect. Then we can complete the following diagram

$$\begin{array}{ccc}
\hat{P} & \xrightarrow{\Psi} & \hat{X}'' \longrightarrow 0 \\
 & \nwarrow{\scriptstyle \varphi} & \uparrow \\
 & & \hat{\Lambda}
\end{array}$$

.

However, $\varrho\psi$ is an epimorphism and ψ is essential. Thus ϱ is an epimorphism (cf. III, 7.1). From the Krull-Schmidt theorem it follows $\hat{P} \cong \hat{\Lambda}$, $\hat{\Lambda}$ being indecomposable as module. The proof of (III, 7.4) now shows that $\hat{X}''$ is indecomposable, since $\hat{X}''/\mathrm{rad}\,\hat{\Lambda}\,\hat{X}''$ is indecomposable. #

4.16 <u>Definition</u>: Let $M \in {}_{\Lambda}\underline{\underline{M}}^{o}$ be given. $Q \in {}_{\Lambda}\underline{\underline{M}}^{o}$ is called <u>M-injective</u> if every diagram with an exact row of Λ-lattices

$$\begin{array}{ccccccccc} 0 & \longrightarrow & M' & \longrightarrow & M & \longrightarrow & M'' & \longrightarrow & 0 \\ & & \downarrow & \swarrow & & & & & \\ & & Q & & & & & & \end{array}$$

can be completed to a commutative diagram.

4.17 <u>Lemma</u>: Let Q be M-injective. If $M = M' \oplus M''$, and $Q = M' \oplus Q''$, then Q'' is M''-injective.

<u>Proof</u>: Given the diagram with an exact row of Λ-lattices

$$\begin{array}{ccccccccc} 0 & \longrightarrow & N & \xrightarrow{\varphi} & M'' & \xrightarrow{\psi} & N'' & \longrightarrow & 0, \\ & & \downarrow{\scriptstyle\sigma} & & & & & & \\ & & Q'' & & & & & & \end{array}$$

we get the diagram with an exact row of Λ-lattices

$$\begin{array}{ccccccccc} 0 & \longrightarrow & N \oplus M' & \xrightarrow{\varphi \oplus 1_{M'}} & M'' \oplus M' & \xrightarrow{\psi \oplus 0_{M'}} & N'' & \longrightarrow & 0 \\ & & {\scriptstyle\sigma \oplus 1_{M'}}\downarrow & \swarrow{\scriptstyle\varrho} & & & & & \\ & & Q'' \oplus M' & & & & & & \end{array}$$

which can be completed by ϱ, since $Q'' \oplus M'$ is $M'' \oplus M'$-injective. But then $\varrho|_{M''}$ completes the original diagram commutatively. #

<u>Exercises § 4</u>:

1.) Let R be a Dedekind domain with quotient field K.

a.) Show that K/R is divisible.

b.) Given two R-lattices $M \supset N$, $KM = KN$. Show that there exist integral ideals $\{\underline{a}_i, \underline{b}_i\}_{1 \le i \le n}$ such that

$$M \cong \oplus_{i=1}^{n} R\underline{a}_i$$

$$N \cong \oplus_{i=1}^{n} R\underline{a}_i\underline{b}_i \; .$$

2.) Prove 4.5!

3.) Drozd-Kirichenko-Roiter [1], Roiter [3]. Let Λ be an R-order in the separable K-algebra A. Let M' be a hypercharacteristic submodule of $M \in {}_{\Lambda}\underline{M}^{o}$. M' is called a D-module if it satisfies the following conditions:

(i) $M' \neq M$, (ii) $M \succ M'$, (iii) if X is a hypercharacteristic submodule of M and $M' + X = M$, then $X = M$.

The largest D-submodule of M is denoted by D(M). Show its existence and uniqueness. Denote by "$\hat{}$" the completion at some fixed $\underline{p} \in$ spec R. Then $\hat{\Lambda}$ is hereditary if and only if $D(\hat{\Lambda}^*) = 0$. (<u>Hint</u>: Show that for $\hat{M} \in {}_{\hat{\Lambda}}\underline{M}^{o}$, every sequence of $\hat{\Lambda}$-lattices

$$0 \longrightarrow \hat{M}' \longrightarrow \hat{M} \longrightarrow \hat{M}'' \longrightarrow 0 \text{ splits}$$

if and only if $D(\hat{M}) = 0$.)

§5 Bass-orders

Λ is a Gorenstein-order if $\Lambda^* = \mathrm{Hom}_R(\Lambda, R)$ is a progenerator, and a Bass-order is an order each overorder of which is Gorenstein. Every lattice over a Bass-order Λ is a direct sum of left Λ-ideals. Moreover, an order is Bass if and only if the full two-sided ideals form a groupoid under proper multiplication.

Commutative Gorenstein-rings and Bass-rings have been investigated by H. Bass [4] (cf. Samuel [1]). Drozd-Kirichenko-Roiter [1] have extended these definitions to orders in separable algebras and they have classified all Bass-orders.

5.1 Definition: Let R be a Dedekind domain with quotient field K and Λ an R-order in the separable finite dimensional K-algebra A. Λ is a Gorenstein-order, if $\Lambda^* \succ \Lambda$ as left Λ-modules; i.e., if we have an epimorphism

$$\tau_{\Lambda^*,\Lambda} : \Lambda^* \otimes_\Lambda \mathrm{Hom}_\Lambda(\Lambda^*, {}_\Lambda\Lambda) \longrightarrow \Lambda,$$

$$\lambda^* \otimes \varphi \longmapsto \lambda^*\varphi,$$

where $\Lambda^* = \mathrm{Hom}_R(\Lambda, R)$.

Λ is called a Bass-order if every R-order in A containing Λ is Gorenstein. We remark that a Gorenstein-order is sometimes also called a quasi-Frobenius order (cf. Endo [1]).

Remark: We point out that $\Lambda^* \succ \Lambda$ as left Λ-lattices is equivalent to ${}_\Lambda\Lambda^*$ being a generator in ${}_\Lambda\underline{\underline{M}}^o$.

5.2 Lemma: Let Λ be a Gorenstein-order in A. Then Λ^* is a progenerator for ${}_\Lambda\underline{\underline{M}}^o$ and for $\underline{\underline{M}}^o_\Lambda$. This shows in particular, that we do not have to distinguish between left and right Gorenstein-orders.

Proof: If suffices to show that

$$_{\hat{\Lambda}_{\underline{\underline{p}}}}\hat{\Lambda}^*_{\underline{\underline{p}}} = \mathrm{Hom}_{\hat{R}_{\underline{\underline{p}}}}({}_{\hat{\Lambda}_{\underline{\underline{p}}}}\hat{\Lambda}_{\underline{\underline{p}}}, \hat{R}_{\underline{\underline{p}}}) \in {}_{\hat{\Lambda}_{\underline{\underline{p}}}}\underline{\underline{M}}^o$$

is a progenerator for every $\underline{\underline{p}} \in$ spec R (cf. IV, 3.1, 3.2); we omit the subscript $\underline{\underline{p}}$. Since ${}_{\hat{\Lambda}}\hat{\Lambda}^* \in {}_{\hat{\Lambda}}\underline{\underline{M}}^o$ is a generator, there exists $\hat{X} \in {}_{\hat{\Lambda}}\underline{\underline{M}}^o$ such that

$$\hat{\Lambda} \oplus \hat{X} \cong ({}_{\hat{\Lambda}}\hat{\Lambda}^*)^{(n)} \quad \text{for some } n \in \underline{\underline{N}}.$$

Let $\{\hat{e}_i\}_{1 \le i \le n}$ be a complete set of orthogonal primitive idempotents of $\hat{\Lambda}$. Then $\hat{\Lambda}\hat{e}_i \cong \hat{\Lambda}\hat{e}_k$ if and only if $\hat{e}_i\hat{\Lambda} \cong \hat{e}_k\hat{\Lambda}$. In fact, $\hat{\Lambda}\hat{e}_i \cong \hat{\Lambda}\hat{e}_k$ if and only if $\bar{\Lambda}\bar{e}_i \cong \bar{\Lambda}\bar{e}_k$, where "-" denotes reduction modulo rad $\hat{\Lambda}$. This follows from the fact that $\hat{\Lambda}\hat{e}_i$ is a projective cover for $\bar{\Lambda}\bar{e}_i$ (cf. III, §7). However, $\bar{\Lambda}$ is a semi-simple $\hat{R}$/rad $\hat{R}$-algebra and so $\bar{\Lambda}\bar{e}_i \cong \bar{\Lambda}\bar{e}_k$ if and only if $\bar{e}_i\bar{\Lambda} \cong \bar{e}_k\bar{\Lambda}$. Now, let $\{\hat{e}_i\}_{1 \le i \le s}$ be the non-equivalent ones among the idempotents $\{\hat{e}_i\}_{1 \le i \le n}$. Then

$${}_{\hat{\Lambda}}\hat{\Lambda} \cong \oplus_{i=1}^{s} \hat{\Lambda}\hat{e}_i^{(\alpha_i)},$$

$$\hat{\Lambda}_{\hat{\Lambda}} \cong \oplus_{i=1}^{s} \hat{e}_i\hat{\Lambda}^{(\beta_i)}.$$

The non-isomorphic indecomposable direct summands of ${}_{\hat{\Lambda}}\hat{\Lambda}^*$ are the $\{(\hat{e}_i\hat{\Lambda})^*\}_{1 \le i \le s}$, and the relation

$$\hat{\Lambda} \oplus \hat{X} \cong ({}_{\hat{\Lambda}}\hat{\Lambda}^*)^{(n)},$$

together with the Krull-Schmidt theorem shows that ${}_{\hat{\Lambda}}\hat{\Lambda}^* \in {}_{\hat{\Lambda}}\underline{\underline{P}}^f$. It remains to show that $\hat{\Lambda}^*_{\hat{\Lambda}} \in \underline{\underline{M}}^o_{\hat{\Lambda}}$ is a generator; but ${}_{\hat{\Lambda}}\hat{\Lambda}^* \in {}_{\hat{\Lambda}}\underline{\underline{P}}^f$ implies that

$${}_{\hat{\Lambda}}\hat{\Lambda}^* \oplus \hat{Y} \cong {}_{\hat{\Lambda}}\hat{\Lambda}^{(n)} \quad \text{for some } \hat{Y} \in {}_{\hat{\Lambda}}\underline{\underline{M}}^o.$$

Passing to the dual lattice we conclude

$$\hat{\Lambda}_{\hat{\Lambda}} \oplus \hat{Y}^* \cong \hat{\Lambda}^{*(n)}_{\hat{\Lambda}},$$

and $\Lambda^*_{\hat{\Lambda}}$ is a generator in $\underline{\underline{M}}^o_{\hat{\Lambda}}$. #

5.3 <u>Theorem</u>: Let $\hat{R}$ be a completion of R and let $\hat{\Lambda}$ be a Gorenstein-order in $\hat{A}$. If $\hat{M} \in {}_{\hat{\Lambda}}\underline{\underline{M}}^o$ is indecomposable, then either

(1) $\hat{M} \in {}_{\hat{\Lambda}}\underline{\underline{P}}^f$ or

(ii) $\hat{M}$ is a lattice over an order $\hat{\Lambda}_1 \underset{\neq}{\supset} \hat{\Lambda}$.

<u>Proof</u>: We assume that $\hat{M}$ is not a lattice for any $\hat{\Lambda}_1$ with $\hat{\Lambda}_1 \underset{\neq}{\supset} \hat{\Lambda}$. If $\hat{M}$ is not faithful, then $\mathrm{ann}_{\hat{A}}(\hat{K}\hat{M}) = \hat{A}\hat{e} \neq 0$ for some central idempotent $\hat{e}$ of $\hat{A}$, and $\hat{M}$ is a faithful $\hat{\Lambda}(1-\hat{e})$-lattice. Because of our hypothesis, $\hat{\Lambda} = \hat{\Lambda}\hat{e} \oplus \hat{\Lambda}(1-\hat{e})$. If $\hat{M} \in {}_{\hat{\Lambda}(1-\hat{e})}\underline{\underline{P}}^f$, then $\hat{M} \in {}_{\hat{\Lambda}}\underline{\underline{P}}^f$. Since with $\hat{\Lambda}$ also $\hat{\Lambda}(1-\hat{e})$ is a Gorenstein-order, we may assume that $\hat{M}$ is faithful. If (ii) does not hold, then

$$\mathrm{End}_{\mathrm{End}_{\hat{\Lambda}}(\hat{M})}(\hat{M}) = \hat{\Lambda},$$

$\hat{K}\hat{M}$ being a progenerator for ${}_{\hat{A}}\underline{\underline{M}}^f$. Thus we have the following chain of natural isomorphisms

$$\hat{M} \otimes_{\mathrm{End}_{\hat{\Lambda}}(\hat{M})} \hat{M}^* \cong (\hat{M} \otimes_{\mathrm{End}_{\hat{\Lambda}}(\hat{M})} \hat{M}^*)^{**} \cong$$

$$\mathrm{Hom}_{\hat{R}}(\hat{M} \otimes_{\mathrm{End}_{\hat{\Lambda}}(\hat{M})} \hat{M}^*, \hat{R})^* \cong \mathrm{Hom}_{\mathrm{End}_{\hat{\Lambda}}(\hat{M})}(\hat{M}, \mathrm{Hom}_{\hat{R}}(\hat{M}^*, \hat{R}))^* \cong$$

$$\cong \mathrm{End}_{\mathrm{End}_{\hat{\Lambda}}(\hat{M})}(\hat{M})^* \cong \hat{\Lambda}^*.$$

Let $\hat{\Lambda}^*$ be generated as left $\hat{\Lambda}$-module by $\{\lambda_i^*\}_{1 \le i \le n}$. If

$$\sigma : \hat{M} \otimes_{\mathrm{End}_{\hat{\Lambda}}(\hat{M})} \hat{M}^* \longrightarrow \Lambda^*$$

is the isomorphism (cf. left $\hat{\Lambda}$-modules) established above, then there exist elements $x_i = \sum_{j=1}^{n_i} m_{ij}^o \otimes m_{ij}^{o*}$ such that $x_i\sigma = \lambda_i^*$. We now define

$$\varrho : \hat{M}^{(\sum_{i=1}^n n_i)} \longrightarrow \hat{\Lambda}^*,$$

$$(m_{ij})_{\substack{1 \le j \le n_i \\ 1 \le i \le n}} \longmapsto \sum_{i=1}^n \left(\sum_{j=1}^{n_i} m_{ij} \otimes m_{ij}^{o*}\right)\sigma .$$

Then ϱ is an epimorphism of left $\hat{\Lambda}$-modules. However, $\hat{\Lambda}^* \in {}_{\hat{\Lambda}}\underline{\underline{P}}^f$ (cf. 5.2) and thus $\hat{\Lambda}^*$ is a direct summand of $\hat{M}^{(m)}$ for some $m \in \underline{\underline{N}}$. Since $\hat{M}$ is indecomposable, $\hat{M} \in {}_{\hat{\Lambda}}\underline{\underline{P}}^f$ by the Krull-Schmidt theorem.

We remark that we actually have shown the following: If $\hat{M}$ is a faithful indecomposable $\hat{\Lambda}$-lattice which is not a lattice over a larger order, then $\hat{M}$ is a progenerator in ${}_{\hat{\Lambda}}\underline{\underline{M}}^{o}$. #

Remark: In the proof of (5.3) we have used heavily the fact that the Krull-Schmidt theorem is valid in ${}_{\hat{\Lambda}}\underline{\underline{M}}^{o}$. From (5.3) we can conclude that for a Bass-order $\hat{\Lambda}$ in $\hat{A}$, every $\hat{\Lambda}$-lattice decomposes into a direct sum of ideals. (Here "ideal" means a $\hat{\Lambda}$-submodule of $\hat{\Lambda}$ - not necessarily full.) However, decomposition is a local property, and so this does not imply that globally, every lattice over a Bass-order decomposes into a direct sum of left ideals. But this is nevertheless true, as shows (5.6).

5.4 Lemma: An R-order Λ in A is a Bass-order if and only if $M > N$ implies $M^* > N^*$ for $M,N \in {}_{\Lambda}\underline{\underline{M}}^{o}$.

Proof: Assume that the condition is satisfied and let $\Lambda_1 \supset \Lambda$. Then $\Lambda_1 > \Lambda_1^*$, Λ_1 being a generator for ${}_{\Lambda_1}\underline{\underline{M}}^{o}$. Hence $\Lambda_1^* > \Lambda_1$ and Λ_1 is a Gorenstein-order. Consequently, Λ is a Bass-order.

Conversely, assume that Λ is a Bass-order and let $M,N \in {}_{\Lambda}\underline{\underline{M}}^{o}$ with $M > N$ be given. In view of (4.14) it suffices to show $\hat{M}^* > \hat{N}^*$ where "$\hat{}$" is the completion at some $\underline{\underline{p}} \in$ spec R. $\hat{M} > \hat{N}$ implies

$$\hat{M} \circ \mathrm{Hom}_{\hat{\Lambda}}(\hat{M},\hat{N}) = \hat{N}$$

and so $\mathrm{ann}_{\hat{A}}(\hat{K}\hat{M}) \subset \mathrm{ann}_{\hat{A}}(\hat{K}\hat{N})$, say $\mathrm{ann}_{\hat{A}}(\hat{K}\hat{M}) = \hat{A}\hat{e}_1$ and $\mathrm{ann}_{\hat{A}}(\hat{K}\hat{N}) = \hat{A}(\hat{e}_1+\hat{e}_2)$ for central idempotents $\hat{e}_1$ and $\hat{e}_2$ in $\hat{A}$. We let $\hat{\Gamma}\hat{e}_1$ be a maximal $\hat{R}$-order in $\hat{A}\hat{e}_1$ containing $\hat{\Lambda}\hat{e}_1$ and define

$$\Lambda_1(\hat{M}) = \{a \in \hat{A}(1-\hat{e}_1) : a\hat{M} \subset \hat{M}\} \oplus \hat{\Gamma}\hat{e}_1,$$

$$\Lambda_1(\hat{N}) = \{a \in \hat{A}(1-\hat{e}_1-\hat{e}_2) : a\hat{N} \subset \hat{N}\} \oplus \Lambda_1(\hat{M})(\hat{e}_1+\hat{e}_2).$$

We put $\hat{\Lambda}_{\hat{M}} = \Lambda_1(\hat{M})(1-\hat{e}_1)$ and $\hat{\Lambda}_{\hat{N}} = \Lambda_1(\hat{N})(1-\hat{e}_1)$. Since $\Lambda_1(\hat{M})$ is a Gorenstein-order, so is $\hat{\Lambda}_{\hat{M}}$ and $\hat{M}$ is a faithful $\hat{\Lambda}_{\hat{M}}$-lattice which is not a lattice over any larger order. Then $\hat{M}^*$ is also a faithful $\hat{\Lambda}_{\hat{M}}$-lattice which is not a lattice over any larger order. As in the proof of (5.3), one shows now that $\hat{\Lambda}^*_{\hat{M}}$ is a direct summand of $\hat{M}^{*(m)}$ for some m. However, $\hat{\Lambda}^*_{\hat{M}}$ is a generator (cf. 5.2) and so $\hat{M}^*$ is a generator for ${}_{\hat{\Lambda}_{\hat{M}}}\underline{\underline{M}}^o$; i.e., $\hat{M}^* > \hat{\Lambda}_{\hat{M}}$. But, $\hat{\Lambda}_{\hat{M}} > \hat{M}$ and $\hat{M} > \hat{N}$ implies $\hat{\Lambda}_{\hat{M}} > \hat{N}$; i.e.,

$$\hat{\Lambda}_{\hat{M}} \circ \mathrm{Hom}_{\hat{\Lambda}_{\hat{M}}}(\hat{\Lambda}_{\hat{M}}, \hat{N}) = \hat{N}.$$

Consequently, $\hat{\Lambda}_{\hat{N}} \supset \hat{\Lambda}_{\hat{M}}$ and $\hat{N} \in {}_{\hat{\Lambda}_{\hat{M}}}\underline{\underline{M}}^o$. Hence $\hat{\Lambda}_{\hat{M}} > \hat{N}^*$. Combining this with $\hat{M}^* > \hat{\Lambda}_{\hat{M}}$, we conclude $\hat{M}^* > \hat{N}^*$. #

5.5 <u>Notation</u>: Let $A = \oplus_{i=1}^n A_i$ be the decomposition of A into simple algebras and let $\{L_i\}_{1\le i\le n}$ be the simple A-modules. For $M \in {}_\Lambda\underline{\underline{M}}^o$ we have $KM \cong \oplus_{i=1}^n L_i^{(\alpha_i)}$ and KM is - up to isomorphism - uniquely determined by $(\alpha_i)_{1\le i\le n}$. We then say that <u>M has signature $(\alpha_i)_{1\le i\le n}$</u>, notation $\mathrm{sig}(M) = (\alpha_i)_{1\le i\le n}$.

5.6 <u>Theorem</u> (Drozd-Kirichenko-Roiter [1], Bass [4]): Let Λ be a Bass-order in A and let $M \in {}_\Lambda\underline{\underline{M}}^o$. Then

$$M \cong I \oplus M' \text{ with } I \in {}_\Lambda\underline{\underline{M}}^o$$

such that $\mathrm{sig}(I) = (\gamma_i)_{1\le i\le n}$ with $\gamma_i = \min(\alpha_i, \lambda_i)$, where $\mathrm{sig}(\Lambda) = (\lambda_i)_{1\le i\le n}$ and $\mathrm{sig}(M) = (\alpha_i)_{1\le i\le n}$.

<u>Proof</u>: Taking the theorem for granted for $\hat{\Lambda}_{\underline{\underline{p}}}$, $\underline{\underline{p}} \in \mathrm{spec}\, R$, we show that it holds globally. We write

$$I(KM) = \oplus_{i=1}^n L_i^{(\gamma_i)}, \text{ where } (\gamma_i)_{1\le i\le n}$$

is defined in the theorem; similarly for the completion.
Obviously, $\hat{K}_{\underline{\underline{p}}} \otimes_K I(KM) \cong I(\hat{K}_{\underline{\underline{p}}}\hat{M}_{\underline{\underline{p}}})$. Thus

$$\hat{M}_{\underline{\underline{p}}} \cong \hat{I}_{\underline{\underline{p}}} \oplus \hat{M}'_{\underline{\underline{p}}} \text{ with } \hat{K}_{\underline{\underline{p}}}\hat{I}_{\underline{\underline{p}}} \cong \hat{K}_{\underline{\underline{p}}} I(KM)$$

implies the existence of $\Lambda_{\mathfrak{p}}$-lattices $I_{\mathfrak{p}}$ and $M'_{\mathfrak{p}}$ such that

$$\hat{R}_{\mathfrak{p}} \otimes_{R_{\mathfrak{p}}} I_{\mathfrak{p}} \cong \hat{I}_{\mathfrak{p}} \text{ and } \hat{R}_{\mathfrak{p}} \otimes_{R_{\mathfrak{p}}} M'_{\mathfrak{p}} \cong \hat{M}'_{\mathfrak{p}} \quad \text{(cf. IV, 1.9).}$$

But then $M_{\mathfrak{p}} \cong I_{\mathfrak{p}} \oplus M'_{\mathfrak{p}}$ (cf. IV, 1.2). Moreover, according to (IV, 1.8) there exists $I \in {}_{\Lambda}\underline{M}^{o}$ such that $R_{\mathfrak{p}} \otimes_{R} I \cong I_{\mathfrak{p}}$ for every $\mathfrak{p} \in$ spec R. Thus I is a local direct summand of M and according to (VII, 3.8), there exists $I' \in {}_{\Lambda}\underline{M}^{o}$ with $KI' \cong KI$ such that $I' \underset{\cong}{|} M$ and $\mathrm{sig}(I) = \mathrm{sig}(I')$, and the theorem is established under the assumption that it is true for the completions.

We thus may assume that $\hat{R}$ is the completion of R at some $\mathfrak{p} \in$ spec R. We claim that

$$\hat{Q} = \hat{M} \circ \mathrm{Hom}_{\hat{\Lambda}}(\hat{M}, \hat{\Lambda}^*)$$

is $\hat{M}$-injective (cf. 4.16). For this we observe first that ${}_{\hat{\Lambda}}\hat{\Lambda}^*$ is $\hat{M}$-injective for every $\hat{M} \in {}_{\hat{\Lambda}}\underline{M}^{o}$; in fact, given the diagram with an exact row of lattices

$$\begin{array}{ccccccccc} 0 & \longrightarrow & \hat{M}' & \longrightarrow & \hat{M} & \longrightarrow & \hat{M}'' & \longrightarrow & 0 \\ & & \downarrow & & & & & & \\ & & {}_{\hat{\Lambda}}\hat{\Lambda}^* & & & & & & \end{array}$$

we pass over to the dual lattices and obtain the diagram with exact row

$$\begin{array}{ccccccccc} 0 & \longrightarrow & \hat{M}''^* & \longrightarrow & \hat{M}^* & \longrightarrow & \hat{M}'^* & \longrightarrow & 0 \\ & & & & & \nwarrow^{\varphi} & \uparrow & & \\ & & & & & & \hat{\Lambda}_{\hat{\Lambda}} & & \end{array}$$

which can be completed, $\hat{\Lambda}$ being projective. Taking duals once more, we see that φ^* completes the original diagram.

To show that $\hat{Q} = \hat{M} \circ \mathrm{Hom}_{\hat{\Lambda}}(\hat{M}, \hat{\Lambda}^*)$ is $\hat{M}$-injective, we observe that $\hat{Q}$ is a submodule of $\hat{\Lambda}^*$; let $\iota : \hat{Q} \longrightarrow \hat{\Lambda}^*$ be the injection. Then we can complete the diagram

$$\begin{array}{ccccccc} 0 \longrightarrow & \hat{M}' & \xrightarrow{\Psi} & \hat{M} & \longrightarrow & \hat{M}'' & \longrightarrow 0 \\ & \varphi \downarrow & \swarrow_{\sigma} & & & & \\ & \hat{Q} & & & & & \\ & \iota \downarrow & & & & & \\ & \hat{\Lambda}^* & & & & & \end{array}$$

by $\sigma \in \mathrm{Hom}_{\hat{\Lambda}}(\hat{M}, \hat{\Lambda}^*)$. However, $\mathrm{Im}\,\sigma \subset \hat{Q}$ and thus $\hat{Q}$ is $\hat{M}$-injective. We observe that

$$\hat{M} \circ \mathrm{Hom}_{\hat{\Lambda}}(\hat{M}, \hat{Q}) = \hat{Q}$$

and so $\hat{M} > \hat{Q}$. Applying (5.4) we conclude $\hat{M}^* > \hat{Q}^*$, and we can find an exact sequence of right $\hat{\Lambda}$-lattices

$$0 \longrightarrow \hat{X} \longrightarrow \hat{M}^{*(m)} \longrightarrow \hat{Q}^* \longrightarrow 0 \quad (\text{cf. } 4.9).$$

Passing to the dual lattices we obtain the exact sequence

$$0 \longrightarrow \hat{Q} \longrightarrow \hat{M}^{(m)} \longrightarrow \hat{X}^* \longrightarrow 0,$$

which is split, $\hat{Q}$ being $\hat{M}^{(m)}$-injective. It should be observed that the same argument as above shows that $\hat{Q}$ is $\hat{M}^{(m)}$-injective. The Krull-Schmidt theorem then implies

$$\hat{M} \cong \hat{I}_1 \oplus \hat{M}_1, \quad \hat{Q} \cong \hat{I}_1 \oplus \hat{Q}_1,$$

where $\hat{Q}_1$ is $\hat{M}_1$-injective by (4.7).

The signature $(q_i)_{1 \le i \le n}$ of $\hat{Q}$ is easily seen to be $q_i = \lambda_i$ if $\alpha_i \ne 0$ and $q_i = 0$ if $\alpha_i = 0$ (here $\hat{A} = \oplus_{i=1}^{n} \hat{A}_i$, where $\hat{A}_i$ are simple algebras, and $\hat{\Lambda}$ has signature $(\lambda_i)_{1 \le i \le n}$, $\hat{M}$ has signature $(\alpha_i)_{1 \le i \le n}$); it should be observed that $\hat{A}^*_{\hat{A}} \cong {}_{\hat{A}}\hat{A}$, $\hat{A}$ being semi-simple (cf. proof of 5.2). If

$$\mathrm{sig}(\hat{K}\hat{I}_1) = (\beta_i)_{1 \le i \le n}, \text{ then}$$

$$\mathrm{sig}(\hat{K}\hat{M}_1) = (\alpha_i - \beta_i)_{1 \le i \le n} \text{ and}$$

$$\mathrm{sig}(\hat{K}\hat{Q}_1) = (q_i - \beta_i)_{1 \le i \le n}.$$

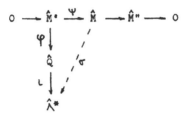

We put $\hat{Q}_1' = \hat{M}_1 \circ \mathrm{Hom}_\Lambda(\hat{M}_1,\hat{Q}_1)$. If $\mathrm{sig}(\hat{K}\hat{Q}_1') = (q_i')_{1\le i\le n}$, then $q_i' = q_i - \beta_i$ for $\alpha_i \neq \beta_i$ and $q_i' = 0$ if $\alpha_i = \beta_i$. Since $\hat{Q}_1$ is $\hat{M}_1$-injective, so is $\hat{Q}_1'$, and we can proceed as above to conclude

$$\hat{M}_1 \cong \hat{I}_2 \oplus \hat{M}_2,\ \hat{Q}_1' \cong \hat{I}_2 \oplus \hat{Q}_2,$$

where $\hat{Q}_2$ is $\hat{M}_2$-injective. After finitely many steps, this proceedure has to stop; i.e., $\hat{Q}_s' = \hat{M}_s \circ \mathrm{Hom}_\Lambda(\hat{M}_s,\hat{Q}_s) = 0$; i.e., $\hat{M}_s$ and $\hat{Q}_s$ do not have a rational component in common. Putting $\hat{I} = \oplus_{i=1}^{s} \hat{I}_i$, we have $\hat{M} \cong \hat{I} \oplus \hat{M}_s$, and $\mathrm{sig}(I) = (\iota_i)_{1\le i\le s}$, where either

$$\iota_i = \alpha_i \text{ and } \lambda_i - \iota_i \geq 0 \text{ or}$$

$$\iota_i = \lambda_i \text{ and } \alpha_i - \iota_i \geq 0;$$

i.e., $\iota_i = \min(\lambda_i, \alpha_i), 1\le i\le n$. Thus $\hat{M} \cong \hat{I} \oplus \hat{X}$ and $\hat{I}$ has the desired properties. #

5.7 <u>Definition</u>: A set G with a partially defined "associative" law of composition is called a <u>groupoid</u>, if for every $g \in G$ there exists exactly one left identity, right identity, left inverse and right inverse. Then the left inverse of an element $g \in G$ is automatically also the right inverse (cf. Ex. 5,4).

If Λ is an R-order in the separable K-algebra A, we consider the two-sided full Λ-ideals under proper multiplication; i.e., if I and J are Λ-ideals, then the product IJ is called proper, if for Λ-ideals $I_1 \supset I$, $J_1 \supset J$ the equality $IJ = I_1J_1$ implies $I = I_1$ and $J = J_1$. With respect to this proper multiplication, every Λ-ideal I has a unique left identity $\Lambda_l(I)$ and a unique right identity $\Lambda_r(I)$. For, if $XI = I$ is a proper product, then $X \subset \Lambda_l(I)$ and $XI = \Lambda_l(I) \cdot I$ implies $X = \Lambda_l(I)$, since the product is proper. Moreover, if I has a left inverse I_l^{-1} and a right inverse, I_j^{-1} then these are uniquely determined and equal. If I has an inverse I^{-1}, then $I^{-1} = \{a \in A : aI \subset \Lambda_r(I)\} = \{a \in A : Ia \subset \Lambda_l(I)\}$.

Obviously, $I^{-1}I = \Lambda_r(I)$ implies $I^{-1} \subset \{a \in A : aI \subset \Lambda_r(I)\} = J_1$, and hence $J_1 I = \Lambda_r(I)$ and the uniqueness of the inverse implies

$$I^{-1} = J_1 = J_r.$$

In (VI, § 8) we have shown that the normal ideals in A form a groupoid, and the next theorem shows that for Bass-orders and only for Bass-orders the two-sided ideals form a groupoid under proper multiplication.

5.8 <u>Theorem</u> (Drozd-Kirichenko-Roiter [1]): Λ is a Bass-order in A if and only if the Λ-ideals form a groupoid under proper multiplication.

<u>Proof</u>: Assume that the Λ-ideals, form a groupoid G_Λ. Then every Λ-ideal has a left and right inverse in the sense of (IV, § 4), whence it is a progenerator (cf. IV, 4.18). In particular, if Λ_1 is an R-order containing Λ, then Λ_1^* is a Λ-ideal; i.e., a progenerator and Λ_1 is a Gorenstein-order. Hence Λ is a Bass-order. Conversely, let Λ be a Bass-order: given a Λ-ideal I, let $\Lambda_1 = \Lambda_l(I)$. As in the proof of (5.3) one shows that ${}_{\Lambda_1}I$ is a generator for ${}_{\Lambda_1}\underline{\underline{M}}^o$, whence it is a projective $\Lambda_r(I)$-module (cf. III, 2.2). A similar argument with the right order shows that I is a progenerator in ${}_{\Lambda_1}\underline{\underline{M}}^o$; whence it has inverses (cf. III, § 1). Hence every Λ-ideal has a left and right inverse, which coincide. If IJ is a proper product, then $J^{-1}I^{-1}$ is a proper product and $J^{-1}I^{-1}IJ = \Lambda_r(J)$ is a proper product. Similarly $IJJ^{-1}I^{-1} = \Lambda_l(I)$, and $\Lambda_l(I) = \Lambda_l(IJ), \Lambda_r(J) = \Lambda_r(IJ)$, the products being proper. Thus the Λ-ideals form a groupoid under proper multiplication. #

5.9 <u>Corollary</u> (Harada [1], Drozd-Kirichenko-Roiter [1]): If Λ is a hereditary R-order in A, then the two-sided Λ-ideals in A form a groupoid under proper multiplication.

<u>Proof</u>: This follows since hereditary orders are Bass-orders. In fact, if Λ is hereditary, then Λ^* is a two-sided projective Λ-lattice, and

thus a progenerator, since $\Lambda = \Lambda_l(\Lambda^*) = \Lambda_r(\Lambda^*)$. However, every order containing Λ is hereditary (cf. 2.5) and so Λ is a Bass-order. #

Exercises §5:

1.) Let A be a semi-simple K-algebra. Show that ${}_A A \cong A^*_A$.

2.) Let Λ be the R-order in $(K)_2$ generated over $\underline{\underline{Z}}_p$ by the matrices

$$\begin{vmatrix} 1 & 0 \\ 0 & 1 \end{vmatrix}, \begin{vmatrix} p & 0 \\ 0 & 0 \end{vmatrix}, \begin{vmatrix} 0 & p \\ 0 & 0 \end{vmatrix}, \begin{vmatrix} 0 & 0 \\ 1 & 0 \end{vmatrix}$$

where p is a rational prime number. Show that Λ is a Bass-order!

3.) Let $\hat{\Lambda}$ be a Bass-order in $\hat{A}$, and let $\hat{M}$ be a faithful $\hat{\Lambda}$-lattice, which is not a lattice over a larger order. Then $\hat{M}$ is a generator for ${}_{\hat{\Lambda}}\underline{\underline{M}}^o$.

4.) Let G be a set with a partially defined internal law of composition $G \times G \longrightarrow G$, $(g_1, g_2) \longmapsto g_1 g_2$, such that:

(i) If ab and (ab)c are defined then bc is defined and (ab)c = a(bc).

(ii) For every $g \in G$ there exists a unique e_1 (resp. e_2) in G such that $ge_1 = g$ (resp. $e_2 g = g$).

(iii) For every $g \in G$ there exists a unique g_1, (resp. g_2) in G such that $g_1 g = e_1$ (resp. $gg_2 = e_2$).

Then G is called a <u>groupoid</u>. Show that the left inverse is equal to the right inverse, and that the left unit of the inverse is the right unit of the original element. Apply these results to the set of Λ-ideals under proper multiplication.

§6 Classification of Bass-orders

In this section we give a local description of all Bass-orders.

We retain the notation of §§4,5.

6.1 Lemma: Bass-orders are invariant under Morita-equivalences.

Proof: Let $E \in {}_{\Lambda}\underline{\underline{M}}^{o}$ be a progenerator and put $\Omega = \mathrm{End}_{\Lambda}(E)$. In view of (5.4) we have to show $M \succ N$ implies $M^* \succ N^*$ for $M, N \in {}_{\Omega}\underline{\underline{M}}^{o}$. But $M \succ N$ if and only if $E \otimes_{\Omega} M \succ E \otimes_{\Omega} N$. Since Λ is a Bass-order, the result follows with the natural isomorphism $\mathrm{Hom}_{\Lambda}(E^*, (E \otimes_{\Omega} M)^*) \cong M^*$. #

Let $\underline{\underline{p}} \in$ spec R be fixed and denote by "^" the $\underline{\underline{p}}$-adic completion.

6.2 Theorem (Drozd-Kirichenko-Roiter [1]): Let $\hat{\Lambda}$ be a Bass-order. For an indecomposable $\hat{\Lambda}$-lattice $\hat{M}$, one of the following cases must occur:

(i) $\mathrm{sig}(M) = (0,\dots,0,2,0,\dots,0)$,

(ii) $\mathrm{sig}(M) = (0,\dots,0,1,0,\dots,0)$,

(iii) $\mathrm{sig}(M) = (0,\dots,0,1,0,\dots,0,1,0,\dots,0)$.

Proof: We recall that the signature of $\hat{M}$, $\mathrm{sig}(\hat{M})$ describes the components of $\hat{K}\hat{M}$ (cf. 5.5). If $\mathrm{ann}_{\hat{\Lambda}}(\hat{K}\hat{M}) = \hat{A}\hat{e}$, for a central idempotent $\hat{e}$ of $\hat{A}$, then $\hat{M}$ is a faithful indecomposable lattice over the Bass-order $\hat{\Lambda}(1-\hat{e})$. It follows from the proof of (5.3) that $\hat{M}$ is a progenerator for the $\hat{R}$-order $\hat{\Lambda}_1 = \{a \in \hat{A}(1-\hat{e}) : a\hat{M} \subset \hat{M}\}$. By (6.1), $\mathrm{End}_{\hat{\Lambda}_1}(\hat{M})$ is a Bass-order, which is completely primary, $\hat{M}$ being indecomposable (cf. VI, 3.2; we recall that a ring is called completely primary if modulo its radical it is a skewfield). Because of Wedderburn's structure theorem, (6.2) follows from the next statement.

6.3 Lemma: If $\hat{\Lambda}$ is a completely primary Bass-order, then one of the following cases must occur:

(i) $\hat{A} = (\hat{D})_2$, $\hat{D}$ a separable skewfield,

(ii) $\hat{A} = \hat{D}$, $\hat{D}$ a separable skewfield,

(iii) $\hat{A} = \hat{D}_1 \oplus \hat{D}_2$, $\hat{D}_i$ separable skewfields, $i=1,2$.

__Proof__: We first establish a lemma which is of interest in itself.

6.4 __Lemma__: Let $\hat{\Lambda}$ be a Bass-order and $\hat{M} \in {}_{\hat{\Lambda}}\underline{\underline{M}}^o$ indecomposable. If $\mathrm{sig}(\hat{\Lambda}) = (\lambda_i)_{1 \le i \le t}$ and $\mathrm{sig}(\hat{M}) = (\alpha_i)_{1 \le i \le t}$, then there exists $n \in \underline{\underline{N}}$ such that $\lambda_i = n\alpha_i$ for all i with $\alpha_i \neq 0$.

__Proof__: Let $\hat{K}\hat{M}$ be a faithful $\hat{A}_1 = \hat{A}/\mathrm{ann}_{\hat{A}}(\hat{K}\hat{M})\hat{A}$-module. Then $\hat{M}$ is a faithful indecomposable $\hat{\Lambda}_1$-module, where $\hat{\Lambda}_1 = \{a \in \hat{A}_1 : a\hat{M} \subset \hat{M}\}$ and thus $\hat{M}$ is a progenerator for ${}_{\hat{\Lambda}_1}\underline{\underline{M}}^o$ (cf. proof of 5.3); i.e., $\hat{\Lambda}_1 \cong \hat{M}^{(m)}$. Then $\hat{A}_1 = \mathrm{End}_{\hat{A}}(\hat{K}\hat{M})_m$, and the statement follows. #

Now we turn to the __proof of (6.3)__: Let

$$\mathrm{sig}(\hat{\Lambda}) = (\lambda_i)_{1 \le i \le t}.$$

(i) Assume that for some i, $\lambda_i \geq 2$, say $\lambda_1 \geq 2$. Then we can find an exact sequence of $\hat{\Lambda}$-lattices (cf. IV, proof of 1.13)

$$0 \longrightarrow \hat{M}' \longrightarrow \hat{\Lambda} \longrightarrow \hat{M}'' \longrightarrow 0$$

with $\mathrm{sig}(\hat{M}') = (1,0,\dots,0)$. Moreover, $\hat{M}''$ is indecomposable by (4.15), $\hat{\Lambda}$ being completely primary. An application of (6.4) shows $\lambda_i = 0$ for $i \geq 2$, and $\lambda_1 = 2$ since the equation $n(\lambda_1 - 1) = 1$ has exactly one positive integral solution $n = 2$, $\lambda_1 = 2$. Thus $\hat{A} = (\hat{D})_2$, $\hat{D}$ a separable skewfield.

(ii) We may therefore assume $\lambda_i = 1$ for all $1 \le i \le t$; i.e., $\hat{A} = \oplus_{i=1}^{t} \hat{D}_i$. If $t \geq 3$ we can find two exact sequences of $\hat{\Lambda}$-lattices

$$0 \longrightarrow \hat{M}' \longrightarrow \hat{\Lambda} \longrightarrow \hat{M}'' \longrightarrow 0, \quad 0 \longrightarrow \hat{N}' \longrightarrow \hat{\Lambda} \longrightarrow \hat{N}'' \longrightarrow 0$$

with $\hat{K}\hat{M}'' \cong \hat{D}_1 \oplus \hat{D}_2$, $\hat{K}\hat{N}'' \cong \hat{D}_2 \oplus \hat{D}_3$. Moreover, $\hat{M}''$ and $\hat{N}''$ are indecomposable by (4.15). We consider

$$\hat{M} = \hat{M}'' \oplus \hat{N}''$$

and apply (5.6), to conclude

$$\hat{M} \cong \hat{I} \oplus \hat{M}_o, \text{ with } \mathrm{sig}(\hat{I}) = (1,1,1,0,\dots,0).$$

But this is a contradiction to the Krull-Schmidt theorem. Thus $t \leq 2$ and either (ii) or (iii) of the theorem must occur. #

6.5 Theorem (Drozd-Kirichenko-Roiter [1]): Let $\hat{\Lambda}$ be a Bass-order which is indecomposable as ring. Then $\hat{\Lambda}$ is of one of the following types:

I.) There exists an indecomposable $\hat{\Lambda}$-lattice $\hat{M}$ with $\mathrm{sig}(\hat{M}) = (1,1,0,\dots,0)$. Then $\hat{A} = \hat{A}_1 \oplus \hat{A}_2$, where $\hat{A}_i$ is a simple $\hat{K}$-algebra, $i = 1,2$. $\hat{\Lambda}$ has signature (λ,λ) and every indecomposable projective $\hat{\Lambda}$-lattice has signature (1,1).

II.) There exists an indecomposable $\hat{\Lambda}$-lattice $\hat{M}$ with $\mathrm{sig}(\hat{M}) = (2,0,\dots,0)$. Then $\hat{A}$ is simple and $\mathrm{sig}(\hat{\Lambda}) = (2\lambda)$. Every indecomposable projective $\hat{\Lambda}$-lattice has signature (2).

III.) Every indecomposable $\hat{\Lambda}$-lattice is irreducible. Then $\hat{A}$ is simple.

Proof: (I) Let $\mathrm{sig}(\hat{M}) = (1,1,0,\dots,0)$, $\mathrm{sig}(\hat{\Lambda}) = (\lambda_i)_{1\leq i\leq t}$. If $\lambda_1 \neq \lambda_2$, we may assume $\lambda_1 > \lambda_2$. Then $\hat{M}^{(\lambda_1)}$ decomposes into a lattice of signature $(\lambda_1,\lambda_2,0,\dots,0)$ and one of signature $(0,\lambda_1-\lambda_2,0,\dots,0)$ by (5.6); a contradiction to the Krull-Schmidt theorem. Thus $\lambda = \lambda_1 = \lambda_2$. If there exists an indecomposable $\hat{\Lambda}$-lattice $\hat{N}$ with $\mathrm{sig}(\hat{N}) = (0,1,0,\dots,0,\underset{i}{1},0,\dots,0)$, then the same argument as above shows $\lambda = \lambda_i$ and the lattice $\hat{N}^{(\lambda)} \oplus \hat{M}^{(\lambda)}$ decomposes into a lattice of signature $(\lambda,\lambda,0,\dots,0,\underset{i}{\lambda},0,\dots,0)$ and one of signature $(0,\lambda,0,\dots,0)$ a contradiction to the Krull-Schmidt theorem. Hence, $t = 2$, since $\hat{\Lambda}$ does not contain non-trivial central idempotents. If $\hat{\Lambda}$ has an indecomposable lattice $\hat{N}$ of type (2,0), then considering the lattice $\hat{M}^{(\lambda-1)} \oplus \hat{N}$ gives a contradiction. Hence $\hat{\Lambda}$ can only have indecomposable lattices of type (1,1), (1,0) or (0,1). Assume that $\hat{P}$ is a projective lattice of type (1,0). This means that $\hat{\Lambda}$ contains an idempotent $\hat{e}$ which is primitive in $\hat{A}$. However, $\hat{M}$ is a faithful indecomposable

$\hat{\Lambda}$-lattice and so it is a progenerator for an order $\hat{\Lambda}_1 \supset \hat{\Lambda}$. Since $\hat{\Lambda}_1$ does not contain idempotents which are primitive in $\hat{A}$, we have obtained a contradiction; i.e., every indecomposable projective $\hat{\Lambda}$-lattice has signature (1,1).

(II) Let $\mathrm{sig}(\hat{M}) = (2,0,\dots,0)$, then λ_1 must be even, say $\lambda_1 = 2\lambda$. If there is an indecomposable $\hat{\Lambda}$-lattice $\hat{N}$ with $\mathrm{sig}(\hat{N}) = (1,1,0,\dots,0)$, then $\hat{M}^{(\lambda-1)2} \oplus \hat{N}^{(3)}$ decomposes into a lattice of signature $(2\lambda, 3)$, a contradiction to the Krull-Schmidt theorem. Hence $\hat{A}$ is simple. The same argument as in the proof of (I) shows that $\hat{\Lambda}$ can not have any irreducible projective lattice. Hence every indecomposable projective lattice must have signature (2).

(III) This is the remaining case (cf. 6.2). #

We <u>remark</u> that the Bass-orders of type III are not necessarily hereditary. In fact, though every indecomposable lattice is irreducible, there may very well exist exact sequence of lattices

$$0 \longrightarrow M' \longrightarrow M \longrightarrow M'' \longrightarrow 0$$

which are not split.

6.6 <u>Lemma</u>: If $\hat{\Lambda}$ is a completely primary Gorenstein-order, then $\hat{M}_1 = \mathrm{Hom}_{\hat{\Lambda}}(\mathrm{rad}\,\hat{\Lambda}, \hat{\Lambda})$ is the unique minimal left and right $\hat{\Lambda}$-ideal in $\hat{A}$ properly containing $\hat{\Lambda}$. If $\hat{\Lambda}$ is not maximal, then $\hat{M}_1$ is an order and $\hat{M}_1 = \Lambda_r(\mathrm{rad}\,\hat{\Lambda}) = \Lambda_1(\mathrm{rad}\,\hat{\Lambda})$.

<u>Proof</u>: It should be observed that for maximal orders the statement is trivial, since completely primary maximal orders are orders in skewfields, and so the result follows from (IV, 5.2).
We therefore assume that $\hat{\Lambda}$ is not maximal. Let $\hat{N} = \mathrm{rad}\,\hat{\Lambda}$; then $\hat{N}$ is the unique maximal left and right $\hat{\Lambda}$-ideal, $\hat{\Lambda}$ being completely primary. Since $\hat{\Lambda}$ is a Gorenstein-order $\hat{\Lambda}^*$ is a progenerator, and because of the Krull-Schmidt theorem we have $\hat{\Lambda} \cong \hat{\Lambda}^*$, $\hat{\Lambda}$ being completely primary. Thus

there exists a regular element $x \in \hat{A}$ such that $\hat{\Lambda}^* = \hat{\Lambda}x$. Obviously, $\hat{N}x$ is the unique maximal submodule of $\hat{\Lambda}^*$ and $\hat{N}x = \hat{N}\hat{\Lambda}^*$. But $\hat{N}\hat{\Lambda}^*$ is a hypercharacteristic submodule of $\hat{\Lambda}^*$. In fact, for $\varphi \in \text{Hom}_{\hat{\Lambda}}(\hat{N}\hat{\Lambda}^*, \hat{\Lambda}^*)$ we have $\varphi' = x\varphi x^{-1} \in \text{Hom}_{\hat{\Lambda}}(\hat{N}, \hat{\Lambda})$ and if $\text{Im}\,\varphi' \not\subset \hat{N}$, then $\text{Im}\,\varphi' + \hat{N} = \hat{\Lambda}$, $\hat{N}$ being maximal. Hence Nakayama's lemma implies $\text{Im}\,\varphi' = \hat{\Lambda}$. But then $\hat{N} \cong \hat{\Lambda}$ and $\hat{\Lambda}$ is hereditary (cf. 2.4). However, then $\hat{A}$ must be a skewfield (cf. § 2) and $\hat{\Lambda}$ is maximal, since every hereditary $\hat{R}$-order in a complete skewfield is maximal (cf. V, 4.10). But this was excluded; i.e., $\text{Im}\,\varphi' \subset \hat{N}$ and consequently $\text{Im}\,\varphi \subset \hat{N}\hat{\Lambda}^*$ and $\hat{N}\hat{\Lambda}^*$ is a hypercharacteristic submodule. With (4.7) we conclude that $\hat{\Lambda}_1 = (\hat{N}\hat{\Lambda}^*)^* = \cong \text{Hom}_{\hat{\Lambda}}(\hat{N}, \hat{\Lambda})$ (cf. 4.4) is the unique minimal $\hat{R}$-order properly containing $\hat{\Lambda}$, since $\hat{N}\hat{\Lambda}^* \neq \hat{\Lambda}^*$. Since $\hat{N}\hat{\Lambda}^*$ is the unique maximal submodule of $\hat{\Lambda}^*$, $\hat{\Lambda}_1$ is at the same time the unique minimal $\hat{\Lambda}$-over-module of $\hat{\Lambda}$.#

6.7 <u>Theorem</u> (Drozd-Kirichenko-Roiter [1]): Let $\hat{\Lambda}$ be a Bass-order which is indecomposable as ring. If there exists a faithful indecomposable projective $\hat{\Lambda}$-lattice $\hat{P}$ such that $\text{End}_{\hat{\Lambda}}(\hat{P}) = \hat{\Omega}$ is not a maximal $\hat{R}$-order, then $\hat{\Lambda} = (\hat{\Omega})_n$ for some n.

<u>Proof</u>: Since $\hat{P}$ is faithful,

$$\hat{\Lambda}_1 = \{a \in \hat{A} : a\hat{P} \subset \hat{P}\}$$

is a Bass-order in $\hat{A}$ containing $\hat{\Lambda}$, and $\hat{P}$ is a progenerator for ${}_{\hat{\Lambda}_1}\underline{\underline{M}}^o$ (cf. proof of 5.3); i.e., $\hat{P}^{(n)} \cong \hat{\Lambda}_1$ and $\hat{\Lambda}_1 \cong (\hat{\Omega})_n$. We may therefore assume $\hat{\Lambda}_1 = (\hat{\Omega})_n \supset \hat{\Lambda}$. We decompose $\hat{\Lambda}$ into indecomposable lattices

$${}_{\hat{\Lambda}}\hat{\Lambda} = \oplus_{i=1}^{t} \hat{\Lambda}\hat{e}_i = \oplus_{i=1}^{t} \hat{P}_i$$

In view of (6.5) all $\hat{\Lambda}\hat{e}_i$ have the same signature. Then

$$\hat{\Omega}_i = \text{End}_{\hat{\Lambda}}(\hat{\Lambda}\hat{e}_i) = \hat{e}_i\hat{\Lambda}\hat{e}_i \text{ and}$$

$\text{End}_{\hat{\Omega}_i}(\hat{\Lambda}\hat{e}_i) = \hat{\Lambda}_1 \supset \hat{\Lambda}$, $\hat{P}_i$ being faithful, $1 \le i \le t$; consequently $t = n$. We then have

$$\hat{\Omega}_i = \hat{e}_i\hat{\Lambda}\hat{e}_i \subset \hat{e}_i\hat{\Lambda}_j\hat{e}_i = \hat{\Omega}_j,$$

since

$$\mathrm{End}_{\hat{\Omega}_i}(\hat{\Lambda}\hat{e}_i) = \oplus_{j=1}^{t} \hat{e}_j\, \mathrm{End}_{\hat{\Omega}_i}(\hat{\Lambda}\hat{e}_i)\hat{e}_j.$$

Thus $\hat{\Omega} = \hat{\Omega}_i$ is the same for all $1 \le i \le t$. Obviously $\hat{\Lambda} = \oplus_{i,j=1}^{n} (\hat{e}_i \hat{\Lambda} \hat{e}_j)$, and we can represent $\hat{\Lambda}$ as a ring of matrices

$$\hat{\Lambda} = (\hat{\Omega}_{ij}) \text{ where } \hat{\Omega}_{ij} = \hat{e}_i \hat{\Lambda} \hat{e}_j;\ 1 \le i,j \le n,$$

as in the proof of (VI, 5.12) it follows that $\hat{\Lambda}$ is uniquely determined by the $\hat{\Omega}$-ideals $\{\hat{\Omega}_{ij}\}_{1 \le i,j \le n}$. Since $\hat{\Lambda} \subset (\hat{\Omega})_n$, all those ideals are integral. But $\hat{P}_1 = \hat{\Lambda}\hat{e}_1$ was indecomposable, and so $\hat{\Omega}$ is a completely primary Bass-order. We shall assume $\hat{\Lambda} \subsetneq (\hat{\Omega})_n$. Then there exists a pair (i,j), $i \neq j$ such that $\hat{\Omega}_{ij} \subset \hat{N}$, where $\hat{N} = \mathrm{rad}\,\hat{\Omega}$. But then also $\hat{\Omega}_{ii}\hat{\Omega}_{ij} \subset \hat{\Omega}_{ij} \subset \hat{N}$, and since $\hat{N}$ is a maximal ideal, $\hat{\Omega}_{ii} \subset \hat{N}$ or $\hat{\Omega}_{ij} \subset \hat{N}$ (observe that maximal ideals are prime). We had assumed $\hat{\Lambda}_1 = \mathrm{End}_{\hat{\Omega}}(\hat{\Lambda}\hat{e}_1) = (\hat{\Omega})_n$; thus $\hat{\Omega}_{ii} = \hat{\Omega}$, $1 \le i \le n$, and consequently $\hat{\Omega}_{ij} \subset \hat{N}$. Let $1 < s$ be the smallest integer such that $\hat{\Omega}_{1s} \subset \hat{N}$. Then $\hat{\Omega}_{1s} \supset \hat{\Omega}_{1k}\hat{\Omega}_{ks}$ implies $\hat{\Omega}_{ks} \subset \hat{N}$ for $k < s$. Moreover, for $1 \le j < s$ we have $\hat{\Omega}_{1j} \supset \hat{\Omega}_{11}\hat{\Omega}_{1j} = \hat{\Omega}$; hence for $1 \le i < s$, $s \le j \le n$ we get

$$\hat{\Omega}_{ij} = \hat{\Omega}_{is}\hat{\Omega}_{sj} \subset \hat{\Omega}_{is} = \hat{N}.$$

Consequently $\hat{\Lambda}$ is contained in the following order

$$\hat{\Lambda} \subset \begin{pmatrix} \hat{\Omega} & \cdots & \hat{\Omega} & & \\ \vdots & & \vdots & \hat{N}_{s-1 \times n-s} & \\ \vdots & & \hat{\Omega} & & \\ \vdots & & \hat{\Omega} & \cdots & \hat{\Omega} \\ \hat{\Omega} & \cdots & \cdots & \cdots & \hat{\Omega} \end{pmatrix} s = \hat{\Lambda}' \subset (\hat{\Omega})_n.$$

But we assume that $\hat{\Omega}$ is not maximal. By (6.6) $\hat{\Omega}$ is contained in a unique minimal over-order $\hat{\Omega}'$ such that $\hat{N}$ is a two-sided $\hat{\Omega}'$-ideal*). Hence

$$\hat{\Lambda}'' = \left(\begin{array}{ccc|c} \hat{\Omega}' & \cdots & \hat{\Omega}' & \hat{N}_{s-1 \times n-s} \\ \vdots & & \vdots & \\ \hline \vdots & & \vdots & \\ \hat{\Omega}' & \cdots & \hat{\Omega}' & \hat{\Omega}_{n-1+1 \times n-s} \end{array}\right)$$

*) Observe that $\hat{\Omega}$ is a Bass-order $(\hat{\Omega})_n$ being one.

is an $\hat{R}$-order containing $\hat{\Lambda}$. Whence it has to be a Bass-order. But in the earlier part of this proof we had seen that for every Bass-order $e_1 \hat{\Lambda}" e_1$ is the same for every i. Thus $\hat{\Lambda}"$ can not be a Bass-order and we have obtained a contradiction; i.e., $\hat{\Lambda} = (\hat{\Omega})_n$. #

6.8 Corollary: If $\hat{\Lambda}$ is a Bass-order of type I,II, indecomposable as ring, then

$$\hat{\Lambda} \cong (\hat{\Omega})_n,$$

where $\hat{\Omega}$ is a completely primary Bass-order of type I or II resp.

Proof: If $\hat{\Lambda}$ is of type I or II, then $\hat{\Lambda}$ has a faithful indecomposable projective lattice $\hat{P}$, which is not irreducible. Hence $\hat{\Omega} = \mathrm{End}_{\hat{\Lambda}}(\hat{P})$ is not maximal, and (6.7) yields the desired result. It should be observed that $\hat{\Omega}$ and $(\hat{\Omega})_n$ are of the same type. #

Remark: In view of (6.8), Bass-orders of type I and II are determined by the class of completely primary Bass-orders of these types.
We now turn to the study of Bass-orders of type III.

6.9 Theorem (Drozd-Kirichenko-Roiter [1]): Let $\hat{\Lambda}$ be a non-hereditary Bass-order of type III, which is indecomposable as ring. Then either $\hat{\Lambda} = (\hat{\Omega}_1)_n$, where $\hat{\Omega}_1$ is a non-maximal Bass-order in the skewfield $\hat{D}$, or $\hat{\Lambda}$ is Morita- equivalent to the order

$$\begin{pmatrix} \hat{\Omega} & \hat{N}^d \\ \hat{\Omega} & \hat{\Omega} \end{pmatrix}$$

where $\hat{\Omega}$ is the maximal $\hat{R}$-order in $\hat{D}$ and $\hat{N} = \mathrm{rad}\, \hat{\Omega}$, with $d \geq 2$; here $\hat{A} = (\hat{D})_n$. Conversely, every such order is a Bass-order of type III.

Proof: Let $\hat{\Lambda} = \oplus_{i=1}^{t} \hat{P}_i$ be the decomposition of $\hat{\Lambda}$ into indecomposable lattices. Then all the $\hat{P}_i$'s are irreducible, $\hat{\Lambda}$ being of type III, and it follows from the proof of (6.7), that $\hat{\Omega}' = \mathrm{End}_{\hat{\Lambda}}(\hat{P}_i)$ is the same for all $1 \leq i \leq t$. If $\hat{\Omega}'$ is not maximal, then we conclude as in the proof of (6.7), that $\hat{\Lambda} = (\hat{\Omega}')_n$. Thus we may assume that $\hat{\Omega}' = \hat{\Omega}$ is the maximal

$\hat{R}$-order in $\hat{D}$.

Also for future use we need the following statement:

6.10 <u>Proposition</u>: Let $\hat{\Lambda}$ be an $\hat{R}$-order in the separable $\hat{K}$-algebra $\hat{A}$. For $\hat{M} \in {}_{\hat{\Lambda}}\underline{\underline{M}}^f$, rad $\hat{M}$ = rad $\hat{\Lambda} \cdot \hat{M}$.

<u>Proof</u>: We recall that rad $\hat{M}$ is the intersection of the maximal left $\hat{\Lambda}$-submodules of $\hat{M}$, and that rad $\hat{\Lambda} \cdot \hat{M} \subset$ rad $\hat{M}$ (cf. I, 4.16); on the other hand, $\hat{M}$/rad $\hat{\Lambda} \cdot \hat{M}$ is the direct sum of simple $\hat{\Lambda}$-modules, and consequently rad($\hat{M}$/rad $\hat{\Lambda} \cdot \hat{M}$) = 0 (cf. III, 5.4). Thus rad $\hat{\Lambda} \cdot \hat{M} \supset$ rad $\hat{M}$; i.e., rad $\hat{\Lambda} \cdot \hat{M}$ = rad $\hat{M}$. #

We now continue with the <u>proof of (6.9)</u>:

Let $\hat{M}$ be a non-projective irreducible $\hat{\Lambda}$-lattice; such a lattice exists, $\hat{\Lambda}$ being non-hereditary. If $\hat{M}$ had only one maximal submodule, then $\hat{M}$/rad$\hat{\Lambda}\hat{M}$ were simple (cf. 6.10) and thus one $\hat{P}_i$ would be a projective cover for $\hat{M}$ (cf. III, 7.4, 7.6); i.e., we had an epimorphism

$$\hat{P}_i \longrightarrow \hat{M} \longrightarrow 0,$$

which would imply $\hat{P}_i \cong \hat{M}$, $\hat{P}_i$ being irreducible; a contradiction; i.e., $\hat{M}$ has at least two maximal submodules, say $\hat{N}_1$ and $\hat{N}_2$. Since $\hat{N}_1 + \hat{N}_2 = \hat{M}$, we have an exact sequence

$$\hat{N}_1 \oplus \hat{N}_2 \longrightarrow \hat{M} \longrightarrow 0;$$

i.e., $\hat{N}_1 \oplus \hat{N}_2 \succ \hat{M}$ (cf. 4.9). Because of (5.4), $\hat{N}_1^* \oplus \hat{N}_2^* \succ M^*$, and we have an exact sequence

$$(\hat{N}_1^* \oplus \hat{N}_2^*)^{(m)} \xrightarrow{\varphi} M^* \longrightarrow 0.$$

Then $\varphi = \oplus_{i=1}^m \varphi_i$, where $\varphi_i : \hat{N}_1^* \oplus \hat{N}_2^* \longrightarrow M$, and $\varphi_i = \psi_i \oplus \chi_i$, where $\psi_i = \varphi_i|_{\hat{N}_1^*}$ and $\chi_i = \varphi_i|_{\hat{N}_2^*}$. Since $\hat{N}_i^* \supset \hat{M}^*$, i=1,2, $\psi_i \in \mathrm{End}_{\hat{\Lambda}}(\hat{N}_1^*) = \hat{\Omega}$ and $\chi_i \in \mathrm{End}_{\hat{\Lambda}}(\hat{N}_2^*) = \hat{\Omega}$. We recall that $\hat{N}_1^*$ is a progenerator for some $\hat{R}$-order $\hat{\Lambda}_1$ containing $\hat{\Lambda}$ and so $\mathrm{End}_{\hat{\Lambda}}(\hat{N}_1^*) = \mathrm{End}_{\hat{\Lambda}_1}(\hat{N}_1^*) = \hat{\Omega}$, $\hat{N}_1^*$ being irreducible, since $\hat{\Omega} = \mathrm{End}_{\hat{\Lambda}}(\hat{P}_i)$ for every indecomposable projective $\hat{\Lambda}$-lattice. Thus Im $\psi_i = \omega_o^{s_i}\hat{N}_1^*$, Im $\chi_i = \omega_o^{t_i}\hat{N}_2^*$, if those maps are different from zero, where $\omega_o\hat{\Omega} = \text{rad}\ \hat{\Omega}$ (cf. IV, §6). (Observe that $\hat{N}_i^*$ are right modules and so maps are written on the left.) Let $s = \min_{\psi_i \neq 0} s_i$,

$t = \min_{\chi_i \neq 0} t_i$. Then $t, s \geq 0$ and

$$\hat{M}^* = \omega_o^s \hat{N}_1^* + \omega_o^t \hat{N}_2^*.$$

Passing to the dual modules, we obtain

$$\hat{M} = \hat{N}_1 \omega_o^{-s} \cap \hat{N}_2 \omega_o^{-t}, \text{ and we must have } s,t \geq 1.$$

However, $\hat{M}\omega_o \subset \text{rad}\ \hat{\Lambda} \cdot \hat{M}$. In fact, if not, there exists a maximal $\hat{\Lambda}$-submodule $\hat{N}$ of $\hat{M}$ such that

$$\hat{N} + \hat{M}\omega_o = \hat{M};\ \text{i.e.,}$$

$\hat{N} \cdot \text{End}_{\hat{\Lambda}}(\hat{M}) = \hat{M}$ by Nakayama's lemma. But $\text{End}_{\hat{\Lambda}}(\hat{N}) = \text{End}_{\hat{\Lambda}}(\hat{M})$. Thus $\hat{N} = \hat{M}$, a contradiction and thus $\hat{M}\omega_o \subset \text{rad}\ \hat{\Lambda} \cdot \hat{M}$. In particular $\hat{N}_1, \hat{N}_2 \supset \hat{M}\omega_o$, and $\hat{N}_1 \cap \hat{N}_2 \supset \hat{M}\omega_o$; i.e., $\hat{N}_1\omega_o^{-1} \cap \hat{N}_2\omega_o^{-1} \supset \hat{M}$. With the relation $\hat{M} = \hat{N}_1\omega_o^{-s} \cap \hat{N}_2\omega_o^{-t}$ we conclude $\hat{N}_1 \cap \hat{N}_2 = \hat{M}\omega_o$, and we have shown that the intersection of any two maximal $\hat{\Lambda}$-submodules of $\hat{M}$ is $\hat{M}\omega_o$, and that $\hat{M}\omega_o = \text{rad}\ \hat{\Lambda} \cdot \hat{M}$. This shows that $\hat{M}/\text{rad}\ \hat{\Lambda} \cdot \hat{M}$ decomposes into the direct sum of two simple $\hat{\Lambda}$-lattices, and we have an exact sequence

$$\hat{P} \oplus \hat{Q} \xrightarrow{\varphi} \hat{M} \longrightarrow 0,$$

where $\hat{P}$ and $\hat{Q}$ are indecomposable projective $\hat{\Lambda}$-lattices. Since $\hat{P} \oplus \hat{Q}$ is a projective cover for $\hat{M}$, one finds readily $\hat{P}\varphi_1 + \hat{Q}\varphi_2 = \hat{M}$ and $\hat{P}\varphi_1 \cap \hat{Q}\varphi_2 = \hat{M}\omega_o$, where $\varphi = \varphi_1 \oplus \varphi_2$. Moreover, $\hat{N}_1 = \hat{P}\varphi_1 + \hat{M}\omega_o$, $\hat{N}_2 = \hat{Q}\varphi_2 + \hat{M}\omega_o$. Let us assume $\hat{P} \cong \hat{Q}$, then $\hat{N}_1/\hat{M}\omega_o \cong \hat{N}_2/\hat{M}\omega_o$; i.e., there exists a map

$$\sigma : \hat{N}_1 \longrightarrow \hat{N}_2 \quad \text{such that}$$

$\hat{N}_1\sigma + \hat{M}\omega_o = \hat{N}_2$, and we have $\hat{N}_1 + \hat{N}_1\sigma + \hat{M}\omega_o = \hat{M}$. By Nakayama's lemma we get $\hat{N}_1 + \hat{N}_1\sigma = \hat{M}$; however $\sigma \in \hat{D}$ and consequently can be written as $u\omega_o^{\varrho}$, where u is a unit in $\hat{\Omega}$ and $\varrho \in \underline{\underline{Z}}$. But then either $\hat{M} = \hat{N}_1$ or $\hat{M} = \hat{N}_1\omega_o^{\varrho}$ if ϱ is negative; yet, none of these cases can happen. Hence $\hat{P} \not\cong \hat{Q}$. At the same time this shows $\hat{N}_1/\hat{M}\omega_o \not\cong \hat{N}_2/\hat{M}\omega_o$. Let $\hat{N}_1/\hat{M}\omega_o = U_1$, $\hat{N}_2/\hat{M}\omega_o = U_2$. From the Jordan-Hölder theorem it follows that for every maximal submodule $\hat{N}$ of $\hat{M}$ we have $\hat{N}/\hat{M}\omega_o \cong U_1$ or $\hat{N}/\hat{M}\omega_o \cong U_2$.

Claim: Given an irreducible $\hat{\Lambda}$-module $\hat{M}_1$, and a maximal submodule $\hat{M}_2$ of $\hat{M}_1$. Then $\hat{M}_1/\hat{M}_2 \cong U_1$ or $\hat{M}_1/\hat{M}_2 \cong U_2$.

Proof: We can embed $\hat{M}_1$ into $\hat{M}$ such that the length of $\hat{M}/\hat{M}_1$ is minimal, say s. We lift a composition series of $\hat{M}/\hat{M}_1$ to a "composition series" between $\hat{M}_1$ and $\hat{M}$

$$\hat{M} \underset{\neq}{\supset} \hat{X}_1 \underset{\neq}{\supset} \hat{X}_2 \underset{\neq}{\supset} \dots \underset{\neq}{\supset} \hat{X}_s \underset{\neq}{\supset} \hat{M}_1 \underset{\neq}{\supset} \hat{M}_2.$$

If s is even, then $\hat{X}_s = \hat{M}\omega_o^{s/2}$, and the result follows from the results established above; if s is odd, then $\hat{M}_1 = \hat{M}\omega_o^{\frac{s+1}{2}}$, and again the statement follows. This proves the claim. In particular it follows from the claim that $\hat{\Lambda}$ has exactly two non-isomorphic projective $\hat{\Lambda}$-lattices, say $\hat{P}_1$, $\hat{P}_2$, and we have a Morita equivalence between $\hat{\Lambda}$ and an order $\hat{\Lambda}_1$ such that $\hat{\Lambda}_1 = \hat{Q}_1 \oplus \hat{Q}_2$, $\hat{Q}_1$ and $\hat{Q}_2$ non-isomorphic irreducible $\hat{\Lambda}_1$-lattices with $\hat{\Omega} = \mathrm{End}_{\hat{\Lambda}_1}(\hat{Q}_1) = \mathrm{End}_{\hat{\Lambda}_1}(\hat{Q}_2)$, and we may assume $\mathrm{End}_{\hat{\Omega}}(\hat{Q}_1) = (\hat{\Omega})_2$. Thus $\hat{\Lambda}_1$ has the form

$$\hat{\Lambda}_1 = \begin{pmatrix} \hat{\Omega} & \hat{I} \\ \hat{\Omega} & \hat{\Omega} \end{pmatrix},$$

where $\hat{I}$ is a two-sided $\hat{\Omega}$-ideal (cf. proof of 6.7). However, $\hat{I} = (\mathrm{rad}\,\hat{\Omega})^d$, $\hat{\Omega}$ being maximal. If $d = 1$, then $\hat{\Lambda}_1$ were hereditary (cf. 2.25) and so $\hat{\Lambda}$ would be hereditary; i.e., $d \geq 2$.
It remains to show that

$$\hat{\Lambda} = \begin{pmatrix} \hat{\Omega} & (\mathrm{rad}\,\hat{\Omega})^d \\ \hat{\Omega} & \hat{\Omega} \end{pmatrix}, \quad d \geq 2,$$

is a Bass-order, where $\hat{\Omega}$ is the maximal $\hat{R}$-order in $\hat{D}$. Let $\hat{\Lambda} = \hat{P} \oplus \hat{Q}$, then $\hat{\Lambda}^* \cong \hat{P}^* \oplus \hat{Q}^*$, and to show that $\hat{\Lambda}^*$ is projective it suffices to prove that both $\hat{P}^*$ and $\hat{Q}^*$ have exactly one maximal submodule; but this is the same as showing that $\hat{P}$ and $\hat{Q}$ have exactly one minimal over-module. We represent $\hat{P}$ and $\hat{Q}$ by means of matrices.

$$\hat{P} = \begin{pmatrix} \hat{\Omega} & 0 \\ \hat{\Omega} & 0 \end{pmatrix}, \quad \hat{Q} = \begin{pmatrix} 0 & (\mathrm{rad}\ \hat{\Omega})^d \\ 0 & \hat{\Omega} \end{pmatrix}, \text{ then}$$

$$\hat{P}_1 = \begin{pmatrix} \hat{\Omega} & 0 \\ (\mathrm{rad}\ \hat{\Omega})^{-1} & 0 \end{pmatrix}, \text{ and } \hat{Q}_1 = \begin{pmatrix} 0 & (\mathrm{rad}\ \hat{\Omega})^{d-1} \\ 0 & \hat{\Omega} \end{pmatrix}$$

are the respective unique minimal over-modules, and thus, $\hat{\Lambda}^*$ is projective. However, $(\hat{P} \oplus \hat{Q}) > \hat{\Lambda}^*$ implies $\hat{P}^* \oplus \hat{Q}^* > \hat{\Lambda}$, and $\hat{\Lambda}^*$ is a generator; i.e., $\hat{\Lambda}$ is a Bass-order, since every over-order of $\hat{\Lambda}$ has the same form. #

In view of (6.8) and (6.9) it remains to clarify the structure of completely primary Bass-orders of type I, II, III and because of (6.5) such orders can only lie in $\hat{D}, (\hat{D})_2$ or $\hat{D}_1 \oplus \hat{D}_2$ where $\hat{D}, \hat{D}_1$ and $\hat{D}_2$ are separable skewfields over $\hat{K}$. We first prove some elementary lemmata:

6.11 <u>Lemma</u>: Let $\hat{\Lambda}$ be a completely primary $\hat{R}$-order in a separable $\hat{K}$-algebra $\hat{A}$ and denote by $\mu_{\hat{\Lambda}}(\hat{M})$ for $\hat{M} \in {}_{\hat{\Lambda}}\underline{M}^f$ the <u>minimal number of generators</u> of $\hat{M}$ as $\hat{\Lambda}$-module. If $\hat{N} = \mathrm{rad}\,\hat{\Lambda}$, then

$$\mu_{\hat{\Lambda}}(\hat{M}) = \dim_{\hat{\Lambda}/\hat{N}}(\hat{M}/\hat{N}\hat{M}).$$

<u>Proof</u>: $\hat{\Lambda}/\hat{N}$ is a skewfield $\bar{S}$, $\hat{\Lambda}$ being completely primary, and we denote by "-" the reduction modulo $\hat{N}$. Since we have an epimorphism $\hat{M} \longrightarrow \bar{M}$, $\mu_{\hat{\Lambda}}(\hat{M}) \geq \mu_{\bar{S}}(\bar{M})$. Conversely, let $\bar{M} = \sum_{i=1}^{n} \bar{S}\, u_i$, where $u_i \in \bar{M}$. We lift the $\{u_i\}_{1\leq i\leq n}$ to elements $\{m_i\}_{1\leq i\leq n}$ in $\hat{M}$. Then $\hat{M} = \sum_{i=1}^{n} \hat{\Lambda} m_i + \hat{N}\hat{M}$, and Nakayama's lemma implies $\hat{M} = \sum_{i=1}^{n} \hat{\Lambda} m_i$; i.e., $\mu_{\hat{\Lambda}}(\hat{M}) \leq \mu_{\bar{S}}(\bar{M})$. #

6.12 <u>Lemma</u>: If $\hat{\Lambda}_1$ is an $\hat{R}$-order containing the completely primary $\hat{R}$-order $\hat{\Lambda}$, then one can always choose 1 as a generator of $\hat{\Lambda}_1$ as $\hat{\Lambda}$-lattice; i.e., $\mu_{\hat{\Lambda}}(\hat{\Lambda}_1) = 1 + \mu_{\hat{\Lambda}}(\hat{\Lambda}_1/\hat{\Lambda})$.

<u>Proof</u>: Let $\hat{\Lambda}_1 = \sum_{i=1}^{n} \hat{\Lambda}\alpha_i$; then $1 = \sum_{i=1}^{n} \lambda_i \alpha_i$. If λ_i is a unit in $\hat{\Lambda}$

for some i, then we can replace α_i by 1. Thus we may assume that $\lambda_i \in \hat{N} = \text{rad}\hat{\Lambda}$ for every i; observe that $\hat{\Lambda}$ is completely primary. But then $1 \in \hat{N}\hat{\Lambda}_1$; i.e., $\hat{\Lambda}_1 = \hat{N}\hat{\Lambda}_1$, a contradiction to Nakayama's lemma. Hence

$$\hat{\Lambda}_1 = \hat{\Lambda} + \sum_{i=2}^{n} \hat{\Lambda}\alpha_i$$

and

$$\mu_{\hat{\Lambda}}(\hat{\Lambda}_1) \geq \mu_{\hat{\Lambda}}(\hat{\Lambda}_1/\hat{\Lambda}) + 1;$$

but obviously,

$$\mu_{\hat{\Lambda}}(\hat{\Lambda}_1) \leq \mu_{\hat{\Lambda}}(\hat{\Lambda}_1/\hat{\Lambda}) + 1.$$

Thus we have equality. #

6.13 Theorem (Drozd-Kirichenko-Roiter [1]): For a completely primary $\hat{R}$-order $\hat{\Lambda}$, the following conditions are equivalent:

(i) $\hat{\Lambda}$ is a Bass-order.

(ii) Every $\hat{\Lambda}$-lattice is the direct sum of (not necessarily full) left $\hat{\Lambda}$-ideals.

(iii) Every $\hat{\Lambda}$-ideal has at most two generators.

Proof: (i) $\Longrightarrow$ (ii). This was proved in (5.6).

(ii) $\Longrightarrow$ (iii). Let $\hat{I}$ be a left $\hat{\Lambda}$-ideal with more than two generators; i.e., $\mu_{\hat{\Lambda}}(\hat{I}) = n \geq 3$. By (6.11) $\hat{I}/\hat{N}\hat{I} \cong \bar{S}^{(n)}$ with $n \geq 3$, where $\hat{N} = \text{rad}\hat{\Lambda}$ and $\bar{S} = \hat{\Lambda}/\hat{N}$. But then we conclude from (III, 7.6) that the projective cover of $\hat{I}$ is $\hat{\Lambda}^{(n)}$, $\hat{\Lambda}$ being indecomposable as lattice, and the exact sequence

$$0 \longrightarrow \hat{X} \xrightarrow{\varphi} \hat{\Lambda}^{(n)} \xrightarrow{\psi} \hat{I} \longrightarrow 0$$

yields the exact sequence

$$0 \longrightarrow \text{Hom}_{\hat{\Lambda}}(\hat{I}, \hat{\Lambda}) \xrightarrow{\psi^*} \hat{\Lambda}^{(n)} \longrightarrow \text{Im}\,\varphi^* \longrightarrow 0$$

of right $\hat{\Lambda}$-modules. However $\text{Hom}_{\hat{\Lambda}}(\hat{I}, \hat{\Lambda})$ is a right ideal, and so Im φ^* can not be an ideal since $n \geq 3$. But then Im φ^* decomposes, say $\text{Im}\,\varphi^* = \hat{X}_1 \oplus \hat{X}_2$, $\hat{X}_i \neq 0$, $i=1,2$ by (ii). (Observe that (ii) is also valid for right $\hat{\Lambda}$-lattices, since $\hat{M}$ is a left ideal if and only if $\text{Hom}_{\hat{R}}(\hat{M},\hat{R})$ is a right ideal, and $\hat{M}$ decomposes if and only if $\text{Hom}_{\hat{R}}(\hat{M},\hat{R})$

decomposes.) Since Im φ^* decomposes, so does its projective cover (cf. 4.15), and the commutative diagram

$$\begin{array}{ccccccc} 0 \longrightarrow \mathrm{Hom}_{\hat\Lambda}(\hat I, \hat\Lambda) & \xrightarrow{\psi^*} & \hat\Lambda^{(n)} & \longrightarrow & \mathrm{Im}\,\varphi^* & \longrightarrow & 0 \\ & & \wr\!\wr & & \wr\!\wr & & \\ & & \hat P_1 \oplus \hat P_2 & \xrightarrow{\alpha\oplus\beta} & \hat X_1 \oplus \hat X_2 & \longrightarrow & 0 \end{array}$$

shows that $\mathrm{Hom}_{\hat\Lambda}(\hat I, \hat\Lambda) = \tilde I_1 \oplus \tilde I_2$ decomposes. If $\tilde I_1 = 0$, then α is an isomorphism, and we have the natural map $\delta_{\hat I} : \hat I \longrightarrow \mathrm{Hom}_{\hat\Lambda}(\mathrm{Hom}_{\hat\Lambda}(\hat I, \hat\Lambda), \hat\Lambda)$; $(x\,\delta_I)\sigma = x\sigma$, $\sigma \in \mathrm{Hom}_{\hat\Lambda}(\hat I, \hat\Lambda)$ (cf. I, 2.11). For a module $\hat Y$ we denote $\mathrm{Hom}_{\hat\Lambda}(\hat Y, \hat\Lambda)$ by $\hat Y^+$. Then we get the commutative diagram

$$\begin{array}{ccccc} 0 \longrightarrow \hat X_1^+ \oplus \hat X_2^+ \longrightarrow & \hat P_1^+ \oplus \hat P_2^+ & \xrightarrow{\gamma} & \tilde I_2^+ & \\ & \wr\!\wr & & \uparrow \delta_{\hat I} & \\ & \hat\Lambda^{(n)} & \xrightarrow{\varphi} & \hat I & \longrightarrow 0, \end{array}$$

with $\gamma : \hat P_1^+ \longrightarrow 0$. However, $\delta_{\hat I}$ is a monomorphism, since there exists $0 \neq r \in \hat R$ such that $r\hat I \subset \hat\Lambda$. Hence we get an epimorphism $\varphi_1 : \hat\Lambda^{(m)} \longrightarrow \hat I$, with $m < n$, a contradiction, since $n = \mu_{\hat\Lambda}(\hat I)$.

Hence $\hat I_1$ and $\hat I_2$ are different from zero. We now distinguish two cases:

(α) If $\hat A = \hat D$ is a skewfield, then we have obtained a contradiction, since $\mathrm{Hom}_{\hat\Lambda}(\hat I, \hat\Lambda)$ is an ideal in $\hat D$, and so it is indecomposable.

(β) If $\hat A = \hat D_1 \oplus \hat D_2$ or $\hat A = (\hat D)_2$, then $\hat I_1$ and $\hat I_2$ are irreducible $\hat\Lambda$-lattices. Since $n \geq 3$, either $\hat P_1$ or $\hat P_2$ is of the form $\hat\Lambda^{(m)}$ with $m \geq 2$, say $\hat P_1 \cong \hat\Lambda^{(m)}$. But then $\hat X_1 \cong \hat P_1/\hat I_1$ is not an ideal, and so it decomposes non-trivially (cf. 5.6), $\hat X_1 = \hat X_1' \oplus \hat X_1''$, and as above one shows that then $\hat I_1$ decomposes non-trivially, a contradiction. (Observe that $\hat P_1$ is the projective cover for $\hat X_1$.)

(iii) $\Longrightarrow$ (i): As in the proof of (6.6), one shows that $\mathrm{Hom}_{\hat\Lambda}({}_{\hat\Lambda}\hat N, \hat\Lambda) = \hat\Lambda_1$ is an overring of $\hat\Lambda$, $\hat N = \mathrm{rad}\,\hat\Lambda$, if $\hat\Lambda$ is not hereditary; but if $\hat\Lambda$ is hereditary, $\hat\Lambda$ is a Bass-order. To show that $\hat\Lambda$ is Gorenstein, we need to prove that $\hat\Lambda^* = \mathrm{Hom}_{\hat R}(\hat\Lambda, \hat R)$ is projective. (We point out, that the following conditions are equivalent: $\hat\Lambda^*$ is a generator and $\hat\Lambda^*$ is projective (cf. proof of 5.2).) In view of the last part of the proof of (6.9), it suffices to show that $\hat{\Lambda}$ has a unique minimal overorder. We shall demonstrate that $\hat{\Lambda}_1$ is this order. $\mu_{\hat{\Lambda}}(\hat{\Lambda}_1) = 2$ by (iii); it can not be equal to 1, since then $\hat{\Lambda}$ would be hereditary. By (6.12) $\mu_{\hat{\Lambda}}(\hat{\Lambda}_1/\hat{\Lambda}) = 1$, and since $\hat{N}\hat{\Lambda}_1 \subset \hat{N}$, $\hat{\Lambda}_1/\hat{\Lambda}$ is isomorphic to the simple $\hat{\Lambda}$-module $U = \hat{\Lambda}/\hat{N}$. This shows that $\hat{\Lambda}_1$ is a minimal overorder of $\hat{\Lambda}$. However,

$$(\hat{N} \cdot \hat{\Lambda}^*)^* \overset{\text{nat}}{\cong} \text{Hom}_{\hat{\Lambda}}(\hat{N}, \hat{\Lambda}) = \hat{\Lambda}_1 \text{ (cf. (4.4))},$$

and by (4.7), $\hat{N} \cdot \hat{\Lambda}^*$ is a maximal submodule of $\hat{\Lambda}^*$. But $\hat{N} \cdot \hat{\Lambda}^* = \text{rad}\, \hat{\Lambda}^*$ (cf. 6.10), and so $\hat{N} \cdot \hat{\Lambda}^*$ is the unique maximal submodule of $\hat{\Lambda}^*$ and $\hat{\Lambda}_1$ is the unique minimal overorder of $\hat{\Lambda}$; i.e., $\hat{\Lambda}$ is Gorenstein. In this way we can continue as long as the overorders are completely primary. Let $\hat{\Lambda}_o$ be an overorder of $\hat{\Lambda}$ that decomposes, $\hat{\Lambda}_o = \hat{\Lambda}_o e_1 \oplus \hat{\Lambda}_o e_2$. But then $\hat{\Lambda}_o^* = e_1 \hat{\Lambda}_o^* \oplus e_2 \hat{\Lambda}_o^*$, and since we may assume $\mu_{\hat{\Lambda}}(\hat{\Lambda}_o^*) = 2$, we have an exact sequence $\hat{\Lambda}^{(2)} \longrightarrow e_1 \hat{\Lambda}_o^* \oplus e_2 \hat{\Lambda}_o^* \longrightarrow 0$, and it follows from the proof of (4.15), that then $\hat{\Lambda}^{(2)}$ decomposes into the projective covers of $e_i \hat{\Lambda}_o^*$, $i=1,2$. This means $\mu_{\hat{\Lambda}}(e_i \hat{\Lambda}_o^*) = 1$ and hence $\mu_{\hat{\Lambda}_o}(\hat{\Lambda}_o^*) = 1$. #

6.14 <u>Theorem</u> (Drozd-Kirichenko-Roiter [1]): If $\hat{\Lambda}$ is a completely primary $\hat{R}$-order in $\hat{D}$ or $\hat{D}_1 \oplus \hat{D}_2$, the following conditions are equivalent:

(i) $\hat{\Lambda}$ is a Bass-order,

(ii) $\hat{\Gamma}/\hat{\Lambda}$ is a cyclic $\hat{\Lambda}$-module, where $\hat{\Gamma}$ is the maximal $\hat{R}$-order in $\hat{A}$ ($\hat{A} = \hat{D}$ or $\hat{A} = \hat{D}_1 \oplus \hat{D}_2$),

(iii) If $\hat{A} = \hat{D}_1 \oplus \hat{D}_2$, then $\hat{\Lambda}$ is a subdirect sum of the maximal $\hat{R}$-orders of $\hat{D}_1$ and $\hat{D}_2$.

<u>Proof</u>: (i) $\Longrightarrow$ (ii). This follows from (6.12) and (6.13).

(ii) $\Longrightarrow$ (i). This we only show in case $\hat{A} = \hat{D}$.

For the other case we show (ii) $\Longrightarrow$ (iii) $\Longrightarrow$ (i).

Let $\hat{\Omega} = \hat{\Gamma}$ be the maximal $\hat{R}$-order in $\hat{A} = \hat{D}$. We shall show that the hypothesis (ii) implies that every $\hat{\Lambda}$-submodule of $\hat{\Omega}/\hat{\Lambda}$ is cyclic. Let

$$\hat{N} = \sum_{i=1}^{s} \hat{\Lambda}\alpha_i, \text{ where } \hat{N} = \text{rad}\,\Lambda,\ \alpha_i \in \hat{\Lambda} \subset \hat{\Omega}.$$

Every element $\alpha \in \hat{\Omega}$ can uniquely be written as $\alpha = u\,\omega_o^{t}$, where u is a unit in $\hat{\Omega}$ and $\omega_o\hat{\Omega} = \text{rad}\,\hat{\Omega}$; in particular, $\alpha_i = u_i\,\omega_o^{t_i}$, $t_i \geq 0$. If t_1 is minimal among the $\{t_i\}_{1\leq i\leq s}$, then

$$\hat{N}\hat{\Omega} = \hat{\Omega}\alpha = \alpha\,\hat{\Omega}, \text{ with } \alpha = \alpha_1 \in \hat{\Lambda},$$

every $\hat{\Omega}$-ideal being two-sided (cf. IV, 5.2). According to (ii), $\hat{\Omega}/\hat{\Lambda}$ is cyclic and this implies $\mu_{\hat{\Lambda}}(\hat{\Omega}) = 2$ (cf. 6.12), say

$$\hat{\Omega} = \hat{\Lambda} + \hat{\Lambda}\beta \text{ and } \hat{N}\hat{\Omega} = \hat{\Omega}\alpha = \hat{\Lambda}\alpha + \hat{\Lambda}\beta\alpha.$$

Thus

$$\hat{X}_i = \hat{\Lambda} + (\hat{N}\hat{\Omega})^i = \hat{\Lambda} + \hat{\Lambda}\alpha^i + \hat{\Lambda}\beta\alpha^i = \hat{\Lambda} + \hat{\Lambda}\beta\alpha^i, \text{ since } \alpha \in \hat{\Lambda}.$$

Hence $\hat{X}_i/\hat{\Lambda}$ is a cyclic $\hat{\Lambda}$-module for every i. Then $\hat{\Omega}/\hat{X}_1 \cong \hat{\Omega}/\hat{\Lambda}/\hat{N}(\hat{\Omega}/\hat{\Lambda})$ and $\hat{\Omega}/\hat{X}_1$ is annihilated by $\hat{N}$. Thus it is cyclic as homomorphic image of a cyclic module. Hence $\hat{X}_1/\hat{\Lambda}$ is the unique maximal submodule of $\hat{\Omega}/\hat{\Lambda}$ (cf. 6.11). Similarly one shows that $\hat{X}_{i+1}/\hat{\Lambda}$ is the unique maximal submodule of $\hat{X}_i/\hat{\Lambda}$. Now, given any $\hat{R}$-order $\hat{\Lambda}_1$ in $\hat{D}$ containing $\hat{\Lambda}$, then the above result shows that $\hat{\Lambda}_1 = \hat{X}_i$ for some i. Then $\mu_{\hat{\Lambda}}(\hat{\Lambda}_1) \leq 2$ and as in the proof of (6.13) we conclude that $\hat{\Lambda}$ is a Bass-order.

(ii) $\Longrightarrow$ (iii). Let $\hat{A} = \hat{D}_1 \oplus \hat{D}_2$, $\hat{\Gamma} = \hat{\Omega}_1 \oplus \hat{\Omega}_2$, where $\hat{\Omega}_i$ is the maximal $\hat{R}$-order in $\hat{D}_i$, i=1,2. Assume that $\hat{\Lambda}$ is a completely primary $\hat{R}$-order in $\hat{A}$, with $\mu_{\hat{\Lambda}}(\hat{\Gamma}) = 2$. By (6.11) $\hat{\Omega}_1/\hat{N}\hat{\Omega}_1$ and $\hat{\Omega}_2/\hat{N}\hat{\Omega}_2$ are simple $\hat{\Lambda}$-modules. Thus $\mu_{\hat{\Lambda}}(\hat{\Omega}_1) = 1$ and $\mu_{\hat{\Lambda}}(\hat{\Omega}_2) = 1$; i.e., $\hat{\Omega}_1 = \hat{\Lambda}e_1$ and $\hat{\Omega}_2 = \hat{\Lambda}e_2$, where e_1 and e_2 are the primitive idempotents in $\hat{A}$ (cf. 6.12). Hence $\hat{\Lambda}$ is a subdirect sum of $\hat{\Omega}_1$ and $\hat{\Omega}_2$.

(iii) $\Longrightarrow$ (i). Since every irreducible $\hat{\Lambda}$-module is either an $\hat{\Omega}_1$- or an $\hat{\Omega}_2$-lattice, $\hat{\Lambda}$ being a subdirect sum of $\hat{\Omega}_1$ and $\hat{\Omega}_2$ (cf. VI, 5.2), there are exactly two non-isomorphic irreducible $\hat{\Lambda}$-lattices, $\hat{\Omega}_1$ and $\hat{\Omega}_2$ (cf. VI, §5). Hence every full left $\hat{\Lambda}$-ideal $\hat{I}$ is an extension of $\hat{\Omega}_i$ by $\hat{\Omega}_j$, i,j = 1,2. If $\hat{I}$ is an extension of $\hat{\Omega}_1$ by $\hat{\Omega}_2$, then

$$\hat{\Omega}_1 = \hat{\Lambda}e_1 \subset \hat{I} \text{ and } \hat{I}/\hat{\Lambda}e_1 \cong \hat{\Lambda}e_2,$$

$\{e_i\}_{i=1,2}$ being the primitive idempotents in $\hat{A}$. Let α be a preimage of e_2 in $\hat{I}$; then

$$\hat{I} = \hat{\Lambda} e_1 + \hat{\Lambda}\alpha,$$

and $\mu_{\hat{\Lambda}}(\hat{I}) \leq 2$. Similarly in the other cases. Hence $\hat{\Lambda}$ is a Bass-order by (6.13). #

6.15 **Theorem**: A completely primary $\hat{R}$-order $\hat{\Lambda}$ in $(\hat{D})_2$, $\hat{D}$ a separable skewfield, is a Bass-order if and only if every irreducible $\hat{\Lambda}$-module is cyclic.

Proof: According to the proof of (6.14, (iii) $\Longrightarrow$ (i)), it suffices to show that for a completely primary Bass-order $\hat{\Lambda}$ in $(\hat{D})_2$, every irreducible lattice is cyclic. Let $\hat{M} \in {}_{\hat{\Lambda}}\underline{\underline{M}}^o$ be irreducible. Then it is a progenerator for its ring of multipliers $\hat{\Lambda}_{\hat{M}} = \{a \in (\hat{D})_2 : a\hat{M} \subset \hat{M}\}$; i.e., $\hat{M} = \hat{\Lambda}_{\hat{M}} e$, where $e \in \hat{\Lambda}_{\hat{M}}$ a primitive idempotent in $(\hat{D})_2$, and it remains to show that for every $\hat{R}$-order $\hat{\Lambda}_1$ in $\hat{A}$ containing $\hat{\Lambda}$, $\mu_{\hat{\Lambda}}(\hat{\Lambda}_1 e) = 1$ for every primitive idempotent of $\hat{e} \in \hat{\Lambda}_1$. According to (6.13), $\mu_{\hat{\Lambda}}(\hat{\Lambda}_1) = 2$ and thus $\hat{\Lambda}_1/\hat{N}\hat{\Lambda}_1 \cong U^{(2)}$, $\hat{N} = \text{rad}\,\hat{\Lambda}$, where $U = \hat{\Lambda}/\hat{N}$ is the simple $\hat{\Lambda}$-module. Thus

$$\hat{\Lambda}_1 e/\hat{N}\hat{\Lambda}_1 e \cong U$$

and (6.11) implies $\mu_{\hat{\Lambda}}(\hat{\Lambda}_1 e) = 1$. Consequently, $\hat{M}$ is cyclic. #

Notation: We turn first to the description of Bass-orders in a separable skewfield $\hat{D}$. For the remainder of this section we shall assume that <u>$\hat{R}/\text{rad}\,\hat{R}$ is a finite field</u>.

6.16 **Lemma**: Let $\hat{\Lambda}$ be a Bass-order in $\hat{D}$. There exists a unique ascending chain of orders

$$\hat{\Lambda} = \hat{\Lambda}_s \subsetneq \hat{\Lambda}_{s-1} \subsetneq \cdots \subsetneq \hat{\Lambda}_o = \hat{\Omega},$$

$\hat{\Lambda}_i = \hat{\Lambda} + (\text{rad}\,\hat{\Lambda})^i\hat{\Omega}$, $i > 0$, where $\hat{\Omega}$ is the maximal $\hat{R}$-order in $\hat{D}$. Moreover $\hat{\Lambda}_i/\hat{\Lambda}_{i+1}$ is isomorphic to the simple $\hat{\Lambda}$-module.

Proof: This follows from the proof of (6.14 (ii) $\Longrightarrow$ (i)). #

6.17 __Lemma:__ With the notation of (6.16), let $\hat{N} = \text{rad}\,\hat{\Lambda}$ and put $F_i = \hat{\Lambda}_i/\hat{N}\hat{\Lambda}_i$, $0 \le i \le s-1$ and $F = \hat{\Lambda}/\hat{N}$. Then F is a field and F_i is a two-dimensional vectorspace over F, $0 \le i \le s-1$. Moreover, as ring $F_i = F[r_i]^{\vee}$, $r_i^2 = 0, 1 \le i \le s-1$ and $F_o = F[r_o]$, where either $r_o^2 = 0$ or F_o is a two-dimensional extension field of F. However, in general F_i is not an F-algebra, since F does not necessarily lie in the center of F_i; still, $Fr_i = r_iF$.

__Proof:__ There are no finite skewfields (cf. III, 6.7) and so $F = \hat{\Lambda}/\hat{N}$ is a field. Moreover, $\hat{N}\hat{\Omega}$ is a two-sided $\hat{\Omega}$-ideal and $\hat{\Lambda}_i\hat{N} = [\hat{\Lambda} + (\hat{N}\hat{\Omega})^i]\hat{N} = \hat{N}\hat{\Lambda}_i$, $i > 0$, is a two-sided $\hat{\Lambda}_i$-ideal. Thus F_i is a ring, $0 \le i \le s-1$. Since it is annihilated by $\hat{N}$, it is an F-module. From (6.6) it follows that $\hat{N}_i = \text{rad}\,\hat{\Lambda}_i$ is a two-sided $\hat{\Lambda}_{i-1}$-ideal, $1 \le i \le s-1$, and that $\hat{\Lambda}_{i-1}$ is the left and right order of $\hat{N}_i$. Since $\hat{\Lambda}_{i-1}$ is a Bass-order, $\hat{N}_i$ is a progenerator for the category of $\hat{\Lambda}_{i-1}$-lattices; i.e., $\hat{N}_i \cong \hat{\Lambda}_{i-1}$. If $\hat{N}_i = \hat{N}_{i-1}$, then $\hat{\Lambda}_{i-1}$ is maximal, since there are no non-maximal hereditary $\hat{R}$-orders in $\hat{D}$. Thus, except possibly for $i = 0$, $\hat{N}_i \ne \hat{N}_{i-1}$. However, $\hat{N}_i = \hat{N} + (\hat{N}\hat{\Omega})^i$ for $i > 0$. In fact,

$$\hat{\Lambda}_i/(\hat{N} + \hat{N}^i\hat{\Omega}) = [\hat{\Lambda} + (\hat{N} + \hat{N}^i\hat{\Omega})]/(\hat{N} + \hat{N}^i\hat{\Omega}) \text{ by } (6.16)$$
$$\cong \hat{\Lambda}/[\hat{\Lambda} \cap (\hat{N} + \hat{N}^i\hat{\Omega})] = \hat{\Lambda}/\hat{N},$$

$\hat{N}$ being the unique maximal ideal in $\hat{\Lambda}$. Thus $\hat{N}_i = \hat{N} + \hat{N}^i\hat{\Omega}$, since we obviously have the inclusion $(\hat{N} + N^i\hat{\Omega}) \subset \hat{N}_i$, $\hat{\Lambda}_i$ being completely primary. On the other hand

$$\mu_{\hat{\Lambda}}(\hat{\Lambda}_i) \le 2 \text{ implies } \dim_F(\hat{\Lambda}_i/\hat{N}\hat{\Lambda}_i) \le 2$$

by (6.11). Hence for $0 \le i \le s-1$, $F_i = \hat{\Lambda}_i/\hat{N}\hat{\Lambda}_i$ is a two-dimensional F-vectorspace, which is a ring, and for $1 \le i \le s-1$, F_i has a one dimensional radical; i.e., $F_i \cong F[r_i]$, $r_i^2 = 0$, $Fr_i = r_iF$, $1 \le i \le s-1$.

For $i = 0$, we can have either $\text{rad}\,\hat{\Omega} = \hat{N}\hat{\Omega}$, in which case F_o is a two-dimensional extension field of F, or $\text{rad}\,\hat{\Omega} \supsetneq \hat{N}\hat{\Omega}$ and $F_o = F[r_o]$, $r_o^2 = 0$, $Fr_o = r_oF$. #

6.18 __Corollary:__ If F_o is a field, then

$$\hat{\Lambda}_i = \hat{\Lambda} + (\text{rad } \hat{\Omega})^i, \ [F_o : F] = 2.$$

Proof: This follows immediately, since rad $\hat{\Omega} = \hat{N}\hat{\Omega}$ in case F_o is a field. #

6.19 **Lemma:** If $F_o = F[r_o]$, $r_o^2 = 0$, then

$$\Lambda_i = \Lambda + (\text{rad } \hat{\Omega})^{2i},$$

and if $\hat{\pi} = \hat{\varepsilon}\,\omega_o^t$ where $\hat{\pi}\hat{R}$ = rad $\hat{R}$ and $\omega_o\hat{\Omega}$ = rad $\hat{\Omega}$ for a unit $\hat{\varepsilon}$ in $\hat{\Omega}$ and for an odd integer t, then $s \leq (t-1)/2$, where s is the number of orders containing $\hat{\Lambda}$.

Proof: If F_o is not a field, then $\hat{N}\hat{\Omega} = (\text{rad } \hat{\Omega})^2 = \omega_o^2\hat{\Omega}$, and for $i > 0$ we have

$$\hat{\Lambda}_i = \hat{\Lambda} + \hat{\Omega}\,\omega_o^{2i}.$$

In addition, it should be observed that

$$\hat{\Omega} = \hat{\Lambda} + \hat{\Omega}\,\omega_o.$$

If now $\hat{\pi} = \hat{\varepsilon}\omega_o^t$ for some unit $\hat{\varepsilon}$ in $\hat{\Omega}$ and an odd integer t, then $\hat{\Omega}\hat{N} = \omega_o^2\hat{\Omega}$ implies the existence of an element $\omega_o^2\hat{\varepsilon}_1 \in \hat{N}$, $\hat{\varepsilon}_1$ a unit in $\hat{\Omega}$. Writing $(t-1) = 2\tau$, we conclude

$$\omega_o^{t-1}\hat{\Omega} = (\omega_o^2\,\hat{\varepsilon}_1)^\tau\hat{\Lambda} + \hat{\Omega}\,\omega_o^t .$$

However, $(\omega_o^2\,\hat{\varepsilon}_1)^\tau\hat{\Lambda} \subset \hat{\Lambda}$ and thus

$$\omega_o^{t-1}\hat{\Omega} \subset \hat{\Lambda} + \hat{\Omega}\,\omega_o^t = \hat{\Lambda} + \hat{\pi}\hat{\Omega} \subset \hat{\Lambda} + \hat{\pi}\hat{\Lambda} + \omega_o\hat{\pi}\hat{\Omega};$$

i.e., $\hat{\Lambda} + \omega_o^{t-1}\hat{\Omega} \subset \hat{\Lambda} + \omega_o^{t+1}\hat{\Omega}$. Moreover,

$$\hat{\Lambda} + \omega_o^{2i}\hat{\Omega} = \hat{\Lambda} + \omega_o^{(2i+1)}\hat{\Omega}.$$

To show this we use induction on i. For $i = 0$, we have $\hat{\Omega} = \hat{\Lambda} + \omega_o\hat{\Omega}$. Assume now

$$\hat{\Lambda} + \omega_o^{2(i-1)}\hat{\Omega} = \hat{\Lambda} + \omega_o^{2i-1}\hat{\Omega}.$$

If $\hat{\Lambda} + \omega_o^{2i}\hat{\Omega} \supsetneq \hat{\Lambda} + \omega_o^{2i+1}\hat{\Omega}$, then

$$\hat{\Lambda} + \omega_0^{2i+1}\hat{\Omega} = \hat{\Lambda} + \omega_0^{2i+2}\hat{\Omega}, \text{ and}$$

$(\hat{\Lambda} + \omega_0^{2i-1}\hat{\Omega})/(\hat{\Lambda} + \omega_0^{2i+1}\hat{\Omega})$ must be isomorphic to $F^{(2)}$ as $\hat{\Lambda}$-module; i.e.,

$$2 = \dim_F [(\hat{\Lambda} + \omega_0^{2i-1}\hat{\Omega})/(\hat{\Lambda} + \omega_0^{2i+1}\hat{\Omega})] =$$

$$= \dim_F [(\hat{\Lambda} + \omega_0^{2i-1}\hat{\Omega})/\hat{\Lambda}/\hat{N}\{(\hat{\Lambda} + \omega_0^{2i-1}\hat{\Omega})/\hat{\Lambda}\}]$$

$$= \mu_{\hat{\Lambda}}[(\hat{\Lambda} + \omega_0^{2i-1}\hat{\Omega})/\hat{\Lambda}] = \mu_{\hat{\Lambda}}[(\hat{\Lambda} + \omega_0^{2(i-1)}\hat{\Omega})/\hat{\Lambda}] = 1,$$

a contradiction.

Returning to the above situation, we therefore have

$$\hat{\Lambda} + \omega_0^{t-1}\hat{\Omega} \subset \hat{\Lambda} + \omega_0^{t+1}\hat{\Omega} = \hat{\Lambda} + \omega_0^{t+2}\hat{\Omega}.$$

Next,

$$\hat{\Lambda} + \omega_0^{t+2}\hat{\Omega} = \hat{\Lambda} + \hat{\pi}\omega_0^{2}\hat{\Omega} \subset \hat{\Lambda} + \omega_0^{3}\hat{\pi}\hat{\Omega} = \hat{\Lambda} + \omega_0^{t+3}\hat{\Omega}.$$

Continuing this way, we conclude for $t > 2s$,

$$\hat{\Lambda} + \omega_0^{t-1}\hat{\Omega} \subset \hat{\Lambda} + \omega_0^{2s}\hat{\Omega} \subset \hat{\Lambda}.$$

Hence $s \leq (t-1)/2$, if t is odd. #

<u>Remark</u>: (6.18) and (6.19) give necessary conditions for the existence of Bass-orders in $\hat{D}$. We shall show next that these conditions are also sufficient for the existence of Bass-orders.

6.20 <u>Lemma</u>: Let $\hat{\Omega}$ be the maximal $\hat{R}$-order in $\hat{D}$ and assume that $\hat{\Omega}/\text{rad}\,\hat{\Omega} = F_0$ contains a subfield F with $(F_0 : F) = 2$. Then there exists a chain of Bass-orders $\hat{\Lambda}_1 = \hat{\Lambda}_1(F)$

$$\hat{\Omega} \supsetneq \hat{\Lambda}_1 \supsetneq \hat{\Lambda}_2 \supsetneq \cdots$$

each of which satisfies (6.16) and (6.17). In addition, $\hat{\Lambda}_{i+1}$ is the minimal $\hat{\Lambda}_i$ - overmodule of $\hat{\Lambda}_i$.

<u>Proof</u>: Let $\varphi : \hat{\Omega} \longrightarrow F_0$ be the canonical epimorphism and consider

$$\hat{\Lambda}_1 = \varphi^{-1}(F).$$

Then $\hat{\Lambda}_1$ is an $\hat{R}$-order in $\hat{D}$ and $\hat{\Omega}/\hat{\Lambda}_1 \cong F$ is a cyclic $\hat{\Lambda}_1$-module, with rad $\hat{\Lambda}_1$ = rad $\hat{\Omega}$. Consequently, $\hat{\Lambda}_1$ is a Bass-order (cf. 6.14). We put $\hat{\Lambda}_1/\text{rad}\,\hat{\Lambda}_1 = F_1 \cong F$. Let $\overline{\Lambda}_1 = \hat{\Lambda}_1/\omega_o\hat{\Lambda}_1$ with $\omega_o\hat{\Omega} = \text{rad}\,\hat{\Omega}$. Then rad $\overline{\Lambda}_1 = \omega_o\hat{\Omega}/\omega_o\hat{\Lambda}_1 \cong \hat{\Omega}/\hat{\Lambda}_1 \cong F$, and as $\hat{\Lambda}_1$-module we have $\overline{\Lambda}_1 = F_1[r]$, $r^2 = 0$, $F_1 r = rF_1$. Let $\varphi_1 : \hat{\Lambda}_1 \longrightarrow \overline{\Lambda}_1$ and put $\hat{\Lambda}_2 = \varphi_1^{-1}(F_1)$, $F_1 \subset \overline{\Lambda}_1$. Then $\hat{\Lambda}_2$ is an $\hat{R}$-order in $\hat{D}$ and $\hat{\Lambda}_1/\hat{\Lambda}_2 \cong F_1$ is a cyclic $\hat{\Lambda}_2$-module. It follows from the proof of (6.13, (iii) $\Longrightarrow$ (i)), that $\hat{\Lambda}_2$ is a Gorenstein-order. According to (6.6), $\hat{\Lambda}_2$ has a unique minimal over-order, which must be $\hat{\Lambda}_1$. Thus $\hat{\Lambda}_2$ is a Bass-order. Continuing this way, we construct a descending chain of Bass-orders with the desired properties. #

6.21 <u>Lemma</u>: Let $\hat{\Omega}$ be the maximal $\hat{R}$-order in $\hat{D}$ and assume $(\text{rad}\,\hat{\Omega})^t = (\text{rad}\,\hat{R})\hat{\Omega}$. For a given prime element $\omega_o \in \hat{\Omega}$ ($\omega_o\hat{\Omega} = \text{rad}\,\hat{\Omega}$), there exists a chain of Bass-orders $\hat{\Lambda}_1(\omega_o) = \hat{\Lambda}_1$

$$\hat{\Omega} \supsetneq \hat{\Lambda}_1 \supsetneq \cdots \supsetneq \hat{\Lambda}_1 \supsetneq \cdots$$

in case t is even and

$$\hat{\Omega} \supsetneq \hat{\Lambda}_1 \supsetneq \cdots \supsetneq \hat{\Lambda}_s, \text{ with } s = (t-1)/2,$$

in case t is odd. These orders satisfy (6.16) and (6.17).

<u>Proof</u>: Let $\omega_o\hat{\Omega} = \text{rad}\,\hat{\Omega}$ and assume $\omega_o^t = \hat{\varepsilon}\hat{\pi}$, $t > 1$ and $\hat{\pi}\hat{R} = \text{rad}\,\hat{R}$, $\hat{\varepsilon}$ a unit in $\hat{\Omega}$. Let $F = \hat{\Omega}/\omega_o\hat{\Omega}$, then $t > 1$ implies that

$$\overline{\Omega} = \hat{\Omega}/\omega_o^2\hat{\Omega}$$

can be considered a finite dimensional $\hat{R}/\hat{\pi}\hat{R}$-algebra. Then $\overline{\Omega} = F + F\alpha$, $\alpha^2 = 0$, $F\alpha = \alpha F$. Let $\hat{\Lambda}_1$ be the preimage of F in $\hat{\Omega}$ under the canonical epimorphism $\hat{\Omega} \longrightarrow \overline{\Omega}$. Then $\hat{\Lambda}_1$ is an $\hat{R}$-order with rad $\hat{\Lambda}_1 = \omega_o^2\hat{\Omega}$ and $\hat{\Omega}/\hat{\Omega}\,\omega_o^2$ is the direct sum of two simple $\hat{\Lambda}_1$-modules, and hence $\hat{\Omega}/\hat{\Lambda}_1$ is a cyclic $\hat{\Lambda}_1$-module (cf. 6.11, 6.12) and $\hat{\Lambda}_1$ is a Bass-order by (6.14).

If t is even or $t > 3$, then $\hat{\varepsilon}^{-1}\omega_o^2\hat{\Lambda}_1 \supset \hat{\pi}\hat{\Lambda}_1$. In fact, for even $t = 2\tau$

$$\hat{\pi}\hat{\Lambda}_1 = \hat{\varepsilon}^{-1}\omega_o^{2\tau}\hat{\Lambda}_1 \subset \hat{\varepsilon}^{-1}\omega_o^2\hat{\Lambda}_1, \text{ since } \omega_o^2 \in \hat{\Lambda}_1.$$

In case of an odd t we observe that $\hat{\Omega}/\hat{\Lambda}_1$ is a simple $\hat{\Lambda}_1$-module. However, $\hat{\Lambda}_1$ is a Bass-order and so $\hat{\Omega}$ is the unique minimal over-order of $\hat{\Lambda}_1$, and consequently $\hat{\Omega} = \hat{\Lambda}_1 + \hat{\Omega}\omega_o$. If now $t \geq 5$, then

$$\hat{\varepsilon}^{-1}\omega_o^2\hat{\Lambda}_1 \supset \omega_o^4\hat{\Omega} = \omega_o^4\hat{\Lambda}_1 + \omega_o^5\hat{\Omega} =$$

$$= \omega_o^4\hat{\Lambda}_1 + \omega_o^5\hat{\Lambda}_1 + \ldots + \hat{\varepsilon}^{-1}\omega_o^t\hat{\Lambda}_1 + \omega_o^{t+1}\hat{\Omega} \supset \hat{\varepsilon}^{-1}\omega_o^t\hat{\Lambda}_1 = \hat{\pi}\hat{\Lambda}_1.$$

Thus $\hat{\varepsilon}^{-1}\omega_o^2\hat{\Lambda}_1 \supset \hat{\pi}\hat{\Lambda}_1$, and we can apply the same construction as above to get an $\hat{R}$-order $\hat{\Lambda}_2$, which is a Gorenstein-order, $\hat{\Lambda}_1/\hat{\Lambda}_2$ being a cyclic $\hat{\Lambda}_2$-module. But $\hat{\Lambda}_1$ is the unique minimal over-order of $\hat{\Lambda}_2$ and so $\hat{\Lambda}_2$ is a Bass-order. If t is even, we obtain this way a descending chain of Bass-orders

$$\hat{\Omega} \underset{\neq}{\supset} \hat{\Lambda}_1 \underset{\neq}{\supset} \ldots \underset{\neq}{\supset} \hat{\Lambda}_i \underset{\neq}{\supset} \ldots .$$

If t is odd, then we only get $\hat{R}$-orders

$$\hat{\Omega} \underset{\neq}{\supset} \hat{\Lambda}_1 \underset{\neq}{\supset} \hat{\Lambda}_2 \underset{\neq}{\supset} \ldots \underset{\neq}{\supset} \hat{\Lambda}_s$$

for $s = (t-1)/2$. #

Remark: This clarifies the structure of Bass-orders in skewfields. Since a completely primary Bass-order in a direct sum of two skewfields is a subdirect sum of the maximal orders, it only remains to characterize the completely primary Bass-orders in $(\hat{D})_2$, $\hat{D}$ a separable skewfield.

6.22 Lemma: If $\hat{\Lambda}$ is a completely primary Bass-order in $(\hat{D})_2$, then a conjugate of $\hat{\Lambda}$ contains $\hat{\Omega}\underline{\underline{E}}$, where $\underline{\underline{E}}$ is the (2x2)-identity matrix.

Proof: Since $\hat{\Lambda}$ is indecomposable it contains a unique minimal over-order $\hat{\Lambda}_1$ (cf. 6.6) (we may assume that $\hat{\Lambda}$ is not maximal); continuing this way, we construct a unique chain of Bass-orders

$$\hat{\Lambda} \subsetneq \hat{\Lambda}_1 \subsetneq \cdots \subsetneq \hat{\Lambda}_k,$$

where $\hat{\Lambda}_1,\ldots,\hat{\Lambda}_{k-1}$ are completely primary and $\hat{\Lambda}_k$ decomposes. If $\hat{M}$ is any irreducible $\hat{\Lambda}$-module, then it is a progenerator over its ring of multipliers, which decomposes; consequently, $\hat{M}$ is an irreducible $\hat{\Lambda}_k$-lattice. Conversely, every irreducible $\hat{\Lambda}_k$-lattice is a $\hat{\Lambda}$-lattice. Since $\hat{\Lambda}$ is a completely primary Bass-order, every irreducible $\hat{\Lambda}$-lattice is cyclic (cf. 6.15) and we claim that $\hat{\Lambda}_k$ has to be hereditary. Let $\hat{M}$ be an irreducible $\hat{\Lambda}$-lattice, then $\hat{M} \cong \hat{\Lambda}\alpha$, for some $\alpha \in (\hat{D})_2$, and we have a non-zero epimorphism

$$\hat{\Lambda}/\mathrm{rad}\,\hat{\Lambda} \longrightarrow \hat{\Lambda}\alpha/(\mathrm{rad}\,\hat{\Lambda})\,\alpha \longrightarrow 0,$$

which shows that $\mathrm{rad}\,\hat{\Lambda} \cdot \hat{M}$ is the unique maximal $\hat{\Lambda}$-submodule of $\hat{M}$. However, $(\mathrm{rad}\,\hat{\Lambda})\alpha$ is also a $\hat{\Lambda}_k$-module, and so, $\hat{M}$ has a unique maximal $\hat{\Lambda}_k$-submodule $\mathrm{rad}\,\hat{\Lambda}_k \cdot \hat{M}$. Hence the projective cover of $\hat{M}$ is an irreducible $\hat{\Lambda}_k$-lattice; i.e., $\hat{\Lambda}_k = \hat{P}_1 \oplus \hat{P}_2$ and we have an epimorphism $\sigma : \hat{P}_1 \longrightarrow \hat{M} \longrightarrow 0$, say. Comparing the dimensions, we find, that σ has to be an isomorphism and every irreducible $\hat{\Lambda}_k$-lattice is projective; i.e., $\hat{\Lambda}_k$ is hereditary. Either $\hat{\Lambda}_k$ is maximal or it is a minimal hereditary $\hat{R}$-order. By (2.22) and (2.24) $\hat{\Lambda}_k$ is conjugate to an order which contains $\underline{\hat{\Omega}\underline{E}}$; extending the conjugation to $\hat{\Lambda}$, we may assume that $\hat{\Lambda}_k$ itself contains $\underline{\hat{\Omega}\underline{E}}$. By induction, we may assume that $\hat{\Lambda}_1$ contains $\underline{\underline{E}\hat{\Omega}}$. Since $\hat{\Lambda}_1$ is the unique minimal overorder of $\hat{\Lambda}$, and since $\hat{\Lambda}$ is indecomposable, $\mathrm{rad}\,\hat{\Lambda}$ is a $\hat{\Lambda}_1$-module, and thus $\underline{\underline{E}\hat{\Omega}}\,\mathrm{rad}\,\hat{\Lambda} \subset \mathrm{rad}\,\hat{\Lambda}$. Now, $\hat{\Lambda}/\mathrm{rad}\,\hat{\Lambda} = U$ is the unique simple $\hat{\Lambda}$-module, and we shall show

$$\underline{\underline{E}\hat{\Omega}} \cdot U \subset U;$$

this obviously would imply $\underline{\hat{\Omega}\underline{E}} \cdot \hat{\Lambda} \subset \hat{\Lambda}$; i.e., $\underline{\underline{E}\hat{\Omega}} \subset \hat{\Lambda}$. Let $\hat{M}$ be an irreducible $\hat{\Lambda}$-lattice. Then $\underline{\hat{\Omega}\underline{E}} \cdot \hat{M} \subset \hat{M}$, on the other hand, since $\hat{M}$ is a cyclic $\hat{\Lambda}$-module, we conclude as above, that $\hat{M}/\mathrm{rad}\hat{\Lambda}\,\hat{M} \cong U$. Since $\underline{\hat{\Omega}\underline{E}}\,\mathrm{rad}\,\hat{\Lambda} \subset \mathrm{rad}\,\hat{\Lambda}$, $\underline{\hat{\Omega}\underline{E}}U \subset U$, which gives the desired result. #

6.23 <u>Theorem</u> (Drozd-Kirichenko [1], Drozd-Kirichenko-Roiter [1]): Let $\hat{D}$ be a skewfield with maximal order $\hat{\Omega}$. The Bass-orders in $(\hat{D})_2$ are precisely those orders $\hat{\Lambda}$ such that no conjugate of $\hat{\Lambda}$ is contained in

$$\hat{\Lambda}_o = \left\{ \begin{pmatrix} \alpha + \omega_o \beta & \omega_o^2 \delta \\ \gamma & \alpha \end{pmatrix}, \alpha, \beta, \gamma, \delta \in \hat{\Omega} \right\} \quad \omega_o \hat{\Omega} = \mathrm{rad}\, \hat{\Omega}.$$

The <u>proof</u> is rather computational, and we refer to the paper of Drozd-Kirichenko [1] (cf. Ex. 6,1). #

<u>Exercise §6</u>:

1.) Let $\hat{R}$ with quotient field $\hat{K}$ be a complete Dedekind domain such that $\hat{R}/\mathrm{rad}\,\hat{R}$ is a finite field. $\hat{D}$ is a separable finite dimensional skewfield over $\hat{K}$, and $\hat{\Omega}$ is the maximal $\hat{R}$-order in $\hat{D}$. We shall classify the Bass-orders in $(\hat{D})_2$ following Drozd-Kirichenko [1]. Let $\omega_o \in \hat{\Omega}$ be such that $\mathrm{rad}\,\hat{\Omega} = \omega_o \hat{\Omega}$. We have seen that every Bass-order in $(\hat{D})_2$ contains $\hat{\Omega}\underline{\underline{E}}_2$ (cf. 6.22), and thus can be considered as $\hat{\Omega}$-lattice.

Show:

(1) <u>Lemma</u>: If $\hat{\Lambda}$ is an $\hat{R}$-order in $(\hat{D})_2$, which is also an $\hat{\Omega}$-lattice, then $\hat{\Lambda}$ is conjugate to an order with $\hat{\Omega}$-basis

$$\begin{pmatrix} \alpha & 0 \\ 0 & 0 \end{pmatrix}, \begin{pmatrix} 1 & 0 \\ 0 & 1 \end{pmatrix}, \begin{pmatrix} 0 & \beta \\ 0 & 0 \end{pmatrix}, \begin{pmatrix} \gamma & \delta \\ 1 & 0 \end{pmatrix}$$

where $\alpha = \omega_o^k$, $\beta = \omega_o^l$, $\gamma = \omega_o^r$, $\delta = \varepsilon \omega_o^m$ with ε a unit in $\hat{\Omega}$ or $\varepsilon = 0$. Moreover, the $\hat{\Omega}$-module spanned by these matrices is a ring if and only if $0 \leq k \leq l \leq k + m$.

(<u>Hint</u>: In $\hat{\Lambda}$ we always can find an $\hat{\Omega}$-basis

$$(1) \qquad \begin{pmatrix} \alpha & 0 \\ 0 & 0 \end{pmatrix}, \begin{pmatrix} \beta_1 & 0 \\ 0 & \beta_2 \end{pmatrix}, \begin{pmatrix} \gamma_1 & \gamma_3 \\ 0 & \gamma_2 \end{pmatrix}, \begin{pmatrix} \delta_1 & \delta_3 \\ \delta_4 & \delta_2 \end{pmatrix}.$$

Show that in this basis, the elements $\alpha, \beta_1, \beta_2, \gamma_1, \gamma_2$ lie in $\hat{\Omega}$. Hence we can assume $\beta_1 = \beta_2 = 1$ and $\gamma_2 = 0$.

Among the orders conjugate to $\hat{\Lambda}$ choose one where $\alpha = \omega_o^k$ and k minimal. But then γ_1 can be made zero, and transformation with

$\begin{pmatrix} \tau_3^{-1} & 0 \\ 0 & 1 \end{pmatrix}$ yields a ring with a basis

$$\begin{pmatrix} \alpha & 0 \\ 0 & 0 \end{pmatrix}, \begin{pmatrix} 1 & 0 \\ 0 & 1 \end{pmatrix}, \begin{pmatrix} 0 & 1 \\ 0 & 0 \end{pmatrix}, \begin{pmatrix} \delta_1 & \delta_3 \\ \delta_4 & \delta_2 \end{pmatrix} .$$

If we put $\delta_4 \delta_3 = \delta$ and $\delta_1 = \gamma$ and $\delta_4 = \beta$; then the transformation with $\begin{pmatrix} 1 & 0 \\ 0 & \beta^{-1} \end{pmatrix}$ yields the desired basis. The condition for $\hat{\Lambda}$ to be a ring is easily verified.)

(ii) The $\hat{R}$-order $\hat{\Lambda}_o$ with the basis

$$\begin{pmatrix} 1 & 0 \\ 0 & 1 \end{pmatrix}, \begin{pmatrix} \omega_o & 0 \\ 0 & 0 \end{pmatrix}, \begin{pmatrix} 0 & \omega_o^2 \\ 0 & 0 \end{pmatrix}, \begin{pmatrix} 0 & 0 \\ 1 & 0 \end{pmatrix}$$

is not Gorenstein. In particular, no order which is conjugately contained in $\hat{\Lambda}_o$ can be Bass. (<u>Hint</u>: Find an irreducible $\hat{\Lambda}_o$-module which is not cyclic.)

(iii) Let for $\underline{\underline{a}} = \begin{pmatrix} a_1 & a_2 \\ a_3 & a_4 \end{pmatrix} \varepsilon (\hat{D})_2$, $tr_{\hat{D}}(\underline{\underline{a}}) = a_1 + a_4$ and put

$$Tr(\underline{\underline{a}}) = Tr_{D/K}(tr_D(\underline{\underline{a}})).$$

<u>Show</u> that for any $\hat{R}$-lattice $\hat{M}$,

$\hat{M}^* = Hom_{\hat{R}}(\hat{M},\hat{R}) \cong \{m^* \varepsilon (\hat{D})_2, Tr(m \cdot m^*) \varepsilon \hat{R}$ for every $m \varepsilon \hat{M}\}$.

(iv) If now $\hat{\Lambda}$ is an $\hat{R}$-order in $(\hat{D})_2$ with the basis as in (i) show that $\hat{\Lambda}^*$ has the basis

$$\begin{pmatrix} 0 & 0 \\ 0 & 1 \end{pmatrix}, \begin{pmatrix} \omega_o^{-k} & -\omega_o^{r-k} \\ 0 & -\omega_o^{-k} \end{pmatrix}, \begin{pmatrix} 0 & -\varepsilon\omega_o^{m-1} \\ \omega_o^{-1} & 0 \end{pmatrix}, \begin{pmatrix} 0 & 1 \\ 0 & 0 \end{pmatrix}.$$

Here we have already used the isomorphism $\hat{\Omega} \cong \hat{\Omega}^*$, since $\hat{\Omega}$ is Bass.

(v) Let now $\hat{\Lambda}$ be an $\hat{R}$-order in $(\hat{D})_2$ with the basis in (i) which is not contained in $\hat{\Lambda}_o$. Then $\hat{\Lambda} \cong \hat{\Lambda}^*$. (<u>Hint</u>: The orders not contained in $\hat{\Lambda}_o$ can be classified according to their bases:

a.) $\begin{pmatrix}1 & 0\\ 0 & 1\end{pmatrix}, \begin{pmatrix}1 & 0\\ 0 & 0\end{pmatrix}, \begin{pmatrix}0 & \omega_o^l\\ 0 & 0\end{pmatrix}, \begin{pmatrix}0 & 0\\ 1 & 0\end{pmatrix}, l \geq 0$

b.) $\begin{pmatrix}1 & 0\\ 0 & 1\end{pmatrix}, \begin{pmatrix}\omega_o & 0\\ 0 & 0\end{pmatrix}, \begin{pmatrix}0 & \omega_o\\ 0 & 0\end{pmatrix}, \begin{pmatrix}0 & 0\\ 1 & 0\end{pmatrix}$

c.) $\begin{pmatrix}1 & 0\\ 0 & 1\end{pmatrix}, \begin{pmatrix}\omega_o^k & 0\\ 0 & 0\end{pmatrix}, \begin{pmatrix}0 & \omega_o^k\\ 0 & 0\end{pmatrix}, \begin{pmatrix}\omega_o^r & \varepsilon\\ 1 & 0\end{pmatrix}, k \geq 1, r \geq 0,$

d.) $\begin{pmatrix}1 & 0\\ 0 & 1\end{pmatrix}, \begin{pmatrix}\omega_o^k & 0\\ 0 & 0\end{pmatrix}, \begin{pmatrix}0 & \omega_o^k\\ 0 & 0\end{pmatrix}, \begin{pmatrix}\omega_o^r & \varepsilon\omega_o\\ 1 & 0\end{pmatrix}, k \geq 1, r \geq 1,$

e.) $\begin{pmatrix}1 & 0\\ 0 & 1\end{pmatrix}, \begin{pmatrix}\omega_o^k & 0\\ 0 & 0\end{pmatrix}, \begin{pmatrix}0 & \omega_o^{k+1}\\ 0 & 0\end{pmatrix}, \begin{pmatrix}\omega_o^r & \varepsilon\omega_o\\ 1 & 0\end{pmatrix}, k \geq 1, r \geq 1.$

There is one more class; but this can be transformed into the class (a). Now check that for these classes, $\hat{\Lambda} \cong \hat{\Lambda}^*$.)

(vi) Classify all Bass-orders in $(\hat{D})_2$. (Show that all overrings of the above rings also belong to the classes a.)... e.).)

CHAPTER X

THE NUMBER OF INDECOMPOSABLE LATTICES OVER ORDERS

§1 Orders with an infinite number of non-isomorphic indecomposable lattices

The problem of the finiteness of the number of non-isomorphic indecomposable Λ-lattices, $n(\Lambda)$, is reduced to the case where Λ is an order over a complete Dedekind domain. The main theorem states that for an order $\hat{\Lambda}$ in a direct sum of complete skewfields $n(\hat{\Lambda}) = \infty$ if $\mu_{\hat{\Lambda}}(\hat{\Gamma}/\hat{\Lambda}) \geq 3$ or if $\mu_{\hat{\Lambda}}(\mathrm{rad}_{\hat{\Lambda}}(\hat{\Gamma}/\hat{\Lambda})) \geq 2$, where $\mu_{\hat{\Lambda}}(X)$ denotes the minimal number of generators of X as $\hat{\Lambda}$-module and $\hat{\Gamma}$ is the unique maximal $\hat{R}$-order in the underlying algebra. Here we have to assume that $\hat{R}$ has a finite residue class field. The proof of the converse of this theorem will take up the remainder of this chapter.

Let K be an $\underline{\underline{A}}$-field (cf. VI, 3.10) with Dedekind domain R, and Λ an R-order in the separable finite dimensional K-algebra A. By $n(\Lambda)$ we denote the number of non-isomorphic indecomposable Λ-lattices, and for $M \in {}_{\Lambda}\underline{\underline{M}}^f$, we write $\mu_{\Lambda}(M)$ for the minimal number of generators of M as Λ-module.

The following theorem of Jones localizes the question of the finiteness of $n(\Lambda)$.

1.1 Theorem (Jones [1]): $n(\Lambda)$ is finite if and only if for every maximal ideal $\underline{\underline{p}}$ of R, dividing the Higman ideal $\underline{\underline{H}}(\Lambda)$, $n(\hat{\Lambda}_{\underline{\underline{p}}})$ is finite.

Proof: Assume that $n(\Lambda) < \infty$. Then the ranks of all indecomposable Λ-lattices are bounded, say by n_o. If $\hat{N} \in {}_{\hat{\Lambda}_{\underline{\underline{p}}}}\underline{\underline{M}}^o$, then we choose a $\hat{\Lambda}_{\underline{\underline{p}}}$-lattice $\hat{X}$, such that there exists $M \in {}_{\Lambda}\underline{\underline{M}}^o$ with $\hat{M}_{\underline{\underline{p}}} \cong \hat{N} \oplus \hat{X}$ (cf. IV, 1.8). If the $\hat{R}$-rank of $\hat{N}$ is larger that n_o, then M decomposes.

$M = \oplus_{i=1}^{s} Y_i$ with rank $Y_i \leq n_o$. Hence $\hat{N} \oplus \hat{X}$ decomposes into modules each of rank $\leq n_o$; by the Krull-Schmidt theorem, $\hat{N}$ decomposes.
The Jordan-Zassenhaus theorem (cf. VI, 3.5, 3.8) now ensures $n(\hat{\Lambda}_{\underline{p}}) < \infty$.
Conversely, assume $n(\hat{\Lambda}_{\underline{p}}) < \infty$ for every $\underline{p} \mid \underline{H}(\Lambda)$, and let $\{\underline{p}_i\}_{1 \leq i \leq s}$ be all maximal ideals that divide $\underline{H}(\Lambda)$ and let $\{\hat{N}_{ij}\}_{1 \leq j \leq s_i}$ be representatives of the non-isomorphic indecomposable $\hat{\Lambda}_{\underline{p}_i}$-lattices. Given $M \in {}_{\Lambda}\underline{M}^{o}$, we have $\hat{M}_{\underline{p}_i} \cong \oplus_{j=1}^{s_i} \hat{N}_{ij}^{(\alpha_{ij}(M))}$. Because of the Krull-Schmidt theorem $\hat{M}_{\underline{p}_i}$ is uniquely determined by $\{\alpha_{ij}(M)\}_{1 \leq j \leq s_i}$, and we have a map

$$\sigma : {}_{\Lambda}\underline{M}^{o} \longrightarrow \underline{Z}^{(\sum_{i=1}^{s} s_i)},$$

$$M \longmapsto (\alpha_{ij}(M))_{\substack{1 \leq i \leq s \\ 1 \leq j \leq s_i}}.$$

As in the proof of (VII, 4.3) one shows that $\operatorname{Im}\sigma$ has only finitely many minimal elements, and it follows from (VIII, 3.8) that M decomposes if $\sigma(M)$ is not minimal.
Thus $n(\Lambda) < \infty$. #

1.2 Lemma: Assume that $\Lambda = \oplus_{i=1}^{n} \Lambda_i$, where Λ_i are R-orders. Then $n(\Lambda) < \infty$ if and only if $n(\Lambda_i) < \infty$, $1 \leq i \leq n$.

Proof: This is clear since an indecomposable Λ-lattice is an indecomposable Λ_i-lattice for some i. #

We now assume that $\hat{R}$ is the completion of R at some maximal ideal $\underline{p}$ of R and that the $\hat{R}$-order $\hat{\Lambda}$ is indecomposable as ring.

1.3 Lemma (Dade [1], Drozd-Roiter [1]): Let $\hat{\Lambda}_1$ be an $\hat{R}$-order containing $\hat{\Lambda}$. Assume that $\hat{I}$ is a full two-sided $\hat{\Lambda}_1$-ideal contained in $\operatorname{rad}\hat{\Lambda}$, and denote $\hat{\Lambda}_1/\hat{I}$ by S. Assume that for every positive integer n, there exists a left $\hat{\Lambda}$-submodule V_n of $S^{(n)}$ satisfying

(i) $SV_n = S^{(n)}$,

(ii) whenever $\overline{\varphi} \in \mathrm{End}_{(\hat{\Lambda}_1, \hat{\Lambda})}(S^{(n)})$ (this indicates $\overline{\varphi} \in \mathrm{End}_{\hat{\Lambda}_1}(S^{(n)})$ is such that $\overline{\varphi}|_{V_n} : V_n \longrightarrow V_n$) is idempotent, then $\overline{\varphi} = 0$ or $\overline{\varphi} = 1$; i.e., $\mathrm{End}_{(\hat{\Lambda}_1, \hat{\Lambda})}(S^{(n)})$ contains only trivial idempotents.

Then $n(\hat{\Lambda}) = \infty$.*)

<u>Proof</u>: Let $\varrho_n : \hat{\Lambda}_1^{(n)} \longrightarrow S^{(n)}$ be the canonical epimorphism and put

$$M_n = (V_n)\varrho_n^{-1}.$$

Then M_n is a $\hat{\Lambda}$-lattice, and (i) implies $\hat{\Lambda}_1 M_n = \hat{\Lambda}_1^{(n)}$. Hence the rank of M_n tends to infinity as $n \longrightarrow \infty$. In particular $M_n \not\cong M_m$ for $m \neq n$. To show $n(\hat{\Lambda}) = \infty$ it suffices to demonstrate that M_n is indecomposable; i.e., $\mathrm{End}_{\hat{\Lambda}}(M_n)$ contains only trivial idempotents. Let $\varphi \in \mathrm{End}_{\hat{\Lambda}}(M_n)$ be idempotent. Then φ induces a homomorphism $\tilde{\varphi} : \hat{\Lambda}_1 M_n = \hat{\Lambda}_1^{(n)} \longrightarrow \hat{\Lambda}_1 M_n = \hat{\Lambda}_1^{(n)}$, and reduction modulo $\hat{I}$ gives an idempotent element $\overline{\varphi} \in \mathrm{End}_{(\hat{\Lambda}_1, \hat{\Lambda})}(S^{(n)})$, which is trivial by (ii); i.e., $\overline{\varphi} = 0$ or $\overline{\varphi} = 1$. If $\overline{\varphi} = 0$, then

$$\tilde{\varphi} : \Lambda_1^{(n)} \longrightarrow \hat{I}\hat{\Lambda}_1^{(n)};$$

but $\tilde{\varphi}$ is idempotent and $\hat{I} \subset \mathrm{rad}\,\hat{\Lambda}$. Thus $\tilde{\varphi} = 0$. If $\overline{\varphi} = 1$, then $\hat{\Lambda}_1^{(n)} = \hat{I}\hat{\Lambda}_1^{(n)} + \mathrm{Im}\,\tilde{\varphi}$ and $\tilde{\varphi} = 1$ by Nakayama's lemma; i.e., M_n is indecomposable and $n(\hat{\Lambda}) = \infty$. #

1.4 <u>Remark</u>: Assume that $\hat{\Lambda}$ is completely primary; i.e., $\hat{\Lambda}/\hat{N} = \underline{\underline{k}}$ is a finite field, where $\hat{N} = \mathrm{rad}\,\hat{\Lambda}$. If $\hat{\Lambda}_1$ is an $\hat{R}$-order in $\hat{A}$ containing $\hat{\Lambda}$ such that $\hat{N}\hat{\Lambda}_1$ is a two-sided $\hat{\Lambda}_1$-ideal, then $S = \hat{\Lambda}_1/\hat{N}\hat{\Lambda}_1$ is a ring and a two-sided $\underline{\underline{k}}$-vectorspace. In order to apply (1.3), we set up V_n in the following form

1.5 $$V_n = \{x_1 + y_1\alpha, x_2 + y_2\alpha + y_1\beta, \dots, x_n + y_n\alpha + y_{n-1}\beta\},$$

where α and β are fixed elements in S and $\{x_i, y_i\}_{1 \le i \le n}$ are arbitrary

---

*) This lemma is valid for any Dedekind domain $\underset{\sim}{R}$ with quotient field K. However, in the sequel we shall assume that $\hat{R}$ has a finite residue class field.

elements in $\underline{\underline{k}}$. Then V_n is a $\hat{\Lambda}$-module, and $SV_n = S^{(n)}$.

Note, that though S is a two-sided $\underline{\underline{k}}$-module, the elements of S do not necessarily commute with the elements of $\underline{\underline{k}}$; i.e., S is not a $\underline{\underline{k}}$-algebra.

Thus, in particular, x_i and y_i do not necessarily commute with α and β.

We assume now that $1, \alpha, \beta$ are linearly independent over $\underline{\underline{k}}$ from the left. Then a left $\underline{\underline{k}}$-basis of V_n is given by

$$e_i = (0,\dots,0,\underset{i}{1},0,\dots,0), 1 \leq i \leq n,$$

and

$$f_i = \alpha e_i + \beta e_{i+1}, 1 \leq i \leq n, \; e_{n+1} = 0.$$

Every element $\varphi \in \mathrm{End}_S(S^{(n)})$ can be represented as an $(n \times n)$ matrix with entries in S, say $\varphi = (\varphi_{ij})$ where $\varphi_{ij} \in S$.

If $\varphi \in \mathrm{End}_{(\hat{\Lambda}_1, \hat{\Lambda})}(S^{(n)})$, then we must have the relations

$\varphi_{ij} = x_{ij} + y_{ij}\alpha + y_{i,j-1}\beta, 1 \leq i,j \leq n, \; y_{i0} = 0, \; x_{ij}, y_{ij} \in \underline{\underline{k}}$, since $(e_i)\varphi \in V_n$.

The conditions $(f_i)\varphi \in V_n$ give rise to a system of linear equations

$$\alpha\varphi_{ij} + \beta\varphi_{i+1,j} = a_{ij} + b_{ij}\alpha + b_{i,j-1}\beta, i,j=1,\dots,n \; b_{i_0} = 0;$$

$$a_{ij}, b_{ij} \in \underline{\underline{k}}.$$

1.6 $$\alpha x_{ij} + \alpha y_{ij}\alpha + \alpha y_{i,j-1}\beta + \beta x_{i+1,j} + \beta y_{i+1,j}\alpha + \beta y_{i+1,j-1}\beta =$$

$$= a_{ij} + b_{ij}\alpha + b_{i,j-1}\beta, \; i,j=1,\dots,n, x_{n+1,j} = y_{n+1,j} = 0.$$

Hence to apply (1.4) we have to show that for an idempotent φ, the system (1.6) has only the trivial solutions $\varphi = 0$ or $\varphi = 1$. For this one has to compute the products.

$\alpha x, \alpha x \alpha, \alpha x \beta, \beta x, \beta x \beta, \beta x \alpha$, for $x \in \underline{\underline{k}}$;

and this computation is in general quite complicated.

1.7 <u>Theorem</u> (Dade [1]): Let $\hat{\Lambda}$ be a completely primary $\hat{R}$-order in $\hat{A}$. Assume that in the decomposition $\hat{A} = \oplus_{i=1}^{s} \hat{A}_i$ of $\hat{A}$ into simple algebras, $s \geq 4$.

Then $n(\hat{\Lambda}) = \infty$.

<u>Proof</u>: Let $\{e_i\}_{1 \leq i \leq s}$ be the corresponding central idempotents, and put $\hat{\Lambda}_1 = \oplus_{i=1}^{s} \hat{\Lambda} e_i$. Then $\hat{N}\hat{\Lambda}e_i = \text{rad}\,\hat{\Lambda}e_i$, where $\hat{N} = \text{rad}\,\hat{\Lambda}$ and $\hat{\Lambda}/\hat{N} = \hat{\Lambda}e_i/\text{rad}\,\hat{\Lambda}e_i$. In fact, the epimorphism $\hat{\Lambda} \to \hat{\Lambda}e_i$ shows $\hat{N}e_i \subset \text{rad}\,\hat{\Lambda}e_i$ (cf. I, Ex. 4,5). On the other hand, we have an epimorphism $\sigma : \hat{\Lambda}/\hat{N} \to \hat{\Lambda}e_i/\hat{N}e_i$; but $\hat{\Lambda}/\hat{N} = \underline{\underline{k}}$ is a field, and so σ must be an isomorphism; i.e., $\text{rad}\,\hat{\Lambda}e_i = \hat{N}e_i$. Hence $\hat{\Lambda}_1/\text{rad}\,\hat{\Lambda}_1 \cong \underline{\underline{k}}^{(s)}$ (as ring). We let $\{\bar{e}_i\}_{1 \leq i \leq s}$ be the corresponding idempotents in $S_1 = \hat{\Lambda}_1/\text{rad}\,\hat{\Lambda}_1$, and for the construction of V_n (cf. 1.5) we take $\alpha = e_1 + e_2$, $\beta = e_2 + e_3$. Then $1, \alpha, \beta$ are linearly independent over $\underline{\underline{k}}$ and $\alpha\beta$ is linearly independent of $1, \alpha, \beta$. Moreover, it should be observed that $1, \alpha, \beta, \alpha\beta$ commute with all elements in $\underline{\underline{k}}$ and $\alpha^2 = \alpha$, $\beta^2 = \beta$, $\alpha\beta = \beta\alpha$. Considering the system (1.6) we obtain

(i) $x_{ij} + y_{ij} = b_{ij}$,

(ii) $a_{ij} = 0$,

(iii) $y_{i,j-1} = -y_{i+1,j}$,

(iv) $x_{i+1,j} + y_{i+1,j-1} = b_{i,j-1}$.

Thus, $y_{nj} = 0, 1 \leq j \leq n-1$ and $x_{nj} = b_{nj}, 1 \leq j \leq n-1$. But $b_{nj} = 0, 1 \leq j \leq n-1$ implies $x_{nj} = 0, 1 \leq j \leq n-1$. And so $x_{nn} = 1$ or $x_{nn} = 0$ since $\varphi^2 = \varphi$.

Now

$$x_{n-1,n-1} + y_{n-1,n-1} = b_{n-1,n-1} \text{ and } x_{n,n} + y_{n,n-1} = b_{n-1,n-1};$$

i.e.,

$$x_{n-1,n-1} + y_{n-1,n-1} = x_{nn}.$$

If $x_{nn} = 1$, then $y_{nn} = 0$ since $1 \cdot \alpha = \alpha$. Consequently, $y_{n-1,n-1} = 0$ and $x_{n-1,n-1} = 1$. If $x_{nn} = 0$, then $x_{n-1,n-1} = -y_{n-1,n-1} = y_{nn}$. But $y_{nn} = 1$ or $y_{nn} = 0$. Since $(x_{n-1,n-1})^2 = x_{n-1,n-1}$, we must have $x_{n-1,n-1} = 0$. Continuing this way, we conclude that

$$(x_{ij}) = \begin{pmatrix} * & \cdots & * \\ & \ddots & \vdots \\ 0 & & * \end{pmatrix} ,$$

where the diagonal entries are either all 1 or all 0. Since $(x_{ij})^2 = (x_{ij})$, we conclude $(x_{ij}) = \underline{\underline{E}}$ or $(x_{ij}) = \underline{\underline{0}}$. However, $\varphi^2 = \varphi$ and so $\varphi = (x_{ij})$ if $(x_{ij}) = \underline{\underline{E}}$. If $(x_{ij}) = \underline{\underline{0}}$, then $(y_{ij})^2 = (y_{ij})$, $(y_{i,j-1})^2 = (y_{i,j-1})$ and $(y_{ij})(y_{i,j-1}) = 0$. If $(y_{ij}) \neq 0$, then $y_{ii} = 1, 1 \leq i \leq n$ and $(y_{i,j-1}) = 0$; i.e., $(y_{ij}) = 0$, a contradiction. Thus $(y_{ij}) = 0$ and hence $(y_{i,j-1}) = 0$. We have therefore shown that for $\varphi^2 = \varphi$, we must have $\varphi = 1$ or $\varphi = 0$; i.e., (1.3) implies $n(\hat{\Lambda}) = \infty$. #

1.8 <u>Lemma</u>: Under the hypotheses of (1.4) assume that one of the following cases occurs:

(i) $1, \alpha, \alpha^2, \beta$ are linearly independent from the left over $\underline{\underline{k}}$, and for every $x \in \underline{\underline{k}}$, $\alpha x \alpha$ is independent from the left of $1, \alpha, \beta$; αx is independent from the left of $1, \alpha x \alpha, \beta$.

(ii) $1, \alpha, \beta$ are linearly independent over $\underline{\underline{k}}$ from the left and for every $x \in \underline{\underline{k}}$, $\alpha x \alpha = \beta x \alpha = \alpha x \beta = \beta x \beta = 0$; moreover αx is independent of $1, \beta$ from the left, βx is independent of $1, \alpha$ from the left.

(iii) $1, \alpha, \beta, \beta\alpha$ are linearly independent over $\underline{\underline{k}}$ from the left; α and $\alpha\beta$ are independent over $\underline{\underline{k}}$ from the right; $1, \alpha\beta$ are independent over $\underline{\underline{k}}$ from the right and from the left, and for every $x \in \underline{\underline{k}}$, $\alpha^2 = \alpha x \alpha = 0, x\beta = \beta x$, moreover, αx is independent from the left of $1, \beta$; *) and β^2 is independent of $\alpha, \beta\alpha, \alpha\beta$ from the left.

Then $n(\hat{\Lambda}) = \infty$.

*) This last condition follows from the indepence of α and β .

<u>Proof</u>: In each of these cases, if $\varphi^2 = \varphi$ is a matrix satisfying (1.6), we have to show $\varphi = 1$ or $\varphi = 0$. The three cases have to be treated separately:

(i) For $i = n$ the system (1.6) yields

$$1.9 \qquad \alpha x_{nj} + \alpha y_{nj}\alpha + \alpha y_{n,j-1}\beta = a_{nj} + b_{nj}\alpha + b_{n,j-1}\beta ;$$

hence $0 = y_{n1} = y_{n2} = \dots = y_{nn}$ and thus

$$\alpha x_{nj} = b_{nj}\alpha,\ a_{nj} = 0 \text{ and } b_{nj} = 0, 1 \le j \le n-1.$$

Hence $x_{nj} = 0, 1 \le j \le n-1$, and $\varphi^2 = \varphi$ implies $x_{nn} = 1$ or $x_{nn} = 0$.

For $i = n-1$, (1.6) implies $0 = y_{n-1,1} = y_{n-1,2} = \dots = y_{n-1,n}$ and thus

$$\alpha x_{n-1,j} + \beta x_{n,j} = a_{n-1,j} + b_{n-1,j}\alpha + b_{n-1,j-1}\beta \text{, and this}$$

implies $x_{n-1,j} = 0$ for $1 \le j \le n-2$ and $x_{n-1,n-1} = 1$ or $x_{n-1,n-1} = 0$,

$\beta x_{nn} = x_{nn}\beta = b_{n-1,n-1}\beta$ and

$$\alpha x_{n-1,n-1} = x_{n-1,n-1}\alpha = b_{n-1,n-1}\alpha ;\ \text{i.e.,}$$

$x_{nn} = x_{n-1,n-1}$.

Continuing this way, we obtain φ in the form

$$\varphi = \begin{pmatrix} 0 & & * \\ & \ddots & \\ 0 & & 0 \end{pmatrix} \quad \text{or } \varphi = \begin{pmatrix} 1 & & * \\ & \ddots & \\ 0 & & 1 \end{pmatrix} .$$

and $\varphi^2 = \varphi$ implies $\varphi = 1$ or $\varphi = 0$.

(ii) In the second case, we obtain from (1.9) for $i = n$: $b_{n,j} = 0$, $1 \le j \le n-1$, and so $x_{nj} = 0$ for $1 \le j \le n-1$, $x_{nn} = 1$ or $x_{nn} = 0$, the y_{nj} are arbitrary.

For $i = n-1$, we get $b_{n-1,j} = 0, 1 \le j \le n-2$; hence $x_{n-1,j} = 0$ for $1 \le j \le n-2$ and $\alpha x_{n-1,n-1} = b_{n-1,n-1}\alpha$; $\beta x_{nn} = x_{nn}\beta = b_{n-1,n-1}\beta$; i.e.,

$x_{n-1,n-1} = x_{nn}$ and $y_{n-1,j}$ is arbitrary, $1 \leq j \leq n$. Thus

$$\varphi = \begin{pmatrix} x & & * \\ & \ddots & \\ 0 & & x \end{pmatrix} + \varphi', \text{ where } x = 0 \text{ or } x = 1$$

and φ' has entries in $\underline{\underline{k}}\alpha + \underline{\underline{k}}\beta$. Now $\varphi^2 = \varphi$ implies $\varphi = 1$ or $\varphi = 0$.

(iii) In the third case we get from (1.9) for $i = n$: $b_{n,j} = 0, 1 \leq j \leq n-1$, and $a_{nj} = 0$, since $\alpha\beta$ and 1 are independent from the right. Hence

$$\alpha(x_{nj} + y_{n,j-1}\beta) = b_{nj}\alpha, 1 \leq j \leq n;$$

since α and $\alpha\beta$ are independent from the right and since β commutes with $\underline{\underline{k}}$, we get $x_{nj} = 0, 1 \leq j \leq n-1$ and $y_{n,j} = 0, 1 \leq j \leq n-2$. Hence

1.10 $\alpha(x_{nn} + y_{n,n-1}\beta) = b_{nn}\alpha$ and y_{nn} is arbitrary.

For $i = n-1$, we get from (1.6)

$$\alpha x_{n-1,j} + \alpha y_{n-1,j-1}\beta + x_{n,j}\beta + y_{n,j}\alpha + y_{n,j-1}\beta^2 = a_{n-1,j} + b_{n-1,j}\alpha + b_{n-1,j-1}\beta.$$

This implies

$$x_{nj}\beta + y_{n,j-1}\beta^2 = b_{n-1,j-1}\beta; \text{ i.e.,}$$

$b_{n-1,j} = 0, 1 \leq j \leq n-2$ and $\alpha(x_{n-1,j} + y_{n-1,j-1}\beta) = 0$, $1 \leq j \leq n-2$. Hence $x_{n-1,j} = 0 = y_{n-1,j-1}, 1 \leq j \leq n-2$, and

1.11 $\alpha(x_{n-1,n-1} + y_{n-1,n-2}\beta) = (b_{n-1,n-1} - \beta y_{n,n-1})\alpha$.

Hence φ has the form

$$\varphi = \begin{pmatrix} & * & \\ \hline *\ldots* & x_{n-1,n-1} + y_{n-1,n-1}\alpha + y_{n-1,n-2}\beta & * \\ 0\ldots0 & y_{n,n-1}\alpha & x_{nn} + y_{nn}\alpha + y_{n,n-1}\beta \end{pmatrix}.$$

Since $\varphi^2 = \varphi$ we get, comparing the (n,n-1)-position:

$$y_{n,n-1}\alpha x_{n-1,n-1} + y_{n,n-1}\alpha y_{n-1,n-2}\beta + x_{nn}y_{n,n-1}\alpha + y_{n,n-1}\beta\alpha = y_{n,n-1}\alpha .$$

Assuming $y_{n,n-1} \neq 0$, we get

$$\alpha(x_{n-1,n-1} + y_{n-1,n-2}\beta) + (x_{n,n} + y_{n,n-1}\beta)\alpha = \alpha.$$

Using (1.11) we conclude

$$(b_{n-1,n-1} - \beta y_{n,n-1} + x_{nn} + y_{n,n-1}\beta)\alpha = \alpha;$$

i.e., $b_{n-1,n-1} + x_{nn} = 0$. From (1.10) we obtain $y_{n,n-1}\beta + 2x_{nn} = 0$. Since 1 and β are independent over $\underline{k}$, we conclude $y_{n,n-1} = 0$, a contradiction to our assumption. Hence $y_{n,n-1} = 0$, and we get

$$b_{nj} = 0, 1 \leq j \leq n-1; \; y_{nj} = 0, 1 \leq j \leq n-1, \; x_{nj} = 0, 1 \leq j \leq n-1.$$

$\alpha x_{nn} = b_{nn}\alpha$, y_{nn} is arbitrary. The condtion $\varphi^2 = \varphi$ implies $x_{nn} = 1$ or $x_{nn} = 0$ and $y_{nn} = 0$. From (1.10) we get for $j = n$, $x_{nn} = b_{n-1,n-1}$.

Continuing this process we get $\varphi = 1$ or $\varphi = 0$. #

Before we state the main theorem of this section, we shall fix some notation.

$\hat{A} = \oplus_{i=1}^{s} \hat{D}_i$, is the direct sum of complete separable skewfields of finite dimension over $\hat{K}$, $\hat{D}_i = \hat{A}e_i$, $1 \leq i \leq s$,

$\hat{\Lambda}$ is a completely primary $\hat{R}$-order in $\hat{A}$,

$\hat{\Gamma}$ is the unique maximal $\hat{R}$-order in $\hat{A}$ (cf. IV, 5.2),

$\hat{N} = \mathrm{rad}\,\hat{\Lambda}$,

$\hat{\Lambda}/\hat{N} = \underline{k}$ is a finite field.

1.12 Theorem (Drozd-Roiter [1], Roggenkamp [9]): If $\mu_{\hat{\Lambda}}(\hat{\Gamma}/\hat{\Lambda}) \geq 3$ or if $\mu_{\hat{\Lambda}}(\mathrm{rad}_{\hat{\Lambda}}(\hat{\Gamma}/\hat{\Lambda})) \geq 2$, then $n(\hat{\Lambda}) = \infty$.

Proof: We may assume $s \leq 3$ (cf. 1.7). Let $\hat{\Gamma} = \oplus_{i=1}^{s} \hat{\Omega}_i$, where $\hat{\Omega}_i$ is the maximal $\hat{R}$-order in $\hat{D}_1$, and $\omega_1\hat{\Omega}_1$ = rad $\hat{\Omega}_1$. We need some auxiliary results:

1.13 $$\mathrm{rad}_{\hat{\Lambda}}(\hat{\Gamma}/\hat{\Lambda}) \cong \hat{\Lambda}_o/\hat{\Lambda}, \text{ where } \hat{\Lambda}_o = \hat{\Lambda} + \hat{N}\hat{\Gamma}.$$

In fact, by (IX, 6.10), $\mathrm{rad}_{\hat{\Lambda}}(\hat{\Gamma}/\hat{\Lambda}) = \hat{N}(\hat{\Gamma}/\hat{\Lambda}) = (\hat{\Lambda} + \hat{N}\hat{\Gamma})/\hat{\Lambda} = \hat{\Lambda}_o/\hat{\Lambda}$. #

1.14 If $\hat{\Lambda}_1$ is an $\hat{R}$-order containing $\hat{\Lambda}$, then rad $\hat{\Lambda}_1 = \hat{\Lambda}_1 \cap \mathrm{rad}\,\hat{\Gamma}$; in particular rad $\hat{\Lambda}_1 \supset$ rad $\hat{\Lambda}$.

To <u>prove</u> this, we observe that $\hat{\Gamma}\cdot \mathrm{rad}\,\hat{\Lambda}_1 e_i \subset \mathrm{rad}\,\hat{\Omega}_i$, $1 \le i \le s$, as follows from Nakayama's lemma, and since rad $\hat{\Omega}_i$ is the unique maximal $\hat{\Omega}_i$-ideal (cf. IV, 5.2). This shows rad $\hat{\Lambda}_1 \subset \hat{\Lambda}_1 \cap \mathrm{rad}\,\hat{\Gamma}$.
<u>Conversely</u>, let $\hat{X} = \hat{\Lambda}_1 \cap \mathrm{rad}\,\hat{\Gamma}$. If $\hat{\Lambda}_1$ is indecomposable, then rad $\hat{\Lambda}_1$ is the unique maximal ideal and $\hat{X} \subset \mathrm{rad}\,\hat{\Lambda}_1$, and so $\hat{X} = \mathrm{rad}\,\hat{\Lambda}_1$. If $\hat{\Lambda}_1$ decomposes, say $\hat{\Lambda}_1 = \hat{\Lambda}' \oplus \hat{\Lambda}''$, then $\hat{\Lambda}' \cdot \mathrm{rad}\,\hat{\Lambda}_1 = \mathrm{rad}\,\hat{\Lambda}'$ (cf. proof of 1.7), and the statement is true for each completely primary summand of $\hat{\Lambda}_1$; hence it is true for $\hat{\Lambda}_1$. #

1.15 $$\mathrm{rad}\,\hat{\Lambda}_o = \hat{N}\hat{\Gamma}, \text{ where } \hat{\Lambda}_o = \hat{\Lambda} + \hat{\Gamma}\hat{N}.$$

$\hat{\Lambda}_o/\hat{\Gamma}\hat{N} \cong \hat{\Lambda}/(\hat{\Lambda} \cap \hat{\Gamma}\hat{N}) = \hat{\Lambda}/\hat{N} = \underline{\underline{k}}$ and so rad $\hat{\Lambda}_o \subset \hat{\Gamma}\hat{N}$. On the other hand, (1.14) implies rad $\hat{\Lambda}_o = \hat{\Lambda}_o \cap \mathrm{rad}\,\hat{\Gamma} \supset \hat{\Lambda}_o \cap \hat{\Gamma}\hat{N} = \hat{N}\hat{\Gamma}$, and so rad $\hat{\Lambda}_o = \hat{N}\hat{\Gamma}$. #

1.16 $$\mu_{\hat{\Lambda}}(\hat{\Gamma}/\hat{\Lambda}) = \tau \text{ implies } \mu_{\hat{\Lambda}}(\mathrm{rad}_{\hat{\Lambda}}(\hat{\Gamma}/\hat{\Lambda})) \le \tau.$$

To prove this, we observe that $\mu_{\hat{\Lambda}}(\hat{\Gamma}/\hat{\Lambda}) = \tau$ means $\mu_{\hat{\Lambda}}(\hat{\Gamma}) = \tau + 1$ (cf. IX, 6.12), and by (1.13), $\mu_{\hat{\Lambda}}(\mathrm{rad}_{\hat{\Lambda}}(\hat{\Gamma}/\hat{\Lambda})) = \mu_{\hat{\Lambda}}(\hat{\Lambda}_o/\hat{\Lambda})$, $\hat{\Lambda}_o = \hat{\Lambda} + \hat{\Gamma}\hat{N}$. Hence we have to show

$$\mu_{\hat{\Lambda}}(\hat{\Gamma}) = t \text{ implies } \mu_{\hat{\Lambda}}(\hat{\Lambda}_o) \le t.$$

If $\mu_{\hat{\Lambda}}(\hat{\Lambda}_o) = t'$, then $\dim_{\underline{\underline{k}}}(\hat{\Lambda}_o/\hat{N}\hat{\Lambda}_o) = t'$ (cf. IX, 6.11); i.e., $\dim_{\underline{\underline{k}}}(\hat{N}\hat{\Gamma}/\hat{N}\hat{\Lambda}_o) = t' - 1$ (cf. 1.15). However $\hat{N}\hat{\Gamma}/\hat{N}\hat{\Lambda}_o = \hat{N}\hat{\Gamma}/(\hat{N} + (\hat{N}\hat{\Gamma})^2)$ is a homomorphic image of $\hat{N}\hat{\Gamma}/(\hat{N}\hat{\Gamma})^2 \cong \hat{\Gamma}/\hat{N}\hat{\Gamma}$. Thus

$t = \mu_{\hat\Lambda}(\hat N\hat\Gamma/(\hat N\hat\Gamma)^2) \geq \mu_{\hat\Lambda}(\hat N\hat\Gamma/(\hat N + (\hat N\hat\Gamma)^2))$.
If we had equality, then $\hat N\hat\Gamma/(\hat N\hat\Gamma)^2 \underset{\hat\Lambda}{\cong} \hat N\hat\Gamma/(\hat N + (\hat N\hat\Gamma)^2)$; i.e., $\hat N \subset (\hat N\hat\Gamma)^2$, hence $\hat\Gamma\hat N \subset (\hat\Gamma\hat N)^2$, a contradiction to Nakayama's lemma. Thus,
$t' - 1 = \mu_{\hat\Lambda}(\hat N\hat\Gamma/(\hat N + (\hat N\hat\Gamma)^2)) \leq t - 1$; i.e.,

$$\mu_{\hat\Lambda}(\hat\Lambda_o) \leq t. \quad \#$$

In order to show $n(\hat\Lambda) = \infty$, we shall apply (1.3). Because of (1.16) the following cases can occur

1.17 (i) $\mu_{\hat\Lambda}(\hat\Gamma/\hat\Lambda) \geq 3$ or
(ii) $\mu_{\hat\Lambda}(\hat\Gamma/\hat\Lambda) = 2$ and $\mu_{\hat\Lambda}(\text{rad}_{\hat\Lambda}(\hat\Gamma/\hat\Lambda) = 2$.

In (i) we shall choose $S = \hat\Gamma/\hat N\hat\Gamma$ and in (ii) we take $S = \hat\Lambda_o/\hat N\hat\Lambda_o$ (cf. 1.3).
Then $\hat N\hat\Lambda_o$ is a two-sided $\hat\Lambda_o$-ideal and we can apply (1.4), and we shall construct elements $1, \alpha, \beta$ of S such that (1.8) is applicable. As pointed out earlier, we have to know the products αx and βx for $x \in \underline{\underline{k}}$.
Altogether, there are 18 cases to be treated. We recall that $s \leq 3$.
We first treat the cases where (1.17,i) occurs.

1.18 If s = 3, then $\hat N\hat\Gamma = \oplus_{i=1}^{3} \omega_i^{s_i}\hat\Omega_i$, $s_i > 0, 1 \leq i \leq 3$. Then $\underline{\underline{k}}_i = \hat\Omega_i/\omega_i\hat\Omega_i$ are finite dimensional extension fields of $\underline{\underline{k}}$, since $\hat N\underline{\underline{k}}_i = 0$. We put $S = \hat\Gamma/\hat N\hat\Gamma = \oplus_{i=1}^{3} \hat\Omega_i/\omega_i^{s_i}\hat\Omega_i$ and

$$\hat\Omega_i/\omega_i^{s_i}\hat\Omega_i = \underline{\underline{k}}_i + \underline{\underline{k}}_i\overline{\omega}_i + \ldots + \underline{\underline{k}}_i\overline{\omega}_i^{s_i-1}, \overline{\omega}_i^{s_i} = 0, 1 \leq i \leq 3.$$

1.19 If s = 2, then $\hat N\hat\Gamma = \hat\Omega_1\omega_1^{s_1} \oplus \hat\Omega_2\omega_2^{s_2}$, $s_i > 0$, i=1,2, and we put $S = \hat\Gamma/\hat N\hat\Gamma = \oplus_{i=1}^{2} \hat\Omega_i/\omega_i^{s_i}\hat\Omega_i. \hat\Omega_i/\omega_i\hat\Omega_i = \underline{\underline{k}}_i$ are extension fields of $\underline{\underline{k}}$, i=1,2 and

$$\hat\Omega_i/\omega_i^{s_i}\hat\Omega_i = \underline{\underline{k}}_i + \underline{\underline{k}}_i\overline{\omega}_i + \ldots + \underline{\underline{k}}_i\overline{\omega}_i^{s_i-1}, \overline{\omega}_i^{s_i} = 0, i=1,2.$$

1.20 If s = 1, then $\hat N\hat\Gamma = \hat\Omega_1\omega_1^{s_1}$, $s_1 > 0$ and we put $S = \hat\Omega_1/\hat N\hat\Omega_1 = \hat\Omega_1/\omega_1^{s_1}\hat\Omega_1$.
$\underline{\underline{k}}_1 = \hat\Omega_1/\omega_1\hat\Omega_1$ is an extension field of $\underline{\underline{k}}$ and

$$\hat{\Omega}_1/\omega_1^{s_1}\hat{\Omega}_1 = \underline{\underline{k}}_1 + \underline{\underline{k}}_1\overline{\omega}_1 + \dots + \underline{\underline{k}}_1\overline{\omega}_1^{s_1-1}, \overline{\omega}_1^{s_1} = 0.$$

Since (1.17,1) holds, we have $\dim_{\underline{\underline{k}}}(S) \geq 4$ in all three cases (cf. IX, 6.11, 6.12). Moreover:

1.) In (1.18) we assume $\hat{N}\hat{\Gamma} = \mathrm{rad}\,\hat{\Gamma}$. Then S is commutative and we may assume $(\underline{\underline{k}}_1 : \underline{\underline{k}}) > 1$, say $\underline{\underline{k}}_1 = \underline{\underline{k}}(\delta)$. Then $1, \alpha = \delta e_1, \alpha^2 = \delta^2 e_1$ and $\beta = e_2$ are linearly independent over $\underline{\underline{k}}$ and they commute with the elements in $\underline{\underline{k}}$. Hence $n(\hat{\Lambda}) = \infty$ by (1.8,1).

2.) In (1.19) we assume $\hat{N}\hat{\Gamma} = \mathrm{rad}\,\hat{\Gamma}$. Then S is commutative and we may assume $(\underline{\underline{k}}_1 : \underline{\underline{k}}) > 1$, say $\underline{\underline{k}}_1 = k(\delta)$. Then $1, \alpha = \delta e_1$ and $\alpha^2 = \delta^2 e_1$ are linearly independent over $\underline{\underline{k}}$. Since $\dim_{\underline{\underline{k}}}(S) \geq 4$, there exists $\beta \in S$ which is independent of $1, \alpha, \alpha^2$. Moreover, $1, \alpha, \alpha^2, \beta$ commute with the elements in $\underline{\underline{k}}$. Hence $n(\hat{\Lambda}) = \infty$ by (1.8,1).

3.) In (1.20) we assume $\hat{N}\hat{\Gamma} = \mathrm{rad}\,\hat{\Gamma}$. Then S is commutative and $(\underline{\underline{k}}_1 : \underline{\underline{k}}) \geq 4$, say $\underline{\underline{k}}_1 = \underline{\underline{k}}(\delta)$. Then $1, \alpha = \delta, \alpha^2$, and $\beta = \alpha^3$ are linearly independent over $\underline{\underline{k}}$ and they commute with the elements in $\underline{\underline{k}}$. Hence $n(\hat{\Lambda}) = \infty$ by (1.8,1).

Thus we may assume rad $S \neq 0$.
If for some i we have $s_i \geq 3$ (cf. 1.18, 1.19, 1.20), then:

4,5.) If $s_1 \geq 3$, then in (1.18) and (1.19) $1, \alpha = \overline{\omega}_1 e_1, \alpha^2 = \overline{\omega}_1^2 e_1$, and $\beta = e_2$ are linearly independent over $\underline{\underline{k}}$. Moreover, $\alpha x \beta = \beta x \alpha = 0$ for every $x \in \underline{\underline{k}}$, and $\alpha x \alpha = x'\alpha^2, \alpha x = x''\alpha$, $x, x', x'' \in \underline{\underline{k}}$. In addition, α^2

is independent of $1, \alpha, \beta$. Hence $n(\hat{\Lambda}) = \infty$ by (1.8,i).

6.) If $s_1 \geq 4$ in (1.20), then $1, \alpha = \overline{\omega}_1, \alpha^2 = \overline{\omega}_1^2, \beta = \overline{\omega}_1^3$ are linearly independent over $\underline{\underline{k}}$, and $\alpha x \alpha = x'\alpha^2, \beta x = x''\beta, \alpha x = x'''\alpha$, $x, x', x'', x''' \in \underline{\underline{k}}$; moreover, $x\alpha^2$ is independent of $1, \alpha, \beta$. Hence $n(\hat{\Lambda}) = \infty$ by (1.8,i).

7.) If $s_1 = 3$ in (1.20), then $(\underline{\underline{k}}_1 : \underline{\underline{k}}) > 1$, say $\underline{\underline{k}}_1 = \underline{\underline{k}}(\delta)$, and $1, \alpha = \overline{\omega}_1, \alpha^2 = \overline{\omega}_1^2, \beta = \delta$ are linearly independent over $\underline{\underline{k}}$, and $\alpha x \alpha = x'\alpha^2, \alpha x \beta = x''\alpha, \alpha x = x'''\alpha, x, x' \in \underline{\underline{k}}, x'', x''' \in \underline{\underline{k}}_1$. Moreover, $x\alpha^2$ is independent of $1, \alpha, \beta$. Hence $n(\hat{\Lambda}) = \infty$ by (1.8,i).

We thus may assume $s_i \leq 2$ for all i. However, $s_i = 1$ for all i can not occur since rad $S \neq 0$. Therefore we shall assume $s_1 = 2$.

8.) In case of (1.18); i.e., $s = 3$, the elements

$$1, \alpha = (e_2 + \overline{\omega}_1 e_1), \alpha^2 = e_2, \beta = e_3$$

are linearly independent over $\underline{\underline{k}}$ from the left, and $\alpha x \alpha = x\alpha^2, x \in \underline{\underline{k}}$ is independent of $1, \alpha, \beta$; $\alpha x \beta = 0, x \in \underline{\underline{k}}$ and αx is independent of $1, \alpha^2, \beta$. Hence $n(\hat{\Lambda}) = \infty$ by (1.8,i).

9.) If in (1.19), $s_2 = 2$, then

$$1, \alpha = \overline{\omega}_1 e_1, \beta = \overline{\omega}_2 e_2$$

are linearly independent over $\underline{\underline{k}}$. Moreover,

$$\alpha x \alpha = \alpha x \beta = \beta x \alpha = \beta x \beta = 0 \text{ for } x \in \underline{\underline{k}}.$$

$\alpha x = x'\alpha$ is independent of $1, \beta$ and $\beta x = x''\beta$ is independent of $1, \alpha$ for $x \in \underline{\underline{k}}, x' \in \underline{\underline{k}}_1, x'' \in \underline{\underline{k}}_2$. Hence $n(\hat{\Lambda}) = \infty$ by (1.8,ii).

10.) If in (1.19), $s_2 = 1$ and $(\underline{\underline{k}}_1 : \underline{\underline{k}}) > 1$, say $\underline{\underline{k}}_1 = \underline{\underline{k}}(\delta)$, then

$$1, \alpha = \overline{\omega}_1 e_1, \beta = \delta e_1, \beta\alpha = \delta\overline{\omega}_1 e_1$$

are linearly independent over $\underline{\underline{k}}$ from the left. α and $\alpha\beta$ are independent from the right over $\underline{\underline{k}}$; $1, \alpha\beta$ are independent. $\alpha^2 = \alpha x \alpha = 0, x\beta = \beta x$, $x \in \underline{\underline{k}}$ and αx is independent of $1, \beta$ from the left, and β^2 is independent of $\alpha, \beta\alpha, \alpha\beta$ from the left. Hence $n(\hat{\Lambda}) = \infty$ by (1.8,iii).

11.) If in (1.19) $s_2 = 1$, $\underline{\underline{k}}_1 = \underline{\underline{k}}$ and $(\underline{\underline{k}}_2 : \underline{\underline{k}}) > 1$, say $\underline{\underline{k}}_2 = \underline{\underline{k}}(\delta)$, then

$$1, \alpha = (\bar{\omega}_1 e_1 + e_2), \alpha^2 = e_2, \beta = e_2 \delta$$

are independent over $\underline{\underline{k}}$ from the left. $\alpha x \alpha = x\alpha^2$ is independent of $1, \alpha, \beta$. Moreover, αx is independent of $1, \alpha^2, \beta$. Hence $n(\hat{\Lambda}) = \infty$ by (1.8,i).

12.) If in (1.20) $s_1 = 2$, then $(\underline{\underline{k}}_1 : \underline{\underline{k}}) > 1$, say $\underline{\underline{k}}_1 = \underline{\underline{k}}(\delta)$, and

$$1, \alpha = \bar{\omega}_1, \beta = \delta, \beta\alpha = \delta\bar{\omega}_1$$

are linearly independent over $\underline{\underline{k}}$ from the left. α and $\alpha\beta$ are independent from the right over $\underline{\underline{k}}$; $1, \alpha\beta$ are independent and $\alpha^2 = \alpha x \alpha = 0$, $x\beta = \beta x, x \in \underline{\underline{k}}$ and αx is independent of $1, \beta$; β^2 is independent of $\alpha, \beta\alpha, \alpha\beta$ from the left. Hence $n(\hat{\Lambda}) = \infty$ by (1.8,iii).

These are all possible cases that can occur if we assume (1.17,i) to hold. From now on we assume (1.17,ii); i.e., $\mu_\lambda(\hat{\Gamma}/\hat{\Lambda}) = 2$ and $\mu_\lambda(\hat{\Gamma}/\hat{N}\hat{\Gamma}) = 2$.

We put $S = \hat{\Gamma}/\hat{N}\hat{\Gamma}$ and $T = \hat{\Lambda}_o/\hat{N}\hat{\Lambda}_o$, where $\hat{\Lambda}_o = \hat{\Lambda} + \hat{N}\hat{\Gamma}$. Then $T = (\hat{\Lambda} + \hat{\Gamma}\hat{N})/(\hat{N} + \hat{N}^2\hat{\Gamma})$; moreover, $\dim_{\underline{\underline{k}}}(S) = \dim_{\underline{\underline{k}}}(T) = 3$. We shall compute $\operatorname{rad}(T) = \hat{N}\hat{\Gamma}/(\hat{N} + \hat{N}^2\hat{\Gamma})$ (cf. 1.15); obviously, $(\operatorname{rad} T)^2 = 0$. From the proof of (1.15) we conclude $\dim_{\underline{\underline{k}}}(\operatorname{rad} T) = 2$ and $\dim_{\underline{\underline{k}}}(\hat{N}\hat{\Gamma}/\hat{N}^2\hat{\Gamma}) = \dim_{\underline{\underline{k}}}(\hat{\Gamma}/\hat{N}\hat{\Gamma}) = 3$.

We have an epimorphism

$$\varphi : \hat{N}\hat{\Gamma}/\hat{N}^2\hat{\Gamma} \longrightarrow \hat{N}\hat{\Gamma}/(\hat{N}+\hat{N}^2\hat{\Gamma}) \cong \hat{N}\hat{\Gamma}/\hat{N}^2\hat{\Gamma} \,/\, (\hat{N}+\hat{N}^2\hat{\Gamma})/\hat{N}^2\hat{\Gamma}.$$

which is two-sided $\underline{\underline{k}}$-linear. If y_1, y_2, y_3 is a $\underline{\underline{k}}$-basis for $\hat{N}\hat{\Gamma}/\hat{N}^2\hat{\Gamma}$, then two of the elements $\{y_i\varphi\}_{1\le i\le 3}$ must form a $\underline{\underline{k}}$-basis for $\hat{N}\hat{\Gamma}/(\hat{N}+\hat{N}^2\hat{\Gamma})$.

13.) If (1.18) occurs; i.e., $s = 3$, then rad $\hat{\Gamma} = \hat{N}\hat{\Gamma}$ and

$$\hat{N}\hat{\Gamma}/\hat{N}^2\hat{\Gamma} = \underline{\underline{k}}\,\overline{\omega}_1 e_1 \oplus \underline{\underline{k}}\,\overline{\omega}_2 e_2 \oplus \underline{\underline{k}}\,\overline{\omega}_3 e_3, \; \overline{\omega}_i^2 = 0, 1\le i\le 3.$$

We then may assume that

$$1, \alpha = (\overline{\omega}_1 e_1)\varphi; \; \beta = (\overline{\omega}_2 e_2)\varphi$$

are in T and they are linearly independent over $\underline{\underline{k}}$. We have $\alpha x\alpha = \beta x\alpha = \alpha x\beta = \beta x\beta = 0$, and $\alpha x = x'\alpha$, $\beta x = x''\beta$, $x, x', x \in \underline{\underline{k}}$. Hence $n(\hat{\Lambda}) = \infty$ by (1.8,ii).

If (1.19) occurs; i.e., $s = 2$, then we can have the situation

$$\hat{N}\hat{\Gamma}/\hat{N}^2\hat{\Gamma} = \underline{\underline{k}}_1\,\overline{\omega}_1 e_1 \oplus \underline{\underline{k}}\,\overline{\omega}_2 e_2, \; \overline{\omega}_i^2 = 0, i=1,2,$$

$(\underline{\underline{k}}_1 : \underline{\underline{k}}) = 2$, $\underline{\underline{k}}_1 = \underline{\underline{k}}(\delta)$. The elements $\overline{\omega}_1 e_1, \delta\overline{\omega}_1 e_1$ and $\overline{\omega}_2 e_2$ are linearly independent over $\underline{\underline{k}}$ in $\hat{N}\hat{\Gamma}/\hat{N}^2\hat{\Gamma}$. We always must have $(\overline{\omega}_2 e_2)\varphi \ne 0$. Assume $(\overline{\omega}_2 e_2)\varphi = 0$. Then Ker$\varphi$ is a $\hat{\Gamma}$-module, and so Im φ is a $\hat{\Gamma}$-module; i.e., $\hat{\Gamma}(\hat{N} + \hat{N}^2\hat{\Gamma}) \subset \hat{N} + \hat{N}^2\hat{\Gamma}$, and $\hat{\Gamma}\hat{N} \subset \hat{N} + \hat{N}^2\hat{\Gamma}$ and Nakayama's lemma implies $\hat{\Gamma}\hat{N} = \hat{N}$; i.e., $\hat{\Lambda}_o = \hat{\Lambda}$. But we had assumed $\mu_\lambda(\hat{\Lambda}_o/\hat{\Lambda}) = 2$. Thus either $(\overline{\omega}_1 e_1)\varphi \ne 0$ and $(\overline{\omega}_2 e_2)\varphi \ne 0$ or $(\overline{\omega}_2 e_2)\varphi \ne 0$ and $(\delta\,\overline{\omega}_1 e_1)\varphi \ne 0$.

14.) If $(\overline{\omega}_1 e_1)\varphi \ne 0$ and $(\overline{\omega}_2 e_2)\varphi \ne 0$, then

$$1, \alpha = (\overline{\omega}_1 e_1)\varphi, \beta = (\overline{\omega}_2 e_2)\varphi$$

are in T and linearly independent over $\underline{\underline{k}}$ from the left and from the right. Moreover

$$\alpha x\alpha = \alpha x\beta = \beta x\alpha = \beta x\beta = 0;$$

αx is independent of $1, \beta$; $x \in \underline{\underline{k}}$ and $\beta x = x'\beta$, $x' \in \underline{\underline{k}}$. Hence $n(\hat{\Lambda}) = \infty$ by (1.8,ii).

15.) If $(\delta\overline{\omega}_1 e_1)\varphi \ne 0$ and $(\overline{\omega}_2 e_2)\varphi \ne 0$, then we take

$$1, \alpha = (\delta\overline{\omega}_1 e_1)\varphi, \beta = (\overline{\omega}_2 e_2)\varphi,$$

and again by (1.8,ii), $n(\hat{\Lambda}) = \infty$.

16.) However if case (1.19) occurs, we also can have the situation

$$\hat{N}\hat{\Gamma}/\hat{N}^2\hat{\Gamma} = (\underline{\underline{k}}\,\overline{\omega}_1^2 + \underline{\underline{k}}\,\overline{\omega}_1^3)e_1 \oplus \underline{\underline{k}}\,\overline{\omega}_1 e_2,\ \overline{\omega}_2^2 = 0,\ \overline{\omega}_1^4 = 0.$$

The elements $\overline{\omega}_1^2 e_1, \overline{\omega}_1^3 e_1, \overline{\omega}_2 e_2$ are linearly independent, and in all possible cases we may choose $1, \alpha, \beta \in T$ such that the hypotheses of (1.8,ii) are satisfied; i.e., $n(\hat{\Lambda}) = \infty$.

17.) If (1.20) occurs; i.e., $s = 1$, then we can have the situation

$$\hat{N}\hat{\Gamma}/\hat{N}^2\hat{\Gamma} = \underline{\underline{k}}_1\,\overline{\omega}_1,\ \overline{\omega}_1^2 = 0 \text{ and } (\underline{\underline{k}}_1 = \underline{\underline{k}}(\delta) : \underline{\underline{k}}) = 3.$$

We observe first that we may choose $\omega_1 \in \hat{N}$. In fact, $\hat{\Gamma}\hat{N} = \hat{\Gamma}\omega_1$. We write $\omega_1 = \sum_{i=1}^{t} \gamma_i n_i, n_i \in \hat{N}, \gamma_i \in \hat{\Gamma}$. Not all i can lie in $\hat{\Gamma}\hat{N}^2$, say $n_1 = \varepsilon_1 \omega_1 + \omega_1^2 \gamma_0$, where ε_1 is a unit in $\hat{\Gamma}$. Then $\hat{\Gamma}\hat{N} = \hat{\Gamma} n_1$, and we may replace ω_1 by n_1; consequently we can assume $\omega_1 \in \hat{N}$. Then

$$\hat{N}\hat{\Gamma}/\hat{N}^2\hat{\Gamma} = \underline{\underline{k}}\,\overline{\omega}_1 + \underline{\underline{k}}\,\delta\overline{\omega}_1 + \underline{\underline{k}}\,\delta^2\overline{\omega}_1,$$

where $\overline{\omega}_1 \in (\hat{N} + \hat{N}^2\hat{\Gamma})/(\hat{N}^2\hat{\Gamma})$. Since $\dim_{\underline{\underline{k}}}((\hat{N} + \hat{N}^2\hat{\Gamma})/\hat{N}^2\hat{\Gamma}) = 1$, we have $(\hat{N} + \hat{N}^2\hat{\Gamma})/(\hat{N}^2\hat{\Gamma}) = \underline{\underline{k}}\,\overline{\omega}_1$. Moreover, this is a two-sided $\underline{\underline{k}}$-module, and so $\underline{\underline{k}}\,\overline{\omega}_1 = \overline{\omega}_1\underline{\underline{k}}$. But then also $\underline{\underline{k}}(\delta\overline{\omega}_1) = (\delta\overline{\omega}_1)\underline{\underline{k}}$. Hence

$$\hat{N}\hat{\Gamma}/(\hat{N} + \hat{N}^2\hat{\Gamma}) = \underline{\underline{k}}(\delta\overline{\omega}_1) \oplus \underline{\underline{k}}(\delta^2\overline{\omega}_1).$$

We choose

$$1, \alpha = \delta\,\overline{\omega}_1, \beta = \delta^2\,\overline{\omega}_1, \text{ in } T$$

which are linearly independent over $\underline{\underline{k}}$ from the left. Moreover, $\alpha x \alpha = \alpha x \beta = \beta x \alpha = \beta x \beta = 0, x \in \underline{\underline{k}}$ and $x\alpha = \alpha x', x, x' \in \underline{\underline{k}}$; $x\beta = \beta x''$, $x, x'' \in \underline{\underline{k}}$. Hence $n(\hat{\Lambda}) = \infty$ by (1.8,ii).

18.) If (1.20) occurs, we can also have

$$\hat{N}\hat{\Gamma}/\hat{N}^2\hat{\Gamma} = \underline{\underline{k}}\,\overline{\omega}_1^3 + \underline{\underline{k}}\,\overline{\omega}_1^4 + \underline{\underline{k}}\,\overline{\omega}_1^5,\ \overline{\omega}_1^6 = 0.$$

Then we may choose $1, \alpha, \beta$ in T such that they satisfy (1.8,ii); hence $n(\hat{\Lambda}) = \infty$.

These are all the cases that can occur, and we have proved theorem (1.12). #

Exercises §1:

1.) Let R be a Dedekind domain and K its quotient field. Let A be a finite dimensional K-algebra, the radical of which is not zero. If Λ

§2 Separation of the three different cases

We sketch the proof of the main theorem (cf. 2.1 below) of this chapter, and reduce the proof of (2.1) to treating three different cases (2.15).

We keep the notation and terminology of § 1: in particular, $\hat{R}$ has a finite residue class field.

$\hat{A} = \oplus_{i=1}^{n} \hat{D}_i$, where $\hat{D}_i$ is a separable skewfield over the completion $\hat{K}$ of the $\underline{\underline{A}}$-field K, $1 \leq i \leq n$,

$\hat{\Gamma} = \oplus_{i=1}^{n} \hat{\Omega}_i$ is the unique maximal $\hat{R}$-order in $\hat{A}$,

$\hat{\Lambda}$ is a fixed completely primary $\hat{R}$-order in $\hat{A}$,

$\hat{N} = \text{rad}\, \hat{\Lambda}$,

$\mu_{\hat{\Lambda}}(\hat{X})$ denotes the minimal number of generators of $\hat{X}$ as left $\hat{\Lambda}$-module.

We shall prove the following statement:

2.1 Theorem (Drozd-Roiter [1], Jacobinski [2], Roggenkamp [8,9]): $n(\hat{\Lambda})$, the number of non-isomorphic indecomposable $\hat{\Lambda}$-lattices, is finite if and only if

(i) $\mu_{\hat{\Lambda}}(\hat{\Gamma}/\hat{\Lambda}) \leq 2$ and

(ii) $\mu_{\hat{\Lambda}}(\text{rad}_{\hat{\Lambda}}(\hat{\Gamma}/\hat{\Lambda})) \leq 1$.

2.2 Corollary: Let $A = \oplus_{i=1}^{n} D_i$ be a separable K-algebra which is the direct sum of skewfields, and let Λ be an R-order in A such that for every maximal ideal $\underline{\underline{p}}$ of R, dividing the Higman ideal $\underline{\underline{H}}(\Lambda)$, $\hat{D}_{i_{\underline{\underline{p}}}}$ decomposes into a direct sum of skewfields, $1 \leq i \leq n$, and let Γ be the unique maximal R-order in A. Then $n(\Lambda) < \infty$ if and only if $\mu_{\Lambda}(\Gamma/\Lambda) \leq 2$ and $\mu_{\Lambda}(\text{rad}_{\Lambda}(\Gamma/\Lambda)) \leq 1$.

Proof (of the Corollary): Because of (1.1), $n(\Lambda) < \infty$ if and only if $n(\hat{\Lambda}_{\underline{\underline{p}}}) < \infty$ for all $\underline{\underline{p}} | \underline{\underline{H}}(\Lambda)$; moreover, since every idempotent in $\hat{A}_{\underline{\underline{p}}}$ is central, $n(\hat{\Lambda}_{\underline{\underline{p}}}) < \infty$ if and only if $n(\hat{\Lambda}_{\underline{\underline{p}},i}) < \infty$, where $\hat{\Lambda}_{\underline{\underline{p}},i}$ are the completely primary summands of $\hat{\Lambda}_{\underline{\underline{p}}}$ (cf. 1.2). Hence, taking (2.1) for granted, it remains to show

$$\mu_\Lambda(\Gamma/\Lambda) \leq 2 \text{ and } \mu_\Lambda(\text{rad}_\Lambda(\Gamma/\Lambda)) \leq 1 \text{ if and only if}$$

$$\mu_{\hat{\Lambda}_{\underline{p},i}}(\hat{\Gamma}_{\underline{p},i}/\hat{\Lambda}_{\underline{p},i}) \leq 2 \text{ and } \mu_{\hat{\Lambda}_{\underline{p},i}}(\text{rad}_{\hat{\Lambda}_{\underline{p},i}}(\hat{\Gamma}_{\underline{p},i}/\hat{\Lambda}_{\underline{p},i})) \leq 1$$

for all $\underline{p} \mid \underline{H}(\Lambda)$ and all i. Obviously, $\mu_\Lambda(\Gamma/\Lambda) \leq 2$ implies $\mu_{\hat{\Lambda}_{\underline{p},i}}(\hat{\Gamma}_{\underline{p},i}/\hat{\Lambda}_{\underline{p},i}) \leq 2$ for all $\underline{p}$ and all i. Conversely, if for all $\underline{p},i$ $\mu_{\hat{\Lambda}_{\underline{p},i}}(\hat{\Gamma}_{\underline{p},i}/\hat{\Lambda}_{\underline{p},i}) \leq 2$, then $\mu_{\hat{\Lambda}_{\underline{p}}}(\hat{\Gamma}_{\underline{p}}/\hat{\Lambda}_{\underline{p}}) \leq 2$ for all $\underline{p} \mid \underline{H}(\Lambda)$. Let $\underline{p}_1, \dots, \underline{p}_s$ be all the maximal ideals dividing $\underline{H}(\Lambda)$. We write $I = \Gamma/\Lambda$; then $\hat{I}_{\underline{p}} = \hat{\Gamma}_{\underline{p}}/\hat{\Lambda}_{\underline{p}}$, and if $\hat{I}_{\underline{p}_i} = \hat{\Lambda}_{\underline{p}_i}\alpha_i + \hat{\Lambda}_{\underline{p}_i}\beta_i$, then we may assume that $\alpha_i, \beta_i \in I$. Moreover, by the Chinese remainder theorem (I, 7.7), we can find $\alpha, \beta \in I$ such that

$$\alpha \equiv \alpha_i \bmod(\underline{p}_i),$$

$$\beta \equiv \beta_i \bmod(\underline{p}_i), 1 \leq i \leq s.$$

We consider $J = \Lambda\alpha + \Lambda\beta$. Then

$$\hat{J}_{\underline{p}_i} + \underline{p}_i \hat{I}_{\underline{p}} = \hat{I}_{\underline{p}_i} \text{ and by Nakayama's lemma,}$$

$\hat{J}_{\underline{p}_i} = \hat{I}_{\underline{p}_i}, 1 \leq i \leq s$. However, for $\underline{p} \neq \underline{p}_i, 1 \leq i \leq s$, $\hat{J}_{\underline{p}} \subset \hat{I}_{\underline{p}} = 0$. And so $J = \oplus_{i=1}^{s} \hat{J}_{\underline{p}_i} = \oplus_{i=1}^{s} \hat{I}_{\underline{p}_i} = I$. Hence $\mu_\Lambda(\Gamma/\Lambda) \leq 2$, and it remains to show that $\text{rad}_\Lambda I = \oplus_{\underline{p},i} \text{rad}_{\hat{\Lambda}_{\underline{p},i}} \hat{I}_{\underline{p},i}$. We have

$\hat{I}_{\underline{p}} = \oplus_i \hat{I}_{\underline{p},i}$ and obviously $\text{rad}_{\hat{\Lambda}_{\underline{p}}} \hat{I}_{\underline{p}} = \oplus_i \text{rad}_{\hat{\Lambda}_{\underline{p},i}}(\hat{I}_{\underline{p},i})$. Therefore it suffices to prove $\text{rad}_\Lambda(I) = \oplus_{\underline{p} \mid \underline{H}(\Lambda)} \text{rad}_{\hat{\Lambda}_{\underline{p}}}(\hat{I}_{\underline{p}})$. If J' is a maximal submodule of I, then either $\hat{J}'_{\underline{p}} = \hat{I}_{\underline{p}}$ or $\hat{J}'_{\underline{p}}$ is a maximal submodule of $\hat{I}_{\underline{p}}$. However, $J' = \oplus_{\underline{p} \mid \underline{H}(\Lambda)} \hat{J}'_{\underline{p}}$, and so $\hat{J}'_{\underline{p}_i} = \hat{I}_{\underline{p}_i}$ for all i except one.

Hence $\mathrm{rad}_{\Lambda}(I) \supset \bigoplus_{i=1}^{s} \mathrm{rad}_{\hat{\Lambda}_{\underline{p}_i}}(\hat{I}_{\underline{p}_i})$. Conversely, given a maximal submodule $\hat{J}_i$ of $\hat{I}_{\underline{p}_i}$, then $J_i = \hat{J}_i \oplus (\bigoplus_{l \neq i} \hat{I}_{\underline{p}_l})$ is a maximal submodule of I; thus $\mathrm{rad}_{\Lambda}(I) \subset \bigoplus_{i=1}^{s} \mathrm{rad}_{\hat{\Lambda}_{\underline{p}_i}}(\hat{I}_{\underline{p}_i})$. #

__Remark 1:__ In (2.2) it suffices to require that $\hat{D}_{i_{\underline{p}}}$ stays a skewfield for all $\underline{p}$ for which $\hat{\Lambda}_{\underline{p}}$ is not a Bass-order. In fact for a Bass-order $\hat{\Lambda}_1$, $n(\hat{\Lambda}_1) < \infty$ (cf. IX, 5.6) by the Jordan-Zassenhaus theorem.

__Remark 2:__ (2.1) has be proved independently by Drozd-Roiter [1] and Jacobinski [2] in case A is commutative; however Jacobinski has obtained other conditions involving ramification indices (cf. Remark § 12). The generalization to direct sums of skewfields is due to Roggenkamp [8,9]. In view of (1.12) we only have to prove one direction of (2.1), namely the difficult one. Since this proof is rather involved, we shall __sketch the proof first:__ The central role in the proof is played by Bass-orders (cf. IX, §§ 5,6). Since $\hat{\Lambda}$ (in 2.1) is completely primary, we may assume $n \leq 3$ (cf. 1.7), for the proof of the sufficiency. We have to treat the three cases separately.

2.3 (i) $\hat{A} = \hat{D}$ is a complete skewfield,

(ii) $\hat{A} = \hat{D}_1 \oplus \hat{D}_2$ is the direct sum of two complete skewfields,

(iii) $\hat{A} = \hat{D}_1 \oplus \hat{D}_2 \oplus \hat{D}_3$ is the direct sum of three complete skewfields.

In (2.12) we shall show that we may assume that

$$\hat{\Lambda}_1 = \{a \in \hat{A} : a\hat{N} \subset \hat{\Lambda}\} = \{a \in \hat{A} : \hat{N}a \subset \hat{\Lambda}\}$$

is a Bass-order containing $\hat{\Lambda}$ (cf. 2.10).

In the case (2.3,i) we associate with $\hat{M} \in {}_{\hat{\Lambda}}\underline{\underline{M}}^{o}$ the exact sequence

2.4 $$0 \longrightarrow \hat{N}\hat{M} \longrightarrow \hat{M} \longrightarrow \hat{M}/\hat{N}\hat{M} \longrightarrow 0,$$

where $\hat{N}\hat{M}$ is a lattice over the Bass-order $\hat{\Lambda}_1$ and $\hat{M}/\hat{N}\hat{M} \cong \underline{k}^{(m)}$, where

$\underline{\underline{k}} = \hat{\Lambda}/\hat{N}$ is a finite field, since $\hat{R}$ has a finite residue class field, $\hat{K}$ being an $\underline{\underline{A}}$-field.

In the case (2.3,ii) we choose a primitive idempotent e_2 of $\hat{A}$ such that $\hat{\Lambda}(1-e_2)$ is a maximal $\hat{R}$-order and $\hat{\Lambda}e_2$ is a Bass-order. With $\hat{M} \in {}_{\hat{\Lambda}}\underline{\underline{M}}^o$ we associate the exact sequence

2.5 $$0 \longrightarrow \hat{M} \cap \hat{A}e_2\hat{M} \longrightarrow \hat{M} \longrightarrow \hat{M}/(\hat{M} \cap \hat{A}e_2\hat{M}) \longrightarrow 0.$$

Then $\hat{M} \cap \hat{A}e_2\hat{M} \in {}_{\hat{\Lambda}e_2}\underline{\underline{M}}^o$ is a lattice over a Bass-order and $\hat{M}/(\hat{M} \cap \hat{A}e_2\hat{M}) \cong \hat{\Lambda}(1-e_2)^{(m)}$.

In the case (2.3,iii) we choose an idempotent $e = e_1 + e_2$, the sum of two orthogonal primitive idempotents such that $\hat{\Lambda}(1-e)$ is a maximal $\hat{R}$-order and $\hat{\Lambda}e$ is a Bass-order. With $\hat{M} \in {}_{\hat{\Lambda}}\underline{\underline{M}}^o$ we associate the exact sequence

2.6 $$0 \longrightarrow \hat{M} \cap \hat{A}e\hat{M} \longrightarrow \hat{M} \longrightarrow \hat{M}/(\hat{M} \cap \hat{A}e_2\hat{M}) \longrightarrow 0.$$

Then $\hat{M} \cap \hat{A}e\hat{M} \in {}_{\hat{\Lambda}e}\underline{\underline{M}}^o$ is a lattice over a Bass-order and

$$\hat{M}/(\hat{M} \cap \hat{A}e_2\hat{M}) \cong \hat{\Lambda}(1-e)^{(m)}.$$

This leads the way for a possible proof of (2.1):

In each case we have associated with $\hat{M} \in {}_{\hat{\Lambda}}\underline{\underline{M}}^o$ an exact sequence

$$E_{\hat{M}} : 0 \longrightarrow \hat{M}' \longrightarrow \hat{M} \longrightarrow \hat{M}'' \longrightarrow 0,$$

where $\hat{M}'$ is a lattice over a Bass-order $\hat{\Lambda}'$; i.e.,

$$\hat{M}' \cong \oplus_{i=1}^{s} \hat{N}_i'^{(s_i)}$$

with $\hat{N}_i' \not\cong \hat{N}_j'$ for $i \neq j$ and each $\hat{N}_i'$ is a projective lattice over some R-order $\hat{\Lambda}_t'$ containing $\hat{\Lambda}'$. Moreover, in all three cases, $\hat{M}'$ is a characteristic submodule of $\hat{M}$; i.e., for $\varphi \in \mathrm{End}_{\hat{\Lambda}}(\hat{M})$, $\varphi|_{\hat{M}'} : \hat{M}' \longrightarrow \hat{M}'$. In addition $\hat{M}'' \cong S''^{(m)}$ where S'' is a ring which is at the same time a $\hat{\Lambda}$-module, and $\mathrm{Hom}_{\hat{\Lambda}}(\hat{M}'',\hat{M}') = 0$.

Thus, the equivalence class $[E_{\hat{M}}]$ of the sequence $E_{\hat{M}}$ lies in

$$\mathrm{Ext}^1_{\hat\Lambda}(S''^{(m)}, \oplus_{i=1}^{s} \hat N_i'^{(s_i)}) \overset{nat}{\cong} \oplus_{i=1}^{s} \mathrm{Ext}^1_{\hat\Lambda}(S''^{(m)}, \hat N_i'^{(s_i)}) \overset{nat}{\cong}$$

$$\oplus_{i=1}^{s} (\mathrm{Ext}^1_{\hat\Lambda}(S'', \hat N_i'))_{m \times s_i},$$

where $(X)_{m \times s_i}$ denotes the set of $(m \times s_i)$ matrices with entries in X. Hence we have a bijection between the classes $[E_{\hat M}]$ and $(m \times \sum_{i=1}^{s} s_i)$ matrices $X_{\hat M}$

$$[E_{\hat M}] \longleftrightarrow \underline{\underline{X}}_{\hat M} = (\underline{\underline{X}}_1, \dots, \underline{\underline{X}}_s), \underline{\underline{X}}_i \in (\mathrm{Ext}^1_{\hat\Lambda}(S'', \hat N_i'))_{m \times s_i}.$$

However, $\mathrm{Ext}^1_{\hat\Lambda}(\hat M'', \hat M')$ is an $[\mathrm{End}_{\hat\Lambda}(\hat M''), \mathrm{End}_{\hat\Lambda}(\hat M')]$-bimodule (cf. II, 5.7), and the elements in $\mathrm{End}_{\hat\Lambda}(\hat M'')$ may be considered as $(m \times m)$ matrices $\underline{\underline{Z}}$ and the elements in $\mathrm{End}_{\hat\Lambda}(\hat M')$ as $(\sum_{i=1}^{s} s_i \times \sum_{i=1}^{s} s_i)$-matrices $\underline{\underline{Y}}$. To decide when $\hat M$ decomposes we can use a lemma of Heller-Reiner [3], which reduces the problem to the decomposition of matrices:

2.7 $\hat M$ decomposes if and only if $\underline{\underline{Z}}\underline{\underline{X}}_{\hat M}\underline{\underline{Y}}$ decomposes for some invertible matrices $\underline{\underline{Z}}$ in $\mathrm{End}_{\hat\Lambda}(\hat M'')$ and $\underline{\underline{Y}}$ in $\mathrm{End}_{\hat\Lambda}(\hat M')$.

Now we can list the steps that have to be taken to prove (2.1):

1.) Find all $\hat R$-orders $\hat\Lambda_t'$ that contain the Bass-order $\hat\Lambda'$; there are only finitely many.

2.) For every t, find all non-isomorphic indecomposable $\hat\Lambda_t'$-lattices; there are only finitely many, say $\{\hat M_{it}\}$.

3.) Compute $\mathrm{Ext}^1_{\hat\Lambda}(S'', \hat M_{it})$ explicitely.

4.) Compute $\mathrm{Hom}_{\hat\Lambda}(\hat M_{it}, \hat M_{jt'})$ explicitely.

5.) Compute the action:

$$\mathrm{Hom}_{\hat\Lambda}(\hat M_{it}, \hat M_{jt'}) : \mathrm{Ext}^1_{\hat\Lambda}(S'', \hat M_{it}) \longrightarrow \mathrm{Ext}^1_{\hat\Lambda}(S'', \hat M_{jt'}),$$

and the operation of $\mathrm{End}_{\hat\Lambda}(S'')$ on $\mathrm{Ext}^1_{\hat\Lambda}(S'', \hat M_{it})$.

6.) Characterize the matrices

$$\underline{\underline{X}} \in \mathrm{Ext}^1_{\hat{\Lambda}}(\hat{M}'', \hat{M}')$$

which actually do correspond to exact sequences of the type $E_{\hat{M}}$.

7.) Decompose these matrices under $\underline{\underline{Z X_{\hat{M}} Y}}$.

8.) Show that the number of indecomposable matrices, that thus occur, is finite.

We shall now prove (2.1) following the above steps. First some general facts.

We assume that $\hat{\Lambda}$ is completely primary $\mu_{\hat{\Lambda}}(\hat{\Gamma}/\hat{\Lambda}) \leq 2$ and $\mu_{\hat{\Lambda}}(\mathrm{rad}_{\hat{\Lambda}}(\hat{\Gamma}/\hat{\Lambda})) \leq 1$.

2.8 <u>Lemma</u>: $\hat{A}$ has at most 3 idempotents and $\hat{\Lambda}_o = \hat{\Lambda} + \hat{N}\hat{\Gamma}$ has at most two idempotents.

<u>Proof</u>: $\mu_{\hat{\Lambda}}(\hat{\Gamma}/\hat{\Lambda}) \leq 2$ implies $\mu_{\hat{\Lambda}}(\hat{\Gamma}) \leq 3$ (cf. IX, 6.12). Hence $\dim_{\underline{\underline{k}}}(\hat{\Gamma}/\hat{N}\hat{\Gamma}) \leq 3$ (cf. IX, 6.11). However, $\hat{N}\hat{\Gamma} \subset \mathrm{rad}\,\hat{\Gamma}$ (cf. 1.14) and the method of lifting idempotents shows that $\hat{\Gamma}$ can have at most three idempotents. Since $\hat{\Gamma}$ is maximal, $\hat{A}$ has at most three idempotents. Similarly one shows that $\hat{\Lambda}_o$ has at most 2 idempotents. #

2.9 <u>Lemma</u>: If $\hat{\Lambda}$ is a Gorenstein order, then $\hat{\Lambda}_1 = \Lambda_1(\hat{N})$ is an $\hat{R}$-order in $\hat{A}$ satisfying

$$\mu_{\hat{\Lambda}_1}(\hat{\Gamma}/\hat{\Lambda}_1) \leq 2 \text{ and } \mu_{\hat{\Lambda}_1}(\mathrm{rad}_{\hat{\Lambda}_1}(\hat{\Gamma}/\hat{\Lambda}_1)) \leq 1. \text{ *)}$$

<u>Proof</u>: If $\hat{\Lambda}$ is Gorenstein, we may assume $\hat{\Lambda} \neq \hat{\Gamma}$ and so $\hat{\Lambda}_1 = \Lambda_r(\hat{N})$ is the unique minimal over order of $\hat{\Lambda}$ (cf. IX, 6.6). We obviously have $\mu_{\hat{\Lambda}_1}(\hat{\Gamma}/\hat{\Lambda}_1) \leq 2$ and it remains to show $\mu_{\hat{\Lambda}_1}(\mathrm{rad}_{\hat{\Lambda}_1}(\hat{\Gamma}/\hat{\Lambda}_1)) \leq 1$. If $\hat{\Lambda}_1$ is completely primary, then $\mathrm{rad}_{\hat{\Lambda}_1}(\hat{\Gamma}/\hat{\Lambda}_1) = (\hat{\Lambda}_1 + \hat{\Gamma}\,\mathrm{rad}\,\hat{\Lambda}_1)/\hat{\Lambda}_1$ (cf. 1.13) and $\mathrm{rad}\,\hat{\Lambda}_1 \subset \hat{N}\hat{\Gamma}$. In fact, $\hat{\Lambda}_o = \hat{\Lambda} + \hat{N}\hat{\Gamma} \supset \hat{\Lambda}_1$ and hence by (1.14), $\mathrm{rad}\,\hat{\Lambda}_o \supset \mathrm{rad}\,\hat{\Lambda}_1$; i.e., $\hat{N}\hat{\Gamma} \supset \mathrm{rad}\,\hat{\Lambda}_1$ (cf. 1.15), if $\hat{\Lambda}_o \supsetneq \hat{\Lambda}$. But $\hat{\Lambda}_o = \hat{\Lambda}$ implies $\hat{N}\hat{\Gamma} = \hat{N}$ and $\hat{\Lambda}_1 = \hat{\Gamma}$; but then $\hat{\Lambda}_1$ obviously satisfies

*) The reader should always distinguish between $1 = \ell$ and 1 = one!!!

the above conditions. Thus $(\hat{\Lambda}_1 + \hat{\Gamma}\,\mathrm{rad}\,\hat{\Lambda}_1)/\hat{\Lambda}_1 = \hat{\Lambda}_o/\hat{\Lambda}_1$ is the homomorphic image of $\hat{\Lambda}_o/\hat{\Lambda}$, and so $\mu_{\hat{\Lambda}_1}(\mathrm{rad}_{\hat{\Lambda}_1}(\hat{\Gamma}/\hat{\Lambda}_1)) \leq 1$. If $\hat{\Lambda}_1$ decomposes, then $\hat{\Lambda}$ is a subdirect sum of $\hat{\Lambda}_1$; i.e., $\hat{\Lambda}_1 = \hat{\Lambda}e_1 \oplus \hat{\Lambda}(1-e_1)$ where e_1 is a primitive idempotent in Λ_1, and rad $\hat{\Lambda}_1 = \hat{N}e_1 \oplus \hat{N}(1-e_1)$. Consequently,

$$\mu_{\hat{\Lambda}_1}(\mathrm{rad}_{\hat{\Lambda}_1}(\hat{\Gamma}/\hat{\Lambda}_1)) \leq 1. \qquad \#$$

We recall from the proof of (IX, 6.13):

2.10 If for a completely primary $\hat{R}$-order $\hat{\Lambda}'$ with rad $\hat{\Lambda}' = \hat{N}'$, we have $\mu^r_{\hat{\Lambda}'}(\Lambda_1(\hat{N}')/\hat{\Lambda}') \leq 1$ or $\mu_{\hat{\Lambda}'}(\hat{\Lambda}_r(\hat{N}')/\hat{\Lambda}') \leq 1$, then $\hat{\Lambda}'$ is a Gorenstein order. (Here $\underline{\mu}^r(-)$ denotes the minimal number of generators of a right module. In view of (IX, 5.2) an order is left Gorenstein if and only if it is right Gorenstein.)

<u>We shall now assume that $\hat{\Lambda}$ is not Gorenstein.</u>

2.11 <u>Lemma</u>: Every left $\hat{\Lambda}$-submodule of $\hat{\Lambda}_o/\hat{\Lambda}$ is cyclic; here $\hat{\Lambda}_o = \hat{\Lambda} + \hat{N}\hat{\Gamma}$.

<u>Proof</u>: Since every $\hat{\Gamma}$-ideal is two-sided and principal, we have $\hat{N}\hat{\Gamma} = \hat{\Gamma}\alpha$ for some regular element α in $\hat{\Gamma}$. But, what is more, we may even choose $\alpha \in \hat{N}$. We shall prove this only in case $\hat{A} = \hat{D}_1 \oplus \hat{D}_2 \oplus \hat{D}_3$, the other cases being similar. Then $\hat{\Gamma} = \oplus_{i=1}^{3} \hat{\Gamma}e_i$. $\hat{N}$ is a finitely generated $\hat{\Lambda}$-lattice, say

$$\hat{N} = \sum_{j=1}^{s} \hat{\Lambda}\beta_j.$$

Then there are elements β_{j_i}, $i=1,2,3$ such that $\hat{\Gamma}\alpha e_i = \hat{\Gamma}\beta_{j_i}$, $1 \leq i \leq 3$. If $j_1 = j_2 = j_3$ then $\hat{\Gamma}\hat{N} = \hat{\Gamma}\beta_{j_1}$; if $j_1 = j_2 \neq j_3$ then we choose a unit $u \in \hat{\Lambda}$ such that $(\beta_{j_1} + u\beta_{j_3})e_i \neq 0$, $1 \leq i \leq 3$; then $\hat{\Gamma}\hat{N} = \hat{\Gamma}(\beta_{j_1} + u\beta_{j_3})$. If $j_1 \neq j_2 \neq j_3 \neq j_1$, there exist units u_1, u_2 such that

$$(\beta_{j_1} + u_1\beta_{j_2} + u_2\beta_{j_3})e_i \neq 0, 1 \leq i \leq 3;$$

then $\hat{\Gamma}\hat{N} = \hat{\Gamma}(\beta_{j_1} + u_1\beta_{j_2} + u_3\beta_{j_3})$ (cf. Ex. 2,1).

Now, $\mathrm{rad}_{\hat\Lambda}(\hat\Gamma/\hat\Lambda) = \hat\Lambda_o/\hat\Lambda$ and we can write

$$\hat\Lambda_o = \hat\Lambda + \hat\Lambda\beta \text{ (cf. IX, 6.11, 6.12)},$$

since $\mu_{\hat\Lambda}(\hat\Lambda_o/\hat\Lambda) \leq 1$; and $\mu_{\hat\Lambda}(\hat\Gamma/\hat\Lambda) \leq 2$ implies

$$\hat\Gamma = \hat\Lambda + \hat\Lambda\gamma_1 + \hat\Lambda\gamma_2.$$

Thus

$$(\hat N\hat\Gamma)^i = \hat\Gamma\alpha^i = \hat\Lambda\alpha^i + \hat\Lambda\gamma_1\alpha^i + \hat\Lambda\gamma_2\alpha^i,$$

and

$$\begin{aligned}\hat\Lambda + (\hat N\hat\Gamma)^i &= \hat\Lambda + \hat\Lambda\alpha^i + \hat\Lambda\gamma_1\alpha^i + \hat\Lambda\gamma_2\alpha^i \\ &= \hat\Lambda + \hat\Lambda\gamma_1\alpha^i + \hat\Lambda\gamma_2\alpha^i = \hat\Lambda + (\hat\Lambda + \hat N\hat\Gamma)\alpha^{i-1} \\ &= \hat\Lambda + \hat\Lambda\beta\alpha^{i-1}, \text{ since } \alpha \in \hat N.\end{aligned}$$

Consequently, $\hat X_i = [\hat\Lambda + (\hat N\hat\Gamma)^i]/\hat\Lambda$ is a cyclic $\hat\Lambda$-submodule of $\hat\Lambda_o/\hat\Lambda$. For the proof of (2.11) it suffices to show that these are the only submodules of $\hat\Lambda_o/\hat\Lambda = \hat X_1$. But $\hat N\hat X_i = \hat X_{i+1}$, and so $\hat N\hat X_i = \mathrm{rad}_{\hat\Lambda}\hat X_i$ is the unique maximal submodule of $\hat X_i$. Thus $\{\hat X_i\}$ are the only submodules of $\hat\Lambda_o/\hat\Lambda$. It should be observed that $\hat X_{i_o+j} \subset \hat\Lambda$ for some i_o. #

2.12 <u>Theorem</u>: If $\hat\Lambda$ is not a Gorenstein-order, then $\Lambda_r(\hat N) = \Lambda_1(\hat N) = \hat\Lambda_1$ is a Bass-order in $\hat A$, and

$$\mu_{\hat\Lambda}(\hat\Gamma) = 3, \ \hat\Lambda_1 \not\subset \hat\Lambda + \hat N\hat\Gamma, \ \dim_{\underline{k}}(\hat\Lambda_1/\hat N) = 3.$$

<u>Proof</u>: In view of (2.10) we may assume

$\mu_{\hat\Lambda}(\Lambda_r(\hat N)/\hat\Lambda) \geq 2$, and we write $\hat\Lambda_r = \Lambda_r(\hat N)$.

If $\hat\Lambda_r \subset \hat\Lambda + \hat N\hat\Gamma$, then $\hat\Lambda_r/\hat\Lambda$ is a cyclic $\hat\Lambda$-module by (2.11), a contradiction to our assumption. If $\mu_{\hat\Lambda}(\hat\Gamma/\hat\Lambda) = 1$, then $\hat\Lambda$ is a Bass-order by (IX, 6.14); thus we may assume $\mu_{\hat\Lambda}(\hat\Gamma/\hat\Lambda) = 2$, and so

$$\dim_{\underline{k}}(\hat\Gamma/[\hat N\hat\Gamma + \hat\Lambda_r]) < \dim_{\underline{k}}(\hat\Gamma/[\hat N\hat\Gamma + \hat\Lambda]) = \mu_{\hat\Lambda}(\hat\Gamma/\hat\Lambda) = 2,$$

since $\hat{\Lambda}_r \not\subset \hat{\Lambda} + \hat{N}\hat{\Gamma}$, where $\underline{\underline{k}} = \hat{\Lambda}/\hat{N}$. Hence

2.13
$$\dim_{\underline{\underline{k}}}(\hat{\Gamma}/[\hat{N}\hat{\Gamma} + \hat{\Lambda}_r]) \leq 1; \text{ i.e.,}$$

$$1 \geq \mu_{\hat{\Lambda}}(\hat{\Gamma}/\hat{\Lambda}_r) \geq \mu_{\hat{\Lambda}_r}(\hat{\Gamma}/\hat{\Lambda}_r).$$

If $\hat{\Lambda}_r$ is indecomposable, it is a Bass-order by (IX, 6.14). If it decomposes, say $\hat{\Lambda}_r = \hat{\Lambda}_1 \oplus \hat{\Lambda}_2$, then

$$1 \geq \mu_{\hat{\Lambda}_r}(\hat{\Gamma}/\hat{\Lambda}_r) = \max\{\mu_{\hat{\Lambda}_1}(\hat{\Gamma}\hat{\Lambda}_1/\hat{\Lambda}_1), \mu_{\hat{\Lambda}_2}(\hat{\Gamma}\hat{\Lambda}_2/\hat{\Lambda}_2)\},$$

and we conclude that the primary components of $\hat{\Lambda}_r$ are Bass-orders, whence $\hat{\Lambda}_r$ is a Bass-order. To show that $\hat{\Lambda}_1 = \Lambda_1(\hat{N})$ is also a Bass-order, it suffices to show - in view of the previous part - that $\hat{\Lambda}_1 \not\subset \hat{\Lambda} + \hat{N}\hat{\Gamma}$. But, if $\hat{\Lambda}_1 \subset \hat{\Lambda} + \hat{N}\hat{\Gamma}$, then $\hat{\Lambda}_1 = \hat{\Lambda} + (\hat{N}\hat{\Gamma})^i$ for some i (cf. proof of 2.11) and hence $\hat{N}\hat{\Lambda}_1 = \hat{\Lambda}_1\hat{N} \subset \hat{N}$, and $\hat{N}$ is a two-sided $\hat{\Lambda}_1$-module and $\hat{\Lambda}_1/\hat{\Lambda}$ is annihilated on both sides by $\hat{N}$. As left submodule of $\hat{\Lambda}_o/\hat{\Lambda}$ it is cyclic; i.e., as left module it is isomorphic to $\underline{\underline{k}}$. Consequently, it is isomorphic to $\underline{\underline{k}}$ as right module; i.e., $\mu_{\hat{\Lambda}}(\hat{\Lambda}_1/\hat{\Lambda}) = 1$ and $\hat{\Lambda}$ is Gorenstein by (2.11). It remains to show $\hat{\Lambda}_r = \hat{\Lambda}_1$. (2.13) implies either $\hat{\Gamma} = \hat{N}\hat{\Gamma} + \hat{\Lambda}_r$; i.e., $\hat{\Gamma} = \hat{\Lambda}_r$ by Nakayama's lemma, in which case $\hat{\Lambda}_r = \hat{\Lambda}_1 = \hat{\Gamma}$ or $\hat{N}\hat{\Gamma} + \hat{\Lambda}_r$ and $\hat{N}\hat{\Gamma} + \hat{\Lambda}_1$ are maximal $\hat{\Lambda}$-submodules of $\hat{\Gamma}$. (2.13) then implies

$$\dim_{\underline{\underline{k}}}([\hat{\Lambda}_r + \hat{N}\hat{\Gamma}]/[\hat{\Lambda} + \hat{N}\hat{\Gamma}]) = 1; \text{ so}$$

$$\dim_{\underline{\underline{k}}}([\hat{\Lambda}_r + \hat{N}\hat{\Gamma}]/\hat{N}\hat{\Gamma}/[\hat{\Lambda} + \hat{N}\hat{\Gamma}]/\hat{N}\hat{\Gamma}) = 1; \text{ so}$$

$$\dim_{\underline{\underline{k}}}([\hat{\Lambda}_r + \hat{N}\hat{\Gamma}]/\hat{N}\hat{\Gamma}) = 2, \text{ since } \hat{\Lambda} + \hat{N}\hat{\Gamma} \neq \hat{N}\hat{\Gamma}, \text{ so}$$

$$\dim_{\underline{\underline{k}}}(\hat{\Lambda}_r/[\hat{\Lambda}_r \cap \hat{N}\hat{\Gamma}]) = 2, \text{ so}$$

$$\dim_{\underline{\underline{k}}}(\hat{\Lambda}_r/N/[\hat{\Lambda}_r \cap \hat{N}\hat{\Gamma}]/\hat{N}) = 2.$$

But

$$[\hat{\Lambda}_r \cap \hat{N}\hat{\Gamma}]/\hat{N} \cong [\hat{\Lambda}_r \cap \hat{N}\hat{\Gamma}]/([\hat{\Lambda}_r \cap \hat{N}\hat{\Gamma}] \cap \hat{\Lambda}) \cong [(\hat{\Lambda}_r \cap \hat{N}\hat{\Gamma}) + \hat{\Lambda}]/\hat{\Lambda}$$

is a cyclic $\hat{\Lambda}$-module by (2.11) as submodule of $[\hat{\Lambda} + \hat{N}\hat{\Gamma}]/\hat{\Lambda}$.

If $\hat{\Lambda}_r \cap \hat{N}\hat{\Gamma} = \hat{N}$, then $\dim_{\underline{\underline{k}}}(\hat{\Lambda}_r/\hat{N}) = 2 = \mu_{\hat{\Lambda}}(\hat{\Lambda}_r)$ implies that $\hat{\Lambda}$ is Gorenstein, a contradiction. Thus $\dim_{\underline{\underline{k}}}(\hat{\Lambda}_r/\hat{N}) = 3$ and $\hat{\Lambda}_r \cap \hat{N}\hat{\Gamma}$ is a minimal $\hat{\Lambda}$-overmodule of $\hat{N}$ which is a two-sided $\hat{\Lambda}_r$-module. Now, $\hat{N}$ is a projective right $\hat{\Lambda}_r$-module, say $\hat{N} = \alpha\hat{\Lambda}_r$. (Observe that $\hat{K}\hat{N} \cong \hat{A}$ and that each indecomposable projective right $\hat{\Lambda}_r$-module occurs with multiplicity 1 in $\hat{\Lambda}_r$; hence it follows from the Krull-Schmidt theorem that $\hat{N} \cong \hat{\Lambda}_r$.) But then $\hat{\Lambda}_1 = \alpha\hat{\Lambda}_r\alpha^{-1}$; and if we can show $\hat{\Lambda}_r\hat{N} \subset \hat{N}$, then $\hat{\Lambda}_r = \hat{\Lambda}_1$. Let us therefore assume $\hat{\Lambda}_r\hat{N} \neq \hat{N}$. Then $\hat{\Lambda}_r\hat{N} = \hat{\Lambda}_r \cap \hat{N}\hat{\Gamma}$, and

$$\dim_{\underline{\underline{k}}}(\hat{\Lambda}_r/\hat{\Lambda}_r\hat{N}) = 2 = \mu^r_{\hat{\Lambda}}(\hat{\Lambda}_r).$$

Consequently $\mu^r_{\hat{\Lambda}}(\hat{\Lambda}_r/\hat{\Lambda}) = 1$; i.e., $\overline{Y} = \hat{\Lambda}_r/\hat{\Lambda} = x\hat{\Lambda}$ for some $x \in \overline{Y}$. On the other hand $\overline{Y}$ is a two-sided $\hat{\Lambda}$-module, and so $0 \neq \overline{X} = \hat{\Lambda}x \subset \overline{Y}$. Since $\hat{N}\overline{Y} \neq 0$, $\overline{X} \cong \underline{\underline{k}}$ as cyclic left $\hat{\Lambda}$-module. In addition $\overline{X}\hat{\Lambda} = \overline{Y}$. Taking the preimage of $\overline{X}$ with respect to the caconical epimorphism $\hat{\Lambda}_r \longrightarrow \hat{\Lambda}_r/\hat{\Lambda}$, we get a minimal left $\hat{\Lambda}$-overmodule $\hat{X}$ of $\hat{\Lambda}$ such that $\hat{X}\hat{\Lambda} = \hat{\Lambda}_r$. There are two possibilities, either $\hat{\Lambda}_1 \cap \hat{X} = \hat{X}$ or $\hat{\Lambda}_1 \cap \hat{X} = \hat{\Lambda}$; $\hat{X}$ being minimal. In the first case, $\hat{X} \subset \hat{\Lambda}_1 \cap \hat{\Lambda}_r$ and $\hat{\Lambda}_r = \hat{X}\hat{\Lambda} \subset \hat{\Lambda}_1 \cap \hat{\Lambda}_r$; i.e., $\hat{\Lambda}_1 = \hat{\Lambda}_r$, a contradiction to our assumption. Hence $\hat{\Lambda}_1 \cap \hat{X} = \hat{\Lambda}$ and

$$\hat{\Lambda}_1/\hat{\Lambda} = \hat{\Lambda}_1/(\hat{X} \cap \hat{\Lambda}_1) \cong (\hat{\Lambda}_1 + \hat{X})/\hat{\Lambda}_1 = (\hat{\Lambda}_1 + \hat{X}\hat{\Lambda})/\hat{\Lambda}_1 = (\hat{\Lambda}_1 + \hat{\Lambda}_r)/\hat{\Lambda}_1,$$

since $\hat{\Lambda}_1/\hat{\Lambda}$ is a two-sided $\hat{\Lambda}$-module. But then

$$\hat{N}(\hat{\Lambda}_1/\hat{\Lambda}) \cong \hat{N}([\hat{\Lambda}_1 + \hat{\Lambda}_r]/\hat{\Lambda}_1) = 0$$

and $\hat{N}\hat{\Lambda}_1 \subset \hat{\Lambda}$; i.e., $\hat{N}\hat{\Lambda}_1 = \hat{N}$ and $\hat{\Lambda}_1 \subset \hat{\Lambda}_r$; i.e., $\hat{\Lambda}_1 = \hat{\Lambda}_r$ a contradiction to our assumption. Hence we must have $\hat{\Lambda}_1 = \hat{\Lambda}_r$. #

2.14 __Corollary:__ If $\Lambda_1(\hat{N}) = \hat{\Lambda}_1$ is completely primary, then

$$\hat{\Lambda} + \operatorname{rad}\hat{\Lambda}_1 = \alpha^{-1}\hat{\Lambda}\alpha + \operatorname{rad}\hat{\Lambda}_1, \text{ and } \alpha^{-1}\hat{N}\alpha = \hat{N},$$

where $\hat{\Lambda}_1\alpha = \hat{N}$.

__Proof:__ In the proof of (2.13) we have shown that $\alpha^{-1}\hat{\Lambda}_1\alpha = \hat{\Lambda}_1$ and since $\hat{N} = \hat{\Lambda}_1\alpha = \alpha\hat{\Lambda}_1$, as is easily seen, we have $\hat{N} = \alpha^{-1}\hat{N}\alpha$. Hence

$\underline{\underline{k}}_1 = \alpha^{-1}\hat{\Lambda}\alpha/\hat{N} \cong \hat{\Lambda}/\hat{N} = \underline{\underline{k}}$. Moreover, we have two unitary ring monomorphisms

$$\varphi : \underline{\underline{k}} \longrightarrow \hat{\Lambda}_1/\mathrm{rad}\,\hat{\Lambda}_1 = \underline{\underline{k}}_2 ,$$

$$\psi : \underline{\underline{k}}_1 \longrightarrow \hat{\Lambda}_1/\mathrm{rad}\,\hat{\Lambda}_1, \quad x \qquad \alpha^{-1}(x)\varphi\alpha,$$

since $\hat{N} = \mathrm{rad}\,\hat{\Lambda}_1 \cap \hat{\Lambda}$ (cf. 1.14). Thus $\mathrm{Im}\,\varphi$ and $\mathrm{Im}\,\psi$ are subfields of $\underline{\underline{k}}_2$ both with the same number of elements. But $\underline{\underline{k}}_2$ is a finite field and hence $\mathrm{Im}\,\varphi = \mathrm{Im}\,\psi$; i.e., $\hat{\Lambda} + \mathrm{rad}\,\hat{\Lambda}_1 = \alpha^{-1}\hat{\Lambda}\alpha + \mathrm{rad}\,\Lambda_1$. #

2.15 We <u>summarize</u>: The following cases can occur:

(i) $\hat{A} = \hat{D}_1$,

(ii) $\hat{A} = \hat{D}_1 \oplus \hat{D}_2$,

(iii) $\hat{A} = \hat{D}_1 \oplus \hat{D}_2 \oplus \hat{D}_3$,

where $\{\hat{D}_i\}_{1 \le i \le 3}$ are skewfields.

We may assume

1.) $\hat{\Lambda}$ is not a Gorenstein-order,

2.) $\mu_{\hat{\Lambda}}(\hat{\Gamma}/\hat{\Lambda}) = 2$,

3.) $\mu_{\hat{\Lambda}}(\hat{\Lambda}_o/\hat{\Lambda}) = 1$, $\hat{\Lambda}_o = \hat{\Lambda} + \hat{N}\hat{\Gamma}$ and $\hat{\Lambda}_o$ contains at most two idempotents,

4.) $\hat{\Lambda}_1 = \hat{\Lambda}_r$ is a Bass-order with $\dim_{\underline{\underline{k}}}(\hat{\Lambda}_1/\hat{\Lambda}) = 3$ and $\hat{\Lambda}_1 \not\subset \hat{\Lambda} + \hat{N}\hat{\Gamma}$
$\hat{N} = \hat{\Lambda}_1\alpha = \alpha\hat{\Lambda}_1$ and $\alpha^{-1}\hat{N}\alpha = \hat{N}$.

5.) If $\hat{\Lambda}_1$ is completely primary then

$$\hat{\Lambda} + \mathrm{rad}\,\hat{\Lambda}_1 = \alpha^{-1}\hat{\Lambda}\alpha + \mathrm{rad}\,\hat{\Lambda}_1.$$

6.) Either $\hat{\Lambda}_1 = \hat{\Gamma}$ or $\hat{\Lambda}_1 + \hat{N}\hat{\Gamma}$ is a maximal $\hat{\Lambda}$-submodule of $\hat{\Gamma}$.

<u>Exercise §2</u>:

1.) Fill in the details in the proof of (2.11).

§3 The case $\hat{A} = \hat{D}$

We keep the notation of the previous sections and assume that $\hat{A} = \hat{D}$ is a skewfield.

3.1 Proposition: $\hat{\Lambda}_1/\hat{N} = S$ is a three-dimensional $\underline{\underline{k}}$-vectorspace, and one of the following cases must occur:

(i) $\hat{\Lambda}_1 = \hat{\Gamma}$ and S is a three-dimensional extension field of $\underline{\underline{k}}$.

(ii) $\hat{\Lambda}_1 = \hat{\Gamma}$ and $S = \underline{\underline{k}}_1[r]$, $r = \omega + \hat{N}$, where $\omega\hat{\Gamma} = \text{rad}\,\hat{\Gamma}$; $r^3 = 0$, $\underline{\underline{k}}_1 r = r\underline{\underline{k}}_1$, $\underline{\underline{k}}_1 = \hat{\Gamma}/\omega\hat{\Gamma} \cong \underline{\underline{k}}$ and $\underline{\underline{k}}$ acts as $\underline{\underline{k}}_1$ on S.

(iii) $\hat{\Lambda}_1$ is a maximal $\hat{\Lambda}$-submodule of $\hat{\Gamma}$ and $S = \underline{\underline{k}}_2[r]$, $r = \omega^2 + \hat{N}$, $r^3 = 0$, $\underline{\underline{k}}_2 r = r\underline{\underline{k}}_2$, $\underline{\underline{k}}_2 = \hat{\Lambda}_1/\text{rad}\,\hat{\Lambda}_1 \cong \underline{\underline{k}}$ and $\underline{\underline{k}}$ acts as $\underline{\underline{k}}_2$ on S.

Proof: From (2.15, 4.6.) we conclude that S is a three-dimensional $\underline{\underline{k}}$-vectorspace and either $\hat{\Lambda}_1 = \hat{\Gamma}$ or $\hat{\Lambda}_1 + \hat{N}\hat{\Gamma}$ is a maximal submodule of $\hat{\Gamma}$. If $\hat{\Lambda}_1 = \hat{\Gamma}$ then it may happen that $\hat{N} = \text{rad}\,\hat{\Gamma}$. Then S is an extension field of $\underline{\underline{k}}$, say $S = \underline{\underline{k}}(r)$. If $\text{rad}\,\hat{\Gamma} \neq \hat{N}$, then $\hat{N} = \hat{\Gamma}\omega^s$ and $\dim_{\underline{\underline{k}}} \hat{\Gamma}/\hat{N}\hat{\Gamma} = 3$ implies $s = 3$. Putting $r = \omega + \hat{N}$, we get

$$S = \underline{\underline{k}}_1 1 + \underline{\underline{k}}_1 r + \underline{\underline{k}}_1 r^2,\ r^3 = 0,\ \underline{\underline{k}}_1 r = r\underline{\underline{k}}_1,$$

where $\underline{\underline{k}}_1 = \hat{\Gamma}/\omega\hat{\Gamma} \cong \underline{\underline{k}}$, and $\underline{\underline{k}}$ acts as $\underline{\underline{k}}_1$ on S.

We now assume $\hat{\Lambda}_1 \neq \hat{\Gamma}$; in that case rad $S \neq 0$, $\hat{N}$ being a principal $\hat{\Lambda}_1$-module. Since $\hat{\Lambda}_1/\text{rad}\,\hat{\Lambda}_1$ is a field, $\dim_{\underline{\underline{k}}}(\text{rad}\,\hat{\Lambda}_1/\hat{N}) = 2$ and $\hat{\Lambda}_1/\text{rad}\,\hat{\Lambda}_1 = \underline{\underline{k}}_1 \cong \underline{\underline{k}}$. Moreover, $\hat{\Lambda}_1$ being a non-maximal Bass-order, it is contained in a unique minimal over-module $\hat{\Lambda}_1$ which is an order (cf. IX, 6.6), and $\hat{\Lambda}_1$ is the left and right ring of multipliers of rad $\hat{\Lambda}_1$.*) Under the isomorphism $\hat{\Lambda}_1 \xrightarrow{\sim} \hat{N}$, $\hat{\Lambda}_1$ is mapped onto the unique minimal $\hat{\Lambda}_1$-over-module $\hat{N}_1$ of $\hat{N}$ (cf. 2.15, 4.) and $\hat{N}_1/\hat{N} \cong \underline{\underline{k}}$, since $\hat{\Lambda}_1/\text{rad}\,\hat{\Lambda}_1 \cong k$. Thus we have the chain of inclusions

$$\hat{\Lambda}_1 \underset{\neq}{\supset} \hat{\Lambda}_1 \underset{\neq}{\supset} \text{rad}\,\hat{\Lambda}_1 \underset{\neq}{\supset} \hat{N}_1 \underset{\neq}{\supset} \hat{N} = \hat{\Lambda}_1\alpha,$$

and the factormodule of each consecutive pair is isomorphic to $\underline{\underline{k}}$.

*) Note the difference between $\hat{\Lambda}_1$ and $\hat{\Lambda}_1$!

Since $\hat{\Lambda}_1$ is a Bass-order, rad $\hat{\Lambda}_1$ is a projective $\hat{\Lambda}_1$-lattice, say rad $\hat{\Lambda}_1 = \hat{\Lambda}_1\beta$. Thus we have

$$\hat{N}_1\alpha^{-1}/\hat{\Lambda}_1 \cong \hat{N}_1\alpha^{-1}/\hat{N}\alpha^{-1} \cong \hat{N}_1/\hat{N} \cong \underline{\underline{k}},$$

and so $\hat{N}_1\alpha^{-1} = \hat{\Lambda}_1$, $\hat{\Lambda}_1$ being the unique minimal over-module of $\hat{\Lambda}_1$.

However,

$$\underline{\underline{k}} \cong \text{rad}\,\hat{\Lambda}_1/\hat{N}_1 \cong (\text{rad}\,\hat{\Lambda}_1)\beta^{-1}/\hat{N}_1\beta^{-1} \cong \hat{\Lambda}_1/\hat{N}_1\beta^{-1}$$

implies $\hat{N}_1\beta = \text{rad}\,\hat{\Lambda}_1$, $\hat{N}_1$ being a $\hat{\Lambda}_1$-module. But then

$$\text{rad}\,\hat{\Lambda}_1 = \hat{\Lambda}_1\alpha\beta^{-1} \text{ is } \hat{\Lambda}_1 \text{ projective},$$

and so $\hat{\Lambda}_1 = \hat{\Gamma}$, since every hereditary order in $\hat{A}$ is maximal. This shows that $\hat{\Lambda}_1$ is a maximal $\hat{\Lambda}$-submodule of $\hat{\Gamma}$ and hence $\hat{N}\hat{\Gamma} \subset \hat{\Lambda}_1$ (cf. 2.15, 6.). Moreover, $\dim_{\underline{\underline{k}}}(\hat{\Gamma}/\hat{N}\hat{\Gamma}) = 3$ implies $\hat{N}_1 = \hat{N}\hat{\Gamma} = \hat{\Gamma}\omega^3$. Consequently $\hat{\Gamma}/\omega\hat{\Gamma} \cong \underline{\underline{k}}$ and we have the chain of submodules

3.2 $$\hat{N} \subsetneq \hat{N}\hat{\Gamma} = \hat{\Gamma}\omega^3 = \hat{N}_1 \subsetneq \hat{\Gamma}\omega^2 = \text{rad}\,\hat{\Lambda}_1 \subsetneq \hat{\Lambda}_1 \subsetneq \hat{\Gamma}.$$

We claim

3.3 $$\hat{\Gamma}\omega^3 = \hat{N} + \hat{\Lambda}_1\omega^4 \text{ and } \hat{\Gamma}\omega^2 = \hat{N} + \hat{\Lambda}_1\omega^2.$$

$\hat{\Gamma}\omega^4 \subsetneq \hat{N}$, is impossible since $\underline{\underline{k}} \cong \hat{\Gamma}\omega^3/\hat{\Gamma}\omega^4$. But then $\hat{\Gamma}\hat{N} = \hat{\Gamma}\omega^3$ implies $\hat{N} + \hat{\Lambda}_1\omega^4 = \hat{N}\hat{\Gamma}$, since $\hat{N}\hat{\Gamma}$ is the unique minimal over-module of $\hat{N}$. Obviously $\hat{\Lambda}_1\omega^2 + \hat{N} \supsetneq \hat{N}\hat{\Gamma}$, and we must have $\hat{\Lambda}_1\omega^2 + \hat{N} = \text{rad}\,\hat{\Lambda}_1$, since the above chain is the unique chain of submodules such that every module is maximal in the following one. Thus, denoting $\omega^2 + \hat{N}$ by r, we get

$$S = \underline{\underline{k}}_2 \cdot 1 + \underline{\underline{k}}_2 r + \underline{\underline{k}}_2 r^2,\ r^3 = 0,\ \underline{\underline{k}}_2 r^2 = r^2\underline{\underline{k}}_2,$$

$\underline{\underline{k}}_2 = \hat{\Lambda}_1/\text{rad}\,\hat{\Lambda}_1 \cong \underline{\underline{k}}$; and it remains to establish $\underline{\underline{k}}_2 r = r\underline{\underline{k}}_2$. For this it suffices to show that

$$\omega^{-2}\Lambda_1\omega^2 \equiv \hat{\Lambda}_1 \bmod \hat{N}.$$

But as above one shows $\omega^2\hat{\Lambda}_1 + \hat{N} = \text{rad}\,\hat{\Lambda}_1$, since all modules in the chain (3.2) are two-sided $\hat{\Lambda}_1$-modules. Thus $\omega^2\hat{\Lambda}_1 + \hat{N} = \hat{\Lambda}_1\omega^2 + \hat{N}$; i.e., $\underline{\underline{k}}_2 r = r\underline{\underline{k}}_2$. (It should be observed that $\underline{\underline{k}}_2 = \hat{\Lambda}_1/\text{rad}\,\hat{\Lambda}_1$ and $\omega^2(\text{rad}\,\hat{\Lambda}_1)\omega^{-2} = \text{rad}\,\hat{\Lambda}_1$.) #

3.4 Lemma: In all cases of (3.1) we have

$$\text{Ext}^1_{\hat{\Lambda}}(\underline{\underline{k}}, {}_{\hat{\Lambda}}\hat{\Lambda}_1) \overset{\text{nat}}{\cong} \alpha^{-1}\hat{\Lambda}_1/\hat{\Lambda}_1 = T,$$

where

$$T = \underline{\underline{k}}\,\bar{\alpha}^{-1} + \underline{\underline{k}}\,\bar{\alpha}^{-1}r + \underline{\underline{k}}\,\bar{\alpha}^{-1}r^2, \text{ and } \underline{\underline{k}}\,\bar{\alpha}^{-1} = \bar{\alpha}^{-1}\underline{\underline{k}}.$$

(This is to be taken cum grano salis, since one has to distinguish the three cases and use $\underline{\underline{k}}_1$ for $\underline{\underline{k}}$ and keep in mind the different meanings of r.) Moreover T is a $(\underline{\underline{k}},S)$-bimodule.

Proof: The exact sequence of $\hat{\Lambda}$-modules

$$0 \longrightarrow \hat{N} \longrightarrow \hat{\Lambda} \longrightarrow \underline{\underline{k}} \longrightarrow 0$$

induces the exact sequence

$$0 \longrightarrow \text{Hom}_{\hat{\Lambda}}(\underline{\underline{k}}, {}_{\hat{\Lambda}}\hat{\Lambda}_1) \longrightarrow \text{Hom}_{\hat{\Lambda}}(\hat{\Lambda}, {}_{\hat{\Lambda}}\hat{\Lambda}_1) \longrightarrow \text{Hom}_{\hat{\Lambda}}(\hat{N}, {}_{\hat{\Lambda}}\hat{\Lambda}_1) \longrightarrow \text{Ext}^1_{\hat{\Lambda}}(\underline{\underline{k}}, {}_{\hat{\Lambda}}\hat{\Lambda}_1) \longrightarrow 0;$$

i.e.,

$$\text{Ext}^1_{\hat{\Lambda}}(\underline{\underline{k}}, {}_{\hat{\Lambda}}\hat{\Lambda}_1) \overset{\text{nat}}{\cong} \text{Hom}_{\hat{\Lambda}}(\hat{N}, {}_{\hat{\Lambda}}\hat{\Lambda}_1)/\hat{\Lambda}_1 \overset{\text{nat}}{\cong} \alpha^{-1}\hat{\Lambda}_1/\hat{\Lambda}_1 = T.$$

We then have two isomorphisms

$$\sigma_1 : S = \hat{\Lambda}_1/\hat{N} \longrightarrow \hat{\Lambda}_1\alpha^{-1}/\hat{\Lambda}_1 = T \quad (\text{cf. } 2.15)$$

of left $\hat{\Lambda}_1$-modules and

$$\sigma_r : S = \hat{\Lambda}_1/\hat{N} \longrightarrow \alpha^{-1}\hat{\Lambda}_1/\hat{\Lambda}_1 = T$$

of right $\hat{\Lambda}_1$-modules, which give T the structure of a (S,S)-bimodule. However, it should be observed that $\sigma_1 \neq \sigma_r$, in general.

The remainder of the statements is clear, since in all three cases of (3.1) $\hat{\Lambda}_1\alpha = \alpha\hat{\Lambda}_1$ (cf. 2.15) and $(\text{rad}\,\hat{\Lambda}_1)\alpha = \alpha(\text{rad}\,\hat{\Lambda}_1)$. By $\bar{\bar{\alpha}}$ we denote $\alpha^{-1} + \hat{\Lambda}_1$. Thus in the cases (ii,iii) we have $\underline{\underline{k}}\,\bar{\bar{\alpha}} = \bar{\bar{\alpha}} \cdot \underline{\underline{k}}$; in case (i) this follows from (2.15, 5.), since then rad $\hat{\Lambda}_1 = \hat{N}$. #

3.5 <u>Lemma</u>: In case (3.1, iii) we have

$$\mathrm{Ext}^1_{\hat{\Lambda}}(\underline{\underline{k}}, {}_{\hat{\Lambda}}\hat{\Gamma}) \overset{\mathrm{nat}}{\cong} \alpha^{-1}\hat{\Gamma}/\hat{\Gamma} = T'.$$

If we put $r' = \omega + \hat{N}\hat{\Gamma}$, and $S' = \hat{\Gamma}/\hat{N}\hat{\Gamma}$, then

$$S' = \underline{\underline{k}}_3 1' + \underline{\underline{k}}_3 r' + \underline{\underline{k}}_3 r'^2, r'^3 = 0, \underline{\underline{k}}_3 r' = r'\underline{\underline{k}}_3,$$

where

$$\underline{\underline{k}}_3 = \hat{\Gamma}/\omega\hat{\Gamma} \cong \underline{\underline{k}}.$$

Moreover, putting $\bar{\bar{\alpha}}' = \alpha^{-1} + \hat{\Gamma}$, we have

$$T' = \underline{\underline{k}}_3\bar{\bar{\alpha}}'1' + \underline{\underline{k}}_3\bar{\bar{\alpha}}'r' + \underline{\underline{k}}_3\bar{\bar{\alpha}}'r'^2.$$

$\underline{\underline{k}}$ acts as $\underline{\underline{k}}_3$ on S' and T', and T' is an S'-bimodule.

<u>Proof</u>: As in (3.4) one shows

$$\mathrm{Ext}^1_{\hat{\Lambda}}(\underline{\underline{k}}, {}_{\hat{\Lambda}}\hat{\Gamma}) \overset{\mathrm{nat}}{\cong} \mathrm{Hom}_{\hat{\Lambda}}(\hat{N}, {}_{\hat{\Lambda}}\hat{\Gamma})/\hat{\Gamma} \overset{\mathrm{nat}}{\cong} \alpha^{-1}\hat{\Gamma}/\hat{\Gamma},$$

since $\hat{\Gamma}\hat{N} = \hat{\Gamma}\alpha$. The remainder of the statement is clear. #

3.6 <u>Lemma</u>: In (3.1,iii), S' is a two-sided S-module, and the natural injection

$$\tilde{\varphi} : \hat{\Lambda}_1 \longrightarrow \hat{\Gamma}$$

induces a $(\underline{\underline{k}},S)$-bimodule-homomorphism

$$\begin{aligned} \varphi : T &\longrightarrow T', \\ \bar{\bar{\alpha}}1 &\longmapsto \bar{\bar{\alpha}}'1', \\ \bar{\bar{\alpha}}r &\longmapsto \bar{\bar{\alpha}}'r'^2, \\ \bar{\bar{\alpha}}r^2 &\longmapsto 0. \end{aligned}$$

<u>Proof</u>: $\tilde{\varphi}$ induces a $(\underline{\underline{k}},S)$-bimodule-homomorphism

$$\varphi : T = \alpha^{-1}\hat{\Lambda}_1/\hat{\Lambda}_1 \longrightarrow \alpha^{-1}\hat{\Gamma}/\hat{\Gamma} = T',$$

$$\alpha^{-1}\lambda + \hat{\Lambda}_1 \longmapsto \alpha^{-1}\lambda + \hat{\Gamma}.$$

Recalling that $r = \omega^2 + \hat{N} = \omega_1^2 + \hat{\Lambda}_1$, $r' = \omega + \hat{\Gamma}\hat{N} = \omega + \hat{\Gamma}\alpha$, the statement follows immediately with (3.4, 3.5). #

3.7 <u>Lemma</u>: In (3.1,iii), the left $\hat{\Lambda}_1$-homomorphism

$$\tilde{\psi} : \hat{\Gamma} \longrightarrow \hat{\Lambda}_1, \quad \gamma \longmapsto \gamma\,\omega^2,$$

induces a left $\underline{\underline{k}}$-homomorphism

$$\begin{aligned} \psi : \quad T' &\longrightarrow T, \\ \bar{\bar{\alpha}}'1' &\longmapsto \bar{\bar{\alpha}} r, \\ \bar{\bar{\alpha}}' r' &\longmapsto 0, \\ \bar{\bar{\alpha}}' r'^2 &\longmapsto \bar{\bar{\alpha}} r^2 k \qquad \text{for some } 0 \neq k \in \underline{\underline{k}}. \end{aligned}$$

<u>Proof</u>: $\tilde{\psi}$ induces the left $\underline{\underline{k}}$-homomorphism

$$\begin{aligned} \psi : T' = \alpha^{-1}\hat{\Gamma}/\hat{\Gamma} &\longrightarrow \alpha^{-1}\hat{\Lambda}_1/\hat{\Lambda}_1 = T, \\ \alpha^{-1}\gamma + \hat{\Gamma} &\longmapsto \alpha^{-1}\gamma\,\omega^2 + \hat{\Lambda}_1, \end{aligned}$$

thus

$$\begin{aligned} \psi : \bar{\bar{\alpha}}'1' &\longmapsto \bar{\bar{\alpha}} \cdot (\omega^2 + \hat{\Lambda}_1) = \bar{\bar{\alpha}} r, \\ \bar{\bar{\alpha}}' r' = \alpha^{-1}\omega + \Gamma &\longmapsto \alpha^{-1}\omega^3 + \hat{\Lambda}_1, \end{aligned}$$

and it remains to show $\alpha^{-1}\omega^3 \in \hat{\Lambda}_1$. Since $\hat{\Gamma}\alpha = \hat{\Gamma}\omega^3$, we can write $\alpha^{-1} = u\,\omega^{-3}$ where u is a unit in $\hat{\Gamma}$. If $\bar{u}$ denotes the image of u in $\hat{\Gamma}/\hat{N}\hat{\Gamma}$, then

$$\psi : \bar{\bar{\alpha}}' \cdot \bar{u}^{-1} r' \longmapsto 1 + \hat{\Lambda}_1 = 0.$$

But ψ is a $\underline{\underline{k}}$-homomorphism and $\underline{\underline{k}}$ acts on $\hat{\Gamma}/\hat{N}\hat{\Gamma}$ as $\underline{\underline{k}}_3 = \hat{\Gamma}/\mathrm{rad}\,\hat{\Gamma}$, and consequently, there exists $\bar{u}' \in \underline{\underline{k}}_3$ such that $\bar{u}'\,\bar{\bar{\alpha}}' = \bar{\bar{\alpha}}' \cdot \bar{u}^{-1}$ and

$$(\bar{\bar{\alpha}}' \cdot r')\psi = \bar{u}'^{-1}(\bar{u}'\bar{\bar{\alpha}}' \cdot r')\psi = \bar{u}'^{-1} \cdot 0 = 0.$$

Thus

$$\begin{aligned} \psi &: \bar{\bar{\alpha}}' \cdot r' \longmapsto 0, \text{ and} \\ \psi &: \bar{\bar{\alpha}}' \cdot r'^2 = \alpha^{-1}\omega^2 + \hat{\Gamma} \longmapsto \bar{\bar{\alpha}} \cdot \omega^4 + \hat{\Lambda}_1, \end{aligned}$$

and a similar argument as above shows $\alpha^{-1}\omega^4 + \hat{\Lambda}_1 = \bar{\bar{\alpha}} \cdot r'^2 \cdot k$ for some $0 \neq k \in \underline{\underline{k}}$. #

3.8 <u>Remark</u>: In the cases (i,ii) of (4.1) the only indecomposable $\hat{\Lambda}_1$-lattice is $\hat{\Lambda}_1$, and in case (iii) the non-isomorphic indecomposable $\hat{\Lambda}_1$-lattices are $\hat{\Lambda}_1$ and $\hat{\Gamma}$.

3.9 <u>Remark</u>: (i) Every element $\tilde{\sigma} \in \mathrm{Hom}_{\hat{\Lambda}}(\hat{\Lambda}_1, {}_{\hat{\Lambda}}\hat{\Gamma})$ induces a left $\underline{\underline{k}}$-homomorphism

$$\sigma : \mathrm{Ext}^1_{\hat{\Lambda}}(\underline{\underline{k}}, {}_{\hat{\Lambda}}\hat{\Lambda}_1) \longrightarrow \mathrm{Ext}^1_{\hat{\Lambda}}(\underline{\underline{k}}, {}_{\hat{\Lambda}}\hat{\Gamma}).$$

However, $\mathrm{Hom}_{\hat{\Lambda}}(\hat{\Lambda}_1, {}_{\hat{\Lambda}}\hat{\Gamma}) \overset{\mathrm{nat}}{\cong} \hat{\Gamma}$, and $\tilde{\sigma}$ can be written uniquely as $\tilde{\sigma} = \tilde{\varphi}\gamma$ for some $\gamma \in \hat{\Gamma}$. Hence $\sigma = \varphi s'$ for some $s' \in S' = \hat{\Gamma}/\hat{N}\hat{\Gamma}$; and every $s' \in S'$ can occur.

(ii) Every $\tilde{\tau} \in \mathrm{Hom}_{\hat{\Lambda}}(\hat{\Gamma}, {}_{\hat{\Lambda}}\hat{\Lambda}_1) \overset{\mathrm{nat}}{\cong} \hat{\Gamma}\omega^2$ induces a left $\underline{\underline{k}}$-homomorphism

$$\tau : \mathrm{Ext}^1_{\hat{\Lambda}}(\underline{\underline{k}}, {}_{\hat{\Lambda}}\hat{\Gamma}) \longrightarrow \mathrm{Ext}^1_{\hat{\Lambda}}(\underline{\underline{k}}, {}_{\hat{\Lambda}}\hat{\Lambda}_1).$$

However, $\tilde{\tau} = \gamma\tilde{\psi}$ and thus $\tau = s'\psi$ for some $s' \in S'$; and every s' can occur.

§ 4 The case $\hat{A} = \hat{D}_1 \oplus \hat{D}_2$

We keep the notation of the previous sections and assume now that $\hat{A} = \hat{D}_1 \oplus \hat{D}_2$ is the direct sum of two skewfields.

4.1 **Lemma**: Let $\hat{\Gamma} = \hat{\Omega}_1 \oplus \hat{\Omega}_2$, $\hat{\Omega}_i = \hat{\Gamma} e_i$, where $\hat{\Omega}_i$ is the maximal $\hat{R}$-order in $\hat{D}_i$, $i=1,2$. Then we have - if necessary after renumbering:

(i) $\mu_{\hat{\Lambda}}(\hat{\Omega}_1) = 2$ and $\mu_{\hat{\Lambda}}(\hat{\Omega}_2) = 1$,

(ii) $\hat{\Lambda}_1 = \hat{\Lambda} e_1 \neq \hat{\Omega}_1$ and $\hat{\Lambda}_1$ is a Bass-order; *)

(iii) $\hat{\Lambda}_2 = \hat{\Lambda} e_2 = \hat{\Omega}_2$.

Proof: $\mu_{\hat{\Lambda}}(\hat{\Gamma}) = 3 = \dim_{\underline{k}}(\hat{\Gamma}/\hat{N}\hat{\Gamma}) = \dim_{\underline{k}}(\hat{\Omega}_1/\hat{N}\hat{\Omega}_1) + \dim_{\underline{k}}(\hat{\Omega}_2/\hat{N}\hat{\Omega}_2) =$

$$= \mu_{\hat{\Lambda}}(\hat{\Omega}_1) + \mu_{\hat{\Lambda}}(\hat{\Omega}_2).$$

Hence we may assume $\mu_{\hat{\Lambda}}(\hat{\Omega}_1) = 2$ and $\mu_{\hat{\Lambda}}(\hat{\Omega}_2) = 1$. However, all idempotents are two-sided and thus

$$\mu_{\hat{\Lambda}_1}(\hat{\Omega}_1) = 2 \text{ and } \mu_{\hat{\Lambda}_2}(\hat{\Omega}_2) = 1; \text{ i.e.,}$$

$\hat{\Lambda}_2 = \hat{\Omega}_2$, $\hat{\Lambda}_1 \neq \hat{\Omega}_1$ and $\hat{\Lambda}_1$ is a Bass-order (cf. IX, 6.14). #

4.2 **Lemma**: If $\hat{\Lambda}_1 = \Lambda_1(\hat{N})$ is indecomposable, then $\hat{\Lambda}_1$ is a subdirect sum of $\hat{\Omega}_1$ and $\hat{\Omega}_2$. If $\hat{\Lambda}_1$ decomposes, then $\hat{N} = \hat{N}e_1 \oplus \hat{N}e_2$.

Proof: The first statement follows from (IX, 6.14), $\hat{\Lambda}_1$ being a Bass-order. If $\hat{\Lambda}_1$ decomposes, so does $\hat{N}$, since it is isomorphic to $\hat{\Lambda}_1$ (cf. 2.15). #

4.3 **Lemma**: If $\hat{\Lambda}_1$ decomposes, then

$$\hat{\Lambda} \cap \hat{\Omega}_1 = \text{rad}\, \hat{\Lambda}_1 = \hat{N}e_1$$

and the $\hat{R}$-orders containing $\hat{\Lambda}_1$ are linearly ordered.

Proof: $\hat{\Lambda} \cap \hat{\Omega}_1 = \hat{N} \cap \hat{\Omega}_1 = \hat{N}e_1$. However

$\hat{\Lambda}_1 / \hat{N}e_1 = \hat{\Lambda}/(\hat{\Lambda} \cap \hat{D}_2)/\hat{N}(\hat{\Lambda}/[\hat{\Lambda} \cap \hat{D}_2]) \cong \hat{\Lambda}/[\hat{N} + (\hat{\Lambda} \cap \hat{D}_2)] = \hat{\Lambda}/\hat{N}$;

and this shows $\hat{N}e_1 = \hat{N}_1 = \text{rad}\, \hat{\Lambda}_1$; observe that $\hat{\Lambda}_1 = \hat{\Lambda} e_1 = \hat{\Lambda}/(\hat{\Lambda} \cap \hat{D}_2)$.

*) Observe the difference between $\hat{\Lambda}_1$ and $\hat{\Lambda}_1$!

However, $\mu_{\hat{\Lambda}_1}(\hat{\Omega}_1/\hat{\Lambda}_1) = 1$ implies that the $\hat{\Lambda}_1$-modules containing $\hat{\Lambda}_1$ are linearly ordered (cf. proof of 2.11) and these orders are given by

$$\hat{\Omega}_1 = \hat{\Sigma}_s \supsetneqq \hat{\Sigma}_{s-1} \supsetneqq \cdots \supsetneqq \hat{\Sigma}_1 = \hat{\Lambda}_1$$

where $\hat{\Sigma}_i = \hat{\Lambda}_1 + (\hat{N}_1\hat{\Omega}_1)^{s+1-i}$, $1 < i \leq s$. Then $\hat{\Sigma}_i/\hat{\Sigma}_{i-1} \cong \underline{\underline{k}}$, $1<i\leq s$. #

4.4 <u>Proposition</u>: Assume that $\hat{\Lambda}_1$ decomposes. If we put $S_i = \hat{\Sigma}_i/\hat{N}_1\hat{\Sigma}_i$, where $\hat{N}_1 = \hat{N}e_1$, then $\hat{N}_1\hat{\Sigma}_i$ is a two-sided $\hat{\Sigma}_i$-ideal, and we have

(i) $S_1 = \underline{\underline{k}}_1 \cong \underline{\underline{k}}$,

(ii) $S_i = \underline{\underline{k}}_i[b_i]$, $b_i^2 = 0$, $\underline{\underline{k}}_i b_i = b_i \underline{\underline{k}}_i$, $\underline{\underline{k}}_i \cong \underline{\underline{k}}$, $1 < i < s$,

(iii) S_s is either a quadratic extension of $\underline{\underline{k}}$ or $S_s = \underline{\underline{k}}_s[b_s]$, $b_s^2 = 0$, $\underline{\underline{k}}_s b_s = b_s \underline{\underline{k}}_s$, $\underline{\underline{k}}_s \cong \underline{\underline{k}}$; $\underline{\underline{k}}$ acts as $\underline{\underline{k}}_1$ on S_i, $1\leq i\leq s$.

<u>Proof</u>: In the proof of (4.3) we have shown $\hat{\Sigma}_1/\hat{N}_1\hat{\Sigma}_1 = \hat{\Lambda}_1/\hat{N}_1 = \underline{\underline{k}}_1 \cong \underline{\underline{k}}$. We observe that $\hat{\Sigma}_2 = \Lambda_1(\hat{N}_1) = \Lambda_r(\hat{N}_1)$ and $\hat{\Sigma}_2$ being Bass, we have an element $\beta \in \hat{N}_1$ such that $\hat{\Sigma}_2\beta = \beta\hat{\Sigma}_2 = \hat{N}_1$. But then $\beta\hat{\Sigma}_3 = \hat{\Sigma}_3\beta =$ $=$ rad $\hat{\Sigma}_2$ unless $s = 2$; in fact, since rad $\hat{\Sigma}_2$ is a two-sided $\hat{\Sigma}_3$-module (cf. IX, 6.6), and since it is the unique maximal ideal, $\hat{\Sigma}_2\beta \subsetneqq \hat{\Sigma}_3\beta \subset$ rad $\hat{\Sigma}_2$. On the other hand $\dim_{\underline{\underline{k}}}(\hat{\Sigma}_2/\hat{N}_1\hat{\Sigma}_2) = 2$, since $\mu_{\hat{\Lambda}_1}(\hat{\Sigma}_2) = 2$, $\hat{\Sigma}_2/\hat{\Lambda}_1$ being a cyclic $\hat{\Lambda}_1$-module. Thus $\hat{\Sigma}_3\beta =$ rad $\hat{\Sigma}_2 =$ $= \beta\hat{\Sigma}_3$. Observe that $\hat{\Sigma}_2\beta \neq$ rad $\hat{\Sigma}_2$, since $\hat{\Sigma}_2$ is not maximal. A similar argument shows

$$\text{rad } \hat{\Sigma}_i = \hat{\Sigma}_{i+1}\beta = \beta\hat{\Sigma}_{i+1} \text{ for } 2\leq i\leq s-1.$$

However, for $i = s$, we can have either rad $\hat{\Sigma}_s = \beta\hat{\Sigma}_s$, in which case S_s is a field, or rad $\hat{\Sigma}_s \neq \beta\hat{\Sigma}_s$. Now the statements of (4.4) follow easily. #

4.5 <u>Lemma</u>: Assume that $\hat{\Lambda}_1$ decomposes, then

$$\text{Ext}^1_{\hat{\Lambda}}(\hat{\Lambda}_2, {}_{\hat{\Lambda}}\hat{\Sigma}_1) = \underline{\underline{k}}_1 \cong \underline{\underline{k}},$$

$$\mathrm{Ext}^1_{\hat{\Lambda}}(\hat{\Lambda}_2, {}_{\hat{\Lambda}}\hat{\Sigma}_1) = \beta^{-1}\hat{\Sigma}_i/\hat{\Sigma}_i = T_i \cong S_i, 2 \le i \le s.$$

A $\underline{\underline{k}}_i$-basis for S_i is given by $\bar{\bar{\beta}}_i \cdot 1_i, \bar{\bar{\beta}}_i \cdot b_i$, moreover $\underline{\underline{k}}_i \bar{\bar{\beta}}_i = \bar{\bar{\beta}}_i \cdot \underline{\underline{k}}_i$, where $\bar{\bar{\beta}}_i$ is the coset $\beta^{-1} + \hat{\Sigma}_i$. $\underline{\underline{k}}$ acts as $\underline{\underline{k}}_i$ on T_i, except in case S_s is a field. However, if S_s is a field, $T_s = \beta^{-1}\hat{\Omega}_1/\hat{\Omega}_1 = \bar{\bar{\beta}} \cdot S_s$, where S_s is a two-dimensional extension field of $\underline{\underline{k}}$, and here too $\bar{\bar{\beta}} \cdot \underline{\underline{k}} = \underline{\underline{k}}\,\bar{\bar{\beta}}$.

<u>Proof</u>: We have the exact sequence of $\hat{\Lambda}$-lattices

$$0 \to \hat{\Lambda} \cap \hat{D}_1 \to \hat{\Lambda} \to \hat{\Lambda}/(\hat{\Lambda} \cap \hat{D}_1) \to 0.$$

But $\hat{\Lambda} \cap \hat{D}_1 = \hat{N}_1$ and $\hat{\Lambda}/(\hat{\Lambda} \cap \hat{D}_1) = \hat{\Lambda}_2$; so we get the exact sequence

$$0 \to \hat{N}_1 \to \hat{\Lambda} \to \hat{\Lambda}_2 \to 0,$$

which induces the exact sequence

$$0 \to \mathrm{Hom}_{\hat{\Lambda}}(\hat{\Lambda}_2, {}_{\hat{\Lambda}}\hat{\Sigma}_i) \to \mathrm{Hom}_{\hat{\Lambda}}(\hat{\Lambda}, {}_{\hat{\Lambda}}\hat{\Sigma}_i) \to \mathrm{Hom}_{\hat{\Lambda}}(\hat{N}_1, {}_{\hat{\Lambda}}\hat{\Sigma}_i) \to \mathrm{Ext}^1_{\hat{\Lambda}}(\hat{\Lambda}_2, {}_{\hat{\Lambda}}\hat{\Sigma}_i) \to 0.$$

However, $\mathrm{Hom}_{\hat{\Lambda}}(\hat{\Lambda}_2, {}_{\hat{\Lambda}}\hat{\Sigma}_i) = 0$, and so

$$\mathrm{Ext}^1_{\hat{\Lambda}}(\hat{\Lambda}_2, {}_{\hat{\Lambda}}\hat{\Sigma}_i) \overset{\text{nat}}{\cong} \mathrm{Hom}_{\hat{\Lambda}}(\hat{N}_1, {}_{\hat{\Lambda}}\hat{\Sigma}_i)/\hat{\Sigma}_i = \mathrm{Hom}_{\hat{\Sigma}_i}(\hat{\Sigma}_i\hat{N}_1, {}_{\hat{\Lambda}}\hat{\Sigma}_i)/\hat{\Sigma}_i.$$

For $i = 1$, we get

$$\mathrm{Ext}^1_{\hat{\Lambda}}(\hat{\Lambda}_2, {}_{\hat{\Lambda}}\hat{\Sigma}_1) \overset{\text{nat}}{\cong} \hat{\Sigma}_1/\hat{\Lambda}_1 \cong \underline{\underline{k}},$$

and for $2 \le i \le s$,

$$\mathrm{Ext}^1_{\hat{\Lambda}}(\hat{\Lambda}_2, {}_{\hat{\Lambda}}\hat{\Sigma}_i) \cong \beta^{-1}\hat{\Sigma}_i/\hat{\Sigma}_i, \text{ where}$$

β is such that $\hat{N}_1 = \hat{\Sigma}_2\beta$ (cf. proof of 4.4). It remains to show that $\underline{\underline{k}}_i \bar{\bar{\beta}}_i = \bar{\bar{\beta}}_i \cdot \underline{\underline{k}}_i$; but this is clear since $\beta\hat{\Sigma}_i\beta^{-1} = \hat{\Sigma}_i$ and $\beta(\mathrm{rad}\,\hat{\Sigma}_i)\beta^{-1} = \mathrm{rad}\,\hat{\Sigma}_i, 2 \le i \le s$. This also holds if S_s is not a field. However, if S_s is a field; i.e., $\mathrm{rad}\,\hat{\Omega}_1 = \beta\hat{\Omega}_1$, then one shows as in the proof of (2.15, 5.), that $\underline{\underline{k}}_{s-1}\bar{\bar{\beta}} = \bar{\bar{\beta}} \cdot \underline{\underline{k}}_{s-1}$ with $\underline{\underline{k}}_{s-1} = \hat{\Sigma}_{s-1}/\mathrm{rad}\,\hat{\Sigma}_{s-1}$. But $\underline{\underline{k}}_{s-1} = \underline{\underline{k}}$ if S_s is considered as extension field of $\underline{\underline{k}}$. #

4.6 __Proposition:__ Assume that $\hat{\Lambda}_1$ decomposes. Then the homomorphisms

$$\widetilde{\varphi}_{ij} : \hat{\Sigma}_i \longrightarrow \hat{\Sigma}_j, \quad 1 \le i < j \le s,$$

induce $(\underline{\underline{k}}, S_i)$-bimodule homomorphisms

$$\begin{aligned} \varphi_{ij} : T_i &\longrightarrow T_j, \\ \bar{\bar{\beta}}_i \cdot 1_i &\longmapsto \bar{\bar{\beta}}_j \cdot 1_j, \\ \bar{\bar{\beta}}_i \cdot b_i &\longmapsto 0. \text{ *)} \end{aligned}$$

__Proof:__ Obviously, T_i and T_j are $(\underline{\underline{k}}, S_i)$-bimodules, and the homomorphisms φ_{ij} are bimodule homomorphisms.

$$\varphi_{ij} : \bar{\bar{\beta}}_i \cdot 1_i = \beta^{-1} + \hat{\Sigma}_i \longrightarrow \beta^{-1} + \hat{\Sigma}_j = \bar{\bar{\beta}}_j \cdot 1_j$$

and it remains to show that

$$\beta^{-1} \text{rad } \hat{\Sigma}_i \subset \hat{\Sigma}_j \text{ if } i < j; \text{ i.e.,}$$

rad $\hat{\Sigma}_i \subset \beta \hat{\Sigma}_j$; but in the proof of (4.4) we have shown:

$$\text{rad } \hat{\Sigma}_i = \beta \hat{\Sigma}_{i+1} \subset \beta \Sigma_j \text{ for } 1 \le i < s. \qquad \#$$

4.7 __Proposition:__ Assume that $\hat{\Lambda}_1$ decomposes. Then the left $\hat{\Lambda}$-homomorphisms

$$\begin{aligned} \widetilde{\varphi}_{ji} : \hat{\Sigma}_j &\longrightarrow \hat{\Sigma}_i, \quad 1 \le i < j \le s, \\ \sigma &\longmapsto \sigma \beta^{j-i} \end{aligned}$$

induce a left $\underline{\underline{k}}$-homomorphism

$$\begin{aligned} \varphi_{ji} : T_j &\longrightarrow T_i, \\ \bar{\bar{\beta}}_j \cdot 1_j &\longmapsto 0, \\ \bar{\bar{\beta}}_j \cdot 1_j &\longmapsto \bar{\bar{\beta}}_i \cdot b_i k, \text{ for some } 0 \ne k \in \underline{\underline{k}}. \end{aligned}$$

__Proof:__ We first observe that

---

*) We recall: $\hat{N}_i = \hat{\Sigma}_2 \beta$, $\bar{\bar{\beta}}_i = \beta^{-1} + \hat{\Sigma}_i$, $T_i = \beta^{-1} \hat{\Sigma}_i / \hat{\Sigma}_i$, and b_i as in (4.4).

$$\operatorname{Hom}_{\hat{\Sigma}_i}(\hat{\Sigma}_{i+1}, {}_{\hat{\Sigma}_i}\hat{\Sigma}_i) = \operatorname{rad} \hat{\Sigma}_i = \beta\hat{\Sigma}_{i+1}.$$

Thus

$$\tilde{\varphi}_{i+1,i} : \hat{\Sigma}_{i+1} \longrightarrow \hat{\Sigma}_i,$$

induces a left $\underline{\underline{k}}$-homomorphism

$$\varphi_{i+1,i} : \quad T_{i+1} \longrightarrow T_i,$$
$$\beta^{-1}\sigma + \hat{\Sigma}_{i+1} \longmapsto \beta^{-1}\sigma\beta + \hat{\Sigma}_i.$$

Obviously

$$\varphi_{i+1,i} : \beta^{-1}1 + \hat{\Sigma}_{i+1} \longmapsto 0.$$

Next we observe that as a preimage of b_i we can take an element $\beta\sigma_{i+1}$, where $\sigma_{i+1} \in \hat{\Sigma}_{i+1} \setminus \hat{\Sigma}_i$. Thus

$$\varphi_{i+1,i} : \bar{\bar{\beta}}_{i+1} b_{i+1} = \beta^{-1}\beta\sigma_{i+2} + \hat{\Sigma}_{i+1} \longmapsto \beta^{-1}\beta\sigma_{i+2}\beta + \hat{\Sigma}_i.$$

If $\sigma_{i+2}\beta \in \hat{\Sigma}_i$, then it can not be a unit in $\hat{\Sigma}_i$. Hence $\hat{\Sigma}_{i+1}\sigma_{i+2}\beta \subset \hat{\Sigma}_{i+1}\beta$; i.e., $\sigma_{i+2} \in \hat{\Sigma}_{i+1}$, a contradiction. Hence $(\beta^{-1}(\beta\sigma_{i+2}\beta) + \hat{\Sigma}_i) \neq 0$. Moreover, $(\beta\sigma_{i+2}\beta + \hat{\Sigma}_i\beta)$ is not a unit in S_i, and thus

$$\beta^{-1}(\beta\sigma_{i+2}\beta) + \hat{\Sigma}_i = \bar{\bar{\beta}}_i \cdot b_i k, \text{ for some } 0 \neq k \in \underline{\underline{k}}.$$

For the general case, we claim

$$\operatorname{Hom}_{\hat{\Sigma}_i}(\hat{\Sigma}_j, {}_{\hat{\Sigma}_i}\hat{\Sigma}_i) = \beta^{j-i}\hat{\Sigma}_j.$$

To show this, we use induction on $j-i$. For $j-i = 1$, the result has just been established, and

$$\operatorname{Hom}_{\hat{\Sigma}_i}(\hat{\Sigma}_j, {}_{\hat{\Sigma}_i}\hat{\Sigma}_i) = \operatorname{Hom}_{\hat{\Sigma}_i}(\hat{\Sigma}_j, {}_{\hat{\Sigma}_i}\hat{\Sigma}_{i+1}\beta) =$$
$$= \operatorname{Hom}_{\hat{\Sigma}_{i+1}}(\hat{\Sigma}_j, {}_{\hat{\Sigma}_{i+1}}\hat{\Sigma}_{i+1})\beta = \beta^{j-i}\hat{\Sigma}_j.$$

Then we have $\tilde{\varphi}_{j,i} = \tilde{\varphi}_{j,i+1}\tilde{\varphi}_{i+1,i}$ and hence $\varphi_{j,i} = \varphi_{j,i+1}\tilde{\varphi}_{i+1,i}$, and the statements of (4.7) follow easily. #

4.8 __Remark:__ The maps

$$\tilde{\sigma}_{ij} : \hat{\Sigma}_i \longrightarrow \hat{\Sigma}_j \text{ induce maps } \sigma_{ij} : S_i \longrightarrow S_j,$$

and these maps can be factored through φ_{ij}.

4.9 __Theorem:__ If $\hat{\Lambda}_1$ decomposes, we have:

(i) The $\hat{R}$-orders containing the Bass-order $\hat{\Lambda}_1 = \hat{\Lambda}e_1$ are linearly ordered

$$\hat{\Lambda}_1 = \hat{\Sigma}_1 \underset{\neq}{\subset} \hat{\Sigma}_2 \underset{\neq}{\subset} \cdots \underset{\neq}{\subset} \hat{\Sigma}_s = \hat{\Omega}_1,$$

$\hat{\Lambda}_2 = \hat{\Lambda}e_2 = \hat{\Omega}_2$, and $\hat{N}_1 = \hat{N}e_1 = \text{rad}\,\hat{\Lambda}e_1$.

(ii) The non-isomorphic indecomposable $\hat{\Lambda}_1 \oplus \hat{\Lambda}_2$-lattices are $\hat{\Lambda}_2$, $\hat{\Sigma}_i, 1 \leq i \leq s$.

(iii) $\hat{\Sigma}_1/\hat{N}_1 \cong \underline{k}$, $\hat{\Sigma}_i/\hat{\Sigma}_i\hat{N}_1 = \underline{\underline{k}}_i[b_i]$, $1 < i < s$, $b_i^2 = 0$, $\underline{\underline{k}}_i b_i = b_i \underline{\underline{k}}_i$, $\underline{\underline{k}}_i = \hat{\Sigma}_i/\text{rad}\,\hat{\Sigma}_i \cong \underline{\underline{k}}$, and either $\hat{\Sigma}_s/\hat{\Sigma}_s\hat{N}_s = \underline{\underline{k}}_s[b_s]$, $b_s^2 = 0$, $\underline{\underline{k}}_s b_s = b_s\underline{\underline{k}}_s$, $\underline{\underline{k}}_s = \hat{\Sigma}_s/\text{rad}\,\hat{\Sigma}_s \cong \underline{\underline{k}}$, or $\hat{\Sigma}_s/\hat{\Sigma}_s\hat{N}_s$ is a quadratic field extension of $\underline{\underline{k}}$. $S_i = \hat{\Sigma}_i/\hat{N}_1\hat{\Sigma}_i$.

(iv) $\text{Ext}^1_{\hat{\Lambda}}(\hat{\Lambda}_2, \hat{\Sigma}_1) = T_1$, where $T_1 \cong \underline{\underline{k}}$,

$$T_i = \beta^{-1}\hat{\Sigma}_i/\hat{\Sigma}_i = \underline{\underline{k}}_i\bar{\bar{\beta}}_i \cdot 1_i + \underline{\underline{k}}_i\bar{\bar{\beta}}_i \cdot b_i, 1 < i \leq s,$$

where $\bar{\bar{\beta}}_i \cdot \underline{\underline{k}}_i = \underline{\underline{k}}_i\bar{\bar{\beta}}_i$.

(v) T_i is a right S_i-module and $\text{Hom}_{\hat{\Lambda}}(\hat{\Sigma}_i, \hat{\Sigma}_j)$ acts as

$$\varphi_{ij}S_j : \text{Ext}^1_{\hat{\Lambda}}(\hat{\Lambda}_2, {}_{\hat{\Lambda}}\hat{\Sigma}_i) \longrightarrow \text{Ext}^1_{\hat{\Lambda}}(\hat{\Lambda}_2, {}_{\hat{\Lambda}}\hat{\Sigma}_j) \text{ for } i < j$$

and as

$$S_i\varphi_{ij} : \text{Ext}^1_{\hat{\Lambda}}(\hat{\Lambda}_2, {}_{\hat{\Lambda}}\hat{\Sigma}_i) \longrightarrow \text{Ext}^1_{\hat{\Lambda}}(\hat{\Lambda}_2, {}_{\hat{\Lambda}}\hat{\Sigma}_j) \text{ for } i > j.$$

(vi) T_i is a left $\hat{\Lambda}_2$-module, and $\hat{\Lambda}_2$ acts as $\hat{\Lambda}_2/\text{rad}\,\hat{\Lambda}_2 \cong \underline{\underline{k}}$ on $T_i, 1 \leq i \leq s$.

__Proof:__ In view of the previous theorems we need only to prove (vi).

One sees easily that

$$\hat{\Lambda}_2/\hat{N}_2 \cong \hat{\Lambda}/(\hat{\Lambda} \cap \hat{D}_1)/\hat{N}(\hat{\Lambda}/[\hat{\Lambda} \cap \hat{D}_1]) \cong \hat{\Lambda}/[\hat{N} + (\hat{\Lambda} \cap \hat{D}_1)] \cong \hat{\Lambda}/\hat{N} = \underline{\underline{k}},$$

and it suffices to show $\omega_2 \mathrm{Ext}^1_{\hat{\Lambda}}(\hat{\Lambda}_2, {}_{\hat{\Lambda}}\hat{\Sigma}_1) = 0$, where $\omega_2 \hat{\Lambda}_2 = \mathrm{rad}\, \hat{\Lambda}_2$. But $\omega_2 \hat{\Gamma} \subset \hat{\Lambda}$ and hence the right multiplication with ω_2 is a projective endomorphism of $\hat{\Gamma}_2$ (cf. V, § 2, 4.7, 4.10). Thus $\omega_2 \mathrm{Ext}^1_{\hat{\Lambda}}(\hat{\Lambda}_2, {}_{\hat{\Lambda}}\hat{\Sigma}_1) = 0$, (observe $\hat{\Lambda}_2 \oplus \hat{\Omega}_1 = \hat{\Gamma}$). #

We now study the case where $\hat{\Lambda}_1$ is indecomposable.

4.9 __Lemma__: If $\hat{\Lambda}_1$ is indecomposable, we have:

(i) $\hat{\Lambda}_1 e_1 = \hat{\Omega}_1$, $\hat{\Lambda}_1 e_2 = \hat{\Omega}_2$,

(ii) $\hat{\Omega}_1$ is the minimal over-order of the Bass-order $\hat{\Lambda}_1 = \hat{\Lambda} e_1$ and $\hat{N} e_1 = \mathrm{rad}\, \hat{\Lambda}_1$.

__Proof__: This follows readily from (4.2 and IX, 6.6). #

4.10 __Proposition__: Assume that $\hat{\Lambda}_1$ is indecomposable. Then we have:

(i) $\hat{\Omega}_1/\mathrm{rad}\, \hat{\Omega}_1 \cong \underline{\underline{k}}$,

(ii) $\mathrm{rad}\, \hat{\Lambda}_1 = \hat{N} e_1 = \hat{\Omega}_1 \omega_1^2$, $\omega_1 \hat{\Omega}_1 = \mathrm{rad}\, \hat{\Omega}_1$,

(iii) $\hat{I} = \hat{\Lambda} \cap \hat{D}_1 = \hat{\Omega}_1 \omega_1^3$.

__Proof__: We have $2 = \mu_{\hat{\Lambda}}(\hat{\Omega}_1) = \dim_{\underline{\underline{k}}}(\hat{\Omega}_1/\hat{N}\hat{\Omega}_1) = \dim_{\underline{\underline{k}}}(\hat{\Omega}_1/\hat{N} e_1)$. Hence $\hat{N} e_1 = \hat{\Omega} \omega_1^s$ with $s \leq 2$. Moreover,

$$\hat{I} = \hat{\Lambda} \cap \hat{D}_1 = \hat{\Lambda} \cap \hat{\Omega}_1 = \hat{N} \cap \hat{\Omega}_1$$

and so $\hat{I}$ is a $\hat{\Lambda}_1$-module. But $1 \varepsilon \hat{\Lambda}_1$ acts on $\hat{I}$ as e_1, and so $\hat{I}$ is a $\hat{\Lambda}_1 e_1 = \hat{\Omega}_1$ module; i.e., $\hat{I} = \hat{\Omega}_1 \omega_1^t$. If $t = 1$ then $\hat{N} e_1 = \hat{I}$ and hence $\hat{N} e_1 \subset \hat{N}$ and $\hat{N}$ decomposes, a contradiction to the fact that $\hat{\Lambda}_1$ is indecomposable. Thus $t \geq 2$. We have

$$\hat{N} e_1/\hat{I} = \hat{N}_1 e_1/(\hat{\Lambda} \cap \hat{N}_1 e_1) = (\hat{N}_1 e_1 + \hat{\Lambda})/\hat{\Lambda}$$

and $(\hat{N}_1 e_1 + \hat{\Lambda})/\hat{\Lambda}$ is cyclic as submodule of $(\hat{\Lambda} + \hat{N}\hat{\Gamma})/\hat{\Lambda}$ (cf. 2.11);

hence

$$1 \geq \mu_{\hat{\Lambda}}(\hat{N}e_1/\hat{I}) = \dim_{\underline{k}}(\hat{N}e_1/(\hat{N}^2 e_1 + \hat{I}),$$

If $s = 1$, then $\dim_{\underline{k}}(\hat{\Omega}_1/\mathrm{rad}\ \hat{\Omega}_1) = 2$ and we have a contradiction. Thus $s = 2$ and $t = 3$. #

4.11 <u>Lemma</u>: $T_1 = \mathrm{Ext}^1_{\hat{\Lambda}}(\hat{\Lambda}_2, {}_{\hat{\Lambda}}\hat{\Omega}_1) \overset{\text{nat}}{\cong} \omega_1^{-3}\hat{\Omega}_1/\hat{\Omega}_1$ and $T_2 = \mathrm{Ext}^1_{\hat{\Lambda}}(\hat{\Lambda}_2, {}_{\hat{\Lambda}}\hat{\Lambda}_1) \overset{\text{nat}}{\cong} \omega_1^{-1}\hat{\Omega}_1/\hat{\Lambda}_1$.

<u>Proof</u>: We have the exact sequence

$$0 \longrightarrow \hat{I} \longrightarrow \hat{\Lambda} \longrightarrow \hat{\Lambda}_2 \longrightarrow 0,$$

which shows

$$\mathrm{Ext}^1_{\hat{\Lambda}}(\hat{\Lambda}_2, {}_{\hat{\Lambda}}\hat{\Omega}_1) \cong \mathrm{Hom}_{\hat{\Lambda}}(\hat{I}, \hat{\Omega}_1)/\hat{\Omega}_1 = \omega_1^{-3}\hat{\Omega}_1/\hat{\Omega}_1,$$

and

$$\mathrm{Ext}^1_{\hat{\Lambda}}(\hat{\Lambda}_2, {}_{\hat{\Lambda}}\hat{\Lambda}_1) \cong \mathrm{Hom}_{\hat{\Lambda}_1}(\hat{I}, \hat{\Lambda}_1)/\hat{\Lambda}_1 = \omega_1^{-1}\hat{\Omega}_1/\hat{\Lambda}_1. \quad \#$$

4.12 <u>Lemma</u>: As right $\hat{\Lambda}_1$-module, T_1 can be generated by the elements $\{\omega_1^{-3} + \hat{\Omega}_1, \omega_1^{-2} + \hat{\Omega}_1, \omega_1^{-1} + \hat{\Omega}_1\}$ and T_2 can be generated as right $\hat{\Lambda}_1$-module by the elements $\{\omega_1^{-1} + \hat{\Lambda}_1, \omega_1 + \hat{\Lambda}_1\}$.

<u>Proof</u>: Since $\hat{\Omega}_1$ is the minimal $\hat{\Lambda}_1$-over-module of $\hat{\Lambda}_1$, we must have $\hat{\Lambda}_1 + \omega_1\hat{\Lambda}_1 = \hat{\Omega}_1$ and a composition series of T_1 is given by

$$\omega_1^{-3}\hat{\Omega}_1/\hat{\Omega}_1 \underset{\neq}{\supset} \omega_1^{-2}\hat{\Omega}_1/\hat{\Omega}_1 \underset{\neq}{\supset} \omega^{-1}\hat{\Omega}_1/\hat{\Omega}_1 \underset{\neq}{\supset} 0$$

and $\omega_1^{-1}\hat{\Omega}_1/\hat{\Omega}_1 = (\omega_1^{-1}\hat{\Lambda}_1 + \hat{\Lambda}_1)/\hat{\Omega}_1$,

$$\omega_1^{-2}\hat{\Omega}_1/\omega_1^{-1}\hat{\Omega}_1 = (\omega_1^{-2}\hat{\Lambda}_1 + \omega_1^{-1}\hat{\Lambda}_1)/(\omega_1^{-1}\hat{\Lambda}_1 + \hat{\Lambda}_1),$$

$$\omega_1^{-3}\hat{\Omega}_1/\hat{\Omega}_1 = (\omega_1^{-3}\hat{\Lambda}_1 + \omega_1^{-2}\hat{\Lambda}_1)/(\omega_1^{-2}\hat{\Lambda}_1 + \omega_1^{-1}\hat{\Lambda}_1).$$

Hence every element in T_1 can be written as

$$s = \bar{\bar{\omega}}_1^3 \cdot k_1 + \bar{\bar{\omega}}_1^2 \cdot k_2 + \bar{\bar{\omega}}_1 \cdot k_3,\ k_i \in \underline{\underline{k}}_1 = \hat{\Lambda}_1/\hat{N}_1,$$

where $\bar{\bar{\omega}}_1$ is the coset $\omega_1^{-1} + \hat{\Omega}_1$, $\bar{\bar{\omega}}_1^2 = \omega_1^{-2} + \hat{\Omega}_1$ and $\bar{\bar{\omega}}_1^3 = \omega_1^{-3} + \hat{\Omega}_1$. Similarly $\omega_1^{-1}\hat{\Omega}_1/\hat{\Lambda}_1$ has a composition series

$$(\omega_1^{-1}\hat{\Lambda}_1 + \hat{\Lambda}_1)/\hat{\Lambda}_1 \underset{\neq}{\supset} (\hat{\Lambda}_1 + \omega_1\hat{\Lambda}_1)/\hat{\Lambda}_1 \underset{\neq}{\supset} 0$$

and every element in T_2 can be written as

$$x = \bar{\bar{\omega}}_1 \cdot k_1 + \bar{\omega}_1 k_2, \; k_1 \in \underline{\underline{k}}_1 = \hat{\Lambda}_1/\hat{N}_1, \text{ where}$$

$\bar{\omega}_1 = \omega_1 + \hat{\Lambda}_1$ and $\bar{\bar{\omega}}_1 = \omega_1^{-1} + \hat{\Lambda}_1$. Moreover, these expressions are unique.#

4.13 **Lemma**: The maps of left $\hat{\Lambda}_1$-modules

$$\tilde{\varphi}_{o1} : \hat{\Lambda}_1 \longrightarrow \hat{\Omega}_1,$$
$$\lambda \longmapsto \lambda,$$

and

$$\tilde{\varphi}_{1o} : \hat{\Omega}_1 \longrightarrow \hat{\Lambda}_1,$$
$$\lambda \longmapsto \lambda\omega_1^2,$$

induce a left $\underline{\underline{k}}$-homomorphism

$$\varphi_{o1} : T_2 \longrightarrow T_1,$$
$$\bar{\bar{\omega}}_1 \cdot k_1 + \bar{\omega}_1 k_2 \longmapsto \bar{\bar{\omega}}_1 \cdot k_1,$$

and a left $\underline{\underline{k}}$-homomorphism

$$\varphi_{1o} : T_1 \longrightarrow T_2$$
$$\bar{\bar{\omega}}_1^3 \cdot k_1 + \bar{\bar{\omega}}_1^2 \cdot k_2 + \bar{\bar{\omega}}_1 \cdot k_3 \longmapsto \bar{\bar{\omega}}_1^3 \cdot k_1 \bar{\bar{\omega}}_1^{+2} + \bar{\bar{\omega}}_1 \cdot k_3 \bar{\omega}_1^{+2} = \bar{\bar{\omega}}_1 \cdot k_1' + \bar{\omega}_1 k_3'$$

some $k_1', k_3' \in \underline{\underline{k}}_1$.

Moreover, $\omega_1\hat{\Lambda}_1 = \hat{\Lambda}_1\omega_1$, and k_1', k_3' are different from zero if and only if k_1 and k_3 are different from zero.

Proof: It only remains to show $\omega_1\hat{\Lambda}_1 = \hat{\Lambda}_1\omega_1$. But in the proof of (IX, 6.21) we have shown that for a properly choosen ω_1, $\hat{\Lambda}_1$ is the inverse image of $\hat{\Omega}_1/\text{rad}\,\hat{\Omega}_1$ as submodule of $\hat{\Omega}_1/(\text{rad}\,\hat{\Omega}_1)^2$ and so conjugation with ω_1 induces an automorphism of $\underline{\underline{k}}_1 = \hat{\Lambda}_1/\hat{N}_1$; in particular $\omega_1\hat{\Lambda}_1 = \hat{\Lambda}_1\omega_1$ since $\hat{N}_1$ is a two-sided $\hat{\Omega}_1$-module. #

4.14 <u>Remark</u>: The reader should keep in mind the different meanings of $\overline{\overline{\omega}}_1$ in T_1 and of $\overline{\omega}_1$ in T_2.

4.15 <u>Lemma</u>: T_1 and T_2 are right $\underline{\underline{k}}_1$-modules and if $\omega_2\hat{\Lambda}_2 = \mathrm{rad}\,\hat{\Lambda}_2$ then $\hat{\Lambda}_2$ acts as $\hat{\Lambda}_2/\omega_2^2\hat{\Lambda}_2$ on T_1 and T_2 as follows:

(i) on T_1: $\omega_2(\overline{\overline{\omega}}_1^{\,3}\cdot k_1 + \overline{\overline{\omega}}_1^{\,2}\cdot k_2 + \overline{\overline{\omega}}_1\cdot k_3) = \overline{\overline{\omega}}_1\cdot k_1$.

(ii) on T_2: $\omega_2(\overline{\overline{\omega}}_1\cdot k_1 + \overline{\omega}_1 k_2) = \overline{\omega}_1 k_1$.

<u>Proof</u>: We have the exact sequence

$$E_o : 0 \longrightarrow \hat{I} \longrightarrow \hat{\Lambda} \longrightarrow \hat{\Lambda}/\hat{I} \longrightarrow 0,$$

and we remark that $\hat{\Lambda}/\hat{I} = \hat{\Lambda}e_2 = \hat{\Lambda}_2$. The commutative diagram with exact rows

$$\begin{array}{cccccccccc}
E_o & : & 0 & \longrightarrow & \hat{I} & \longrightarrow & \hat{\Lambda} & \longrightarrow & \hat{\Lambda}/\hat{I} & \longrightarrow 0 \\
\downarrow & & & & \uparrow 1_{\hat{I}} & & \uparrow \psi & & \uparrow \varphi & \\
\omega_2 E_o & : & 0 & \longrightarrow & \hat{I} & \longrightarrow & \hat{X} & \longrightarrow & \hat{\Lambda}/\hat{I} & \longrightarrow 0
\end{array}$$

where φ is multiplication with ω_2 on the right, induces the commutative diagram

$$\begin{array}{ccccccc}
0 \longrightarrow & \mathrm{Hom}_{\hat{\Lambda}}(\hat{\Lambda},{}_{\hat{\Lambda}}\hat{\Omega}_1) & \longrightarrow & \mathrm{Hom}_{\hat{\Lambda}}(\hat{I},{}_{\hat{\Lambda}}\hat{\Omega}_1) & \longrightarrow & \mathrm{Ext}^1_{\hat{\Lambda}}(\hat{\Lambda}_2,{}_{\hat{\Lambda}}\hat{\Omega}_1) & \longrightarrow 0 \\
& \downarrow i & & \downarrow \psi^* & & \downarrow \varphi^* & \\
0 \longrightarrow & \mathrm{Hom}_{\hat{\Lambda}}(\hat{X},{}_{\hat{\Lambda}}\hat{\Omega}_1) & \longrightarrow & \mathrm{Hom}_{\hat{\Lambda}}(\hat{I},{}_{\hat{\Lambda}}\hat{\Omega}_1) & \xrightarrow{\sigma} & \mathrm{Ext}^1_{\hat{\Lambda}}(\hat{\Lambda}_2,{}_{\hat{\Lambda}}\hat{\Omega}_1). &
\end{array}$$

The commutativity of this diagram shows

$$\omega_2\mathrm{Ext}^1_{\hat{\Lambda}}(\hat{\Lambda}_2,{}_{\hat{\Lambda}}\hat{\Omega}_1) = \mathrm{Im}\,\sigma \cong \mathrm{Hom}_{\hat{\Lambda}}(\hat{I},{}_{\hat{\Lambda}}\hat{\Omega}_1)/\mathrm{Hom}_{\hat{\Lambda}}(\hat{X},{}_{\hat{\Lambda}}\hat{\Omega}_1).$$

The construction of $\omega_2 E_o$ shows

$$\hat{X} = \{(\lambda, \lambda_2) : \lambda e_2 = \lambda_2\omega_2, \lambda \in \hat{\Lambda}, \lambda_2 \in \hat{\Lambda}_2\}.$$

But

$$N' = \{\lambda \in \hat{\Lambda} : (\lambda, \lambda_2) \in \hat{X} \text{ for some } \lambda_2 \in \hat{\Lambda}_2\} = \hat{N}.$$

Hence we have

$$\mathrm{Hom}_{\hat\Lambda}(\hat X, {}_{\hat\Lambda}\hat\Omega_1) = \mathrm{Hom}_{\hat\Lambda}(\hat N, {}_{\hat\Lambda}\hat\Omega_1) = \mathrm{Hom}_{\hat\Lambda_1}(\hat N_1, {}_{\hat\Lambda}\hat\Omega_1) = \omega_1^{-2}\hat\Omega_1,$$

and so

$$\text{(i)}\qquad \omega_2 \mathrm{Ext}^1_{\hat\Lambda}(\hat\Lambda_2, {}_{\hat\Lambda}\hat\Omega_1) = \omega_1^{-3}\hat\Omega_1/\omega_1^{-2}\hat\Omega_1,$$

and we get

$$\omega_2(\bar{\bar\omega}_1^3 \cdot k_1 + \bar{\bar\omega}_1^2 \cdot k_2 + \bar{\bar\omega}_1 \cdot k_3) = \bar{\bar\omega}_1 \cdot k_1.$$

This shows also $\omega_2^2 T_1 = 0$, and thus the elements in $\hat\Lambda_2$ act as $\hat\Lambda_2/\omega_2^2\hat\Lambda_2$ and the formula (i) determines the structure of T_1 as $\hat\Lambda_2$-module.

Similarly one shows that

$$\omega_2 \mathrm{Ext}^1_{\hat\Lambda}(\hat\Lambda_2, {}_{\hat\Lambda}\hat\Lambda_1) = \omega_1^{-1}\hat\Omega_1/\hat\Omega_1, \text{ and we get}$$

$$\omega_2(\bar\omega_1 \cdot k_1 + \bar\omega_1 k_2) = \bar\omega_1 k_1. \qquad \#$$

4.16 __Remark:__ If $\hat\Lambda_1$ is indecomposable, then the only $\hat R$-order containing the Bass-order $\hat\Lambda_1 \oplus \hat\Lambda_2$ is $\hat\Omega_1 \oplus \hat\Lambda_2$, and the only non-isomorphic indecomposable $\hat\Lambda_1 \oplus \hat\Lambda_2$-lattices are $\hat\Lambda_1, \hat\Lambda_2, \hat\Omega_1$. $\hat\Omega_1$ acts on the right on $T_1 = \mathrm{Ext}^1_{\hat\Lambda}(\hat\Lambda_2, {}_{\hat\Lambda}\hat\Omega_1)$ by multiplication with the elements in $\hat\Omega_1/\omega_1^3\hat\Omega_1$. $\hat\Lambda_1$ acts on the right on $T_2 = \mathrm{Ext}^1_{\hat\Lambda}(\hat\Lambda_2, {}_{\hat\Lambda}\hat\Lambda_1)$. The morphisms (on the right)

$$\mathrm{Ext}^1_{\hat\Lambda}(\hat\Lambda_2, {}_{\hat\Lambda}\hat\Lambda_1) \longrightarrow \mathrm{Ext}^1_{\hat\Lambda}(\hat\Lambda_2, {}_{\hat\Lambda}\hat\Omega_1)$$

are given by $\varphi_{01}(\hat\Omega_1/\omega_1^3\hat\Omega_1)$ and the maps

$$\mathrm{Ext}^1_{\hat\Lambda}(\hat\Lambda_2, {}_{\hat\Lambda}\hat\Omega_1) \longrightarrow \mathrm{Ext}^1_{\hat\Lambda}(\hat\Lambda_2, {}_{\hat\Lambda}\hat\Lambda_1)$$

are given by $(\hat\Omega_1/\omega_1^3\hat\Omega_1)\varphi_{10}$. We put $S = \hat\Omega_1/\omega_1^3\hat\Omega_1$.

§ 5 The case $\hat{A} = \hat{D}_1 \oplus \hat{D}_2 \oplus \hat{D}_3$

We keep the notation of the previous sections and assume now that $\hat{A}$ is the direct sum of three skewfields $\hat{A} = \hat{D}_1 \oplus \hat{D}_2 \oplus \hat{D}_3$. The maximal $\hat{R}$-order in $\hat{D}_i$ is $\hat{\Omega}_i$ with rad $\hat{\Omega}_i = \omega_i\hat{\Omega}_i$, and the central idempotents in $\hat{A}$ are denoted by e_i, $1 \leq i \leq 3$.

5.1 Lemma: $\hat{\Lambda} e_i = \hat{\Omega}_i$, $1 \leq i \leq 3$.

Proof: This follows immediately from $\mu_{\hat{\Lambda}}(\hat{\Gamma}) = 3$ (cf. proof of 4.1). #

5.2 Lemma: $\hat{\Omega}_i \omega_i = \hat{N} e_i$, $\hat{\Omega}_i/\hat{\Omega}_i\omega_i = \underline{\underline{k}}_i \cong \underline{\underline{k}}$, $1 \leq i \leq 3$.

Proof:

$\underline{\underline{k}}_1 = \hat{\Omega}_1/\hat{N}\hat{\Omega}_1 \cong \hat{\Lambda}/(\hat{\Lambda} \cap [\hat{D}_2 \oplus \hat{D}_3])/\hat{N}(\hat{\Lambda}/[\hat{\Lambda} \cap \{\hat{D}_2 \oplus \hat{D}_3\}]) \cong \hat{\Lambda}/\hat{N} = \underline{\underline{k}}$.

Since $\hat{N}\hat{\Omega}_i = \hat{N}e_i$, the statements follow. #

5.3 Lemma: $\hat{\Lambda}_1$ decomposes and we may assume $\hat{\Lambda}_1 = \hat{\Omega}_1 \oplus \hat{\Sigma}$, where $\hat{\Sigma}$ is a Bass-order in $\hat{D}_1 \oplus \hat{D}_2$.*)

Proof: It follows from (IX, § 6) that $\hat{\Lambda}_1$ decomposes into two Bass-orders, and the statement follows from (5.1). #

5.4 Lemma: $\hat{\Sigma}_o = \hat{\Lambda}(e_1+e_2) = \hat{\Lambda}/(\hat{\Lambda} \cap \hat{\Omega}_3)$ is an indecomposable Bass-order in $\hat{\Omega}_1 \oplus \hat{\Omega}_2$.

Proof: $2 = \mu_{\hat{\Lambda}}(\hat{\Omega}_1 \oplus \hat{\Omega}_2) = \mu_{\hat{\Sigma}_o}(\hat{\Omega}_1 \oplus \hat{\Omega}_2)$ and so $\hat{\Sigma}_o$ is a Bass-order by (IX, 6.14) since $\hat{\Sigma}_o$ can not decompose. If $\hat{\Sigma}_o$ would decompose, then $\mu_{\hat{\Lambda}}(\hat{\Gamma}) = 2$, a contradiction. #

5.5 Lemma: The unique minimal $\hat{\Sigma}_o$-over-module of $\hat{\Sigma}_o$ is $\hat{\Omega}_1 \oplus \hat{\Omega}_2$.

Proof: rad $\hat{\Sigma}_o = \hat{N}(e_1+e_2) = \hat{N}e_1 \oplus \hat{N}e_2$ by (5.3). Hence the ring of multipliers of rad $\hat{\Sigma}_o$ is $\hat{\Omega}_1 \oplus \hat{\Omega}_2$. #

5.6 Lemma: $\hat{I} = \hat{\Lambda} \cap (\hat{\Omega}_1 \oplus \hat{\Omega}_2) \in {}_{\hat{\Omega}_1 \oplus \hat{\Omega}_2}\underline{\underline{M}}^o$; moreover,

---

*) Observe the difference between 1 and i!

$$\hat{I} = (\hat{\Omega}_1 \oplus \hat{\Omega}_2)\alpha \text{ with } \alpha = (\omega_1, \omega_2^d, 0) \text{ for some } d \geq 1.$$

<u>Proof</u>: $\hat{I} = \hat{N} \cap (\hat{\Omega}_1 \oplus \hat{\Omega}_2)$ is an $(\hat{\Omega}_1 \oplus \hat{\Omega}_2)$-lattice and since $\hat{\Lambda}_1 = \Lambda_1(\hat{N}) = \hat{\Omega}_1 \oplus \hat{\Sigma}$, we get $\alpha = (\omega_1, \omega_2^d, 0)$. #

5.7 <u>Lemma</u>: We have

$$\mathrm{Ext}^1_{\hat{\Lambda}}(\hat{\Omega}_3, {}_{\hat{\Lambda}}\hat{\Omega}_1) \overset{nat}{\cong} \omega_1^{-1}\hat{\Omega}_1/\hat{\Omega}_1 = T_1 \cong \underline{\underline{k}},$$

$$\mathrm{Ext}^1_{\hat{\Lambda}}(\hat{\Omega}_3, {}_{\hat{\Lambda}}\hat{\Omega}_2) \overset{nat}{\cong} \omega_2^{-d}\hat{\Omega}_2/\hat{\Omega}_2 = T_2,$$

$$\mathrm{Ext}^1_{\hat{\Lambda}}(\hat{\Omega}_3, {}_{\hat{\Lambda}}\hat{\Sigma}_o) \overset{nat}{\cong} (\hat{\Omega}_1 \oplus \omega_2^{1-d}\hat{\Omega}_2)/\hat{\Sigma}_o = T_o.$$

<u>Proof</u>: The exact sequence with canonical homomorphisms

$$0 \longrightarrow \hat{I} \longrightarrow \hat{\Lambda} \longrightarrow \hat{\Omega}_3 \longrightarrow 0$$

implies

$$\mathrm{Ext}^1_{\hat{\Lambda}}(\hat{\Omega}_3, {}_{\hat{\Lambda}}\hat{X}) \overset{nat}{\cong} \mathrm{Hom}_{\hat{\Lambda}}(\hat{I}, {}_{\hat{\Lambda}}\hat{X})/\hat{X},$$

where $X = \hat{\Omega}_1, \hat{\Omega}_2, \hat{\Sigma}_o$.

(i) For $\hat{X} = \hat{\Omega}_1$, we have

$$\mathrm{Ext}^1_{\hat{\Lambda}}(\hat{\Omega}_3, {}_{\hat{\Lambda}}\hat{\Omega}_1) \overset{nat}{\cong} \omega_1^{-1}\hat{\Omega}_1/\hat{\Omega}_1.$$

(ii) For $\hat{X} = \hat{\Omega}_2$,

$$\mathrm{Ext}^1_{\hat{\Lambda}}(\hat{\Omega}_3, {}_{\hat{\Lambda}}\hat{\Omega}_2) \overset{nat}{\cong} \omega_2^{-d}\hat{\Omega}_2/\hat{\Omega}_2.$$

(iii) For $\hat{X} = \hat{\Sigma}_o$,

$$\mathrm{Hom}_{\hat{\Lambda}}(\hat{I}, \hat{\Sigma}_o) = \mathrm{Hom}_{\hat{\Lambda}}(\hat{I}, \mathrm{rad}\, \hat{\Sigma}_o),$$

since $\hat{\Sigma}_o$ is indecomposable. Thus

$$\mathrm{Ext}^1_{\hat{\Lambda}}(\hat{\Omega}_3, {}_{\hat{\Lambda}}\hat{\Sigma}_o) \overset{nat}{\cong} (\hat{\Omega}_1 \oplus \omega_2^{1-d}\hat{\Omega}_2)/\hat{\Sigma}_o. \quad \#$$

5.8 <u>Remark</u>:

$$T_2 = \bar{\bar{\omega}}_2^d \cdot \underline{\underline{k}}_2 + \dots + \bar{\bar{\omega}}_2 \cdot \underline{\underline{k}}_2, \quad \underline{\underline{k}}_2 = \hat{\Omega}_2/\omega_2\hat{\Omega}_2$$

where $\bar{\bar{\omega}}_2^1 = \omega_2^{-1} + \hat{\Omega}_2$; and these elements form a two-sided $\underline{\underline{k}}$-basis, since T_2 is a two-sided $\hat{\Omega}_2$-module. Moreover, $\hat{\Omega}_2 = \hat{\Sigma}_o e_2$ and

$\hat{\Omega}_2/\omega_2\hat{\Omega}_2 \cong \hat{\Sigma}_o/\text{rad}\,\hat{\Sigma}_o$; i.e., T_2 is also a two-sided $\underline{\underline{k}}_o$-module, $\underline{\underline{k}}_o = \hat{\Sigma}_o/\text{rad}\,\hat{\Sigma}_o$.

A composition series of T_o as right $\hat{\Sigma}_o$-module is given by

$$(\hat{\Sigma}_o e_1 + \omega_2^{1-d}\hat{\Sigma}_o)/\hat{\Sigma}_o \supsetneq \cdots \supsetneq (\hat{\Sigma}_o e_1 + \omega_2^{-1}\hat{\Sigma}_o)/\hat{\Sigma}_o \supsetneq (\hat{\Sigma}_o e_1 + \hat{\Sigma}_o e_2)/\hat{\Sigma}_o \supsetneq 0.$$

But $(\hat{\Sigma}_o e_1 + \hat{\Sigma}_o\omega_2^{-1})/\hat{\Sigma}_o = (\hat{\Sigma}_o + \hat{\Sigma}_o\omega_2^{-1})/\hat{\Sigma}_o$, and so every element in T_o can uniquely be written as

$$x = \bar{\bar{\omega}}_2^{d-1}k_{1-d} + \ldots + \bar{\bar{\omega}}_2\cdot k_{-1} + \bar{e}_2 k_o,$$

where $k_i \in \underline{\underline{k}}_o$. But since $\underline{\underline{k}}_o\bar{e}_2 = \underline{\underline{k}}_2$, we may also assume that $k_i \in \underline{\underline{k}}_2$. Moreover it should be observed that conjugation with ω_2 induces an automorphism on $\underline{\underline{k}}_2$ - and also on $\underline{\underline{k}}_o$ (by the argument of 4.13).

5.9 <u>Proposition</u>: There are non-zero maps

$$\text{Ext}^1_\Lambda(\hat{\Omega}_3,{}_{\hat{\Lambda}}\hat{\Omega}_1) \longrightarrow \text{Ext}^1_\Lambda(\hat{\Omega}_3,{}_{\hat{\Lambda}}\hat{\Omega}_2), \text{ or}$$

$$\text{Ext}^1_\Lambda(\hat{\Omega}_3,{}_{\hat{\Lambda}}\hat{\Omega}_2) \longrightarrow \text{Ext}^1_\Lambda(\hat{\Omega}_3,{}_{\hat{\Lambda}}\hat{\Omega}_1).$$

However, the left $\hat{\Lambda}$-homomorphisms

$$\tilde{\varphi}_{1o} : \hat{\Omega}_1 \longrightarrow \hat{\Sigma}_o;\; e_1 \longmapsto e_1\omega_1,$$

$$\tilde{\varphi}_{2o} : \hat{\Omega}_2 \longrightarrow \hat{\Sigma}_o;\; e_2 \longmapsto e_2\omega_2,$$

$$\tilde{\varphi}_{o1} : \hat{\Sigma}_o \longrightarrow \hat{\Omega}_1;\; \lambda \longmapsto \lambda e_1,$$

$$\tilde{\varphi}_{o2} : \hat{\Sigma}_o \longrightarrow \hat{\Omega}_2;\; \lambda \longmapsto \lambda e_2,$$

induce $\underline{\underline{Z}}$-homomorphisms

$$\varphi_{1o} : T_1 \longrightarrow T_o,$$

$$\bar{\bar{\omega}}_1\cdot k_1 \longmapsto \bar{\bar{\omega}}_1\cdot k_1\bar{\omega}_1 \in e_2\underline{\underline{k}}_o;$$

$$\varphi_{2o} : T_2 \longrightarrow T_o,$$

$$\sum_{i=1}^d \bar{\bar{\omega}}_2^i\cdot k_i \longmapsto \sum_{i=1}^d \bar{\bar{\omega}}_2^i\cdot k_i\bar{\omega}_2, \text{ where}$$

$\bar{\bar{\omega}}_2^i\cdot k_i\bar{\omega}_2 \in \bar{\bar{\omega}}_2^{i-1}\underline{\underline{k}}_o$, and right $\hat{\Lambda}$-homomorphisms

$$\varphi_{o1} : T_o \longrightarrow T_1,$$

$$\sum_{i=o}^{d-1} \bar{\bar{\omega}}_2^i k_i \longmapsto 0;$$

$$\varphi_{o2} : T_o \longrightarrow T_2$$

$$\sum_{i=o}^{d-1} \bar{\bar{\omega}}_2^i \cdot k_i \longmapsto \sum_{i=1}^{d-1} \bar{\bar{\omega}}_2^i \cdot k_i .$$

The proof follows readily from (5.8). #

5.10 Lemma: T_o, T_1 and T_2 are left $\hat{\Omega}_3$-modules,

(i) $\hat{\Omega}_3$ acts as T_1 as $\hat{\Omega}_3/\omega_3\hat{\Omega}_3 \cong \underline{\underline{k}}$.

(ii) $\hat{\Omega}_3$ acts on T_2 as $\hat{\Omega}_3/\omega_3^d\hat{\Omega}_3$, where

$$\omega_3(\sum_{i=1}^{d} \bar{\bar{\omega}}_2^i \cdot k_i) = \sum_{i=2}^{d} \bar{\bar{\omega}}_2^{i-1} k_i .$$

(iii) $\hat{\Omega}_3$ acts on T_o as $\hat{\Omega}_3/\omega_3^d\hat{\Omega}_3$, where

$$\omega_3(\sum_{i=o}^{d-1} \bar{\bar{\omega}}_2^i \cdot k_i) = \sum_{i=1}^{d-1} \bar{\bar{\omega}}_2^{i-1} k_i .$$

Proof: This is proved as the similar statement of (4.15). #

5.11 Remark: The only $\hat{R}$-order containing $\hat{\Sigma}_o = \hat{\Lambda}(e_1 + e_2)$ is $\hat{\Omega}_1 \oplus \hat{\Omega}_2$ and the non-isomorphic indecomposable $\hat{\Sigma}_o$-lattices are $\hat{\Sigma}_o$, $\hat{\Omega}_1$ and $\hat{\Omega}_2$. $\mathrm{End}_{\hat{\Lambda}}({}_{\hat{\Lambda}}\hat{\Omega}_1)$ acts as $\mathrm{Ext}^1_{\hat{\Lambda}}(\hat{\Omega}_3, {}_{\hat{\Lambda}}\hat{\Omega}_1) = T_1$ as right multiplication with the elements in $\hat{\Omega}_1/\omega_1\hat{\Omega}_1$. $\mathrm{End}_{\hat{\Lambda}}({}_{\hat{\Lambda}}\hat{\Omega}_2)$ acts on $\mathrm{Ext}^1_{\hat{\Lambda}}(\hat{\Omega}_3, {}_{\hat{\Lambda}}\hat{\Omega}_2) = T_2$ as right multiplication with the elements in $\hat{\Omega}_2/\omega_2^d\hat{\Omega}_2$, and $\mathrm{End}_{\hat{\Lambda}}({}_{\hat{\Lambda}}\hat{\Sigma}_o) = \hat{\Sigma}_o$ acts via right multiplication with the elements in

$$\hat{\Sigma}_o/[\hat{\Sigma}_o \cap (e_1 + \omega_2^{d-1})\hat{\Sigma}_o].$$

(The latter statement follows since $(\hat{\Omega}_1 + \hat{\Omega}_1\omega_2^{d-1}) = \hat{\Sigma}_o + \omega_2^{d-1}\hat{\Sigma}_o$.) There are no induced maps $\mathrm{Ext}^1_{\hat{\Lambda}}(\hat{\Omega}_3, {}_{\hat{\Lambda}}\hat{\Omega}_1) \rightleftarrows \mathrm{Ext}^1_{\hat{\Lambda}}(\hat{\Omega}_3, {}_{\hat{\Lambda}}\hat{\Omega}_2)$. $\mathrm{Hom}_{\hat{\Lambda}}(\hat{\Omega}_1, {}_{\hat{\Lambda}}\hat{\Sigma}_o)$ acts as $(\hat{\Omega}_1/\omega_1\hat{\Omega}_1)\varphi_{1o}$ on

$$\mathrm{Ext}^1_{\hat{\Lambda}}(\hat{\Omega}_3, {}_{\hat{\Lambda}}\hat{\Omega}_1) \longrightarrow \mathrm{Ext}^1_{\hat{\Lambda}}(\hat{\Omega}_3, {}_{\hat{\Lambda}}\hat{\Sigma}_o).$$

$\mathrm{Hom}_{\hat{\Lambda}}(\hat{\Omega}_2, {}_{\hat{\Lambda}}\hat{\Sigma}_o)$ acts as $(\hat{\Omega}_2/\omega_2^d\hat{\Omega}_2)\varphi_{2o}$ on

$$\mathrm{Ext}^1_{\hat{\Lambda}}(\hat{\Omega}_3, {}_{\hat{\Lambda}}\hat{\Omega}_2) \longrightarrow \mathrm{Ext}^1_{\hat{\Lambda}}(\hat{\Omega}_3, {}_{\hat{\Lambda}}\hat{\Sigma}_o).$$

$\mathrm{Hom}_{\hat{\Lambda}}(\hat{\Sigma}_o, {}_{\hat{\Lambda}}\hat{\Omega}_1)$ induces only the zero map.

$\mathrm{Hom}_{\hat{\Lambda}}(\hat{\Sigma}_o, {}_{\hat{\Lambda}}\hat{\Omega}_2)$ acts as $\varphi_{o2}(\hat{\Omega}_2/\omega_2^d\hat{\Omega}_2)$ on

$$\mathrm{Ext}^1_{\hat{\Lambda}}(\hat{\Omega}_3, {}_{\hat{\Lambda}}\hat{\Sigma}_o) \longrightarrow \mathrm{Ext}^1_{\hat{\Lambda}}(\hat{\Omega}_3, {}_{\hat{\Lambda}}\hat{\Omega}_2).$$

§6 Reduction of the proof of (2.1) to the decomposition of matrices

We keep the notation of the previous sections, and we reduce the problem, of deciding when a $\hat{\Lambda}$-lattice $\hat{M}$ is indecomposable, to the problem of decomposing certain matrices.

6.1 <u>Theorem</u> (Heller-Reiner [3]): Given an exact sequence of left $\hat{\Lambda}$-modules

$$E_{\hat{M}} : 0 \longrightarrow \hat{M}' \longrightarrow \hat{M} \longrightarrow \hat{M}'' \longrightarrow 0$$

such that

(i) $\hat{M}'$ is a characteristic submodule of $\hat{M}$; i.e., $\varphi|_{\hat{M}'} : \hat{M}' \longrightarrow \hat{M}'$, for every $\varphi \in \operatorname{End}_{\hat{\Lambda}}(\hat{M})$,

(ii) $\operatorname{Hom}_{\hat{\Lambda}}(\hat{M}'',\hat{M}') = 0$,

(iii) $\hat{M}'' \cong \hat{N}''^{(m)}$, where $\hat{N}''$ is an indecomposable left $\hat{\Lambda}$-module, and $\hat{M}' \cong \oplus_{i=1}^{s} \hat{N}_i^{(s_i)}$, where $\{\hat{N}_i\}_{1 \le i \le s}$ are non-isomorphic indecomposable $\hat{\Lambda}$-lattices. Then

$$\operatorname{Ext}^1_{\hat{\Lambda}}(\hat{M}'',\hat{M}') \cong \oplus_{i=1}^{s} \left(\operatorname{Ext}^1_{\hat{\Lambda}}(\hat{N}'',\hat{N}_i)\right)_{m \times s_i},$$

and the congruence class $[E_{\hat{M}}]$ of $E_{\hat{M}}$ corresponds to a matrix

$$\underline{\underline{X}}_{\hat{M}} = (\underline{\underline{X}}_i)_{1 \le i \le s}, \quad \underline{\underline{X}}_i \in \left(\operatorname{Ext}^1_{\hat{\Lambda}}(\hat{N}'',\hat{N}_i)\right)_{m \times s_i}.$$

Moreover,

$\hat{M}$ decomposes if and only if $\alpha \underline{\underline{X}}_{\hat{M}} \beta$ decomposes for some $\hat{\Lambda}$-automorphism α of $\hat{M}''$ and β of $\hat{M}'$.

<u>Proof</u>: (i) Replacing $E_{\hat{M}}$ by $\alpha E_{\hat{M}} \beta$, where α and β are automorphisms, does not change the isomorphism class of $\hat{M}$. In fact, we have the commutative diagram

$$\begin{array}{lccccccccc}
E_{\hat{M}} & : 0 \longrightarrow & \hat{M}' & \longrightarrow & \hat{M} & \longrightarrow & \hat{M}'' & \longrightarrow 0 \\
 & & \beta \downarrow & & \varphi \downarrow & & \downarrow 1_{\hat{M}''} & \\
E_{\hat{M}}\beta & : 0 \longrightarrow & \hat{M}' & \longrightarrow & \hat{X} & \longrightarrow & \hat{M}'' & \longrightarrow 0 \\
 & & \beta^{-1} \downarrow & & \psi \downarrow & & \downarrow 1_{\hat{M}''} & \\
E_{\hat{M}}\beta\beta^{-1} & : 0 \longrightarrow & \hat{M}' & \longrightarrow & \hat{Y} & \longrightarrow & \hat{M}'' & \longrightarrow 0,
\end{array}$$

and since $[E_{\hat{M}}] = [E_{\hat{M}}\beta\beta^{-1}]$, $\varphi\psi$ is an isomorphism; hence φ is a monomorphism. Since $[E_{\hat{M}}\beta] = [E_{\hat{M}}\beta(\beta^{-1}\beta)]$ we conclude that ψ is also a monomorphism and so φ is an isomorphism. The general case is done similarly.

(ii) Every automorphism γ of $\hat{M}$ gives rise to a commutative diagram

$$\begin{array}{ccccccccc} 0 & \longrightarrow & \hat{M}' & \longrightarrow & \hat{M} & \longrightarrow & \hat{M}'' & \longrightarrow & 0 \\ & & \downarrow{\scriptstyle\beta} & & \downarrow{\scriptstyle\gamma} & & \downarrow{\scriptstyle\alpha^{-1}} & & \\ 0 & \longrightarrow & \hat{M}' & \longrightarrow & \hat{M} & \longrightarrow & \hat{M}'' & \longrightarrow & 0, \end{array}$$

where α and β are automorphisms (cf. Ex. 6,1). We define a map

$$\Phi : \mathrm{End}_{\hat{\Lambda}}(\hat{M}) \longrightarrow \mathrm{End}_{\hat{\Lambda}}(\hat{M}') \oplus \mathrm{End}_{\hat{\Lambda}}(\hat{M}''),$$
$$\varphi \longmapsto (\varphi|_{\hat{M}'}, \varphi^*),$$

where φ^* is induced from the homomorphism

$$\tilde{\varphi}^* : \hat{M}/\hat{M}' \longrightarrow \hat{M}/\hat{M}',$$
$$m+\hat{M}' \longmapsto m\varphi + \hat{M}'.$$

$\varphi|_{\hat{M}'}$ and φ^* are well-defined since $\hat{M}'$ is a characteristic submodule of $\hat{M}$. Φ is then a monomorphism. In fact, if $(\varphi|_{\hat{M}'}, \varphi^*) = 0$ we define

$$\psi : \hat{M}/\hat{M}' \longrightarrow \hat{M}',$$
$$m+M' \longmapsto m\varphi .$$

But $\mathrm{Hom}_{\hat{\Lambda}}(\hat{M}'',\hat{M}') = 0$ implies $\psi = 0$; i.e., $\varphi = 0$ and Φ is monic.
It is clear that an automorphism γ of $\hat{M}$ gives rise to two automorphisms α and β completing the above diagram, Φ being monic.

(iii) If $[\alpha E_{\hat{M}}\beta]$ decomposes, we may as well assume that $[E_{\hat{M}}]$ decomposes (cf. (i)). But then $[E_{\hat{M}}] = [E_1 \oplus E_2]$ and $\hat{M}$ decomposes.

(iv) Assume that $\hat{M}$ decomposes, say $\hat{M} = \hat{M}_1 \oplus \hat{M}_2$, $\hat{M}_i \neq 0, i=1,2$, and let $\pi_i : \hat{M} \longrightarrow \hat{M}_i$ be the corresponding projections. Then we obtain the commutative diagram

$$\begin{array}{ccccccccc} 0 & \longrightarrow & \hat{M}' & \xrightarrow{\sigma} & \hat{M} & \xrightarrow{\tau} & \hat{M}'' & \longrightarrow & 0 \\ & & \downarrow & & \downarrow & & \downarrow & & \\ 0 & \longrightarrow & \hat{M}_1'\oplus\hat{M}_2' & \xrightarrow{\sigma_1\oplus\sigma_2} & \hat{M}_1\oplus\hat{M}_2 & \xrightarrow{\tau_1\oplus\tau_2} & \hat{M}_1''\oplus\hat{M}_2'' & \longrightarrow & 0 \end{array}$$

and $E_{\hat{M}}$ decomposes. According to the Krull-Schmidt theorem,

$$\hat{M}'_j \cong \oplus_{i=1}^{s} \hat{N}_i^{(s'_{ij})},\ s'_{i1} + s'_{i2} = s_i \text{ and}$$

$$\hat{M}''_i \cong \hat{N}''^{(m_1)},\ m_1 + m_2 = m.$$

Consequently, there exist automorphisms α of $\hat{M}''$ and β of $\hat{M}'$ such that $[\alpha E_{\hat{M}} \beta] = \alpha X_{\hat{M}} \beta$ decomposes because of (ii). #

6.2 Remark: We point out, that α and β can be represented as invertible matrices: $\alpha \in \mathrm{Aut}_{\hat{\Lambda}}(\hat{M}'')$ corresponds to an element $Y \in GL(m, \mathrm{End}_{\hat{\Lambda}}(\hat{N}''))$ and β can be represented as an invertible $((\sum_{i=1}^{s} s_i) \times (\sum_{i=1}^{s} s_i))$-matrix Z where Z has a block-decomposition,

$$Z = (Z_{ij})_{1 \leq i, j \leq s} \text{ with } Z_{ij} \in (\mathrm{Hom}_{\hat{\Lambda}}(\hat{N}'_i, \hat{N}'_j))_{s_i \times s_j}.$$

Hence we have to decompose $X_{\hat{M}}$ under the operations $YX_{\hat{M}}Z$.

6.A: The case $\hat{\Lambda} = \hat{D}$.

If $\hat{\Lambda}$ is a skewfield we can associate with $\hat{M} \in {}_{\hat{\Lambda}}M^{o}$ the exact sequence

$$E_{\hat{M}} : 0 \longrightarrow \hat{N}\hat{M} \longrightarrow \hat{M} \longrightarrow \hat{M}/\hat{N}\hat{M} \longrightarrow 0,$$

where $\hat{N} = \mathrm{rad}\,\hat{\Lambda}$. We have:

(i) $\hat{N}\hat{M}$ is a characteristic submodule of $\hat{M}$,

(ii) $\mathrm{Hom}_{\hat{\Lambda}}(\hat{M}/\hat{N}\hat{M}, \hat{N}\hat{M}) = 0$, since $\hat{M}/\hat{N}\hat{M} \cong k^{(m)}$, $k = \hat{\Lambda}/\mathrm{rad}\,\hat{\Lambda}$.

Hence we may apply (6.1):

6.A.I: $\hat{\Lambda}_1 = \Lambda_1(\hat{N}) = \hat{\Gamma}$ and $S = \hat{\Lambda}_1/\hat{N}$ is a three-dimensional extension field of k. (This is (3.1,i).)

$\hat{N}\hat{M} \cong \hat{\Gamma}^{(s)}$ and the matrix $X_{\hat{M}}$ has entries in $T = \alpha^{-1}\hat{\Gamma}/\hat{\Gamma}$, where α is such that $\hat{\Gamma}\alpha = \mathrm{rad}\,\hat{\Gamma}$. The matrix Y has entries in k and the matrix Z has entries in $S = k(r)$. Moreover, conjugation with α induces an automorphism of k (cf. 2.15,5.). Therefore we may assume w.l.o.g. that $S = T = \hat{\Lambda}_1/\hat{N}$. We have to decompose $X_{\hat{M}}$ under $YX_{\hat{M}}Z$.

6.A.II: $\hat{\Lambda}_1 = \hat{\Gamma}$ and $S = \hat{\Lambda}_1/\hat{N} = k_1[r]$, $r^3 = 0$, $k_1 r = r k_1$, $k_1 = \hat{\Gamma}/\mathrm{rad}\,\hat{\Gamma} \cong k$.

(This is case (3.1,ii).)

$\hat{N}\hat{M} \cong \hat{\Gamma}^{(s)}$; the matrix $\underline{\underline{X}}_{\hat{M}}$ has entries in $T = \alpha^{-1}\hat{\Gamma}/\hat{\Gamma} = \bar{\bar{\alpha}} S$,

$\underline{\underline{k}}_1 \bar{\bar{\alpha}} = \bar{\bar{\alpha}} \cdot \underline{\underline{k}}_1$. The matrix $\underline{\underline{Y}}$ has entries in $\underline{\underline{k}}_1$ and the matrix $\underline{\underline{Z}}$ has entries in S. Again we may assume $T = S = \hat{\Lambda}_1/\hat{N}$ and $\underline{\underline{k}} = \underline{\underline{k}}_1$. We have to decompose $\underline{\underline{X}}_{\hat{M}}$ under $\underline{\underline{Y}}\underline{\underline{X}}_{\hat{M}}\underline{\underline{Z}}$.

6.A.III: $\hat{\Lambda}_1 \neq \hat{\Gamma}$, $S = \hat{\Lambda}_1/\hat{N} = \underline{\underline{k}}_2[r]$, $r^3 = 0$, $\underline{\underline{k}}_2 r = r\underline{\underline{k}}_2$, $\underline{\underline{k}}_2 = \hat{\Lambda}_1/\mathrm{rad}\,\hat{\Lambda}_1$, $S' = \hat{\Gamma}/\hat{N}\hat{\Gamma} = \underline{\underline{k}}_3[r']$, $r'^3 = 0$, $\underline{\underline{k}}_3 r = r\underline{\underline{k}}_3$, $\underline{\underline{k}}_3 = \hat{\Gamma}/\mathrm{rad}\,\hat{\Gamma}$. (This is the case (3.1,iii)).

$\hat{N}\hat{M} \cong \hat{\Lambda}_1^{(s_1)} \oplus \hat{\Gamma}^{(s_2)}$, $\underline{\underline{X}}_{\hat{M}} = (\underline{\underline{X}}_1, \underline{\underline{X}}_2)$, where $\underline{\underline{X}}_1$ has entries in $T = \alpha^{-1}\hat{\Lambda}_1/\hat{\Lambda}_1 = \bar{\bar{\alpha}} \cdot S$, $\underline{\underline{k}}_2 \bar{\bar{\alpha}} = \bar{\bar{\alpha}} \cdot \underline{\underline{k}}_2$; $\underline{\underline{X}}_2$ has entries in $T' = \alpha^{-1}\hat{\Gamma}/\hat{\Gamma} = \bar{\bar{\alpha}} \cdot S'$, $\underline{\underline{k}}_3 \bar{\bar{\alpha}} = \bar{\bar{\alpha}} \cdot \underline{\underline{k}}_3$. $\underline{\underline{Y}}$ has entries in $\underline{\underline{k}}$; however, $\underline{\underline{k}}$ acts as $\underline{\underline{k}}_2$ on S and as $\underline{\underline{k}}_3$ on S'. Thus we may assume $\underline{\underline{k}} = \underline{\underline{k}}_2 = \underline{\underline{k}}_3$ and

$$S = T = \hat{\Lambda}_1/\hat{N}, \quad S' = T' = \hat{\Gamma}/\hat{N}\hat{\Gamma}.$$

The matrix $\underline{\underline{Z}}$ has the form

$$\underline{\underline{Z}} = \begin{pmatrix} \underline{\underline{Z}}_{11} & \underline{\underline{Z}}_{12} \\ \underline{\underline{Z}}_{21} & \underline{\underline{Z}}_{22} \end{pmatrix}, \text{where}$$

$\underline{\underline{Z}}_{11}$ has entries in S,

$\underline{\underline{Z}}_{22}$ has entries in S',

$\underline{\underline{Z}}_{12}$ has entries in $\varphi S'$, where

$$\varphi : S \longrightarrow S',$$
$$1 \longmapsto 1';\ r \longmapsto r'^2;\ r^2 \longmapsto 0,$$

$\underline{\underline{Z}}_{21}$ has entries in $S'\psi$, where

$$\psi : S' \longrightarrow S,$$
$$1' \longmapsto r,\ r' \longmapsto 0,\ r'^2 \longmapsto r^2 k \text{ for}$$

some $0 \neq k \in \underline{\underline{k}}$.

φ is a $(\underline{\underline{k}},S)$-bimodule homomorphism and ψ is a left $\underline{\underline{k}}$-homomorphism.

$\underline{\underline{Y}}$ has entries in $\underline{\underline{k}}$.

We have to decompose $\underline{\underline{X}}_{\hat{M}}$ under $\underline{\underline{Y}}\underline{\underline{X}}_{\hat{M}}\underline{\underline{Z}}$.

6.B: The case $A = \hat{D}_1 \oplus \hat{D}_2$.

If $\hat{A}$ is the direct sum of two skewfields, we associate with $\hat{M} \in {}_{\hat{\Lambda}}\underline{\underline{M}}^{o}$ the exact sequence

$$0 \longrightarrow \hat{M} \cap \hat{D}_1\hat{M} \longrightarrow \hat{M} \longrightarrow \hat{M}/(\hat{M} \cap \hat{D}_1\hat{M}) \longrightarrow 0.$$

We have:

(i) $\hat{M} \cap \hat{D}_1\hat{M}$ is a lattice over the Bass-order $\hat{\Lambda}_1 = \hat{\Lambda}e_1$ and $\hat{M}/(\hat{M} \cap \hat{D}_1\hat{M}) = \hat{M}e_2$ is an $\hat{\Omega}_2$-lattice. Obviously $\hat{M} \cap \hat{D}_1\hat{M}$ is a characteristic submodule of $\hat{M}$, since it is an $\hat{R}$-pure submodule.

(ii) $\mathrm{Hom}_{\hat{\Lambda}}(\hat{M}e_2, \hat{M} \cap \hat{D}_1\hat{M}) = 0$, and $\hat{M}e_2 \cong \hat{\Omega}_2^{(m)}$.

Hence we may apply (6.1):

6.B.I: $\hat{\Lambda}_1$ decomposes.

By (4.3),

$$\hat{M} \cap \hat{D}_1\hat{M} \cong \oplus_{i=1}^{s} \hat{\Sigma}_i^{(s_i)}.$$

$S_1 = \underline{\underline{k}}_1 = \hat{\Lambda}_1/\mathrm{rad}\,\hat{\Lambda}_1$

$S_i = \hat{\Sigma}_i/\hat{N}\hat{\Sigma}_i = \underline{\underline{k}}_i[b_i]$, $b_i^2 = 0$, $\underline{\underline{k}}_i b_i = b_i \underline{\underline{k}}_i$,

$\underline{\underline{k}}_i = \hat{\Sigma}_i/\mathrm{rad}\,\hat{\Sigma}_i \cong \underline{\underline{k}}$, $1 < i < s$,

S_s is either a two-dimensional extension field of $\underline{\underline{k}}$ or $S_s = \underline{\underline{k}}_s[b_s]$, $b_s^2 = 0$, $\underline{\underline{k}}_s b_s = b_s \underline{\underline{k}}_s$. $\underline{\underline{k}}_s = \hat{\Omega}_1/\mathrm{rad}\,\hat{\Omega}_1 \cong \underline{\underline{k}}$.

But $\underline{\underline{k}}$ acts on S_i as does $\underline{\underline{k}}_i$ and so we can replace $\underline{\underline{k}}_i$ by $\underline{\underline{k}}$.

In the matrix

$$\underline{\underline{X}}_{\hat{M}} = (\underline{\underline{X}}_1, \dots, \underline{\underline{X}}_s),$$

$\underline{\underline{X}}_1$ has entries in $\underline{\underline{k}}$,

$\underline{\underline{X}}_i$ has entries in $T_i = \bar{\bar{\beta}} \cdot S_i$, $1 < i \leq s$, and $\bar{\bar{\beta}} \cdot \underline{\underline{k}} = \underline{\underline{k}}\bar{\bar{\beta}}$, except perhaps in case S_s is a field, where $\underline{\underline{X}}_s$ has entries in $\beta^{-1}\hat{\Omega}_1/\hat{\Omega}_1$. But conjugation with $\bar{\bar{\beta}}$ induces an automorphism of $\underline{\underline{k}}$ (4.5). Hence, w.l.o.g. we may assume $T_1 = S_1$.

The matrix $\underline{\underline{Z}}$ has the form

$$\underline{\underline{Z}} = (\underline{\underline{Z}}_{ij})_{1\leq i,j\leq s}, \text{ and}$$

$\underline{\underline{Z}}_{ii}$ has entries in S_i, $1\leq i\leq s$,

$\underline{\underline{Z}}_{ij}$ has entries in $\varphi_{ij}S_j$, $1\leq i < j\leq s$, where

$$\varphi_{ij} : S_i \longrightarrow S_j,$$
$$1_i \longmapsto 1_j,\ b_i \longmapsto 0;$$

φ_{ij} are $(\underline{\underline{k}}, S_1)$ bimodule homomorphisms.

$\underline{\underline{Z}}_{ij}$ has entries in $S_i\varphi_{ij}$, $1\leq j < i\leq s$, where

$$\varphi_{ij} : S_i \longrightarrow S_j,$$
$$1_i \longmapsto 0,\ b_i \longmapsto b_j k \text{ for some } 0 \neq k \in \underline{\underline{k}}.$$

φ_{ij} are left $\underline{\underline{k}}$-homomorphisms.

This follows from (4.9).

$\underline{\underline{Y}}$ has entries in $\underline{\underline{k}}$.

<u>6.B.II:</u> $\hat{\Lambda}_1$ <u>is indecomposable.</u>

Then

$$\hat{M} \cap \hat{D}_1\hat{M} \cong \hat{\Lambda}_1^{s_1} \oplus \hat{\Omega}_1^{s_2},$$

and in the matrix

$$\underline{\underline{X}}_{\hat{M}} = (\underline{\underline{X}}_1, \underline{\underline{X}}_2),$$

$\underline{\underline{X}}_1$ has entries in $\omega_1^{-3}\hat{\Omega}_1/\hat{\Omega}_1 = T_1$,

$\underline{\underline{X}}_2$ has entries in $\omega_1^{-1}\hat{\Omega}_1/\hat{\Lambda}_1 = T_2$.

$T_1 = \{\bar{\bar{\omega}}_2^3\cdot k_1 + \bar{\bar{\omega}}_2^2\cdot k_2 + \bar{\bar{\omega}}_2\cdot k_3 : k_i \in \underline{\underline{k}}_1 = \hat{\Lambda}_1/\hat{N}_1\}$ with $\bar{\bar{\omega}}_2\cdot\underline{\underline{k}}_1 = \underline{\underline{k}}_1\bar{\bar{\omega}}_2$.

$T_2 = \{\bar{\bar{\omega}}_1\cdot k_1 + \bar{\bar{\omega}}_1 k_2 : k_i \in \underline{\underline{k}}_1\}$, $\bar{\bar{\omega}}_1\cdot\underline{\underline{k}}_1 = \underline{\underline{k}}_1\bar{\bar{\omega}}_1$.

In

$$\underline{\underline{Z}} = \begin{pmatrix} \underline{\underline{Z}}_{11} & \underline{\underline{Z}}_{12} \\ \underline{\underline{Z}}_{21} & \underline{\underline{Z}}_{22} \end{pmatrix}$$

$\underline{\underline{Z}}_{11}$ has entries in $S_1 = \hat{\Omega}_1/\omega_1^3\hat{\Omega}_1$,

$\underline{\underline{Z}}_{22}$ has entries in $S_2 = \hat{\Lambda}_1/(\hat{\Lambda}_1 \cap \omega_1\hat{\Lambda}_1)$.

$\underline{\underline{Z}}_{12}$ has entries in $\varphi_{o1}S_1$ (cf. 4.16),

$\underline{\underline{Z}}_{21}$ has entries in $S_1\varphi_{1o}$ (cf. 4.16).

However, $\underline{\underline{k}}\,\overline{\overline{\omega}}_1 = \overline{\overline{\omega}}_1\cdot\underline{\underline{k}}$ and

$$S_2 = \hat{\Lambda}_1/(\hat{\Lambda}_1 \cap \hat{\Lambda}_1\omega_1) \overset{nat}{\cong} (\hat{\Lambda}_1 + \hat{\Lambda}_1\omega_1)/\omega_1\hat{\Lambda}_1 \cong$$
$$\cong \omega_1^{-1}\hat{\Omega}_1/\hat{\Lambda}_1,$$

and $T_2 \cong \overline{\overline{\omega}}_1\cdot S_2$. Hence we may assume $T_1 = S_1$, $T_2 = S_2$ and then the maps are given as follows:

$$\varphi_{o1} : S_1 \longrightarrow S_2,$$
$$\sum_{i=o}^{2} \overline{\omega}_1^{i}k_i \longmapsto 1k'_o + \overline{\omega}_1^{2}k'_2,$$

where $k_o, k_2 \neq 0$ if and only if $k'_o, k'_2 \neq 0$.

$$\varphi_{1o} : S_2 \longrightarrow S_1,$$
$$1k_o + \overline{\omega}_1^{2}k_2 \longmapsto \overline{\omega}_1^{2}k_o.$$

Both maps are left $\underline{\underline{k}}$-homomorphisms.

$\underline{\underline{Y}}$ has entries in $\hat{\Omega}_2$, and the action of $\hat{\Omega}_2$ on S_1 is given via $\hat{\Omega}_2/\omega_2^2\hat{\Omega}_2$

$$\omega_2(\sum_{i=o}^{2} \overline{\omega}_1^{i}k_i) = \overline{\omega}_1^{2}k_2 \text{ on } S_1$$
$$\omega_2(1\cdot k_o + \overline{\omega}_1^{2}k_1) = \overline{\omega}_1^{2}k_o \text{ on } S_2.$$

$\hat{\Omega}_2/\text{rad}\,\hat{\Omega}_2$ acts as $\underline{\underline{k}}$ on S_i, $i=1,2$.

This follows from (4.12-4.16).

6.C: The case $\hat{A} = \hat{D}_1 \oplus \hat{D}_2 \oplus \hat{D}_3$.

If A is the direct sum of three skewfields we associate with $\hat{M} \in {}_{\hat{\Lambda}}\underline{\underline{M}}^o$ the exact sequence

$$0 \longrightarrow \hat{M} \cap (\hat{D}_1 \oplus \hat{D}_2)\hat{M} \longrightarrow \hat{M} \longrightarrow e_3\hat{M} \longrightarrow 0.$$

We have:

(i) $\hat{M}' = \hat{M} \cap (\hat{D}_1 \oplus \hat{D}_2)\hat{M}$ is a lattice over the Bass-order $\hat{\Sigma}_o = \hat{\Lambda}(e_1 + e_2)$, and $e_3\hat{M}$ is a lattice over $\hat{\Omega}_3$. Then $\hat{M}'$ is a characteristic submodule of $\hat{M}$.

(ii) $\mathrm{Hom}_{\hat{\Lambda}}(e_3\hat{M},\hat{M}') = 0$ and $e_3\hat{M} \cong \hat{\Omega}_3^{(m)}$.

Hence we may apply (6.1).

According to (5.11)

$$\hat{M}' = \hat{M} \cap (\hat{D}_1 \oplus \hat{D}_2)\hat{M} \cong \hat{\Sigma}_o^{(s_1)} \oplus \hat{\Omega}_1^{(s_2)} \oplus \hat{\Omega}_2^{(s_3)},$$

and in the matrix

$$\underline{\underline{X}}_{\hat{M}} = (\underline{\underline{X}}_1, \underline{\underline{X}}_2, \underline{\underline{X}}_3)$$

$\underline{\underline{X}}_1$ has entries in $T_o = (\hat{\Omega}_1 \oplus \omega_2^{1-d}\hat{\Omega}_2)/\hat{\Sigma}_o$,

$\underline{\underline{X}}_2$ has entries in $T_1 = \omega_1^{-1}\hat{\Omega}_1/\hat{\Omega}_1 \cong \underline{\underline{k}}$,

$\underline{\underline{X}}_3$ has entries in $T_2 = \omega_2^{-d}\hat{\Omega}_2/\hat{\Omega}_2$.

$T_o = \{ \bar{\bar{\omega}}_2^{d-1} k_{1-d} + \ldots + \bar{\bar{\omega}}_2 \cdot k_1 + \bar{e}_2 k_o : k_i \in \underline{\underline{k}}_o = \hat{\Sigma}_o/\mathrm{rad}\,\hat{\Sigma}_o \}$,

$T_2 = \{ \bar{\bar{\omega}}_2^d k_d + \ldots + \bar{\bar{\omega}}_2 \cdot k_1 : k_i \in \underline{\underline{k}}_2 = \hat{\Omega}_2/\omega_2\hat{\Omega}_2 \}$.

Moreover, we may assume $\underline{\underline{k}}_1 = \underline{\underline{k}}_o = \underline{\underline{k}}$.

In

$$\underline{\underline{Z}} = (\underline{\underline{Z}}_{ij})_{1 \le i,j \le 3},$$

$\underline{\underline{Z}}_{11}$ has entries in $S_o = \hat{\Sigma}_o/(\hat{\Sigma}_o \cap (e_1 + \omega_2^{d-1})(\hat{\Omega}_1 \oplus \hat{\Omega}_2))$,

$\underline{\underline{Z}}_{22}$ has entries in $S_1 = \hat{\Omega}_1/\omega_1\hat{\Omega}_1 = \underline{\underline{k}}$,

$\underline{\underline{Z}}_{33}$ has entries in $S_2 = \hat{\Omega}_2/\omega_2^d\hat{\Omega}_2$.

$\underline{\underline{Z}}_{32} = \underline{\underline{Z}}_{23} = \underline{\underline{Z}}_{12} = 0$,

$\underline{\underline{Z}}_{21}$ has entries in $(\hat{\Omega}_1/\omega_1\hat{\Omega}_1)\varphi_{1o} = S_1\varphi_{1o}$ (cf. 5.9),

$\underline{\underline{Z}}_{31}$ has entries in $(\hat{\Omega}_2/\omega_2^d\hat{\Omega}_2)\varphi_{2o} = S_2\varphi_{2o}$ (cf. 5.9).

However, $\underline{\underline{k}}\,\overline{\overline{\omega}}_2 = \overline{\overline{\omega}}_2\,\underline{\underline{k}}$ and $\overline{(e_1 + \omega_2^{d-1})}\underline{\underline{k}} = \underline{\underline{k}}\overline{(e_1 + \omega_2^{d-1})}$,

$$S_o \overset{nat}{\cong} \hat{\Sigma}_o + (e_1 + \omega_2^{d-1})(\hat{\Omega}_1 \oplus \hat{\Omega}_2)/(e_1 + \omega_2^{d-1})(\hat{\Omega}_1 + \hat{\Omega}_2) \cong T_o.$$

Hence we may assume $T_i = S_i, i=0,1,2$, and then the maps φ_{ij} are given as follows:

$$\varphi_{1o} : S_1 \longrightarrow S_o,$$

$$k \longmapsto \overline{\omega}_2^{d-1}k', \text{ where } k \neq 0 \text{ if and only if } k' \neq 0;$$

$$\varphi_{o2} : k_{1-d} + \ldots + \overline{\omega}_2^{d-1}k_o \longmapsto k_{1-d} + \ldots + \overline{\omega}_2^{d-1-1}k_1;$$

$$\varphi_{2o} : k_o + \ldots + \overline{\omega}_2^{d-1}k_d \longmapsto \overline{\omega}_2 k'_o + \ldots + \overline{\omega}_2^{d-2}k'_{d-1},$$

where $k_i \neq 0$ if and only if $k'_i \neq 0$.

The matrix $\underline{\underline{Y}}$ has entries in $\hat{\Omega}_3$, and its action on S_i is given in (5.10).

<u>Exercise §6</u>:

1.) Under the hypotheses of (6.1), every homomorphism $\varphi: M \longrightarrow M$ gives rise to the commutative diagram

$$\begin{array}{ccccccccc} 0 & \longrightarrow & M' & \longrightarrow & M & \longrightarrow & M'' & \longrightarrow & 0 \\ & & \Big\downarrow{\scriptstyle \varphi'} & & \Big\downarrow{\scriptstyle \varphi} & & \Big\downarrow{\scriptstyle \varphi''} & & \\ 0 & \longrightarrow & M' & \longrightarrow & M & \longrightarrow & M'' & \longrightarrow & 0. \end{array}$$

If φ is an automorphism, then so are φ' and φ''.

§7 Decomposition of the matrix $\underline{\underline{X}}_{\hat{M}}$ of 6.A.I

We have to show that under the operations $\underline{\underline{Y}}\,\underline{\underline{X}}_{\hat{M}}\underline{\underline{Z}}$ where $\underline{\underline{Y}}$ and $\underline{\underline{Z}}$ are invertible, every matrix $\underline{\underline{X}}_{\hat{M}}$ can be decomposed into "indecomposable" ones and that among these indecomposable ones there are only finitely many non-"equivalent" ones. (Here "indecomposable" means indecomposable under the operation $\underline{\underline{Y}}\,\underline{\underline{X}}_{\hat{M}}\underline{\underline{Z}}$; the relation $\underline{\underline{X}}_{M} \sim \underline{\underline{X}}'_{M}$ if $\underline{\underline{X}}'_{\hat{M}} = \underline{\underline{Y}}\,\underline{\underline{X}}_{\hat{M}}\underline{\underline{Z}}$ is obviously an equivalence relation.)

However, in all the cases of § 6 we can decompose $\underline{\underline{X}}_{\hat{M}}$ already with "elementary transformations" into a finite number of non-equivalent ones. Where "elementary transformations" are the following operations (notation: ET).

We recall, that we had associated with $\hat{M}$ an exact sequence

$$0 \longrightarrow \oplus_{i=1}^{s} \hat{N}_i'^{(s_i)} \longrightarrow \hat{M} \longrightarrow \hat{N}''^{(m)} \longrightarrow 0,$$

$\underline{\underline{Y}}$ was an invertible $(m \times m)$-matrix with entries in $\mathrm{End}_{\hat{\Lambda}}(\hat{N}'')$ and $\underline{\underline{Z}} = (\underline{\underline{Z}}_{ij})$ was an invertible matrix, where $\underline{\underline{Z}}_{ij}$ had entries in $\mathrm{Hom}_{\hat{\Lambda}}(\hat{N}_i', \hat{N}_j')$. We define the elementary transformations as follows:

(i) $(\underline{\underline{E}} + s\underline{\underline{E}}_{ij})\underline{\underline{X}}_{\hat{M}}$, $i \neq j$,

where $s \in \mathrm{End}_{\hat{\Lambda}}(\hat{N}'')$. This has the effect of adding the s-fold of the j-th row to the i-th row.

(ii) $\underline{\underline{X}}_{\hat{M}}(\underline{\underline{E}} + s\underline{\underline{E}}_{ij})$ $i \neq j$,

where $s \in \mathrm{Hom}_{\hat{\Lambda}}(\hat{N}_i', \hat{N}_j')$. This has the effect of adding the s-fold of the j-th column to the i-th column.

(iii) Multiplying the i-th row of $\underline{\underline{X}}_{\hat{M}}$ with a unit in $\mathrm{End}_{\hat{\Lambda}}(\hat{N}'')$.

(iv) Multiplying the i-th column of $\underline{\underline{X}}_{\hat{M}}$ with a unit in $\mathrm{End}_{\hat{\Lambda}}(\hat{N}_i')$.

The operations (i) and (iii) are denoted by <u>ETL</u> and the operations (ii) and (iv) by <u>ETR</u>.

<u>Remark</u>: $\mathrm{End}_{\hat{\Lambda}}(\hat{N}'')$ is completely primary and thus the invertible matrix $\underline{\underline{Y}}$ can be represented as a product of ETLs (if one considers ETs as matrices) (cf. VIII, 2.5). And we leave it as an exercise to show that $\underline{\underline{Z}}$ too can be represented as a product of elementary matrices. Hence it

suffices to decompose $X_{\underset{=}{M}}$ by successively applying ETs, since obviously, the ETs represent invertible matrices.

We recall from 6,A,I:

$S = \underline{k}(\alpha)$ is a three dimensional extension field of the finite field $\underline{k}$; $X_{\underset{=}{M}}$ has entries in S, we can apply ETLs with elements in $\underline{k}$ and ETRs with elements in S.

By means of ETRs we obtain $X_{\underset{=}{M}}$ in the following form

$$X_{\underset{=}{M}} = \begin{bmatrix} 1 & & & & \\ & \ddots & & & 0 \\ & & 1 & & \\ & * & & t & 0 \end{bmatrix} \quad m \times s$$

In treating case 6,A,II, we shall encounter a similar situation, where $S = \underline{k}[r]$, $r^3 = 0$. Then by ETs we obtain $X_{\underset{=}{M}}$ in the form

$$X_{\underset{=}{M}} = \begin{bmatrix} 1 & & & & \\ & \ddots & & 0 & \\ & & 1 & & \\ & * & & ** & \\ & & t & & \end{bmatrix} \quad m \times s$$

where $*, **$ denote entries in rad S. To treat both cases together, we prove:

7.1 **Proposition:** If the matrix $\underline{X} = X_{\underset{=}{M}}$ corresponds to an exact sequence

$$E_{\hat{M}} : 0 \longrightarrow \hat{NM} \longrightarrow \hat{M} \longrightarrow \hat{M}/\hat{NM} \longrightarrow 0 \quad (cf. 6,A)$$

then $s = t$.

Proof: The exact sequence

$$\Xi_0 : 0 \longrightarrow \hat{\Gamma}^{(m)} \overset{\varphi_1}{\longrightarrow} \hat{\Lambda}^{(m)} \longrightarrow \underline{k}^{(m)} \longrightarrow 0,$$

where $\varphi_1 : \hat{\Gamma}^{(m)} \underset{\sim}{\longrightarrow} \hat{N}^{(m)}$ implies

$$Ext^1_{\hat{\Lambda}}(\underline{k}^{(m)}, {}_{\hat{\Lambda}}\hat{\Gamma}^{(s)}) \cong (\hat{\Gamma}/\hat{N})_{m \times s}.$$

To the matrix $X_{\underset{=}{M}}$ we pick a matrix $\widetilde{X} \epsilon (\hat{\Gamma})_{m \times s}$ which is a preimage of $\underset{=}{X}_{\hat{M}}$. Then we may assume

$$\underset{=}{\tilde{X}} = \left[\begin{array}{ccc|c} 1 & & & \\ & \ddots & & 0 \\ & & 1 & \\ \hline & * & & ** \end{array}\right] \quad m \times s, \quad t ,$$

where (**) has entries in rad $\hat{\Gamma}$. The sequence $E_{\hat{M}}$ is then given - up to congruence - by $E_o\underset{=}{\tilde{X}}$, $\underset{=}{\tilde{X}} \in \mathrm{Hom}_{\hat{\Lambda}}(\hat{\Gamma}^{(m)}, \hat{\Gamma}^{(s)}_{\hat{\Lambda}})$:

$$\begin{array}{llllllllll} E_o & : 0 & \longrightarrow & \hat{\Gamma}^{(m)} & \xrightarrow{\varphi_1} & \hat{\Lambda}^{(m)} & \longrightarrow & \underset{=}{k}^{(m)} & \longrightarrow & 0 \\ & & & \downarrow \underset{=}{\tilde{X}} & & \downarrow \psi & & \downarrow 1_{\underset{=}{k}^{(m)}} & & \\ E_o\underset{=}{\tilde{X}} & : 0 & \longrightarrow & \hat{\Gamma}^{(s)} & \xrightarrow{\varphi_2} & \hat{M} & \longrightarrow & \underset{=}{k}^{(m)} & \longrightarrow & 0. \end{array}$$

Thus we may assume

$$\hat{M} = (\hat{\Gamma}^{(s)} \oplus \hat{\Lambda}^{(m)})/\hat{M}_o, \text{ where}$$

$$\hat{M}_o = \{(\gamma^{(s)}\underset{=}{\tilde{X}}, -\gamma^{(m)}\varphi_1), \gamma^{(s)} \in \hat{\Gamma}^{(s)}, \gamma^{(m)} \in \hat{\Gamma}^{(m)}\}.$$

If $t < s$, then $\underset{=}{\tilde{X}}$ is not an epimorphism. However, $\underset{=}{X}$ is the matrix of the sequence $E_{\hat{M}}$, and so we must have $\hat{N}\hat{M} = \mathrm{Im}\,\varphi_2$.

But

$$\varphi_2 : \hat{\Gamma}^{(s)} \longrightarrow \hat{M},$$

$$\gamma^{(s)} \longmapsto (\gamma^{(s)}, 0) + \hat{M}_o;$$

i.e., under the above congruence $E_{\hat{M}} \equiv E_o\underset{=}{\tilde{X}}$, $\mathrm{Im}\,\varphi_2 = \hat{\Gamma}^{(s)}$. We have

$$\hat{N}\hat{M} = [(\hat{N}^{(s)} \oplus \hat{N}^{(m)}) + \hat{M}_o]/\hat{M}_o$$

$$= ([(\hat{N}^{(s)} + \hat{\Gamma}^{(m)}\underset{=}{\tilde{X}}) \oplus 0] + \hat{M}_o)/\hat{M}_o, \text{ since } \mathrm{Im}\,\varphi_1 = \hat{N}^{(m)}.$$

But $[(\hat{N}^{(s)} + \hat{\Gamma}^{(m)}\underset{=}{\tilde{X}}) \oplus 0] \cap \hat{M}_o = 0$ implies $\hat{N}^{(s)} + \hat{\Gamma}^{(m)}\underset{=}{\tilde{X}} = \hat{\Gamma}^{(s)}$, a contradiction to Nakayama's lemma. Thus $t \leq s$; i.e., $t = s$. #

Hence we may assume $t = s$ and $m > s$, since for $m = s = t$, $\underset{=}{X}_{\hat{M}}$ splits off a factor 1. $\underset{=}{X} = \underset{=}{X}_{\hat{M}}$ then has the form

$$\underset{=}{X} = \left[\begin{array}{cc} 1 & 0 \\ & \ddots \\ 0 & 1 \\ \hline & * \end{array}\right]^{m \times s} = \left[\begin{array}{c} \underset{=}{E}_s \\ \hline \underset{=}{X}' \end{array}\right] ,$$

and applying ETLs with elements in $\underset{=}{k}$, we may assume

$$\underset{=}{X}' = \alpha \underset{=}{C}_1 + \alpha^2 \underset{=}{C}_2 , \ \underset{=}{C}_1 \in (\underset{=}{k})_{m-s,s} .$$

We then can diagonalize $\underset{=}{C}_1$ by ETLs with elements in $\underset{=}{k}$.

$$\underset{=}{C}_1 = \left[\begin{array}{cc|c} 1 & 0 & \\ & \ddots & * \\ 0 & 1 & \\ \hline & 0 & * \end{array}\right] .$$

We remark that ETRs with elements in $\underset{=}{k}$ can be reversed on the $\underset{=}{E}_s$-part of $\underset{=}{X}$ by ETLs with elements in $\underset{=}{k}$, which leave $\underset{=}{X}'$ invariant. Therefore $\underset{=}{C}_1$ gets the form

$$\underset{=}{C}_1 = \left[\begin{array}{cc|c} 1 & 0 & \\ & \ddots & 0 \\ 0 & 1 & \\ \hline & 0 & 0 \end{array}\right] .$$

Decomposing $\underset{=}{C}_2$ accordingly into blocks, yields

$$\underset{=}{C}_2 = \begin{bmatrix} \underset{=}{C}_{11} & \underset{=}{C}_{12} \\ \underset{=}{C}_{21} & \underset{=}{C}_{22} \end{bmatrix} .$$

7.2 If $\underset{=}{C}_{22} \neq 0$ we can diagonalize $\underset{=}{C}_{22}$ without changing the $\underset{=}{C}_1$-part or the $\underset{=}{E}_s$-part of $\underset{=}{X}$; i.e.,

$$\underline{\underline{X}} = \left[\begin{array}{c|c|c}
\multicolumn{3}{c}{1 \; \cdots \; 1} \\ \hline
* & \multicolumn{2}{c}{\underline{\underline{C}}_{12}\,\alpha^2} \\ \hline
\underline{\underline{C}}_{21}\,\alpha^2 & \begin{matrix}\alpha^2 & & \\ & \ddots & \\ & & \alpha^2\end{matrix} & 0 \\ \cline{2-3}
 & 0 & 0
\end{array}\right]$$

and we can transform $\underline{\underline{X}}$ into

$$\left[\begin{array}{c|c|c}
\multicolumn{3}{c}{1 \; \cdots \; 1} \\ \hline
* & \begin{matrix}0\\ \vdots \\ 0\end{matrix} & * \\ \hline
0 \; \cdots \; 0 & \alpha^2 & 0 \; \cdots \; 0 \\ \hline
* & \begin{matrix}0\\ \vdots \\ 0\end{matrix} & *
\end{array}\right]$$

i.e., $\underline{\underline{X}}$ splits off factor $\begin{vmatrix}1\\ \alpha^2\end{vmatrix}$ and we thus may assume $\underline{\underline{C}}_{22} = 0$.

7.3 Let $\underline{\underline{C}}_{22} = 0$ and $\underline{\underline{C}}_{21} \neq 0$; then we can diagonalize $\underline{\underline{C}}_{21}$, and $\underline{\underline{X}}$ has the form

$$\left[\begin{array}{c|c|c}
\multicolumn{3}{c}{1 \; \cdots \; 1} \\ \hline
\begin{matrix}\alpha & & \\ & \ddots & \\ & & \alpha\end{matrix} & \underline{\underline{D}}_1\alpha^2 & \underline{\underline{D}}_2\alpha^2 \\ \hline
0 & \underline{\underline{E}}\,\alpha + \underline{\underline{D}}_4\,\alpha^2 & \underline{\underline{D}}_3\alpha^2 \\ \hline
\begin{matrix}\alpha^2 & & \\ & \ddots & \\ & & \alpha^2\end{matrix} & 0 & 0
\end{array}\right].$$

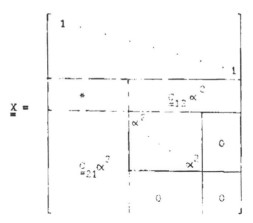

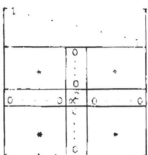

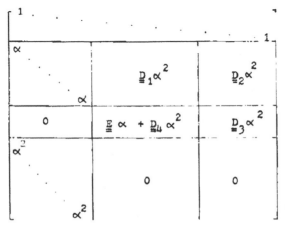

If $\underline{\underline{D}}_2 \neq 0$ we can diagonalize it and $\underline{\underline{X}}$ splits a factor $\begin{pmatrix} 1 & 0 \\ 0 & 1 \\ \alpha & \alpha^2 \\ \alpha^2 & 0 \end{pmatrix}$.

If $\underline{\underline{D}}_2 = 0$ and $\underline{\underline{D}}_1 = 0$, then we get a factor $\begin{pmatrix} 1 \\ \alpha \\ \alpha^2 \end{pmatrix}$.

Hence we assume $D_1 \neq 0$.

We recall that $\underline{\underline{k}}$ is a finite field and that if char $\underline{\underline{k}} = 3$, then the primitive element of S over $\underline{\underline{k}}$ can be chosen to satisfy an equation $X^3 - X - a = 0$ for some $a \in \underline{\underline{k}}$ (cf. Lang [1, p 215, Thm. 1 (Artin-Schreier)]). If 3 is prime to the characteristic of $\underline{\underline{k}}$, we may - replacing S be an isomorphic field if necessary - assume that α satisfies $X^3 - a = 0$ for some $a \in \underline{\underline{k}}$. (This is a consequence of Hilbert's theorem 90, cf. Lang [1, p 213, Thm.; p 214, Thm. 10].) Hence in either case we may assume $\alpha^3 = a\alpha + b$, $a,b \in \underline{\underline{k}}$, $b \neq 0$ and consequently $\alpha^{-1} = c\,\alpha^2 + d$, $c,d \in \underline{\underline{k}}$, $c \neq 0$.

We bring $\underline{\underline{X}}$ into the form

$$\left[\begin{array}{c|c|c|c}
\multicolumn{4}{c}{\begin{matrix}1 & & \\ & \ddots & \\ & & 1\end{matrix}} \\ \hline
\alpha & 0 \cdots 0 & \alpha^2 & 0 \cdots 0 \\ \hline
\begin{matrix}0\\ \vdots\\ 0\end{matrix} & * & \begin{matrix}0\\ \vdots\\ 0\end{matrix} & * \\ \hline
\begin{matrix}0\\ \vdots\\ 0\end{matrix} & 0 & \begin{matrix}\alpha\\ 0\\ \vdots\end{matrix} & * \\ \hline
\alpha^2 & 0 \cdots 0 & 0 & 0 \cdots 0 \\ \hline
\begin{matrix}0\\ \vdots\\ 0\end{matrix} & * & \begin{matrix}0\\ \vdots\\ 0\end{matrix} & 0
\end{array}\right]$$

and we shall show that we can split a factor $\begin{bmatrix} 1 \\ \alpha \end{bmatrix}$ from

$\begin{bmatrix} 1 & 0 \\ 0 & 1 \\ \alpha & \alpha^2 \\ 0 & \alpha \\ \alpha^2 & 0 \end{bmatrix}$ without moving the row before the last one. (This factor does not involve the row before the last one.)

$$\begin{bmatrix} 1 & 0 \\ 0 & 1 \\ \alpha & \alpha^2 \\ 0 & \alpha \\ \alpha^2 & 0 \end{bmatrix} \xrightarrow{\cdot\alpha} \begin{bmatrix} \alpha & 0 \\ 0 & 1 \\ \alpha^2 & \alpha^2 \\ 0 & \alpha \\ \alpha^3 & 0 \end{bmatrix} \Longrightarrow \begin{bmatrix} \alpha & -\alpha \\ 0 & 1 \\ \alpha^2 & 0 \\ 0 & \alpha \\ \alpha^3 & -\alpha^3 \end{bmatrix} \xrightarrow{+b} \begin{bmatrix} \alpha & -\alpha \\ 0 & 1 \\ \alpha^2 & 0 \\ 0 & \alpha \\ \alpha^3 & -a\alpha \end{bmatrix} -a$$

$$\begin{bmatrix} \alpha & -\alpha \\ 0 & 1 \\ \alpha^2 & 0 \\ 0 & \alpha \\ \alpha^3 - a\alpha & 0 \end{bmatrix} = \begin{bmatrix} \alpha & -\alpha \\ 0 & 1 \\ \alpha^2 & 0 \\ 0 & \alpha \\ b & 0 \end{bmatrix} \Longrightarrow \begin{bmatrix} \alpha & -\alpha \\ 0 & 1 \\ \alpha^2 & 0 \\ \alpha & 0 \\ 1 & 0 \end{bmatrix} \xrightarrow{-1} \begin{bmatrix} 0 & -\alpha \\ 0 & 1 \\ \alpha^2 & 0 \\ \alpha & 0 \\ 1 & 0 \end{bmatrix}$$

and hence $\underline{\underline{X}}$ splits a factor $\begin{pmatrix} 1 \\ \alpha \end{pmatrix}$.

7.4 We thus may assume $\underline{\underline{C}}_{22} = \underline{\underline{C}}_{21} = 0$ and $\underline{\underline{C}}_{12} \neq 0$. Diagonalizing $\underline{\underline{C}}_{12}$, we obtain $\underline{\underline{X}}$ in the form

$$\left[\begin{array}{c|c|c|c} \multicolumn{4}{c}{1 \quad \cdots \quad 1} \\ \hline \begin{matrix} \alpha & & \\ & \ddots & \\ & & \alpha \end{matrix} & 0 & \begin{matrix} \alpha^2 & & \\ & \ddots & \\ & & \alpha^2 \end{matrix} & 0 \\ \hline \underline{\underline{D}}_1 \alpha^2 & \underline{\underline{E}}\alpha + \underline{\underline{D}}_2 \alpha^2 & & 0 \end{array}\right].$$

If $\underline{\underline{D}}_1 = 0$ we can split off a factor $\begin{pmatrix} 1 & 0 \\ 0 & 1 \\ \alpha & \alpha^2 \end{pmatrix}$. We diagonalize $\underline{\underline{D}}_1$ and get

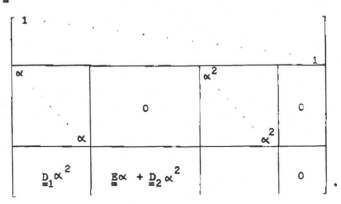

$$\underline{\underline{X}} = \left[\begin{array}{cc|c|c|c}
\multicolumn{5}{c}{1 \quad \cdots \quad 1} \\ \hline
\multicolumn{2}{c|}{\alpha \;\ddots\; \alpha} & 0 & 0 & \alpha^2 \;\ddots\; \alpha^2 \\ \hline
\alpha^2 \;\ddots\; \alpha^2 & 0 & \alpha \;\ddots\; \alpha & 0 & \\ \cline{1-4}
0 & 0 & f(\alpha^2) & \underline{\underline{E}}\alpha + g(\alpha^2) & 0
\end{array}\right].$$

Here $f(\alpha^2)$ and $g(\alpha^2)$ indicate that in these blocks all entries are multiples of α^2. We shall split off a factor $\left|\begin{smallmatrix}1\\ \alpha\end{smallmatrix}\right|$ from the matrix $\begin{bmatrix}1&0&0\\0&1&0\\0&0&1\\ \alpha&0&\alpha^2\\ \alpha^2&\alpha&0\end{bmatrix}$ without moving the middle column, and such that the factor to be split off does not involve the middle column.

$$\begin{bmatrix}1&0&0\\0&1&0\\0&0&1\\ \alpha&0&\alpha^2\\ \alpha^2&\alpha&0\end{bmatrix}_{\cdot\alpha\text{ (col. 2)}},\quad
\begin{bmatrix}1&0&0\\0&\alpha&0\\0&0&1\\ \alpha&0&\alpha^2\\ \alpha^2&\alpha^2&0\end{bmatrix}_{-},\quad
\begin{bmatrix}1&-1&0\\0&\alpha&0\\0&0&1\\ \alpha&-\alpha&\alpha^2\\ \alpha^2&0&0\end{bmatrix},\quad
\begin{bmatrix}1&-1&0\\0&\alpha&0\\0&0&1\\ \alpha&0&\alpha^2\\ \alpha^2&0&0\end{bmatrix}_{\cdot\alpha\text{ (col. 3)}},\quad
\begin{bmatrix}1&-1&0\\0&\alpha&0\\0&0&\alpha\\ \alpha&0&\alpha^3\\ \alpha^2&0&0\end{bmatrix}\quad \alpha^3 = a\alpha + b,$$

$$\begin{bmatrix}1&-1&0\\0&\alpha&0\\0&0&\alpha\\ \alpha&0&b\\ \alpha^2&0&0\end{bmatrix},\quad
\begin{bmatrix}1&-1&0\\0&\alpha&0\\0&\alpha&\alpha\\ \alpha&b&b\\ \alpha^2&0&0\end{bmatrix} b^{-1},\quad
\begin{bmatrix}1+\alpha/b&0&1\\0&\alpha&0\\0&\alpha&\alpha\\ \alpha&b&b\\ \alpha^2&0&0\end{bmatrix}_{-\alpha/b},\quad
\begin{bmatrix}1&0&1\\0&\alpha&0\\0&0&\alpha\\0&b&b\\ \alpha^2&0&0\end{bmatrix} -b$$

$$\begin{bmatrix}1&0&1\\0&\alpha&0\\0&0&\alpha\\-b&b&0\\ \alpha^2&0&0\end{bmatrix} b^{-1},\quad
\begin{bmatrix}0&1&1\\0&\alpha&0\\0&0&\alpha\\-b&b&0\\ \alpha^2&0&0\end{bmatrix}_{-},\quad
\begin{bmatrix}0&0&1\\0&\alpha&0\\0&0&\alpha\\-b&b&0\\ \alpha^2&0&0\end{bmatrix}$$

and X splits off factor $\left|\begin{smallmatrix}1\\ \alpha\end{smallmatrix}\right|$.

7.5 We may assume that $\underline{\underline{C}}_{12} = \underline{\underline{C}}_{21} = \underline{\underline{C}}_{22} = 0$, and $\underline{\underline{X}}$ has the form

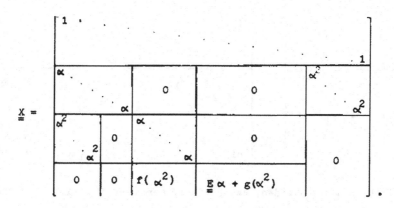

$$\underline{\underline{X}} = \left[\begin{array}{c} \begin{matrix} 1 & & \\ & \ddots & \\ & & 1 \end{matrix} \\ \hline \alpha \underline{\underline{E}} + \underline{\underline{Y}} \alpha^2 \end{array}\right] = \left[\begin{array}{cccc} 1 & & & \\ & \ddots & & \\ & & & 1 \\ \hline \alpha + k_{11}\alpha^2 & k_{12}\alpha^2 & \cdots & k_{1n}\alpha^2 \\ k_{21}\alpha^2 & \ddots & & \vdots \\ \vdots & & \ddots & k_{n-1,n}\alpha^2 \\ k_{n1}\alpha^2 & \cdots & k_{n,n-1}\alpha^2 & \alpha + k_{nn}\alpha^2 \end{array}\right] .$$

Not all k_{1j} and k_{j1}, $1<j\leq n$, can be zero, for then $\underline{\underline{X}}$ splits off factor

$$\left[\begin{array}{c} 1 \\ \alpha + k_{11}\alpha^2 \end{array}\right] \quad , k_{11} \in \underline{\underline{k}}.$$

Say, $k_{21} \neq 0$, $n\geq 2$, then we bring $\underline{\underline{X}}$ to the form

$$\underline{\underline{X}} = \left[\begin{array}{cccc} 1 & & & \\ & \ddots & & \\ & & & 1 \\ \hline \alpha & k'_{12}\alpha^2 & \cdots & k'_{1n}\alpha^2 \\ \alpha^2 & \alpha + k_{22}\alpha^2 & & \vdots \\ 0 & & & \vdots \\ 0 & \vdots & & \vdots \\ \vdots & \vdots & & \vdots \\ 0 & k'_{n2}\alpha^2 & \cdots & \alpha + k'_{nn}\alpha^2 \end{array}\right] .$$

Continuing this way, we bring $\underline{\underline{X}}$ into the "Frobenius-form"

$$\underline{\underline{X}} = \left[\begin{array}{ccccc} 1 & & & & \\ & \ddots & & & \\ & & & & 1 \\ \hline \alpha & & & & \beta_1\alpha^2 \\ \alpha^2 & \alpha & & & \vdots \\ & \alpha^2 & \ddots & & \vdots \\ & & \ddots & \alpha & \beta_{n-1}\alpha^2 \\ & & & \alpha^2 & \alpha + \beta_n\alpha^2 \end{array}\right] .$$

Multiplication of the first column with α yields

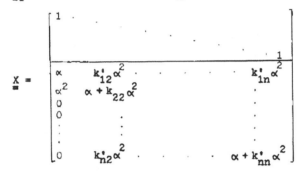

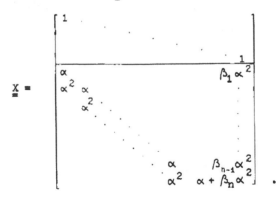

$$\underline{\underline{X}} = \left[\begin{array}{cccccc} \alpha & 1 & \cdot & & & \\ & & & \cdot & \cdot & 1 \\ \hline \alpha^2 & & & & & \beta_1\alpha^2 \\ a_1+b_2\alpha & \alpha & & & & a_1+\beta_2\alpha^2 \\ & \alpha^2 & \cdot & \cdot & & \cdot \\ & & \cdot & \cdot & \cdot & \cdot \\ & & & & & \alpha+\beta_n\alpha^2 \end{array}\right].$$

Multiplying the last column with $0 \neq k \in \underline{k}$, if necessary, we may assume $1-\beta_1 \neq 0$. Thus $\underline{\underline{X}}$ has the form

$$\underline{\underline{X}} = \left[\begin{array}{cccccc} 1 & \cdot & \cdot & & & \\ & & & \cdot & \cdot & 1 \\ \hline \alpha & & & & & \beta_1'\alpha^2 \\ & \alpha & & & & \cdot \\ & \alpha^2 & \cdot & & & \cdot \\ & & \cdot & \cdot & & \cdot \\ & & & & \alpha & \beta_{n-1}\alpha^2 \\ & & & & \alpha^2 & \alpha\gamma+\beta_n\alpha^2 \end{array}\right].$$

This reduction can be done $(n-2)$ times:

$$\underline{\underline{X}} = \left[\begin{array}{cccccc} 1 & \cdot & \cdot & & & \\ & & & \cdot & \cdot & 1 \\ \hline \alpha & & & & & \beta_1\alpha^2 \\ & \alpha & & & & \cdot \\ & & \cdot & & & \cdot \\ & & & & \alpha & \beta_{n-1}\alpha^2 \\ & & & & \alpha^2 & \alpha\gamma+\beta_n\alpha^2 \end{array}\right].$$

(i) If $\beta_{n-1} \neq 0$, then $\underline{\underline{X}}$ decomposes into factors of the form

$$\begin{bmatrix} 1 \\ \alpha \end{bmatrix} \quad \text{and} \quad \begin{bmatrix} 1 & 0 \\ 0 & 1 \\ \alpha & \beta_{n-1}\alpha^2 \\ \alpha^2 & \alpha\gamma+\beta_n\alpha^2 \end{bmatrix}.$$

(ii) If $\beta_{n-1} = 0$, we may assume $\beta_1 \neq 0, \dots, \beta_{n-2} \neq 0$, since otherwise $\underline{\underline{X}}$ splits a factor $\begin{bmatrix} 1 \\ \alpha \end{bmatrix}$. Thus we may assume $n = 3$; i.e.,

$$\underline{\underline{X}} = \begin{bmatrix} 1 & 0 & 0 \\ 0 & 1 & 0 \\ 0 & 0 & 1 \\ \alpha & 0 & \beta_1 \alpha^2 \\ 0 & \alpha & 0 \\ 0 & \alpha^2 & \alpha\gamma + \beta_3 \alpha^2 \end{bmatrix}, \beta_1 \neq 0.$$

<u>Summary</u>: We thus have shown that the degrees of the indecomposable parts of $\underline{\underline{X}}$ are bounded. Since $\underline{\underline{k}}(\alpha)$ is a finite field, there are only finitely many "non-equivalent indecomposable" matrices; i.e., $n(\Lambda) < \infty$ in this case. It is now easy to compute the non-equivalent indecomposable matrices. They are

$$(1), \begin{bmatrix} 1 \\ \alpha \end{bmatrix}, \begin{bmatrix} 1 \\ \alpha \\ \alpha^2 \end{bmatrix}, \begin{bmatrix} 1 & 0 \\ 0 & 1 \\ \alpha & \alpha^2 \end{bmatrix}.$$

<u>Exercises §7</u>:

1.) Decompose the matrix $\underline{\underline{X}}$ p. 70!

§8 <u>Decomposition of the matrix $\underline{\underline{X}}_{\underline{\underline{M}}}^{\Delta}$ of 6.A.II</u>

$S = \underline{\underline{k}}[r]$, $r^3 = 0$, $\underline{\underline{k}}r = r\underline{\underline{k}}$, and $\underline{\underline{X}} = \underline{\underline{X}}_{\underline{\underline{M}}}^{\Delta}$ has entries in S. We decompose $\underline{\underline{X}}$ by applying ETLs with elements in $\underline{\underline{k}}$ and ETRs with elements in S. With the help of ETs we bring $\underline{\underline{X}}$ into the following form

$$\underline{\underline{X}} = \left[\begin{array}{ccc|c} 1 & & & \\ & \ddots & & 0 \\ & & 1 & \\ \hline & * & & ** \end{array}\right]^{m \times s}_{\quad t} .$$

where the blocks $*$ and $**$ have only entries in rad S. According to (7.1) we may assume s = t, and we turn to the decomposition of

$$\underline{\underline{X}} = \left[\begin{array}{ccc} 1 & & \\ & \ddots & \\ & & 1 \\ \hline & \underline{\underline{X}}' & \end{array}\right]^{m \times s} ,$$

where $\underline{\underline{X}}'$ has entries in rad S. If m = s, $\underline{\underline{X}}$ splits off a factor (1) and we may assume m > s.

$$\underline{\underline{X}}' = r\underline{\underline{C}}_1 + r^2\underline{\underline{C}}_2 \; ; \; \underline{\underline{C}}_i \in (\underline{\underline{k}})_{(m-s) \times s}, \; i=1,2.$$

We can diagonalize $\underline{\underline{C}}_1$ without changing the $\underline{\underline{E}}_s$-part of $\underline{\underline{X}}$. This uses essentially the fact $\underline{\underline{k}}r = r\underline{\underline{k}}$; thus

$$\underline{\underline{C}}_1 = \left[\begin{array}{ccc|c} 1 & & & \\ & \ddots & & 0 \\ & & 1 & \\ \hline & 0 & & 0 \end{array}\right] .$$

We write $\underline{\underline{C}}_2$ in an analogous block-decomposition

$$\underline{\underline{C}}_2 = \begin{bmatrix} \underline{\underline{C}}_{11} & \underline{\underline{C}}_{12} \\ \underline{\underline{C}}_{21} & \underline{\underline{C}}_{22} \end{bmatrix} .$$

As in §7, we may assume $\underline{\underline{C}}_{22}=0$ for otherwise $\underline{\underline{X}}$ splits off factor $\begin{bmatrix} 1 \\ r^2 \end{bmatrix}$.

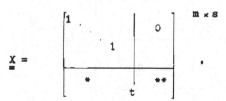

If $\underline{\underline{C}}_{22} = 0$ and $\underline{\underline{C}}_{12} \neq 0$, we can diagonalize $\underline{\underline{C}}_{12}$ and $\underline{\underline{X}}$ has the form

$$\underline{\underline{X}} = \left[\begin{array}{c|c|c} \multicolumn{3}{c}{\begin{matrix}1 & & \\ & \ddots & \\ & & 1\end{matrix}} \\ \hline \begin{matrix}r & & \\ & \ddots & \\ & & r\end{matrix} & 0 & \begin{matrix}r^2 & & \\ & \ddots & \\ & & r^2\end{matrix} \\ \hline \underline{\underline{D}}_1 r^2 & \underline{\underline{E}}r + \underline{\underline{D}}_4 r^2 & \\ \cline{1-2} \underline{\underline{D}}_2 r^2 & \underline{\underline{D}}_3 r^2 & 0 \end{array}\right] .$$

Assuming $\underline{\underline{D}}_2 \neq 0$, we diagonalize it:

$$\underline{\underline{X}} = \left[\begin{array}{c|c|c|c} \multicolumn{4}{c}{\begin{matrix}1 & & \\ & \ddots & \\ & & 1\end{matrix}} \\ \hline \multicolumn{2}{c|}{\begin{matrix}r & & \\ & \ddots & \\ & & r\end{matrix}} & 0 & \begin{matrix}r^2 & & \\ & \ddots & \\ & & r^2\end{matrix} \\ \hline 0 & * & * & \\ \cline{1-3} \begin{matrix}r^2 & & \\ & \ddots & \\ & & r^2\end{matrix} & 0 & 0 & 0 \\ \cline{1-3} 0 & 0 & & \end{array}\right] ,$$

and we obtain a factor $\begin{bmatrix} 1 & 0 \\ 0 & 1 \\ r & r^2 \\ r & 0 \end{bmatrix}$.

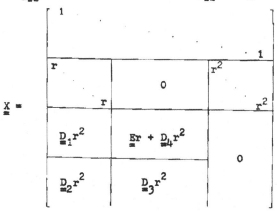

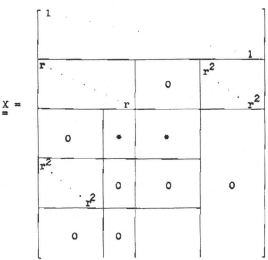

If $\underline{\underline{D}}_2 = 0$ and $\underline{\underline{D}}_1 = 0$, then we get a factor $\begin{bmatrix} 1 & 0 \\ 0 & 1 \\ r & r^2 \end{bmatrix}$, and so we may assume $\underline{\underline{D}}_2 = 0$, $\underline{\underline{D}}_1 \neq 0$. We bring $\underline{\underline{X}}$ into the form

$$\underline{\underline{X}} = \left[\begin{array}{c|c|c|c|c} \multicolumn{5}{c}{\begin{matrix}1 & & \\ & \ddots & \\ & & 1\end{matrix}} \\ \hline \begin{matrix}r & & \\ & \ddots & \\ & & r\end{matrix} & \multicolumn{3}{c|}{0} & \begin{matrix}r^2 & & \\ & \ddots & \\ & & r^2\end{matrix} \\ \hline r^2 \; 0 \cdots 0 & r+kr^2 & \multicolumn{2}{c|}{0 \cdots\cdots 0} & \\ \cline{1-4} 0 \;\; f(r^2) & g(r^2) & \multicolumn{2}{c|}{\underline{\underline{E}}'r + \underline{\underline{D}}_4'r^2} & 0 \\ \cline{1-4} 0 \;\; 0 & \multicolumn{3}{c|}{\underline{\underline{D}}_3 r^2} & \end{array}\right].$$

3 ⌐ (row operation indicated); • r ⟶ 1 (column operation indicated); 2 • r (column operation indicated)

The indicated ETs show that we have a factor $\begin{bmatrix} 1 & -r \\ 0 & 1 \\ r & 0 \end{bmatrix}$, since $r^3 = 0$.

Hence we can assume $\underline{\underline{C}}_{22} = 0 = \underline{\underline{C}}_{12}$, $\underline{\underline{C}}_{21} \neq 0$. Diagonalizing $\underline{\underline{C}}_{21}$ we get

$$\underline{\underline{X}} = \left[\begin{array}{c|c} \multicolumn{2}{c}{\begin{matrix}1 & & \\ & \ddots & \\ & & 1\end{matrix}} \\ \hline \begin{matrix}r & & \\ & \ddots & \\ & & r\end{matrix} & f(r^2) \\ \hline 0 & \underline{\underline{E}}r + g(r^2) \\ \hline \begin{matrix}r^2 & & \\ & \ddots & \\ & & r^2\end{matrix} & 0 \\ \hline 0 & 0 \end{array}\right].$$

Here $f(r^2)$ indicates that the corresponding block has entries in $\underline{\underline{k}}r^2$. The block $f(r^2)$ can be brought into diagonal form. If the diagonal does not go through the entire block, then we obtain a factor $\begin{bmatrix} 1 \\ r \\ r^2 \end{bmatrix}$.

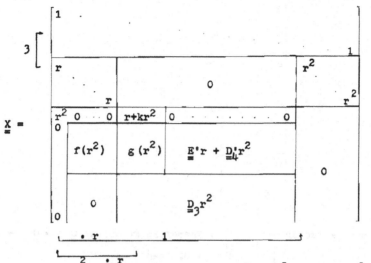

Consequently $\underline{\underline{X}}$ has the form

$$\underline{\underline{X}} = \left[\begin{array}{c|c} \multicolumn{2}{c}{\begin{matrix} 1 & & \\ & \ddots & \\ & & 1 \end{matrix}} \\ \hline \begin{matrix} r & & \\ & \ddots & \\ & & r \end{matrix} & \begin{matrix} r^2 & & \\ & \ddots & \\ & & r^2 \end{matrix} \\ \hline 0 & \begin{matrix} r & & \\ & \ddots & \\ & & r \end{matrix} \\ \hline \begin{matrix} r^2 & & \\ & \ddots & \\ & & r^2 \end{matrix} & 0 \end{array}\right],$$

and we have a factor $\begin{bmatrix} 1 & 0 \\ 0 & 1 \\ r & r^2 \\ 0 & r \\ r^2 & 0 \end{bmatrix}$. Therefore we assume $\underline{\underline{C}}_{12} = \underline{\underline{C}}_{21} = \underline{\underline{C}}_{22} = 0$. Then

$$\underline{\underline{X}} = \left[\begin{array}{cccc} 1 & & & \\ & \ddots & & \\ & & & 1 \\ \hline r+k_{11}r^2 & k_{12}r^2 & \cdots & k_{1n}r^2 \\ k_{21}r^2 & r+k_{22}r^2 & & \vdots \\ \vdots & & & \\ k_{n1}r^2 & \cdots & \cdots & r+k_{nn}r^2 \end{array}\right].$$

If $n = 1$, we get a factor $\begin{bmatrix} 1 \\ r+kr^2 \end{bmatrix}$. Otherwise we bring $\underline{\underline{X}}$ into the "Frobenius-form" (cf. 7.5):

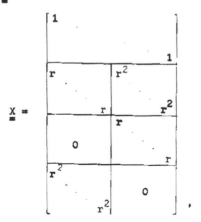

$$\underset{=}{X} = \left[\begin{array}{cccccc} 1 & & & & & \\ & \ddots & & & & \\ & & & & & 1 \\ \hline r & & & & & \beta_1 r^2 \\ r^2 & r & & & & \vdots \\ & r^2 & \ddots & & & \vdots \\ & & \ddots & \ddots & & \\ & & & & r & \beta_{n-1} r^2 \\ & & & & r^2 & r + \beta_n r^2 \end{array}\right]$$

(2; $1 \cdot r$; 3: $\beta_1 r^2$)

By means of the indicated transformations, we obtain:

$$\underset{=}{X} = \left[\begin{array}{c|cccccc} 1 & & & & & & \\ & \ddots & & & & & \\ & & & & & & 1 \\ \hline r & 0 & \cdots & & & 0 & \beta_1 r^2 \\ \hline 0 & r & & & & & \vdots \\ \vdots & r^2 & r & & & & \vdots \\ \vdots & & r^2 & \ddots & & & \vdots \\ \vdots & & & \ddots & & r & \beta_{n-1} r^2 \\ 0 & & & & & r^2 & r + \beta_n r^2 \end{array}\right].$$

This reduction can be done (n - 2) times:

$$\underset{=}{X} = \left[\begin{array}{ccccc} 1 & & & & \\ & \ddots & & & \\ & & & & 1 \\ \hline r & & & & \beta_1 r^2 \\ & \ddots & & & \vdots \\ & & r & & \vdots \\ & & & r & \beta_{n-1} r^2 \\ & & & r^2 & r + \beta_n r^2 \end{array}\right].$$

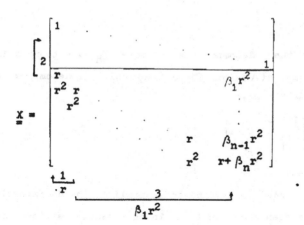

If one of the $\{\beta_i\}_{1 \le i \le n-2}$ is zero, we get a factor $\begin{bmatrix} 1 \\ r \end{bmatrix}$. Hence we may assume $n = 3$.

8.1 The degrees of the indecomposable matrices $\underline{\underline{X}}_{\hat{M}}$ are bounded in case of 6.A,II; hence $n(\hat{\Lambda}) < \infty$. The non-equivalent indecomposable matrices $\underline{\underline{X}}_{\hat{M}}$ can be listed:

$$(1),\ \begin{bmatrix}1\\ r\end{bmatrix},\ \begin{bmatrix}1\\ r^2\end{bmatrix},\ \begin{bmatrix}1\\ r\\ r^2\end{bmatrix},\ \begin{bmatrix}1 & 0\\ 0 & 1\\ r^2 & r\end{bmatrix},\ \begin{bmatrix}1 & 0\\ 0 & 1\\ r & r^2\\ r^2 & 0\end{bmatrix}.$$

These matrices are needed in §9, and so we shall do the decomposition explicitly. We have seen above that all indecomposable matrices must occur as factors of one of the following matrices.

$$(1),\ \begin{bmatrix}1\\ r\end{bmatrix},\ \begin{bmatrix}1\\ r^2\end{bmatrix},\ \begin{bmatrix}1\\ r+kr^2\end{bmatrix},\ \begin{bmatrix}1 & 0\\ 0 & 1\\ r & r^2\\ r^2 & 0\end{bmatrix},\ \begin{bmatrix}1\\ r\\ r^2\end{bmatrix},\ \begin{bmatrix}1 & -r\\ 0 & 1\\ r & 0\end{bmatrix},\ \begin{bmatrix}1 & 0\\ 0 & 1\\ r & r^2\end{bmatrix},$$

$$\begin{bmatrix}1 & 0\\ 0 & 1\\ r & r^2\\ 0 & r\\ r^2 & 0\end{bmatrix} \quad \text{and} \quad \begin{bmatrix}1 & 0 & 0\\ 0 & 1 & 0\\ 0 & 0 & 1\\ r & 0 & \beta_1 r^2\\ 0 & r & \beta_2 r^2\\ 0 & r^2 & r+\beta_3 r^2\end{bmatrix}.$$

(i) $\begin{bmatrix}1 & -r\\ 0 & 1\\ r & 0\end{bmatrix} \Rightarrow \begin{bmatrix}1 & 0\\ 0 & 1\\ r & r^2\end{bmatrix}$.

(ii) $\begin{bmatrix}1\\ r+kr^2\end{bmatrix} \Rightarrow \begin{bmatrix}1\\ r(1+kr)\end{bmatrix} \Rightarrow \begin{bmatrix}(1+kr)^{-1}\\ r\end{bmatrix} \Rightarrow \begin{bmatrix}1+k'r+k''r^2\\ r\end{bmatrix} \Rightarrow \begin{bmatrix}1+k''r^2\\ r\end{bmatrix}$

$\Rightarrow \begin{bmatrix}1\\ r(1-k''r^2)\end{bmatrix} \Rightarrow \begin{bmatrix}1\\ r\end{bmatrix}$.

(iii) $\begin{bmatrix}1 & 0\\ 0 & 1\\ r & r^2\\ 0 & r\\ r^2 & 0\end{bmatrix} \Rightarrow \begin{bmatrix}1 & 0\\ r & 0\\ r^2 & 0\\ 0 & 1\\ 0 & r\end{bmatrix}$ and $\underline{\underline{X}}$ decomposes into the factors $\begin{bmatrix}1\\ r\\ r^2\end{bmatrix}$ and $\begin{bmatrix}1\\ \text{[illegible]}\end{bmatrix}$.

(iv)

$$\underline{\underline{X}} = \begin{bmatrix} 1 & 0 & 0 \\ 0 & 1 & 0 \\ 0 & 0 & 1 \\ r & 0 & \beta_1 r^2 \\ 0 & r & \beta_2 r^2 \\ 0 & r^2 & r+\beta_3 r^2 \end{bmatrix} \begin{matrix} 3 \\ \beta_1 \end{matrix}$$

$$1 - r \qquad 2-\beta_1 r \qquad 4-\beta_1\beta_3 r^2 \quad .$$

If $\beta_1 = 0$, then we have a factor $\begin{bmatrix}1\\r\end{bmatrix}$. We assume $\beta_1 \neq 0$ and obtain a factor $\begin{bmatrix}1\\r\end{bmatrix}$. The remainder has the form

$$\begin{bmatrix} 1 & 0 \\ -r & 1 \\ r & \beta_2 r^2 \\ 0 & \beta_3 r^2+r \end{bmatrix} \begin{matrix} 2 \\ + \end{matrix} \begin{matrix} 3 \\ \beta_2 \end{matrix}$$

$$1-\beta_2 r \qquad 4-\beta_2\beta_3 r^2 \quad ,$$

and $\underline{\underline{X}}$ decomposes into a factor $\begin{bmatrix}1\\r\end{bmatrix}$ and a factor of type (ii).

§9 Decomposition of the matrix 6.A.III

We have shown in (6,A,III):

$$\mathbf{X}_{\hat{M}} = (\mathbf{X}_1, \mathbf{X}_2)$$

$\mathbf{X}_1 \in (\mathbf{k}[r])_{m \times s_1}$, $r^3 = 0$, $\mathbf{k}r = r\mathbf{k}$, $S = \mathbf{k}[r]$,

$\mathbf{X}_2 \in (\mathbf{k}[r'])_{m \times s_2}$, $r'^3 = 0$, $\mathbf{k}r' = r'\mathbf{k}$, $S' = \mathbf{k}[r']$,

$$\mathbf{Z} = \begin{bmatrix} \mathbf{Z}_{11} & \mathbf{Z}_{12} \\ \mathbf{Z}_{21} & \mathbf{Z}_{22} \end{bmatrix}$$

$\mathbf{Z}_{11} \in (\mathbf{k}[r])_{s_1}$, $\mathbf{Z}_{22} \in (\mathbf{k}[r'])_{s_2}$,

$\mathbf{Z}_{12} \in (\varphi\,\mathbf{k}[r'])_{s_1 \times s_2}$, $\mathbf{Z}_{21} \in (\mathbf{k}[r']\psi)_{s_2 \times s_1}$,

where $\varphi : 1 \longmapsto 1'$, $r \longmapsto r'^2$, $r^2 \longmapsto 0$;

$\psi : 1' \longmapsto r$, $r' \longmapsto 0$, $r'^2 \longmapsto r^2 k$ for some $0 \neq k \in \mathbf{k}$.

$\mathbf{Y}$ has entries in $\mathbf{k}$.

By means of ETs we obtain $\mathbf{X} = \mathbf{X}_{\hat{M}}$ in the form

$\mathbf{X} =$

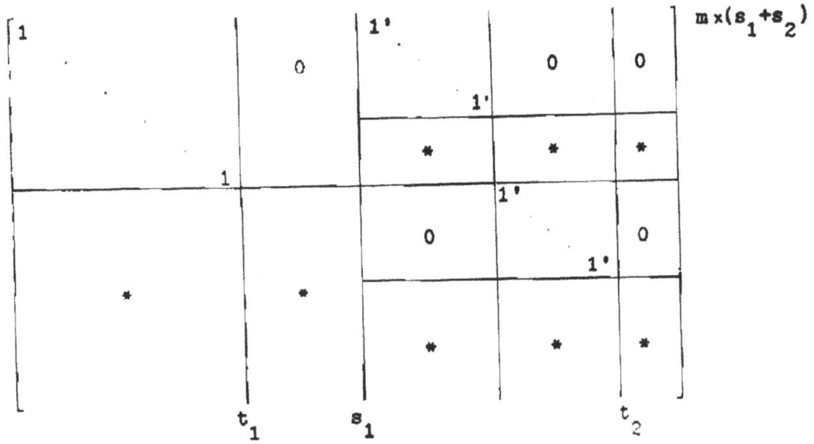

where (*) denotes entries in rad S and rad S' resp. Here we have used the following ETs: ETLs with entries in $(\mathbf{k})_m$ and ETRs with matrices of the form

$$\begin{bmatrix} \underline{\underline{E}}_{s_1} & 0 \\ 0 & \underline{\underline{Z}}_2 \end{bmatrix} \quad \text{and} \quad \begin{bmatrix} \underline{\underline{Z}}_1 & 0 \\ 0 & \underline{\underline{E}}_{s_2} \end{bmatrix},$$

where $\underline{\underline{Z}}_1$ and $\underline{\underline{Z}}_2$ correspond to ETRs with entries in $\underline{\underline{k}}[r]$ and $\underline{\underline{k}}[r']$ resp.

We recall that we consider $\hat{\Lambda}$-lattices $\hat{M}$ in exact sequences of the form

$$E_{\hat{M}} : 0 \longrightarrow \hat{\Lambda}_1^{(s_1)} \oplus \hat{\Gamma}^{(s_2)} \xrightarrow{\sigma} \hat{M} \longrightarrow \underline{\underline{k}}^{(m)} \longrightarrow 0,$$

where $\operatorname{Im}\sigma = \hat{N}\hat{M}$ (cf. 6,A).

9.1 __Proposition:__ If the matrix $\underline{\underline{X}}$ corresponds to an exact sequence $E_{\hat{M}}$, then we must have $t_1 = s_1$, $t_2 = s_2$ and $\hat{\Lambda}_1^{(m)} \underline{\underline{X}}_2 = \hat{\Gamma}^{(s_2)}$.

__Proof:__ Let

$$\pi_1 : \hat{\Lambda}_1^{(s_1)} \oplus \hat{\Gamma}^{(s_2)} \longrightarrow \hat{\Lambda}_1^{(s_1)},$$

$$\pi_2 : \Lambda_1^{(s_1)} \oplus \hat{\Gamma}^{(s_2)} \longrightarrow \hat{\Gamma}^{(s_2)}$$

be the projections corresponding to the above decomposition. Then we obtain the following commutative diagram

$$\begin{array}{ccccccccc} E_{\hat{M}} : 0 & \longrightarrow & \hat{M}_1' \oplus \hat{M}_2' & \xrightarrow{\sigma} & \hat{M} & \longrightarrow & \underline{\underline{k}}^{(m)} & \longrightarrow & 0 \\ & & \downarrow \pi_1 & & \downarrow \varphi_1 & & \downarrow 1_{k^{(m)}} & & \\ E_{\hat{M}}\pi_1 : 0 & \longrightarrow & \hat{M}_1' & \xrightarrow{\sigma_1} & \hat{M}_1 & \xrightarrow{\tau_1} & \underline{\underline{k}}^{(m)} & \longrightarrow & 0, \end{array}$$

with $\hat{M}_1' = \hat{\Lambda}_1^{(s_1)}$, $\hat{M}_2' = \hat{\Gamma}^{(s_2)}$.

Since $\operatorname{Im}\sigma = \hat{N}\hat{M}$ and since π_1 is an epimorphism, we have $\hat{N}\hat{M}_1' \supset \operatorname{Im}\sigma_1$. On the other hand, $\hat{N}\hat{M}_1' \subset \operatorname{Ker}\tau_1$; i.e., $\hat{N}\hat{M}_1' = \operatorname{Im}\sigma_1$. Similarly one shows $\operatorname{Im}\sigma_2 = \hat{N}\hat{M}_2'$ with selfexplanatory notations.

However, it is clear that

$$\underline{\underline{X}}_1 = \left[\begin{array}{ccc|c} 1 & & & \\ & \ddots & & 0 \\ & & 1 & \\ \hline & * & & ** \end{array}\right]\begin{array}{l} m \times s_1 \\ \\ \\ \\ \end{array}$$
$$\qquad\qquad t_1$$

is the matrix corresponding to the sequence $E_{\hat{M}}\pi_i$, $i=1,2$. Now it follows from (7.1), that $s_i = t_i$, $i=1,2$.

As for the second statement, we recall that

$$\mathrm{Ext}^1_{\hat\Lambda}(\underline{\underline{k}}, \hat\Gamma) = E_o \mathrm{Hom}_{\hat\Lambda}(\hat\Lambda_1, \hat\Gamma),$$

where

$$E_o : 0 \longrightarrow \hat\Lambda_1 \xrightarrow{\varphi} \hat\Lambda \longrightarrow \underline{\underline{k}} \longrightarrow 0,$$

$\varphi : \hat\Lambda_1 \xrightarrow{\sim} \hat N$. Since $\mathrm{Hom}_{\hat\Lambda}(\hat\Lambda_1, \hat\Gamma) = \iota\hat\Gamma$, with $\iota : \hat\Lambda_1 \longrightarrow \hat\Gamma$ the injection, we can write every $E \in \mathrm{Ext}^1_{\hat\Lambda}(\underline{\underline{k}}, \hat\Gamma)$ as $E = E_o\iota\gamma$ for some $\gamma \in \hat\Gamma$. To apply this to our situation, we choose a preimage $\tilde{\underline{\underline{X}}}_2 \in (\hat\Gamma)_{m \times s_2}$ of $\underline{\underline{X}}_2$. The sequence corresponding to $\underline{\underline{X}}_2$ is given by

$$\begin{array}{lccccccccc}
E_o^{(m)} & : 0 & \longrightarrow & \hat\Lambda_1^{(m)} & \xrightarrow{\varphi(m)} & \hat\Lambda^{(m)} & \longrightarrow & \underline{\underline{k}}^{(m)} & \longrightarrow & 0 \\
 & & & \downarrow{\scriptstyle \iota^{(m)}\tilde{\underline{\underline{X}}}_2} & & \downarrow{\scriptstyle g} & & \downarrow{\scriptstyle 1_{\underline{\underline{k}}}(m)} & & \\
E = E_o^{(m)}\iota^{(m)}\tilde{\underline{\underline{X}}}_2 & : 0 & \longrightarrow & \hat\Gamma^{(s_2)} & \longrightarrow & \hat M_2 & \longrightarrow & \underline{\underline{k}}^{(m)} & \longrightarrow & 0.
\end{array}$$

We have

$$\hat M_2 = (\hat\Gamma^{(s_2)} \oplus \hat\Lambda^{(m)})/\hat M_o ,$$
$$\hat M_o = \{(\lambda\iota^{(m)}\underline{\underline{X}}_2, -\lambda\varphi^{(m)}) : \lambda \in \hat\Lambda_1^{(m)}\}.$$

We recall

$$\hat\Gamma^{(s_2)} = \hat N \hat\Gamma^{(s_2)} + \hat\Lambda_1^{(m)}\iota^{(m)}\tilde{\underline{\underline{X}}}_2.$$

Nakayama's lemma implies

$$\hat\Gamma^{(s_2)} = \Lambda_1^{(m)}\iota^{(m)}\tilde{\underline{\underline{X}}}_2 \quad \text{or}$$

$$S'^{(s_2)} = S^{(m)}\underline{\underline{X}}_2.$$

Here $S'^{(s_2)}$ and $S^{(m)}$ are considered as spaces of row-vectors, and S' is an S-module via φ; i.e., $11' = 1'$, $r1' = r'^2$, $r^2 1' = 0$.

If $s_1 = 0$ or $s_2 = 0$, then we obtain the indecomposable lattices that correspond to the matrices of (8.2) for r and r' resp. However in $\underline{\underline{X}}_2$ we can only have the matrices $\begin{bmatrix} 1' \\ r' \end{bmatrix}$ and $\begin{bmatrix} 1' \\ r' \\ r'^2 \end{bmatrix}$, since these are the only ones that satisfy $S'^{(s_1)} = S^{(m)}\underline{\underline{X}}_2$ (cf. 9.1) (observe the action of S on S').

We now use the ETs that correspond to the homomorphisms φ and ψ ; i.e., ETs of the following kind

$$\begin{array}{cl} \begin{array}{ccc|ccc} & & & & j & \end{array} & \\ \left[\begin{array}{ccc|ccc} 1 & & & & \vdots & \\ & \ddots & & \cdots & \varphi\alpha' & \cdots \\ & & 1 & & & \\ \hline & & & 1' & & \\ & & & & \ddots & \\ & & & & & 1' \end{array}\right] & \begin{array}{l} (s_1+s_2)\times(s_1+s_2) \\ i \\ s_1 \\ \\ \\ \\ \end{array} \\ \begin{array}{ccc|ccc} & & & s_1 & & \end{array} & \end{array} , \quad \alpha' \in \underline{\underline{k}}[r']$$

and

$$\begin{array}{ccl} & \left[\begin{array}{ccc|ccc} 1 & & & & & \\ & \ddots & & & & \\ & & 1 & & & \\ \hline & \vdots & & 1' & & \\ \cdots & \alpha'\psi & \cdots & & \ddots & \\ & \vdots & & & & 1' \end{array}\right] & \begin{array}{l} (s_1+s_2)\times(s_1+s_2) \\ \\ s_1 \\ \\ \\ \\ \end{array} \\ i & & \\ & \begin{array}{ccc|ccc} & j & & s_1 & & \end{array} & \end{array} , \quad \alpha' \in \underline{\underline{k}}[r'] .$$

This has the effect of applying $\varphi\alpha'$ to the i-th column of $\underline{\underline{X}}_1$ and adding this to the j-th column of $\underline{\underline{X}}_2$; similarly for $\alpha'\psi$.

9.2 <u>Lemma</u>: We may assume $\underline{\underline{X}}$ in the form:

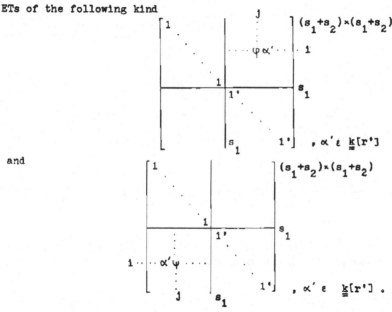

$$\underline{\underline{X}} = \left[\begin{array}{c|c} 1 \cdots 1 & 0 \\ \hline 0 & 1' \cdots 1' \\ \hline \text{rad } S & \text{rad } S' \end{array}\right]_{m\times(s_1+s_2)} \quad (\text{first block column of width } s_1) .$$

Proof: $\underline{\underline{X}}$ has the form:

$$\underline{\underline{X}} = \left[\begin{array}{c|c|c} & 1' \cdots 1' & 0 \\ \cline{2-3} 1 \cdots 1 & \text{rad } S' & \text{rad } S' \\ \hline & 0 & 1' \cdots 1' \\ \cline{2-3} \text{rad } S & * & \text{rad } S' \end{array}\right] .$$

$$\underbrace{\hphantom{\text{rad } S}}_{\varphi\alpha}$$

Elementary transformations with $\varphi\alpha$ yield

$$\underline{\underline{X}} = \left[\begin{array}{c|c|c} & \multicolumn{2}{c}{0} \\ \cline{2-3} 1 \cdots 1 & \multicolumn{2}{c}{f(r')} \\ \hline & 0 & 1' \cdots 1' \\ \cline{2-3} * & * & * \end{array}\right] .$$

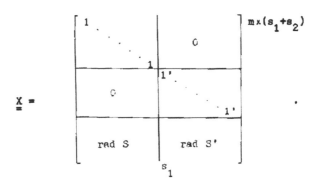

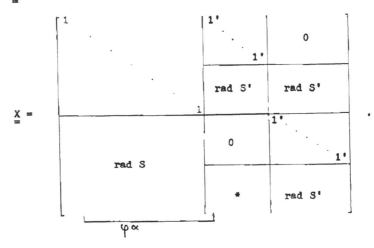

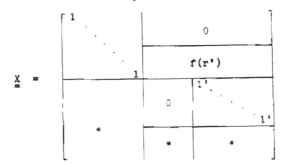

We diagonalize (f(r')) from the left, and we may assume that (*) has entries in rad(S'): otherwise we could enlarge the diagonal with 1'. Hence $\underline{\underline{X}}$ has the form

$\underline{\underline{X}} =$

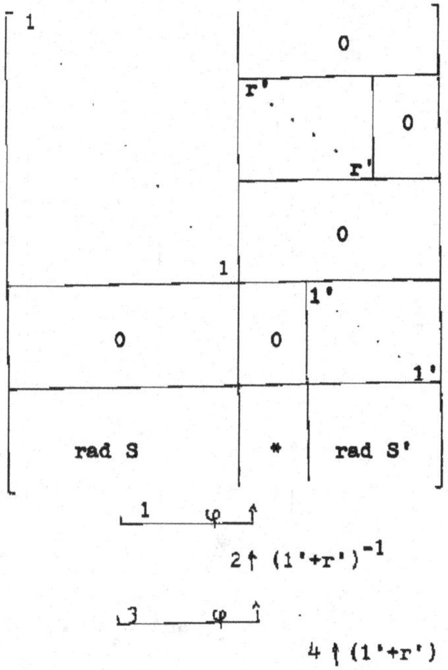

Using the indicated transformations and applying (9.1) once more, we conclude that $\underline{\underline{X}}$ has the following form

$$\underline{\underline{X}} = \begin{bmatrix} \underline{\underline{E}}_{s_1} & 0 \\ 0 & \underline{\underline{E}}_{s_2} \\ \underline{\underline{C}}_1 & \underline{\underline{C}}_2 \end{bmatrix} ,$$

where $\underline{\underline{C}}_1$ has entries in rad(S) and $\underline{\underline{C}}_2$ has entries in rad(S'). #

9.3 <u>Lemma</u>: We can decompose the matrix $\begin{bmatrix} \underline{\underline{E}}_{s_2} \\ \underline{\underline{C}}_2 \end{bmatrix}$ without changing the remainder of $\underline{\underline{X}}$. In this decomposition, there occur only the factors

$$\begin{bmatrix} 1' \\ r' \end{bmatrix} \quad \text{and} \quad \begin{bmatrix} 1' \\ r' \\ r'^2 \end{bmatrix} .$$

The proof is clear with (9.1) and (8.1) and using ψ. #

Therefore we may assume:

$$\underline{\underline{X}} = \left[\begin{array}{c|c} 1 \; \cdots \; 1 & 0 \\ \hline 0 & 1' \; \cdots \; 1' \\ \hline \underline{\underline{C}}_1 & r' \; \cdots \; r' \\ \hline \underline{\underline{C}}_2 & r'^2\varepsilon_1 \; \cdots \; r'^2\varepsilon_2 \end{array}\right]$$

$$\uparrow \psi \qquad r'$$

where either $\varepsilon = 0$ or $\varepsilon = 1$. $\underline{\underline{C}}_1$ and $\underline{\underline{C}}_2$ have entries in rad(S).

9.4 Lemma: Multiplying the columns with r', and using the homomorphism ψ, we may assume that $\underline{\underline{C}}_1$ has entries in $r\underline{\underline{k}}$. Moreover, we can bring $\underline{\underline{X}}$ into the form

$$\underline{\underline{X}} = \left[\begin{array}{cccccc|c} \multicolumn{6}{c|}{1 \; \cdots \; 1} & 0 \\ \hline \multicolumn{6}{c|}{0} & 1' \; \cdots \; 1' \\ \hline \multicolumn{2}{c|}{0} & \multicolumn{4}{c|}{f(r)} & r' \; \cdots \; r' \\ \hline \multicolumn{2}{c|}{0} & \multicolumn{4}{c|}{g(r+r^2)} & r'^2 \; \cdots \; r'^2 \quad 0 \\ \hline r\underline{\underline{E}} & & & r^2\underline{\underline{E}} & & & \\ & r\underline{\underline{E}} & & & r^2\underline{\underline{E}} & & \\ & & r^2\underline{\underline{E}} & & & & 0 \\ r^2\underline{\underline{E}} & & & & & & \end{array}\right]$$

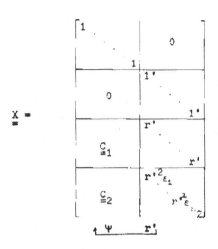

<u>Proof</u>: We can decompose the part $\mathbf{C}_2'$ of $\mathbf{C}_2$ which stands on the same line as those $\epsilon = 0$, and the identity matrix $\mathbf{E}_{s_1}$. Thus the part

$$\left[\begin{array}{cccc|c} 1 & & & & \\ & \cdot & & & \\ & & \cdot & & 0 \\ & & & \cdot & \\ & & & 1 & \\ \hline & & \mathbf{C}_2' & & 0 \end{array}\right]$$

decomposes into a direct sum of the matrices in (8.1); i.e., we obtain $\mathbf{C}_2'$ in the form

$$\left[\begin{array}{c|c|c|c|c|c|c|c} r\mathbf{E} & & & & & & & \\ \hline & r\mathbf{E} & & & & & & \\ \hline & & r\mathbf{E} & & & r^2\mathbf{E} & & \\ \hline & & & r\mathbf{E} & & & r^2\mathbf{E} & \\ \hline & & & & r^2\mathbf{E} & & & \\ \hline & r^2\mathbf{E} & & & & & & \\ \hline & & r^2\mathbf{E} & & & & & \end{array}\right]$$

$\vdash \quad \mathbf{A}_1 \quad \dashv$

If the part $\mathbf{A}_1$ is different from zero, then $\mathbf{X}$ splits a factor $\begin{bmatrix} 1 \\ r \end{bmatrix}$ or $\begin{bmatrix} 1 \\ r \\ r^2 \end{bmatrix}$.

Using ETs we can bring $\mathbf{X}$ to the desired form. #

We write $\mathbf{X}$ in the form

$$
\mathbf{X} = \left[\begin{array}{c|c|c|c|c|c|c|c}
\multicolumn{6}{c|}{1 \ \cdot \ \cdot \ \cdot \ 1} & \multicolumn{2}{c}{} \\
\hline
\multicolumn{6}{c|}{} & \multicolumn{2}{c}{1' \ \cdot \ \cdot \ \cdot \ 1'} \\
\hline
\multicolumn{2}{c|}{} & \mathbf{B}_{11} & \mathbf{B}_{12} & \mathbf{B}_{13} & \mathbf{B}_{14} & r' \cdots r' & \\
\hline
\multicolumn{2}{c|}{} & \mathbf{B}_{21} & \mathbf{B}_{22} & \mathbf{B}_{23} & \mathbf{B}_{24} & & r' \cdots r' \\
\hline
\multicolumn{2}{c|}{} & \mathbf{C}_{1} & \mathbf{C}_{2} & \mathbf{C}_{3} & \mathbf{C}_{4} & r'^2 \cdots r'^2 & \\
\hline
r\mathbf{E} & & & & r^2\mathbf{E} & & \multicolumn{2}{c}{} \\
\cline{1-6}
 & r\mathbf{E} & & r^2\mathbf{E} & & & \multicolumn{2}{c}{} \\
\cline{1-6}
 & & r^2\mathbf{E} & & & & \multicolumn{2}{c}{} \\
\cline{1-6}
r^2\mathbf{E} & & & & & & \multicolumn{2}{c}{}
\end{array}\right],
$$

where $\mathbf{B}_{ij}$ has entries in $\mathbf{kr}$.

Using elementary transformations, we obtain $\mathbf{X}$ as:

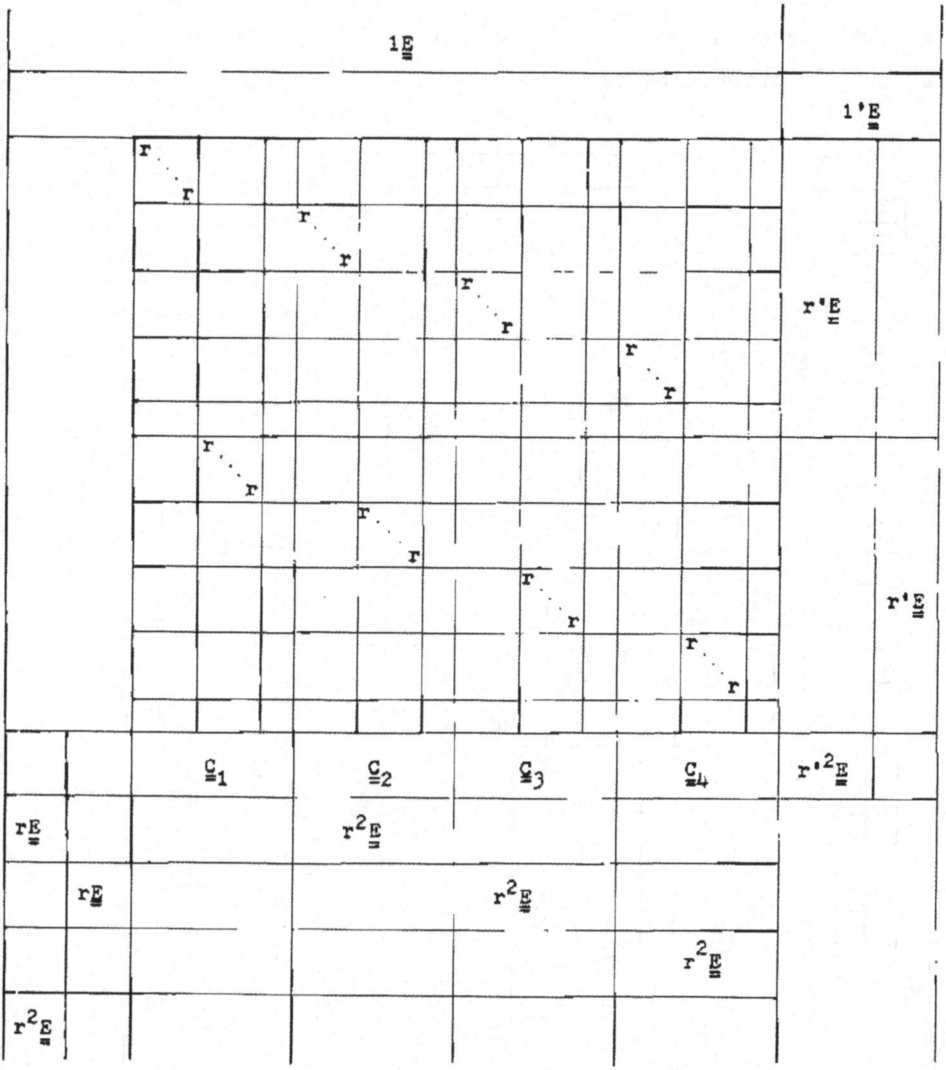
1E
1'E
r
r
r
r
r
r
r
r
r'E
r
r
r
r
r
r
r
r
r'E
C1
C2
C3
C4
r'2E
rE
r2E
rE
r2E
r2E
r2E

Here the matrices $\underline{\underline{C}}_1$ have entries in $\underline{\underline{k}}r^2$, if we assume

(i) The number of r'^2 is minimal; otherwise we would enlarge the lower part of $\underline{\underline{X}}$,

(ii) the number of the r' in $\underline{\underline{B}}_{1j}$ is minimal.

Using ETs we diagonalize $\underline{\underline{X}}$:

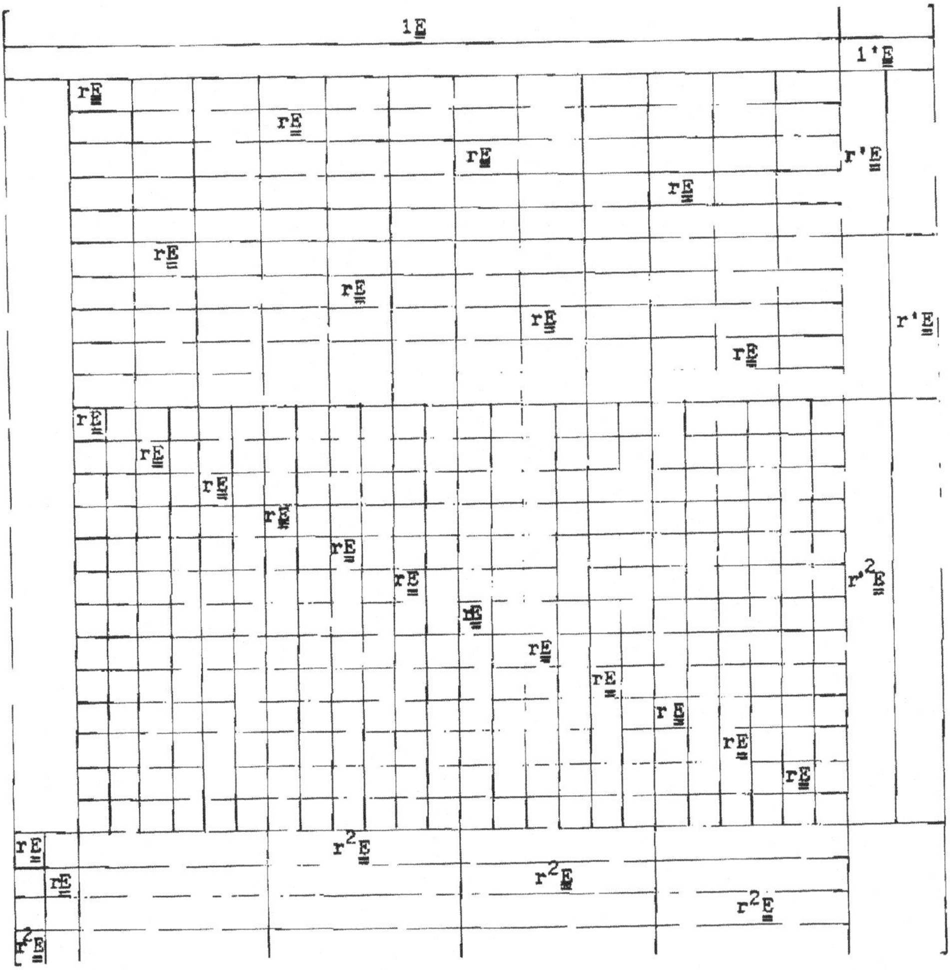

Now it is easily seen that all factors of $\underline{\underline{X}}$ have bounded degree. Thus also in this case $n(\hat{\Lambda}) < \infty$.

§10 Decomposition of the matrix 6,B,I

In 6,B,I we have shown

$$\underline{\underline{X}} = (\underline{\underline{X}}_1,\ldots,\underline{\underline{X}}_s) \text{ with}$$

$\underline{\underline{X}}_1 \in (k)_{m\times s_1}$, $\underline{\underline{k}} = S_1$,

$\underline{\underline{X}}_i \in (\underline{\underline{k}}[b_i])_{m\times s_i}$, $b_i^2 = 0$, $\underline{\underline{k}}b_i = b_i\underline{\underline{k}}$, $1 < i < s$, $k[b_i] = S_i$,

$\underline{\underline{X}}_s \in (\underline{\underline{k}}[b_s])_{m\times s_s}$, where either $b_s^2 = 0$, $\underline{\underline{k}}b_s = b_s\underline{\underline{k}}$, or $\underline{\underline{k}}[b_s]$ is a quadratic field extension of $\underline{\underline{k}}$, $\underline{\underline{k}}[b_s] = S_s$.

$\underline{\underline{Z}} = (\underline{\underline{Z}}_{ij})_{1\leq i,j\leq s}$, where

$\underline{\underline{Z}}_{ij}$ has entries in $\varphi_{ij}S_j$, $1\leq i < j\leq s$,

$$\varphi_{ij} : 1_i \longmapsto 1_j,\ b_i \longmapsto 0.$$

$\underline{\underline{Z}}_{ij}$ has entries in $S_i\varphi_{ij}$, $1\leq j < i\leq s$,

$$\varphi_{ij} : 1_i \longmapsto 0,\ b_i \longmapsto b_j k \text{ for some } k \in \underline{\underline{k}}.$$

$\underline{\underline{Z}}_{ii}$ has entries in S_i for $i > 1$ and $\underline{\underline{Z}}_{11}$ has entries in $\underline{\underline{k}}$.

$\underline{\underline{Y}}$ has entries in $\underline{\underline{k}}$.

By means of ETs we obtain $\underline{\underline{X}}$ in the form

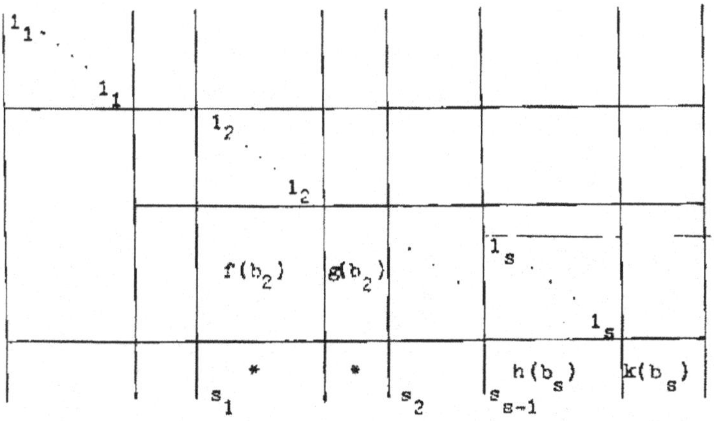

Beginning from the lower left hand corner we conclude that $\underline{\underline{X}}$ decomposes into factors of the following kind: $(1_i), 1<i\leq s$, $(b_i), 1<i\leq s$,

$\begin{bmatrix} 1_i \\ b_i \end{bmatrix}$, $1<i\leq s$. (If $\underline{\underline{k}}[b_s]$ is a field, the case (b_s) does not occur.)

Hence also here, $n(\hat{\Lambda}) < \infty$.

§ 11 Decomposition of the matrix 6.B.II

We had shown in 6,B,II:

$$\underline{\underline{X}}_{\hat{M}} = (\underline{\underline{X}}_1, \underline{\underline{X}}_2),$$

$\underline{\underline{X}}_1 \in (\hat{\Omega}_1/\omega_1^3\hat{\Omega}_1)_{m\times s_1} = (S_1)_{m\times s_1}$,

$\underline{\underline{X}}_2 \in (\hat{\Lambda}_1/(\hat{\Lambda}_1 \cap \omega_1\hat{\Lambda}_1))_{m\times s_2} = (S_2)_{m\times s_2}$,

$$\underline{\underline{Z}} = \begin{bmatrix} \underline{\underline{Z}}_{11} & \underline{\underline{Z}}_{12} \\ \underline{\underline{Z}}_{21} & \underline{\underline{Z}}_{22} \end{bmatrix} ,$$

$\underline{\underline{Z}}_{11}$ has entries in $\omega_1^{-1}\hat{\Omega}_1/\hat{\Lambda}_1 = S_1$,

$\underline{\underline{Z}}_{22}$ has entries in $\omega_1^{-3}\hat{\Omega}_1/\hat{\Omega}_1 = S_2$,

$\underline{\underline{Z}}_{12}$ has entries in $S_1\varphi_{1o}$,

$\underline{\underline{Z}}_{21}$ has entries in $\varphi_{o1}S_1$,

where $\varphi_{o1} : \sum_{i=o}^{2} \bar{\varpi}_1^i k_i \longmapsto 1k_o' + \bar{\varpi}_1^2 k_2'$

$\qquad \varphi_{1o} : 1k_o + \bar{\varpi}_1^2 k_2 \longmapsto \bar{\varpi}_1^2 k_o$.

$\underline{\underline{Y}}$ has entries in $\hat{\Omega}_2/\omega_2^2\hat{\Omega}_2$ with the action

$$\omega_2(\textstyle\sum_{i=o}^{2} \bar{\varpi}_1^i k_i) = \bar{\varpi}_1^2 k_2 \text{ on } S_1,$$

$$\omega_2(1k_o + \bar{\varpi}_1^2 k_2) = \bar{\varpi}_1^2 k_o \text{ on } S_2.$$

By means of ETs we can bring $\underline{\underline{X}}_{\hat{M}}$ into the form

$$\underline{\underline{X}} = \underline{\underline{X}}_{\hat{M}} = \left[\begin{array}{c|c|c|c|c} \underline{\underline{D}}_{11}\bar{\varpi}_1 & \underline{\underline{D}}_{12}\bar{\varpi}_1 & \underline{\underline{D}}_{13}\bar{\varpi}_1 & \underline{\underline{E}}\,1_o & \\ \hline \underline{\underline{D}}_{21}\bar{\varpi}_1 & \underline{\underline{D}}_{22}\bar{\varpi}_1 & \underline{\underline{D}}_{23}\bar{\varpi}_1 & & \underline{\underline{E}}\,r_1 \\ \hline \underline{\underline{E}}\,1_1 & & & & \\ \hline \underline{\underline{D}}_{31}\bar{\varpi}_1 & \underline{\underline{E}}\bar{\varpi}_1^2 + \underline{\underline{D}}_{32}\bar{\varpi}_1 & \underline{\underline{D}}_{33}\bar{\varpi}_1 & & \\ \hline \underline{\underline{D}}_{41}\bar{\varpi}_1 & \underline{\underline{D}}_{42}\bar{\varpi}_1 & \underline{\underline{D}}_{43}\bar{\varpi}_1 & & \end{array}\right] .$$

Here $r_1 = \bar{\omega}_1^2 + \omega_1 \hat{\Lambda}_1$. We now assume that among the equivalent matrices, $\underline{\underline{X}}$ is choosen such that the number of entries different from r_1 and $\bar{\omega}_1^2$ is minimal. Then we must have $\underline{\underline{D}}_{2i} = 0$, $1 \leq i \leq 3$; otherwise one could reduce the number of r_1's. Moreover, $\underline{\underline{D}}_{31} = \underline{\underline{D}}_{33} = 0$ otherwise one could replace one $\bar{\omega}_1^2$ by 0. But then we have to admit elements with ${}_1$ in the (3,2)-block.

If $\underline{\underline{D}}_{43} \neq 0$, we get a factor $(\bar{\omega}_1)$. Hence we may assume

$\underline{\underline{D}}_{43} = 0$, $\underline{\underline{D}}_{42} \neq 0$, $\underline{\underline{D}}_{12} = 0$: $\begin{bmatrix} \bar{\omega}_1^2 + k\,\bar{\omega}_1 \\ \bar{\omega}_1 \end{bmatrix} \sim \begin{bmatrix} \bar{\omega}_1 \\ \bar{\omega}_1^2 \end{bmatrix}$.

$\underline{\underline{D}}_{43} = 0$, $\underline{\underline{D}}_{42} \neq 0$, $\underline{\underline{D}}_{12} = 0$: $\begin{bmatrix} \bar{\omega}_1 & 1_o \\ \bar{\omega}_1^2 & 0 \end{bmatrix}$.

$\underline{\underline{D}}_{43} = 0$, $\underline{\underline{D}}_{42} = 0$, $\underline{\underline{D}}_{41} \neq 0$: $\begin{bmatrix} 0 & 1_o \\ 1_1 & 0^o \\ \omega_1 & 0 \end{bmatrix}$; i.e., (1_o), $\begin{bmatrix} 1_1 \\ \bar{\omega}_1 \end{bmatrix}$.

$\underline{\underline{D}}_{43} = 0$, $\underline{\underline{D}}_{42} = 0$,
$\underline{\underline{D}}_{41} = 0$, $\underline{\underline{D}}_{13} \neq 0$: $(\bar{\omega}_1, 1_o)$.

$\underline{\underline{D}}_{43} = 0$, $\underline{\underline{D}}_{42} = 0$,
$\underline{\underline{D}}_{41} = 0$, $\underline{\underline{D}}_{13} = 0$, $\underline{\underline{D}}_{11} \neq 0$: $\begin{bmatrix} \bar{\omega}_1 & 1_o \\ 1_1 & 0 \end{bmatrix}$.

Besides that we get a factor $\begin{bmatrix} 1_1 & \bar{\omega}_1 \\ \bar{\omega}_1 & 0 \end{bmatrix} \sim \begin{bmatrix} 1_1 & 0 \\ \bar{\omega}_1 & \bar{\omega}_1^2 \end{bmatrix}$, if we can reduce one $\bar{\omega}_1^2$ to zero.

Hence here too we have $n(\hat{\Lambda}) < \infty$.

§12 Decomposition of the matrix 6,C

We have shown in (6,C):

$$\underline{\underline{X}}_{\hat{M}} = (\underline{\underline{X}}_1, \underline{\underline{X}}_2, \underline{\underline{X}}_3),$$

$\underline{\underline{X}}_1 \in (S_o)_{s_1 \times m}$,

$\underline{\underline{X}}_2 \in (S_1)_{s_2 \times m}$,

$\underline{\underline{X}}_3 \in (S_2)_{s_3 \times m}$.

As generating elements we take for

$$S_o : r_{11}, \dots, r_{1d},$$

$$S_1 : r_{21} \ ,$$

$$S_2 : r_{31}, \dots, r_{3d}.$$

The matrix $\underline{\underline{Z}}$ is of the form

$$\underline{\underline{Z}} = \begin{bmatrix} S_o & 0 & \varphi_{o2} S_2 \\ S_1 \varphi_{1o} & S_1 & 0 \\ S_2 \varphi_{21} & 0 & S_2 \end{bmatrix},$$

where

$$\varphi_{1o} : S_1 \longrightarrow S_o, \quad r_{21} \longmapsto r_{1d}$$

$$\varphi_{2o} : S_2 \longrightarrow S_o, \quad r_{3i} \longmapsto r_{1i}, 1 \leq i \leq d$$

$$\varphi_{o2} : S_o \longrightarrow S_2, \quad r_{1i} \longmapsto r_{3,i+1}, 1 \leq i \leq d-1, \ r_{1d} \longmapsto 0.$$

The matrix $\underline{\underline{Y}}$ has entries in $\hat{\Omega}_3$ and ω_3 acts on S_i as follows

$$\Psi_1 : \omega_3 r_{1i} = r_{1i+1}, 1 \leq i \leq d, \ r_{1,d+1} = 0$$

$$\Psi_2 : \omega_3 r_{21} = 0$$

$$\Psi_3 : \omega_3 r_{3i} = r_{3,i+1}, 1 \leq i \leq d, \ r_{3,d+1} = 0.$$

We may diagonalize $\underline{\underline{X}}_1$ by means of ETs:

$$\underline{\underline{X}} = \left[\begin{array}{cccc|c|c} \underline{\underline{E}}r_{11} & & & & & \\ & \underline{\underline{E}}r_{12} & & & & \\ & & \ddots & & \underline{\underline{X}}_2 & \underline{\underline{X}}_3 \\ & & & \underline{\underline{E}}r_{1d} & & \\ & & & 0 & & \end{array}\right].$$

Starting from below, we may diagonalize $\underline{\underline{X}}_2$:

$$\underline{\underline{X}} = \left[\begin{array}{c|c|c|c|c|c|c|c|c|c} \underline{\underline{E}}r_{11} & & & & & & & & \underline{\underline{E}}r_{21} & \underline{\underline{A}}_1 \\ \hline & \underline{\underline{E}}r_{12} & & & & & & \underline{\underline{E}}r_{21} & & \underline{\underline{A}}_2 \\ \hline & & \ddots & & & & \cdot^{\cdot^{\cdot}} & & & \vdots \\ \hline & & & \underline{\underline{E}}r_{1d} & & \underline{\underline{E}}r_{21} & & & & \underline{\underline{A}}_d \\ \hline & & & & \underline{\underline{E}}r_{21} & & & & & \underline{\underline{A}}_{d+1} \end{array}\right].$$

Here $\underline{\underline{A}}_i$ is of the form $f(r_{31}, r_{32}, \ldots, r_{3i})$. We write $\underline{\underline{X}}_3$ in the following selfexplanatory form

$$\underline{\underline{X}}_3 = \begin{bmatrix} \underline{\underline{C}}_{11} \\ \underline{\underline{C}}_{21} \\ \underline{\underline{C}}_{12} \\ \underline{\underline{C}}_{22} \\ \cdot \\ \cdot \\ \cdot \\ \underline{\underline{C}}_{2d} \\ \underline{\underline{C}}_{1,d+1} \\ \underline{\underline{C}}_{2,d+1} \end{bmatrix}.$$

Now we assume that the number of r_{1i} is minimal. Because of the map φ_{31} we must have $\underline{\underline{C}}_{1j} = 0$, for all $j \neq d+1$. Then we can split off the factors (r_{i1}), $1\leq i\leq d$ and (r_{i1}, r_{21}), $1\leq i\leq d$.

Hence only the part

$$\underline{\underline{D}} = \begin{bmatrix} \underline{\underline{E}}r_{21} & \underline{\underline{C}}_{1,d+1} \\ 0 & \underline{\underline{C}}_{2,d+1} \end{bmatrix}$$

remains. We diagonalize $\underline{\underline{C}}_{2,d+1}$ as follows

$$\underline{\underline{C}}_{2,d+1} = \left[\begin{array}{c|c|c|c|c} \underline{\underline{E}}r_{31} & & & & \\ \hline & \underline{\underline{E}}r_{32} & & & \\ \hline & & \ddots & & \\ \hline & & & \underline{\underline{E}}r_{3d} & \\ \hline & & & & \end{array}\right].$$

All columns above $\underline{\underline{E}}r_{31}$ can only contain elements of the form kr_{3j}, $1\leq j\leq i-1$. In particular, we get factors (r_{21}), (r_{31}). If we assume that the number of elements different from zero in $\underline{\underline{C}}_{2,d+1}$ is minimal, then $\underline{\underline{D}}$ must have the form

$$\underline{\underline{D}} = \left[\begin{array}{c|c|c|c|c|c} \underline{\underline{E}}r_{21} & & & & & * \\ \hline & \underline{\underline{E}}r_{31} & & & & \\ \hline & & \underline{\underline{E}}r_{32} & & & \\ \hline & & & \ddots & & \\ \hline & & & & \underline{\underline{E}}r_{3d} & \end{array}\right],$$

and we can split off factors (r_{3i}), $1\leq i\leq d$. It is clear that the remaining factors can only be of the form (r_{21}, r_{31}), $1\leq i\leq d$. Hence also in this case $n(\hat{\Lambda}) < \infty$.

This concludes the proof of (2.1). #

Remark: With (1.1), the main theorem (2.1) gives a necessary and sufficient condition for $n(\Lambda)$ to be finite in case A is commutative. Little is known about general theorems for arbitrary A. Nevertheless, our conditions can be used to solve the problem for $\Lambda = RG$, G a finite group. Using a technique of D.G. Higman [1], one shows $n(RG) < \infty$ if and only if $n(R_{\underline{p}}G) < \infty$ for every maximal ideal p of R with $\underline{p} \mid |G| R$, if and only if $n(\hat{R}_{\underline{p}}G_p) < \infty$ for every $\underline{p} \mid |G| R$, where G_p is a p-Sylowsubgroup of G, $p = \underline{p} \cap \underline{Z}$. But Heller-Reiner [3] have shown that $n(\hat{R}_{\underline{p}}G_p) = \infty$ if G_p is not cyclic. Hence we can restrict ourselves to the case where G_p is commutative, and (2.1) solves the problem. For the sake of completeness, we shall list here the conditions Jacobinski [2] has given for $n(RG)$ to be finite. For every prime $p \in Z$, $p \mid |G|$, let G_p be a p-Sylowsubgroup of G, and let $pR = \prod_{i=1}^{n} \underline{p}_j^{e_i}$, $\underline{p}_j \in$ spec R, (here K is an algebraic number field) and put $e(p) = \max_{1 \le i \le n} e_i$. Then $n(RG) < \infty$ if and only if for every $p \mid |G|$, one of the following conditions is satisfied:

(i) $e(p) = 1$ and G_p is cyclic of order p or p^2,

(ii) $p > 3$, $e(p) = 2$ and G_p is cyclic of order p,

(iii) $p = 3$, $e(p) = 3$ and G_p is cyclic of order p.

BIBLIOGRAPHY

(To the references in this collection, we refer to in the text by e.g. "Smith [7]^{2}".)

BALLEW, D.
1. The module index, projective modules and invertible ideals. Ph.D. Thesis, Univ. of Illinois, 1969.

BANASCHEWSKI, B.
1. Integral group rings of finite groups, Can. Math. Bull. 10 (1967), 635-642.

BOREVICH, Z.I. - D.K. FADDEEV
4. Remarks on orders of cyclic index. Dokl. Akad. Nauk SSSR 164 (1965), 727-728.

BROOKS, J.O.
1. Classification of representation modules over quadratic orders. Ph.D. Thesis, Univ. of Michigan, 1964.

COHN, J.A. - D. Livingstone
1. On the structure of group algebras. Can. J. Math. 17 (1965), 585-593.

COLEMAN, D.B.
1. Idempotents in group rings. Proc. Am. Math. Soc. 17 (1966), 962.

CONNELL, I.G.
1. On the group ring. Can. J. Math. 15 (1963), 650-685.

DADE, E.C.
2. Rings in which no fixed power of ideal classes becomes invertible. Math. Ann. 148 (1962), 65-66.

DADE, E.C. - O. Tauski - H. Zassenhaus
2. On the semi-group of ideal classes in an order of an algebraic number field. Bull. Am. Math. Soc. 67 (1961), 305-308.

DRESS, A.
1. On relative Grothendieck rings. Bull. Am. Math. Soc. 75 (1969), 955.

2. On the decomposition of modules. Bull. Am. Math. Soc. 75 (1969), 984.

3. On integral representations. Bull. Am. Math. Soc. 75 (1969), 1031.

DROZD, Ju.A. - V.V. KIRICHENKO
2. Hereditary orders. Ukrain. Mat. J. 20 (1967), 246-248.

FADDEEV, D.K.
3. On the theory of cubic Z-rings. Trudy Steklov Inst. 80 (1965), 183-187.

4. On the equivalence of systems of integral matrices. Izv. Akad. Nauk SSSR 30 (1966), 449-454.

FADDEEV, D.K.
5. Class numbers of exact ideals for Z-rings. Mat. Zametki 1 (1967), 625-632.

FOSSUM, R.
2. The noetherian different of projective orders. J. reine angew. Math. 224 (1966), 207-218.

FRÖHLICH, A.
4. Ideals in an extension field as modules over the algebraic integers in a finite number field. Math. Zeit. 74 (1960), 29-38.

5. The module structure of Kummer extensions over Dedekind domains. J. reine angew. Math. 209 (1962), 39-53.

GIORGIUTTI, I.
1. Modules projectifs sur les algèbres de groupes finis. C.R. Acad. Sci. Paris 250 (1960), 1419-1420.

GROTHENDIECK, A.
1. Sur quelques points d'algèbre homologique. Tôhoku Math. J. 9 (1957), 119-221.

GUDIVOK, P.M.
6. Representations of finite groups over local number rings. Dopovidi Akad. Nauk Ukrain RSR 8 (1966), 979-981.

HATTORI, A.
2. Semi-simple algebras over a commutative ring. J. Math. Soc. Japan 15 (1963), 404-419.

HELLER, A.
2. Some exact sequences in algebraic K-theory. Topology 3 (1965), 389-408.

HIGMAN, G.
1. The units of group rings. Proc. London Math. Soc. (2), 46 (1940), 231-248.

JENNER, W.E.
2. Block ideals and arithmetics of algebras. Comp. Math. 11 (1953), 187-203.

KAPLANSKI, I.
4. Elementary divisors and modules. Trans. Am. Math. Soc. 66 (1949), 464-491.

KRUGLJAK, S.A.
1. Exact ideals in a second order integral matrix ring. Ukrain. Mat. J. 18 (1966), 58-64.

LARSON, R.
2. Group rings over Dedekind domains I. J. of Algebra 5 (1967), 358-361.

LATIMER, C.G. - C.C. MACDUFFEE
1. A correspondence between classes of ideals and classes of matrices. Ann. of Math. (2), 34 (1933), 313-316.

LEOPOLD, H.
1. Über die Hauptordnung der ganzen Elemente eines abelschen Zahlkörpers. J. reine angew. Math. 201 (1959), 119-149.

LEVY, L.S.

1. Decomposing pairs of modules. Trans. Am. Math. Soc. 122 (1966), 64-80.

MAY, W.

1. Commutative group algebras. Trans. Am. Math. Soc. 136 (1969), 139-149.

NAZAROVA, L.A.

1. Representations of a tetrad. Izv. Akad. Nauk SSR, Ser. Mat. 31 (1967), 1361-1378.

NAZAROVA, L.A. - A.V. ROITER

2. On integral p-adic representations and representations over residue class rings. Ukrain. Math. J. 19 (1967), 125-126.

3. Finitely generated modules over a dyad of a pair of local groups having an abelian normal subgroup of index p. Izv. Akad. Nauk, SSSR, 33 (1969), 65-89.

REINER, I.

17. Maximal orders. Mimeo. notes, Univ. of Illinois (1969).

18. A survey of integral representation theory. Bull. Am. Math. Soc. (1970), (to appear).

ROITER, A.V.

7. On the theory of integral representations of rings. Mat. Zametki 3 (1968), 361-366.

UCHIDA, K.

1. Remarks on Grothendieck groups. Tôhoku Math. J. 19 (1967), 341-348.

INDEX

Corrections Vol. I

p. XVIII; -9 : If no simple component of A is a full matrix ring over a totally definite quoternion algebra, then...

p. I,12; -10 : We have $\varphi^* \psi^* = \mathrm{hom}(\varphi, 1_N)\mathrm{hom}(\psi, 1_N) = \ldots$

p. I,16; -5 : $\Psi(M'')\, \psi_*(F)\sigma = \ldots$

p. I,45; -10 : If A is integral over R, then A_S is integral over R_S

p. I,53; +1 : $\mathfrak{p}$-primary component of M/N.

p. I,60; -2 : ...in $\hat{R}_{\mathfrak{m}}$

p. I,61; 9 : $\hat{R}_{\mathfrak{p}}$ is a...

p. II,50; +9 : ...category $\widetilde{\mathfrak{E}}_R$, where...

p. II,50; +12 : $E(1_{M'}, \sigma, 1_{M''}) = E'$ for...

p. II,54; +3 : ... $\nabla\alpha = (\alpha \oplus \alpha)\nabla$ ...

p. II,55; -8 : ... $\alpha \in \mathrm{Hom}_R(N'', M'')$.

p. III,13; +5 : Omit "and faithfulness"!

p. III,13; +7 : projectives and faithful modules

p. III,23; -3 : $E : 0 \longrightarrow M \longrightarrow X \longrightarrow N \longrightarrow 0, \ldots$

p. IV,7; +4 : ...if there exists $0 \neq N \subset M$...

p. IV,13; -3 : This part of the proof is false: The decomposition of a Λ-module into its torsion-free part and its torsion part is <u>not</u> a decomposition of Λ-modules. The argument here should read: Let $M_{\mathfrak{p}}$ be a generator for every $\mathfrak{p} \in \mathfrak{S}$; i.e.,

$$\tau_{\mathfrak{p}} : M_{\mathfrak{p}} \otimes_{\mathrm{End}_{\Lambda_{\mathfrak{p}}}(M_{\mathfrak{p}})} \mathrm{Hom}_{\Lambda_{\mathfrak{p}}}(M_{\mathfrak{p}}, \Lambda_{\mathfrak{p}}) \longrightarrow \Lambda_{\mathfrak{p}},$$

$$m \otimes \varphi \longmapsto m\varphi$$

is an isomorphism for every $\mathfrak{p} \in \mathfrak{S}$. However, $R_{\mathfrak{p}} \otimes_R$- is a faithful and flat functor on the category of Λ-lattices and by (III,1.2),

$$\mathrm{Hom}_{\Lambda_{\mathfrak{p}}}(X_{\mathfrak{p}}, Y_{\mathfrak{p}}) \cong R_{\mathfrak{p}} \otimes_R \mathrm{Hom}_{\Lambda}(X, Y) \text{ for } X, Y \in {}_{\Lambda}\mathfrak{M}^{\circ}.$$

This implies $\mathrm{Im}(\tau_{\mathfrak{p}}) = (\mathrm{Im}\,\tau)_{\mathfrak{p}}$ and thus

$$\mathrm{Im}\,\tau = \bigcap_{\mathfrak{p} \in \mathfrak{S}} (\mathrm{Im}\,\tau)_{\mathfrak{p}} = \bigcap_{\mathfrak{p} \in \mathfrak{S}} \mathrm{Im}(\tau_{\mathfrak{p}}) = \bigcap_{\mathfrak{p} \in \mathfrak{S}} \Lambda_{\mathfrak{p}} = \Lambda$$

and M is a generator (cf. III, 1.9).

Offsetdruck: Julius Beltz, Weinheim/Bergstr

Lecture Notes in Mathematics

Bisher erschienen/Already published

Vol. 1: J. Wermer, Seminar über Funktionen-Algebren. IV, 30 Seiten. 1964. DM 3,80 / $ 1.10

Vol. 2: A. Borel, Cohomologie des espaces localement compacts d'après. J. Leray. IV, 93 pages. 1964. DM 9,– / $ 2.60

Vol. 3: J. F. Adams, Stable Homotopy Theory. Third edition. IV, 78 pages. 1969. DM 8,– / $ 2.20

Vol. 4: M. Arkowitz and C. R. Curjel, Groups of Homotopy Classes. 2nd revised edition. IV, 36 pages. 1967. DM 4,80 / $ 1.40

Vol. 5: J.-P. Serre, Cohomologie Galoisienne. Troisième édition. VIII, 214 pages. 1965. DM 18,– / $ 5.00

Vol. 6: H. Hermes, Term Logic with Choise Operator. III, 55 pages. 1970. DM 6,– / $ 1.70

Vol. 7: Ph. Tondeur, Introduction to Lie Groups and Transformation Groups. Second edition. VIII, 176 pages. 1969. DM 14,– / $ 3.80

Vol. 8: G. Fichera, Linear Elliptic Differential Systems and Eigenvalue Problems. IV, 176 pages. 1965. DM 13,50 / $ 3.80

Vol. 9: P. L. Ivănescu, Pseudo-Boolean Programming and Applications. IV, 50 pages. 1965. DM 4,80 / $ 1.40

Vol. 10: H. Lüneburg, Die Suzukigruppen und ihre Geometrien. VI, 111 Seiten. 1965. DM 8,– / $ 2.20

Vol. 11: J.-P. Serre, Algèbre Locale. Multiplicités. Rédigé par P. Gabriel. Seconde édition. VIII, 192 pages. 1965. DM 12,– / $ 3.30

Vol. 12: A. Dold, Halbexakte Homotopiefunktoren. II, 157 Seiten. 1966. DM 12,– / $ 3.30

Vol. 13: E. Thomas, Seminar on Fiber Spaces. IV, 45 pages. 1966. DM 4,80 / $ 1.40

Vol. 14: H. Werner, Vorlesung über Approximationstheorie. IV, 184 Seiten und 12 Seiten Anhang. 1966. DM 14,– / $ 3.90

Vol. 15: F. Oort, Commutative Group Schemes. VI, 133 pages. 1966. DM 9,80 / $ 2.70

Vol. 16: J. Pfanzagl and W. Pierlo, Compact Systems of Sets. IV, 48 pages. 1966. DM 5,80 / $ 1.60

Vol. 17: C. Müller, Spherical Harmonics. IV, 46 pages. 1966. DM 5,– / $ 1.40

Vol. 18: H.-B. Brinkmann und D. Puppe, Kategorien und Funktoren. XII, 107 Seiten. 1966. DM 8,– / $ 2.20

Vol. 19: G. Stolzenberg, Volumes, Limits and Extensions of Analytic Varieties. IV, 45 pages. 1966. DM 5,40 / $ 1.50

Vol. 20: R. Hartshorne, Residues and Duality. VIII, 423 pages. 1966. DM 20,– / $ 5.50

Vol. 21: Seminar on Complex Multiplication. By A. Borel, S. Chowla, C. S. Herz, K. Iwasawa, J.-P. Serre. IV, 102 pages. 1966. DM 8,– / $ 2.20

Vol. 22: H. Bauer, Harmonische Räume und ihre Potentialtheorie. IV, 175 Seiten. 1966. DM 14,– / $ 3.90

Vol. 23: P. L. Ivănescu and S. Rudeanu, Pseudo-Boolean Methods for Bivalent Programming. 120 pages. 1966. DM 10,– / $ 2.80

Vol. 24: J. Lambek, Completions of Categories. IV, 69 pages. 1966. DM 6,80 / $ 1.90

Vol. 25: R. Narasimhan, Introduction to the Theory of Analytic Spaces. IV, 143 pages. 1966. DM 10,– / $ 2.80

Vol. 26: P.-A. Meyer, Processus de Markov. IV, 190 pages. 1967. DM 15,– / $ 4.20

Vol. 27: H. P. Künzi und S. T. Tan, Lineare Optimierung großer Systeme. VI, 121 Seiten. 1966. DM 12,– / $ 3.30

Vol. 28: P. E. Conner and E. E. Floyd, The Relation of Cobordism to K-Theories. VIII, 112 pages. 1966. DM 9,80 / $ 2.70

Vol. 29: K. Chandrasekharan, Einführung in die Analytische Zahlentheorie. VI, 199 Seiten. 1966. DM 16,80 / $ 4.70

Vol. 30: A. Frölicher and W. Bucher, Calculus in Vector Spaces without Norm. X, 146 pages. 1966. DM 12,– / $ 3.30

Vol. 31: Symposium on Probability Methods in Analysis. Chairman. D. A. Kappos. IV, 329 pages. 1967. DM 20,– / $ 5.50

Vol. 32: M. André, Méthode Simpliciale en Algèbre Homologique et Algèbre Commutative. IV, 122 pages. 1967. DM 12,– / $ 3.30

Vol. 33: G. I. Targonski, Seminar on Functional Operators and Equations. IV, 110 pages. 1967. DM 10,– / $ 2.80

Vol. 34: G. E. Bredon, Equivariant Cohomology Theories. VI, 64 pages. 1967. DM 6,80 / $ 1.90

Vol. 35: N. P. Bhatia and G. P. Szegö, Dynamical Systems. Stability Theory and Applications. VI, 416 pages. 1967. DM 24,– / $ 6.60

Vol. 36: A. Borel, Topics in the Homology Theory of Fibre Bundles. VI, 95 pages. 1967. DM 9,– / $ 2.50

Vol. 37: R. B. Jensen, Modelle der Mengenlehre. X, 176 Seiten. 1967. DM 14,– / $ 3.90

Vol. 38: R. Berger, R. Kiehl, E. Kunz und H.-J. Nastold, Differentialrechnung in der analytischen Geometrie IV, 134 Seiten. 1967 DM 12,– / $ 3.30

Vol. 39: Séminaire de Probabilités I. II, 189 pages. 1967. DM 14,– / $ 3.90

Vol. 40: J. Tits, Tabellen zu den einfachen Lie Gruppen und ihren Darstellungen. VI, 53 Seiten. 1967. DM 6.80 / $ 1.90

Vol. 41: A. Grothendieck, Local Cohomology. VI, 106 pages. 1967. DM 10,– / $ 2.80

Vol. 42: J. F. Berglund and K. H. Hofmann, Compact Semitopological Semigroups and Weakly Almost Periodic Functions. VI, 160 pages. 1967. DM 12,– / $ 3.30

Vol. 43: D. G. Quillen, Homotopical Algebra. VI, 157 pages. 1967. DM 14,– / $ 3.90

Vol. 44: K. Urbanik, Lectures on Prediction Theory. IV, 50 pages. 1967. DM 5,80 / $ 1.60

Vol. 45: A. Wilansky, Topics in Functional Analysis. VI, 102 pages. 1967. DM 9,60 / $ 2.70

Vol. 46: P. E. Conner, Seminar on Periodic Maps. IV, 116 pages. 1967. DM 10,60 / $ 3.00

Vol. 47: Reports of the Midwest Category Seminar I. IV, 181 pages. 1967. DM 14,80 / $ 4.10

Vol. 48: G. de Rham, S. Maumary et M. A. Kervaire, Torsion et Type Simple d'Homotopie. IV, 101 pages. 1967. DM 9,60 / $ 2.70

Vol. 49: C. Faith, Lectures on Injective Modules and Quotient Rings. XVI, 140 pages. 1967. DM 12,80 / $ 3.60

Vol. 50: L. Zalcman, Analytic Capacity and Rational Approximation. VI, 155 pages. 1968. DM 13.20 / $ 3.70

Vol. 51: Séminaire de Probabilités II. IV, 199 pages. 1968. DM 14,– / $ 3.90

Vol. 52: D. J. Simms, Lie Groups and Quantum Mechanics. IV, 90 pages. 1968. DM 8,– / $ 2.20

Vol. 53: J. Cerf, Sur les difféomorphismes de la sphère de dimension trois (Γ_4 = O). XII, 133 pages. 1968. DM 12,– / $ 3.30

Vol. 54: G. Shimura, Automorphic Functions and Number Theory. VI, 69 pages. 1968. DM 8,– / $ 2.20

Vol. 55: D. Gromoll, W. Klingenberg und W. Meyer, Riemannsche Geometrie im Großen. VI, 287 Seiten. 1968. DM 20,– / $ 5.50

Vol. 56: K. Floret und J. Wloka, Einführung in die Theorie der lokalkonvexen Räume. VIII, 194 Seiten. 1968. DM 16,– / $ 4.40

Vol. 57: F. Hirzebruch und K. H. Mayer, O (n)-Mannigfaltigkeiten, exotische Sphären und Singularitäten. IV, 132 Seiten. 1968. DM 10,80 / $ 3.00

Vol. 58: Kuramochi Boundaries of Riemann Surfaces. IV, 102 pages. 1968. DM 9,60 / $ 2.70

Vol. 59: K. Jänich, Differenzierbare G-Mannigfaltigkeiten. VI, 89 Seiten. 1968. DM 8,– / $ 2.20

Vol. 60: Seminar on Differential Equations and Dynamical Systems. Edited by G. S. Jones. VI, 106 pages. 1968. DM 9,60 / $ 2.70

Vol. 61: Reports of the Midwest Category Seminar II. IV, 91 pages. 1968. DM 9,60 / $ 2.70

Vol. 62: Harish-Chandra, Automorphic Forms on Semisimple Lie Groups X, 138 pages. 1968. DM 14,– / $ 3.90

Vol. 63: F. Albrecht, Topics in Control Theory. IV, 65 pages. 1968. DM 6,80 / $ 1.90

Vol. 64: H. Berens, Interpolationsmethoden zur Behandlung von Approximationsprozessen auf Banachräumen. VI, 90 Seiten. 1968. DM 8,– / $ 2.20

Vol. 65: D. Kölzow, Differentiation von Maßen. XII, 102 Seiten. 1968. DM 8,– / $ 2.20

Vol. 66: D. Ferus, Totale Absolutkrümmung in Differentialgeometrie und -topologie. VI, 85 Seiten. 1968. DM 8,– / $ 2.20

Vol. 67: F. Kamber and P. Tondeur, Flat Manifolds. IV, 53 pages. 1968. DM 5,80 / $ 1.60

Vol. 68: N. Boboc et P. Mustaţă, Espaces harmoniques associés aux opérateurs différentiels linéaires du second ordre de type elliptique. VI, 95 pages. 1968. DM 8,60 / $ 2.40

Vol. 69: Seminar über Potentialtheorie. Herausgegeben von H. Bauer. VI, 180 Seiten. 1968. DM 14,80 / $ 4.10

Vol. 70: Proceedings of the Summer School in Logic. Edited by M. H. Löb. IV, 331 pages. 1968. DM 20,– / $ 5.50

Vol. 71: Séminaire Pierre Lelong (Analyse), Année 1967 – 1968. VI, 19 pages. 1968. DM 14,– / $ 3.90

Bitte wenden / Continued

Vol. 72: The Syntax and Semantics of Infinitary Languages. Edited by J. Barwise. IV, 268 pages. 1968. DM 18,– / $ 5.00

Vol. 73: P. E. Conner, Lectures on the Action of a Finite Group. IV, 123 pages. 1968. DM 10,– / $ 2.80

Vol. 74: A. Fröhlich, Formal Groups. IV, 140 pages. 1968. DM 12,– / $ 3.30

Vol. 75: G. Lumer, Algèbres de fonctions et espaces de Hardy. VI, 80 pages. 1968. DM 8,– / $ 2.20

Vol. 76: R. G. Swan, Algebraic K-Theory. IV, 262 pages. 1968. DM 18,– / $ 5.00

Vol. 77: P.-A. Meyer, Processus de Markov: la frontière de Martin. IV, 123 pages. 1968. DM 10,– / $ 2.80

Vol. 78: H. Herrlich, Topologische Reflexionen und Coreflexionen. XVI, 166 Seiten. 1968. DM 12,– / $ 3.30

Vol. 79: A. Grothendieck, Catégories Cofibrées Additives et Complexe Cotangent Relatif. IV, 167 pages. 1968. DM 12,– / $ 3.30

Vol. 80: Seminar on Triples and Categorical Homology Theory. Edited by B. Eckmann. IV, 398 pages. 1969. DM 20,– / $ 5.50

Vol. 81: J.-P. Eckmann et M. Guenin, Méthodes Algébriques en Mécanique Statistique. VI, 131 pages. 1969. DM 12,– / $ 3.30

Vol. 82: J. Wloka, Grundräume und verallgemeinerte Funktionen. VIII, 131 Seiten. 1969. DM 12,– / $ 3.30

Vol. 83: O. Zariski, An Introduction to the Theory of Algebraic Surfaces. IV, 100 pages. 1969. DM 8,– / $ 2.20

Vol. 84: H. Lüneburg, Transitive Erweiterungen endlicher Permutationsgruppen. IV, 119 Seiten. 1969. DM 10,– / $ 2.80

Vol. 85: P. Cartier et D. Foata, Problèmes combinatoires de commutation et réarrangements. IV, 88 pages. 1969. DM 8,– / $ 2.20

Vol. 86: Category Theory, Homology Theory and their Applications I. Edited by P. Hilton. VI, 216 pages. 1969. DM 16,– / $ 4.40

Vol. 87: M. Tierney, Categorical Constructions in Stable Homotopy Theory. IV, 65 pages. 1969. DM 6,– / $ 1.70

Vol. 88: Séminaire de Probabilités III. IV, 229 pages. 1969. DM 18,– / $ 5.00

Vol. 89: Probability and Information Theory. Edited by M. Behara, K. Krickeberg and J. Wolfowitz. IV, 256 pages. 1969. DM 18,– / $ 5.00

Vol. 90: N. P. Bhatia and O. Hajek, Local Semi-Dynamical Systems. II, 157 pages. 1969. DM 14,– / $ 3.90

Vol. 91: N. N. Janenko, Die Zwischenschrittmethode zur Lösung mehrdimensionaler Probleme der mathematischen Physik. VIII, 194 Seiten. 1969. DM 16,80 / $ 4.70

Vol. 92: Category Theory, Homology Theory and their Applications II. Edited by P. Hilton. V, 308 pages. 1969. DM 20,– / $ 5.50

Vol. 93: K. R. Parthasarathy, Multipliers on Locally Compact Groups. III, 54 pages. 1969. DM 5,60 / $ 1.60

Vol. 94: M. Machover and J. Hirschfeld, Lectures on Non-Standard Analysis. VI, 79 pages. 1969. DM 6,– / $ 1.70

Vol. 95: A. S. Troelstra, Principles of Intuitionism. II, 111 pages. 1969. DM 10,– / $ 2.80

Vol. 96: H.-B. Brinkmann und D. Puppe, Abelsche und exakte Kategorien, Korrespondenzen. V, 141 Seiten. 1969. DM 10,– / $ 2.80

Vol. 97: S. O. Chase and M. E. Sweedler, Hopf Algebras and Galois theory. II, 133 pages. 1969. DM 10,– / $ 2.80

Vol. 98: M. Heins, Hardy Classes on Riemann Surfaces. III, 106 pages. 1969. DM 10,– / $ 2.80

Vol. 99: Category Theory, Homology Theory and their Applications III. Edited by P. Hilton. IV, 489 pages. 1969. DM 24,– / $ 6.60

Vol. 100: M. Artin and B. Mazur, Etale Homotopy. II, 196 Seiten. 1969. DM 12,– / $ 3.30

Vol. 101: G. P. Szegö et G. Treccani, Semigruppi di Trasformazioni Multivoche. VI, 177 pages. 1969. DM 14,– / $ 3.90

Vol. 102: F. Stummel, Rand- und Eigenwertaufgaben in Sobolewschen Räumen. VIII, 386 Seiten. 1969. DM 20,– / $ 5.50

Vol. 103: Lectures in Modern Analysis and Applications I. Edited by C. T. Taam. VII, 162 pages. 1969. DM 12,– / $ 3.30

Vol. 104: G. H. Pimbley, Jr., Eigenfunction Branches of Nonlinear Operators and their Bifurcations. II, 128 pages. 1969. DM 10,– / $ 2.80

Vol. 105: R. Larsen, The Multiplier Problem. VII, 284 pages. 1969. DM 18,– / $ 5.00

Vol. 106: Reports of the Midwest Category Seminar III. Edited by S. Mac Lane. III, 247 pages. 1969. DM 16,– / $ 4.40

Vol. 107: A. Peyerimhoff, Lectures on Summability. III, 111 pages. 1969. DM 8,– / $ 2.20

Vol. 108: Algebraic K-Theory and its Geometric Applications. Edited by R. M. F. Moss and C. B. Thomas. IV, 86 pages. 1969. DM 6,– / $ 1.70

Vol. 109: Conference on the Numerical Solution of Differential Equations. Edited by J. Ll. Morris. VI, 275 pages. 1969. DM 18,– / $ 5.00

Vol. 110: The Many Facets of Graph Theory. Edited by G. Chartrand and S. F. Kapoor. VIII, 290 pages. 1969. DM 18,– / $ 5.00

Vol. 111: K. H. Mayer, Relationen zwischen charakteristischen Zahlen. III, 99 Seiten. 1969. DM 8,– / $ 2.20

Vol. 112: Colloquium on Methods of Optimization. Edited by N. N. Moiseev. IV, 293 pages. 1970. DM 18,– / $ 5.00

Vol. 113: R. Wille, Kongruenzklassengeometrien. III, 99 Seiten. 1970. DM 8,– / $ 2.20

Vol. 114: H. Jacquet and R. P. Langlands, Automorphic Forms on GL (2). VII, 548 pages. 1970. DM 24,– / $ 6.60

Vol. 115: K. H. Roggenkamp and V. Huber-Dyson, Lattices over Orders I. XIX, 290 pages. 1970. DM 18,– / $ 5.00

Vol. 116: Séminaire Pierre Lelong (Analyse) Année 1969. IV, 195 pages. 1970. DM 14,– / $ 3.90

Vol. 117: Y. Meyer, Nombres de Pisot, Nombres de Salem et Analyse Harmonique. 63 pages. 1970. DM 6,– / $ 1.70

Vol. 118: Proceedings of the 15th Scandinavian Congress, Oslo 1968. Edited by K. E. Aubert and W. Ljunggren. IV, 162 pages. 1970. DM 12,– / $ 3.30

Vol. 119: M. Raynaud, Faisceaux amples sur les schémas en groupes et les espaces homogènes. III, 219 pages. 1970. DM 14,– / $ 3.90

Vol. 120: D. Siefkes, Büchi's Monadic Second Order Successor Arithmetic. XII, 130 Seiten. 1970. DM 12,– / $ 3.30

Vol. 121: H. S. Bear, Lectures on Gleason Parts. III, 47 pages. 1970. DM 6,– / $ 1.70

Vol. 122: H. Zieschang, E. Vogt und H.-D. Coldewey, Flächen und ebene diskontinuierliche Gruppen. VIII, 203 Seiten. 1970. DM 16,– / $ 4.40

Vol. 123: A. V. Jategaonkar, Left Principal Ideal Rings. VI, 145 pages. 1970. DM 12,– / $ 3.30

Vol. 124: Séminare de Probabilités IV. Edited by P. A. Meyer. IV, 282 pages. 1970. DM 20,– / $ 5.50

Vol. 125: Symposium on Automatic Demonstration. V, 310 pages. 1970. DM 20,– / $ 5.50

Vol. 126: P. Schapira, Théorie des Hyperfonctions. XI, 157 pages. 1970. DM 14,– / $ 3.90

Vol. 127: I. Stewart, Lie Algebras. IV, 97 pages. 1970. DM 10,– / $ 2.80

Vol. 128: M. Takesaki, Tomita's Theory of Modular Hilbert Algebras and its Applications. II, 123 pages. 1970. DM 10,– / $ 2.80

Vol. 129: K. H. Hofmann, The Duality of Compact Semigroups and C*-Bigebras. XII, 142 pages. 1970. DM 14,– / $ 3.90

Vol. 130: F. Lorenz, Quadratische Formen über Körpern. II, 77 Seiten. 1970. DM 8,– / $ 2.20

Vol. 131: A Borel et al., Seminar on Algebraic Groups and Related Finite Groups. VII, 321 pages. 1970. DM 22,– / $ 6.10

Vol. 132: Symposium on Optimization. III, 348 pages. 1970. DM 22,– / $ 6.10

Vol. 133: F. Topsøe, Topology and Measure. XIV, 79 pages. 1970. DM 8,– / $ 2.20

Vol. 134: L. Smith, Lectures on the Eilenberg-Moore Spectral Sequence. VII, 142 pages. 1970. DM 14,– / $ 3.90

Vol. 135: W. Stoll, Value Distribution of Holomorphic Maps into Compact Complex Manifolds. II, 267 pages. 1970. DM 18,– / $

Vol. 136: M. Karoubi et al., Séminaire Heidelberg-Saarbrücken-Strasbourg sur la K-Théorie. IV, 264 pages. 1970. DM 18,– / $ 5.00

Vol. 137: Reports of the Midwest Category Seminar IV. Edited by S. MacLane. III, 139 pages. 1970. DM 12,– / $ 3.30

Vol. 138: D. Foata et M. Schützenberger, Théorie Géométrique des Polynômes Eulériens. V, 94 pages. 1970. DM 10,– / $ 2.80

Vol. 139: A. Badrikian, Séminaire sur les Fonctions Aléatoires Linéaires et les Mesures Cylindriques. VII, 221 pages. 1970. DM 18,– / $ 5.00

Vol. 140: Lectures in Modern Analysis and Applications II. Edited by C. T. Taam. VI, 119 pages. 1970. DM 10,– / $ 2.80

Vol. 141: G. Jameson, Ordered Linear Spaces. XV, 194 pages. 1970. DM 16,– / $ 4.40

Vol. 142: K. W. Roggenkamp, Lattices over Orders II. V, 388 pages. 1970. DM 22,– / $ 6.10